HIBERNATION AND HYPOTHERMIA, PERSPECTIVES AND CHALLENGES

HIBERNATION AND HYPOTHERMIA, PERSPECTIVES AND CHALLENGES

SYMPOSIUM HELD AT SNOWMASS-AT-ASPEN,
COLORADO, JANUARY 3 - 8, 1971

Edited by

FRANK E. SOUTH
JOHN P. HANNON
JOHN R. WILLIS
ERIC T. PENGELLEY
NORMAN R. ALPERT

ELSEVIER PUBLISHING COMPANY

AMSTERDAM — LONDON — NEW YORK

1972

ELSEVIER PUBLISHING COMPANY
335 JAN VAN GALENSTRAAT
P.O. BOX 211, AMSTERDAM, THE NETHERLANDS

AMERICAN ELSEVIER PUBLISHING COMPANY, INC.
52 VANDERBILT AVENUE
NEW YORK, NEW YORK 10017

This symposium was sponsored by the

SPACE SCIENCES RESEARCH CENTER of the

UNIVERSITY OF MISSOURI, COLUMBIA, MISSOURI

Library of Congress Card Number: 79-188671

ISBN: 0-444-41007-4

With 144 Figures, 20 Plates and 32 Tables.

Copyright © 1972 by Elsevier Publishing Company, Amsterdam

All rights reserved. No part of this publication may be reproduced, stored in a retrieval system, or transmitted in any form or by any means, electronic, mechanical, photocopying, recording, or otherwise, without the prior written permission of the publisher,
Elsevier Publishing Company, Jan van Galenstraat 335, Amsterdam

Printed in The Netherlands

This volume and the symposium from which it is taken is dedicated to the memory of Professor Kenneth C. Fisher, University of Toronto, and Professor Raymond C. Hock, University of Nevada. They are sorely missed by their fellow scientists, students and many friends.

RAYMOND J. HOCK
1918 TO 1970

Ray Hock was a physiologist, naturalist and an outdoorsman who knew well the difficult procedure of using all of nature as a laboratory. His formative scientific years were in company with three of the great physiologists of our time. His now classical Ph.D. dissertation on the subject of hibernation was completed under Donald Griffin's direction. He followed this with a fruitful period of postdoctoral studies under Per Scholander and Lawrence Irving. His research transited the Gulf of St. Lawrence, the Navy Point Barrow Arctic Research Laboratory, Cornell University, the University of Arizona at Tucson, the Arctic Public Health Service Laboratory at Anchorage, the Air Force Aeromedical Laboratory at Fairbanks, the University of California White Mountain Research Station, the Northrup Company at Los Angeles, and finally, the University of Nevada at Las Vegas. His final position was as a colleague of Bruce Dill: rounding out a highly productive scientific career.

For many years Ray was the most knowledgeable scientist in this hemisphere on the subject of the torpor of bears, and also on the subject of life cycles of hibernating Arctic mammals. He was a consultant to Government agencies and to other groups on physiological subjects, particularly hibernation and cold-torpor. He was one of the small party which in 1956 conceived of initiating the symposia on mammalian hibernation. He leaves us, a small band of colleagues, who honor him as best we can with this small dedicatory tribute to a good friend, naturalist, and scientist.

A. R. Dawe

KENNETH C. FISHER
1911 TO 1970

The editors of this volume and those of <u>Hibernation III</u> deeply regret the death on January 22, 1970, of Kenneth C. Fisher, Professor and former Chairman of the Department of Zoology, Faculty of Arts and Science, University of Toronto; Chairman of the organizing committee for the International Conference on Hibernation, Toronto, 1965, and Editor-in-Chief of Hibernation III; an esteemed colleague and friend of all those who had the privilege of meeting him, be it in the field of temperature regulation or in other areas.

Kenneth Fisher was born in Saint John, New Brunswick, November 5, 1911. During his undergraduate career, he had the privilege of attending a summer course in experimental Biology under Dr. L. Irving at the Marine Biological Laboratory, Woods Hole, Massachusetts. That exciting experience determined the course of his future career. He came to the University of Toronto to carry on his post graduate research under Dr. Irving in the Department of Physiology. He obtained his Master of Arts degree in 1934 and two years later was granted the degree of Doctor of Philosophy.

He taught a year at the University of Maine, then returned to the University of Toronto in 1937 as Assistant Professor in the Department of Biology. After a year in this position, he married Dr. Jeanne Manery who also had taken her degree under Dr. Irving. Dr. Fisher was named Professor and Chairman of the Department of Zoology in 1956, and continued in that post for eleven years. He was elected a Fellow of the Royal Society of Canada in 1947 and in 1961 he was honoured by the presentation of a D.Sc. degree by the University of New Brunswick.

For a number of years after joining the Department of Zoology at Toronto, Professor Fisher spent his summers at Woods Hole teaching and continuing his research. His interest in teaching at the undergraduate level and his dedication to the research and training of his graduate students were among his notable contributions to this University. The research problems of his graduate students always involved him deeply. In each case, it was a mutual problem to be solved together, by critical discussion, planning, and investigation.

The field of his interest was broad, encompassing the oxidative metabolism of cells, the physiology of muscle contraction, the biology of animal behavior and the hibernation of animals.

This was a man who served not only as a scholar and teacher but as a model for those following. He will be long remembered and appreciated.

E. Schönbaum

PREFACE

For years those working in the area have discussed and disputed the definition of hibernation as opposed to hypothermia. In spite of the controversy, where it exists, certain characteristics distinguish two phenomena.

Hibernation, as it occurs among certain mammals, is considered to be a naturally recurring torporous state typified by a body temperature which approaches that of the environment and from which the animal can arouse itself - returning to the usual range of mammalian temperature without external aid. In contrast, hypothermia is an artificial or, at times, accidental depression of body temperature from which the animal or human being usually cannot recover without support. Thus, natural hibernation is a controlled physiological process while hypothermia represents a situation in which the normal physiological controls of temperature and metabolism either have been abandoned or overwhelmed.

Apart from the niceties of definition, is the ubiquitous fascination felt for the dramatic ability of some mammals to assume this state which appears to place the animal so close to boarding Charon's ferry. Accompanying the interest is the nagging question as to the inability of other animals, including man, to perform in a similar manner. While surgical and experimental hypothermia approximate hibernation in many respects, there are more serious and, to date, crucial differences between them than similarities. When taken as a whole, we hope that the contributions in this volume will at least help forge a key for unlocking one more gate in the path of man's understanding of nature.

In most instances, scientific symposia and multi-authored books perform the important functions of providing reviews of current progress as well as a forum for discussion and exploration of ideas and concepts. In the last several years work in hibernation and hypothermia has increased to the extent that no single symposium could provide representation to all areas. It was decided that a more useful approach would be to focus on a limited number of "crucial" areas of fundamental importance to the understanding of the phenomena associated with low temperature physiology. The topics chosen for this purpose were: 1. Myocardial Contractility; 2. Membrane-electrolyte Phenomena; 3. Intermediary Metabolism; 4. Central Nervous System and Thermoregulation; and 5. Timing and Synchrony with the Environment. Between three and six investigators were invited to contribute to each of these topical areas. Not only were they asked to present individually authored papers, but they also were requested to devote a significant amount of time to the consideration of the area of their responsibility as related to the problems involved in understanding the physiology of hibernation and hypothermia in general. Each of these groups then were asked to co-author a brief document in which they put together their best thoughts in the form of a "synopsis" or "position paper." We hope that these five highly individualistic and, perhaps, controversial papers will provide not only a summary statement for each of the areas which were considered but, more importantly, an additional challenge and stimulus for future work. Since many of those who acceded to our request to contribute to the symposium and this volume are experts in areas of study tangential to hibernation and hypothermia, they deserve our thanks for attempting to bridge the gap to a new and varied scientific audience.

In the preparation of any volume such as this, acknowledgment of help is not only customary but imperative. In this case, however, it is particularly so. Organization and funding of the symposium was under the sponsorship of the Space Sciences Research Center of the University of Missouri. The Director, Dr. R. H. Schiffman, was constant in his encouragement and support for the project. We are happy to express our particular thanks.

Most of the arrangement for the symposium itself and for the stenographic services afterwards could not have been made without the invaluable assistance and devotion of Mrs. Wanda Wells. The editors and the participants thank her heartily. We are most grateful to Mr. Kenneth Harmon for the patience and skill he contributed to the illustrations for this volume. The cheerfulness and help of Mrs. Donna Wilmsmeyer in typing and, frequently, retyping the manuscripts is acknowledged. To Mrs. Iris Denison fell the task of assisting with the copy editing and caring for many of the details which would have exhausted a less meticulous person. Finally, the preparation of this volume would have been quite impossible without Berna Deane South who undertook the responsibility for most of the copy editing and many of the organizational chores. Not only was this done voluntarily but also with her usual expertise and rare grace and charm.

The editors wish to note that a number of excellent short, voluntary papers were also presented during the symposium itself. It is unfortunate that it is impossible to include them in this volume. Abstracts of the voluntary papers have been published in <u>Cryobiology</u> Volume 8, Number 3 (p. 300-319) 1971. It is also appropriate to mention that several graduate students received awards for their presentations from Mrs. Jean Mannery Fisher, in the name of Professor Kenneth C. Fisher. The awardees were Sue Binkley (University of Texas), Barbara Cannon (Wenner-Grens Institute), William C. Hartner (University of Missouri) and Lloyd Wells (Harvard University).

In addition to Drs. A. R. Dawe and J. L. Shields who helped in many of the organizational details, the editors also wish to thank Drs. C. P. Lyman and E. Schönbaum for their advice, help and encouragement as well as those who served as chairmen and others who volunteered their valuable services.

F.E.S.

PARTICIPANTS

THOMAS F. ALBERT, Department of Biology, Georgetown University, Washington, D.C.
ROLAND C. ALOIA, Department of Life Sciences, University of California, Riverside, California.
NORMAN R. ALPERT, Department of Physiology, University of Vermont, Burlington, Vermont
R. K. ANDJUS, Institute of Physiology, Faculty of Science and Cryobiology Unit, Institute for Biological Research, University of Belgrade, Belgrade, Yugoslavia
SALLY J. ASMUNDSON, Department of Life Sciences, University of California, Riverside, California.
NATALIO BANCHERO, Department of Physiology, University of Colorado Medical School, Denver, Colorado.
DOUGLAS BARNES, Department of Psychology, University of Toronto, Toronto, Ontario, Canada.
GEORGE A. BARTHOLOMEW, Department of Zoology, University of California, Los Angeles, California.
ALEXANDER L. BECKMAN, Department of Physiology, School of Medicine, University of Pennsylvania, Philadelphia, Pennsylvania.
ROBERT BEYER, Department of Zoology, University of Michigan, Ann Arbor, Michigan.
SUE BINKLEY, Department of Zoology, University of Texas, Austin, Texas.
GARY BINTZ, Department of Biology, Eastern Montana College, Billings, Montana.
BARBARA H. BLAKE, Department of Zoology, Drew University, Madison, New Jersey.
J. H. BLAND, University of Vermont, Burlington, Vermont.
RAYMOND BOULOUARD, Laboratorie de Physiologie, Museum National d'Histoire Naturelle, Paris, France.
BAYARD H. BRATTSTROM, Department of Biology, California State College, Fullerton, California.
MARY ANNE BROCK, Gerontology Research Center (NIH), Baltimore City Hospitals, Baltimore, Maryland.
ROBERT W. BULLARD, Indiana University, Bloomington, Indiana.
ROY F. BURLINGTON, Department of Biology, Central Michigan University, Mount Pleasant, Michigan.
BARBARA CANNON, Wenner-Grens Institute, Norrtullsgatan, Stockholm, Sweden.
NADA CHANG, Department of Veterinary Anatomy, University of Missouri, Columbia, Missouri.
WILLIAM CIRKSENA, Department of Nephrology, Walter Reed Army Institute Of Research, Washington, D.C.
A. R. DAWE, Office of Naval Research, Chicago, Illinois.
P. DREIZEN, Downstate Medical Center, Brooklyn, New York.
MELVIN I. DYER, Patuxent Wildlife Research Center, Ohio Field Station, Sandusky, Ohio.
R. R. ELLER, University of Texas at Arlington, Arlington, Texas.
THOMAS E. EMERSON, JR., Department of Physiology, College of Human Medicine, Michigan State University, East Lansing, Michigan.

CECIL ENTENMAN, Institute for Lipid Research, Berkeley, California.
LESLIE L. FANG, University of Illinois, Urbana, Illinois.
J. HOMER FERGUSON, University of Idaho, Moscow, Idaho.
JEANNE MANERY FISHER, Department of Biochemistry, University of Toronto, Toronto, Canada.
G. EDGAR FOLK, Department of Physiology & Biophysics, University of Iowa, Iowa City, Iowa.
DAVID FRENCH, Michigan State University, East Lansing, Michigan.
CHARLES C. GALE, Regional Primate Research Center, University of Washington, Seattle, Washington.
WILLIAM GALSTER, Institute of Arctic Biology, University of Alaska, Fairbanks, Alaska.
CARLOS GITLER, Centro de Investigacion y de Estudios Avanzados, Inst. Politech, Nacional, Mexico City, D.F.
CECILIE A. GOODRICH, Department of Anatomy, Harvard Medical School, Boston, Massachusetts.
WILLIAM E. GRAVES, Syracuse University College of Forestry, Syracuse, New York.
BURT HAMRELL, University of Vermont, Burlington, Vermont.
JAMES HANEGAN, Department of Biology, Eastern Washington State College, Cheney, Washington.
J. P. HANNON, USA Medical Research and Nutrition Lab., Fitzsimons General Hospital, Denver, Colorado.
WILLIAM C. HARTNER, Department of Physiology, University of Missouri, Space Sciences Research Center, Columbia, Missouri.
DAVID HARTSHORNE, Carnegie-Mellon University, Pittsburgh, Pennsylvania.
EDWARD HAWTHORNE, Department of Physiology, Howard University, Washington, D.C.
J. S. HAYWARD, Department of Biology, University of Victoria, Victoria, British Columbia.
J. E. HEATH, Department of Physiology and Biophysics, University of Illinois, Urbana, Illinois.
H. CRAIG HELLER, Scripps Institute of Oceanography, Physiological Research Laboratory, LaJolla, California.
ROBERT E. HENSHAW, Department of Biology, Pennsylvania State University, University Park, Pennsylvania.
CEIL HERMAN, Space Sciences Research Center, University of Missouri, Columbia, Missouri.
JACK HOLLAND, Michigan Technological University, Houghton, Michigan.
FRANCES HUNTER, Department of Physiology, Howard University, Washington, D.C.
KURT JACOBS, Space Sciences Research Center, University of Missouri, Columbia, Missouri.
BERNARD JAROSLOW, Argonne National Laboratory, Argonne, Illinois.
BENGT W. JOHANSSON, Heart Laboratory, Department of Medicine, General Hospital, Malmö, Sweden.
BARBARA KENT, Mt. Sinai Hospital, New York, New York.
GEORGE KLAIN, USAMRNL, Fitzsimons General Hospital, Denver, Colorado.
JAMES KOSKI, Department of Anatomy, University of Michigan, Ann Arbor, Michigan.
PAUL LICHT, Department of Zoology, University of California, Berkeley, California.

RICHARD LUECKE, Department of Chemical Engineering, University of Missouri, Columbia, Missouri.
CHARLES P. LYMAN, Department of Anatomy, Harvard Medical School, Harvard University, Boston, Massachusetts.
G. ROBERT LYNCH, Department of Zoology, University of Iowa, Iowa City, Iowa.
VIRGINIA MAY, Space Sciences Research Center, University of Missouri, Columbia, Missouri.
NICHOLAS MENDLER, Institute of Experimental Surgery, University of Munich, Munich, Germany.
LJUBODRAG MIHAILOVIĆ, Faculty of Medicine, University of Belgrade, Yugoslavia.
ROSE MARY MONACO, Space Sciences Research Center, University of Missouri, Columbia, Missouri.
J. EMIL MORHARDT, Washington University, St. Louis, Missouri.
MARTIN L. MORTON, Department of Biology, Occidental College, Los Angeles, California.
RICHARD MOY, Department of Zoology, University of Montana, Missoula, Montana.
NICHOLAS MROSOVSKY, Department of Zoology, University of Toronto, Toronto, Canada.
ALMA MURPHY, Kirksville College of Osteopathy and Surgery, Kirksville, Missouri.
X. J. MUSACCHIA, Space Sciences Research Center, University of Missouri, Columbia, Missouri.
R. D. MYERS, Department of Psychology, Purdue University, E. Lafayette, Indiana.
RALPH NELSON, Section of Clinical Nutrition, Mayo Clinic, Rochester, Minnesota.
MARGARET C. NEVILLE, Department of Physiology, University of Colorado Medical Center, Denver, Colorado.
FREDERICK OSCHMANN, 117 W. Willow Street, Pomona, California.
JOSEPH A. PANUSKA, Department of Biology, Georgetown University, Washington, D.C.
E. T. PENGELLEY, Department of Life Sciences, University of California, Riverside, California.
PAVA POPOVIC, Department of Physiology, Emory University, Atlanta, Georgia.
V. P. POPOVIC, Department of Physiology, Emory University, Atlanta, Georgia.
RUSSELL PREWITT, Space Sciences Research Center, University of Missouri, Columbia, Missouri.
JOSEFINE RAUCH, Department of Zoology, University of Alberta, Edmonton, Canada.
ROBERT REDDICK, Pathology Department, University of North Carolina Medical School, Chapel Hill, North Carolina.
M. L. RIEDESEL, Department of Biology, University of New Mexico, Albuquerque, New Mexico.
GERALD L. RIGLER, Department of Biology, University of New Mexico, Albuquerque, New Mexico.
EDITH ROSENBERG, Department of Physiology, Howard University, Washington, D.C.
G. D. V. van ROSSUM, Department of Pharmacology, Temple University School of Medicine, Philadelphia, Pennsylvania.

EVELYN SATINOFF, Psychology Department, University of Pennsylvania, Philadelphia, Pennsylvania.
ROBERT H. SCHIFFMAN, Director, Space Sciences Research Center, University of Missouri, Columbia, Missouri.
E. SCHÖNBAUM, University of Toronto, Toronto, Ontario, Canada.
BYRON SCHOTTELIUS, Department of Physiology & Biophysics, University of Iowa, Iowa City, Iowa.
DOROTHY SCHOTTELIUS, Department of Physiology, University of Iowa, Iowa City, Iowa.
GRACE SCOTT, Ramsay-Wright Zoological Lab., University of Toronto, Toronto, Ontario, Canada.
JEROME B. SENTURIA, Department of Biology, Cleveland State University, Cleveland, Ohio.
JAMES SHIELDS, National Institutes of Health, Washington, D.C.
KATHLEEN SHIVERICK, University of Vermont, Burlington, Vermont.
RICHARD C. SIMMONDS, Manned Spacecraft Center - NASA, Houston, Texas.
W. SLEATOR, Department of Physiology & Biophysics, University of Illinois, Urbana, Illinois.
MICHAEL W. SMITH, Institute of Animal Physiology, Babraham, Cambridge, England.
GEORGE SOMERO, Scripps Institute of Oceanography, LaJolla, California.
JOACHIM R. SOMMER, Department of Pathology, Duke University, Durham, North Carolina.
JOSEF SOUHRADA, Department of Anatomy and Physiology, Indiana University, Bloomington, Indiana.
FRANK E. SOUTH, Space Sciences Research Center, University of Missouri, Columbia, Missouri.
EDWARD E. SOUTHWICK, Department of Zoology, Washington State University, Pullman, Washington.
WILMA SPURRIER, Loyola University, Stritch Medical School, Maywood, Illinois.
ROBERT C. STONES, Michigan Technological University, Houghton, Michigan.
F. J. SULLIVAN, Nutrition Lab., Denver, Colorado.
HENRY SWAN, Department of Surgery (Research), 6700 W. Lakeridge Road, Denver, Colorado.
PHILIP TEITELBAUM, Department of Psychology, University of Pennsylvania, Philadelphia, Pennsylvania.
MARTHA THOMPSON, Department of Pathology, University of Missouri, Columbia, Missouri.
KENNETH TORKE, Space Sciences Research Center, University of Missouri, Columbia, Missouri.
IGNATIUS L. TRAPANI, Department of Chemistry, Colorado Mountain College, Glenwood Springs, Colorado.
JANET TWENTE, Space Sciences Research Center, University of Missouri, Columbia, Missouri.
JOHN W. TWENTE, Space Sciences Research Center, University of Missouri, Columbia, Missouri.
RONALD VAN OEVEREN, University of Michigan, Ann Arbor, Michigan.
HEINZ WAHNER, Clinical Pathology, Mayo Clinic, Rochester, Minnesota.
LAWRENCE WANG, University of Alberta, Edmonton, Alberta, Canada.
RICHARD M. WEBSTER, Department of Anatomy, Louisiana State University Medical Center, New Orleans, Louisiana.

RICHARD E. WEEKS, Department of Medicine, Mayo Clinic, Rochester, Minnesota.
LLOYD WELLS, Department of Anatomy, University of Rochester School of Medicine, Rochester, New York.
BERTWELL K. WHITTEN, Michigan Technological University, Houghton, Michigan.
BILL A. WILLIAMS, NASA - Ames Research Center, Moffett Field, California.
DARRELL WILLIAMS, Arctic Health Research Center, Fairbanks, Alaska.
JOHN WILLIS, Department of Physiology & Biophysics, University of Illinois, Urbana, Illinois.
SIDNEY K. WOLFSON, JR., Department of Surgery, Michael Reese Hospital, Chicago, Illinois.
RUTH YOUNG, Metabolic Unit, University of Vermont College of Medicine, Burlington, Vermont.
M. K. YOUSEF, Department of Biological Sciences, University of Nevada, Reno, Nevada.
MARVIN ZATZMAN, Department of Physiology, University of Missouri, Columbia,

NOT IN ATTENDANCE

L. M. N. BACH, Division of Biomedical Sciences, School of Medical Sciences, University of Nevada, Reno, Nevada.
ARNOLD SCHWARTZ, Division of Myocardial Biology, Baylor College of Medicine, Houston, Texas.

CONTENTS

PREFACE

LIST OF PARTICIPANTS

I. METABOLISM IN HIBERNATION AND HYPOTHERMIA

Recent advances in intermediary metabolism of hibernating mammals. Roy F. Burlington. 3

Effects of low temperature on cold sensitive enzymes from mammalian tissues. Robert F. Beyer. 17

Molecular mechanisms of temperature compensation in aquatic poikilotherms. George N. Somero 55

Structured water in biological systems. John H. Bland. . . . 81

A guide for future studies of low temperature metabolic function. John P. Hannon et al. 99

II. MEMBRANES AND ELECTROLYTE PHENOMENA

The significance and analysis of membrane function in hibernation. J. S. Willis, L. S. T. Fang, and Rachel F. Foster 123

The Na^+, K^+-ATPase transport enzyme system. Arnold Schwartz. 149

Cold swelling and energy metabolism in the hypothermic brain of rats and dogs. N. Mendler, H. J. Reulen, and W. Brendel 167

The metabolic coupling of ion transport. G. D. V. van Rossum. 191

Temperature adaptation of transport properties in poikilotherms. M. W. Smith. 219

Analysis of membrane structure. Carlos Gitler. 239

Perspectives on the role of membranes in hibernation and hypothermia. John S. Willis, et al. 281

III. CARDIAC MUSCLE

Ultrastructure of cardiac muscle. J. R. Sommer, R. L. Steere, E. A. Johnson, and P. H. Jewett. 291

Calcium, and the control of muscle activity. D. J. Hartshorne and L. J. Boucher. 357

Chemistry of muscular contraction: myosin.
Paul Dreizen 373

Membrane processes and activation in mammalian cardiac muscle.
William W. Sleator 401

The mechanical properties of the isolated papillary muscle from the thirteen line ground squirrel. Norman R. Alpert, Burt B. Hamrell and William Halpern. 421

Current approaches to the study of the dynamic geometry of the left ventricle in conscious animals.
Edward M. Hawthorne and Frances Kraft-Hunter 457

Excitation and contraction mechanisms in heart muscle during hibernation and hypothermia.
Norman R. Alpert et al. 481

IV. CNS AND THERMOREGULATION

Cortical and subcortical electrical activity in hibernation and hypothermia. Lj. T. Mihailović. 487

Analysis of correlates between levels of consciousness and activity of the central nervous system.
L. M. N. Bach. 535

The role of hypothalamic monoamines in hibernation and hypothermia. R. D. Myers and T. L. Yaksh. 551

A possible model for thermoregulation during deep hibernation.
R. H. Luecke and F. E. South 577

The responsiveness of the preoptic-anterior hypothalamus to temperature in vertebrates. James Edward Heath, Bill A. Williams, Steven H. Mills, and Matthew J. Kluger. 605

Status of the role of the central nervous system and thermoregulation during hibernation and hypothermia.
Frank E. South et al. 629

V. TIMING AND SYNCHRONY OF THE ENVIRONMENT

An analysis of the mechanisms by which mammalian hibernators synchronize their behavioural physiology with the environment. E. T. Pengelley and Sally J. Asmundson 637

Aspects of timing and periodicity of heterothermy.
George A. Bartholomew. 663

Problems in experimentation on timing mechanisms for annual physiological cycles in reptiles. Paul Licht. 681

Problems in the chronobiology of hibernating mammals. E. T. Pengelley et al. 713

Author Index. 717

Subject Index 733

METABOLISM IN HIBERNATION AND HYPOTHERMIA

RECENT ADVANCES IN INTERMEDIARY METABOLISM
OF HIBERNATING MAMMALS

ROY F. BURLINGTON

Department of Biology
Central Michigan University
Mt. Pleasant, Michigan 48858

INTRODUCTION

Hibernating mammals survive unfavorable environmental conditions by existing in a dormant, hypothermic state for extended periods. A number of investigators have become interested in the cellular events associated with this phenomenon; however, the following questions remain only partially answered: How is energy acquired, transformed, and utilized during the seasonal hibernating cycle, and what mechanisms control these processes? What cellular adaptations enable these animals to survive decreased temperatures which are lethal for most homeotherms? Tissue adaptions to cold and energy metabolism in hibernators were reviewed during the preceding Hibernation Symposium (25,33). Current aspects of brown fat metabolism in hibernating species have received adequate review (19,23,24) and will be excluded from the present discussion. The present paper is an attempt to present current information concerning intermediary metabolism in hibernating mammals.

MITOCHONDRIAL METABOLISM

The marked decrease in body temperature and metabolic rate which characterize hibernation are accompanied by changes at the cellular level. Some controversy remains as to the direction of these changes and the problem is compounded by the need to consider temperature effects when *in vitro* systems are utilized and extrapolated to explain physiological phenomena in the living animal. For example, Chaffee, et al. (8) observed a decrease in succinic oxidase activity in liver mitochondria from hibernating hamsters when compared to control nonhibernating animals at 7°C; however, an increased activity was noted when similar comparisions were made at 37°C. On the other hand, a significant decrease in succinic dehydrogenase activity was found at 37°C in heart homogenates from hibernating Citellus tridecemlineatus (34).

More recently, Liu, et al. (20) and Frehn (11) have studied respiration and phosphorylation in liver mitochondria from hibernating hamsters and ground squirrels (C. tridecemlineatus) at temperatures ranging from 5 to 35°C. Succinate and β-hydroxybutyrate oxidation rates were decreased during hibernation. These results agree with the observations of Shug, et al. (22) which showed a decreased capacity for succinate oxidation in liver mitochondria from hibernating golden-mantled ground squirrels (C. lateralis). In support of this observation, the total cytochrome level in mitochondria from hibernating animals was decreased to 1.5 μmoles/mg protein compared to 3.8 μmoles/mg protein in control mitochondria from nonhibernating animals. This change is apparently due to a decrease in a-a_3 cytochromes (Fig. 1). Of further interest in this study was the marked stimulating effect of salicylanilide XIII, a potent uncoupler of oxidative phosphorylation, on the respiration rate of mitochondria from hibernating squirrels (Fig. 2). Thus the respiration rate of mitochondria appears not to be limited by the amount of cytochrome but could possibly be affected by the alteration of the coupling mechanism, extra-mitochondrial factors, or changes in the penetration of nucleotides. In regard to these possibilities, Horwitz, et al. (13) and Horwitz and Nelson (14) found the respiratory rates of washed mitochondria from torpid and aroused bats (Myotis a. austroriparius) were not signifi-

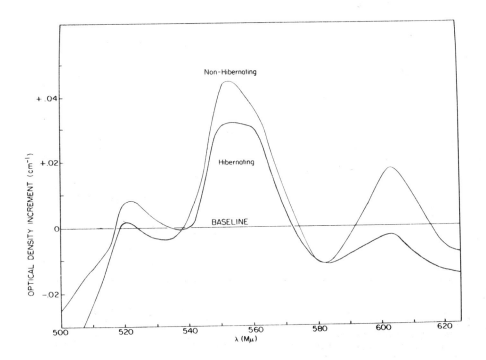

FIG. 1. Difference Spectra for the respiratory components of ground squirrel liver mitochondria. Oxidized (potassium ferricyanide) minus Reduced (sodium hydrosulfite). From Shug, et al. (22).

cantly different at a variety of temperatures. However, at 5° and 21°C, succinate oxidation, with and without ADP, was significantly decreased in unwashed mitochondria from torpid animals. This important observation lends credence to the hypothesis that the metabolic characteristics of mitochondria are influenced by extramitochondrial factors which in turn could be affected by the physiological changes taking place during hibernation. Whether or not these data can be associated with the ability to hibernate remains open to question.

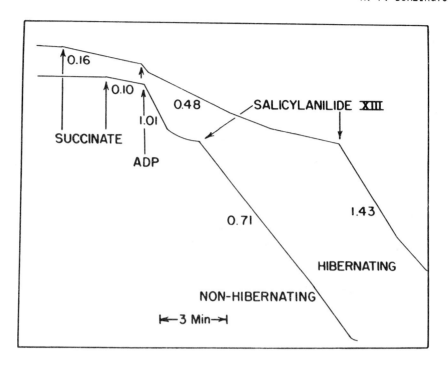

FIG. 2. Control of respiration and response to salicylanilide XIII by liver mitochondria from hibernating and nonhibernating ground squirrels. Numbers represent respiration rate in μmoles of oxygen/mg protein/hr. The oxygen concentration at the time of succinate addition was 0.24 M. From: Shug, et al. (22).

INTERMEDIARY METABOLISM

The deposition of fat in the summer and fall and the subsequent utilization of this energy depot for survival during the winter has long been recognized as a major feature of the hibernating cycle in many mammals. Understanding the complex interrelations of lipid, carbohydrate and protein metabolism which are related to the above events remains a challenge which is being accepted by an increasing number of investigators.

During hibernation, an R. Q. of 0.7 indicates that lipid catabolism is the major source of energy (25). The rate of lipolysis is apparently geared for a reduced but constant supply of energy at the temperatures encountered during hibernation. The capacity for lipogenesis is decreased markedly during the winter period (10). Support for this hypothesis was recently provided by Whitten and Klain (30) who studied glucose and acetate metabolism and the activity of several NADP specific dehydrogenases in hepatic tissue from hibernating, arousing and normothermic ground squirrels. Compared to normothermic animals, the incorporation of glucose-^{14}C-UL and acetate-1-^{14}C into fatty acids, glycerol-glyceride and non-saponifiable lipids was markedly decreased in tissue from hibernators when measured at 37°C. The activities of hexose monophosphate shunt dehydrogenases as well as NADP specific malic dehydrogenase and citrate cleavage enzyme were also substantially reduced during hibernation. A decreased lipogenesis from glucose and acetate was also observed in white adipose tissue from hibernating animals (17). Tashima, et al. (26) found in vivo lipogenesis from glucose-U-^{14}C to be severely restricted during hibernation. These investigators also presented data to support the concept that noncarbohydrates are the primary source of energy during hibernation. With the exception of studies with brown adipose tissue, the mechanisms which control seasonal changes in lipid metabolism have received little attention even though this area would seem to be a major area of concern for the investigator interested in the intermediary metabolism.

Protein synthesis is also drastically reduced during the hibernating season (21). Incorporation of methionine (methyl - ^{14}C) by liver microsomes at 37°C is significantly lower in tissues from hibernating or aroused ground squirrels compared to normothermic squirrels (29). More recently, Whitten, et al. (31) demonstrated a disaggregation of hepatic ribosomes during the winter season (Fig. 3) and no difference in cell sap enzyme activities (28). These changes are accompanied by a decreased protein synthetic capacity in a highly purified cell-free system. Thus, ribosome disaggregation is respon-

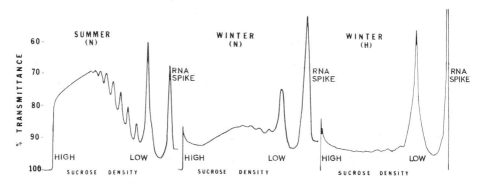

FIG. 3. Profile of hepatic polyribosomes from active summer, active winter, and hibernating winter ground squirrels showing a decreased number of high molecular weight aggregates in winter animals. From: Whitten, et al. (31).

sible for the decreased protein synthetic capacity in hibernation. Although lipid and protein synthesis are severely depressed in the hibernating animal, recent evidence supports the concept that carbohydrate anabolism as well as catabolism continues at a decreased but measurable rate. The half-life of plasma glucose was established as 35 hours in hibernating golden-mantled ground squirrels (26).

Some controversy exists as to the changes in blood glucose levels during hibernation (25). Twente and Twente (27) reviewed earlier data on plasma glucose in hibernators. Blood glucose did not decline in hibernating C. lateralis when supplied with food (27); however, in more recent studies a significant decline was observed in fasting, hibernating C. tridecemlineatus (6) and C. undulatus plesuis (12). Plasma glucose changes can be induced by a variety of experimental conditions including the method of blood withdrawal and handling, and differences in the physiological status of the animal when blood is withdrawn. Although the presence of food in the hibernaculum may have some bearing on glucose values, most investigators agree that food consumption is negligible during the short periods between hibernation (25). If this statement is true, it must follow that plasma glucose is maintained from glycogenolysis or gluconeogenesis. More recent studies have shown that liver glycogen is depleted during the actual hibernating period (6,12,26). Tashima, et al. (26) estimated that the difference in liver glycogen between active and hibernating ground squirrels could maintain plasma glucose for 4.5 days; however, this species is known to remain in continuous hibernation for up to 8 days. Thus, gluconeogenesis must be the source of some plasma glucose during the hibernation period. In support of this proposal, the capacity for gluconeogenesis from a variety of substrates was found to be present at 5°C in kidney cortex slices from C. tridecemlineatus (6). Furthermore, this capacity was significantly enhanced in hibernating and aroused squirrels compared to control nonhibernating animals as measured at 5 or 40°C. Evidence to support these data was supplied by Klain and Whitten (15). In this study the capacity for $C^{14}O_2$ incorporation into glycogen and glucose was enhanced at 6° or 37°C in hepatic tissue from hibernating or aroused ground squirrels.

Blood glucose is apparently near normal during arousal of fasted squirrels (6,12,26), yet the utilization of plasma glucose is markedly increased compared to rates in hibernation (26). Galster and Morrison have shown a replenishment of liver glycogen within five hours after the arousal of fasted ground squirrels (12). Current evidence supports the results of previous data wherein an R. Q. of 1.0 during arousal was thought to reflect an increased oxidative metabolism of glucose (25). Daudova and Stepanova (9) found a markedly increased liver and muscle hexokinase activity during arousal. These results support Tashima's observation of a fifteen fold increase in the rate of plasma glucose disappearance and an apparent increased glycolysis during arousal (26). Based on these data, an increased capacity for gluconeogenesis during hibernation would be an important source of glucose to meet the energy requirements imposed by arousal.

In addition to the increased metabolism of carbohydrate during arousal, current evidence also suggests an increased catabolism of lipids and proteins. During arousal, the ratio of $^{14}CO_2$ (from glucose-^{14}C-UL) to total CO_2 was less than that in active control animals (26). This data was thought to reflect the catabolism of substrate other than carbohydrate during arousal. In 1968, both Klain and Whitten (16) and Kristoffersson (18) showed a significant elevation of plasma amino acids in arousing C. tridecemlineatus and Erinaceus europaeus respectively. A significant increase in alanine catabolism was evident in arousing ground squirrels when measured in vitro at 6 or 37°C and compared to values in tissues from hibernating animals (29). Further studies at 5, 15, 25 and 35°C have demonstrated a transitional increase in both alanine and leuine catabolism during arousal (32).

TEMPERATURE, ENZYMES, AND METABOLIC CONTROL

Evidence to date suggests that high energy phosphates (PC and ATP) are maintained at near normal levels during hibernation despite reduced temperatures (25). Since minimal amounts of energy must be available to maintain physiological integrity during hibernation, it follows that the enzyme systems involved must remain viable at low temperatures. Certain enzymes in hibernators appear to have a lower effective energy of activation than similar enzymes in nonhibernating mammals. Essentially, this hypothesis as developed by South (cf. 25) states that the structural integrity of enzymes in hibernating mammals is maintained at low temperatures. The net result is an effective enzyme activity in the hibernator which allows maintenance of overall metabolic homeostasis in the cell at low temperatures. Recent reports support this hypotheses (4,5).

Although this characteristic of enzymes in hibernators presumably exists throughout the year, in some cases it may be seasonally induced and exist only during the hibernation period (2). Brain microsomal preparations from hibernating hedgehogs were found to have a Na,K,Mg-ATPase which was less sensitive to low temperature than the enzyme from nonhibernating hedgehogs. In contrast, the properties of Mg ATPase from both hibernating and nonhibernating hedgehogs were similar to those from the laboratory rat. The temperature resistant form of Na,K,Mg-ATPase may be synthesized prior to hibernation and under the influence of adrenal cortical hormones.

Isozymes may exist which enable hibernating species to survive low temperatures; however, this possibility has received little attention. Isozyme distribution of lactic dehydrogenase was studied in the ground squirrel (7) and bat (3). In both studies, possible shifts of isozyme patterns toward LDH-5 (the "anaerobic form") were observed in tissues from animals in hibernation. Any conclusions based on these data will remain tenuous until the physiological role of isozymes is more fully understood.

A number of enzyme activities have been shown to increase or decrease during hibernation but perhaps the most striking changes have recently been observed in enzymes within less than 2 hours after the initiation of arousal. Compared to enzyme activities in tissues from hibernating animals assayed at the same temperature, the following activities were found to be increased significantly in tissues from arousing animals: hexokinase (9), succinic dehydrogenase (34), lactic dehydrogenase (6), glutamic-oxaloacetic-transaminase (6), glutamic-pyruvate-transaminase (6), and arginase (29). Increasing _in vivo_ temperatures during arousal will increase the activity of enzymes markedly; however, the above increases, when measured at the same temperature _in vitro_, requires further consideration. Increased enzyme synthesis may be partially responsible for this phenomenon but it seems more reasonable to invoke other possible mechanisms to explain such short term changes.

It has recently been appreciated that enzyme kinetic behavior can be modulated by a metabolite or "effector". Adenine nucleotides (ATP, ADP, AMP) as well as metabolic intermediates themselves (citrate, glucose-6-phosphate) can significantly and immediately modify the affinity of an enzyme for its substrate (1). Thus a "positive" effector substantially increases the rate of an enzymatically catalyzed reaction at a constant substrate concentration whereas a "negative" effector has the opposite effect. For example, AMP will inhibit fructose-1, 6-diphosphatase, which catalyzes the conversion of fructose diphosphate to fructose-6-phosphate. The same compound will promote the activity of phosphofructokinase, an enzyme which catalyzes the above reaction in the reverse direction (1). It is entirely conceivable that the apparent rapid changes in enzyme activities in aroused hibernators are a reflection of "effectors" which may be present in the tissue preparations used for enzyme determinations.

In summary, research concerning intermediary metabolism in hibernating species is entering a new era. Recent advances toward the understanding of complex metabolic control systems in mammals must be considered to reach meaningful conclusions about the nature of cellular metabolism in hibernators. The very nature of the changes which occur throughout the seasons imply that intermediary metabolism in mammalian hibernators is influenced by subtle, dynamically balanced control mechanisms which operate throughout a wide range of temperatures.

REFERENCES

1. ATKINSON, D. E. Biological control at the molecular level. Science 150: 851-857, 1965.

2. BOWLER, K., and C. J. DUNCAN. The temperature characteristics of brain microsomal ATPases of the hedgehog: changes associated with hibernation. Physiol. Zool. 42: 211-219, 1969.

3. BRUSH, A. H. Response of isozymes to torpor in the bat, Eptesicus fuscus. Comp. Biochem. Physiol. 27: 113-120, 1968.

4. BURLINGTON, R. F., and J. E. WIEBERS. Anaerobic glycolysis in cardiac tissue from a hibernator and non-hibernator as effected by temperature and hypoxia. Comp. Biochem. Physiol. 17: 183-189, 1966.

5. BURLINGTON, R. F., and J. E. WIEBERS. The effect of temperature on glycolysis in brain and skeletal muscle from a hibernator and non-hibernator. Physiol. Zool. 40: 201-206, 1967.

6. BURLINGTON, R. F., and G. J. KLAIN. Gluconeogenesis during hibernation and arousal from hibernation. Comp. Biochem. Physiol. 22: 701-708, 1967.

7. BURLINGTON, R. F., and J. H. SAMPSON. Distribution and activity of lactic dehydrogenase isozymes in tissues from a hibernator and non-hibernator. Comp. Biochem. Physiol. 25: 185-192, 1968.

8. CHAFFEE, R. R. J., F. L. HOCH, and C. P. LYMAN. Mitochondrial oxidative enzymes and phosphorylations in cold exposure and hibernation. Amer. J. Physiol. 201: 29-32, 1961.

9. DAUDOVA, G. M., and N. G. STEPANOVA. Hexokinase and glucokinase of cell fractions of liver and muscles in ground squirrels during hibernation and arousal. Federation Proc. 25, Part II: T273-T276, 1966.

10. DENYES, A., and J. D. CARTER. Utilization of acetate-1-C^{14} by hepatic tissue from cold-exposed and hibernating hamsters. Amer. J. Physiol. 200: 1043-1046, 1961.

11. FREHN, J. L., and J. A. THOMAS. Effects of cold exposure and hibernation on the liver, brown fat, and testes from golden hamsters. J. Exptl. Zool. 170: 107-116, 1969.

12. GALSTER, W. A., and P. MORRISON. Cyclic changes in carbohydrate concentrations during hibernation in the arctic ground squirrel. Amer. J. Physiol. 218: 1228-1232, 1970.

13. HORWITZ, B. A., L. NELSON, and V. P. POPOVIC. Effect of temperature on oxidative phosphorylation in hibernators and nonhibernators. J. Appl. Physiol. 22: 639-644, 1967.

14. HORWITZ, B. A., and L. NELSON. Effect of temperature on mitochondrial respiration in a hibernator (Myotis austroriparius) and a non-hibernator (Rattus rattus). Comp. Biochem. Physiol. 24: 385-394, 1968.

15. KLAIN, G. J., and B. K. WHITTEN. Carbon dioxide fixation during hibernation and arousal from hibernation. Comp. Biochem. Physiol. 25: 363-366, 1968.

16. KLAIN, G. J., and B. K. WHITTEN. Plasma free amino acids in hibernation and arousal. Comp. Biochem. Physiol. 27: 617-619, 1968.

17. KLAIN, G. J., and G. B. ROGERS. Seasonal changes in adipose tissue lipogenesis in the hibernator. Intern. J. Biochem. 1: 248-250, 1970.

18. KRISTOFFERSSON, R., and S. BROBERG. Free amino acids in blood serum of hedgehogs in deep hypothermia and after spontaneous arousals. Experientia 24: 148-150, 1968.

19. LINDBERG, O., editor. Brown Adipose Tissue. New York: Elsevier, 1970.

20. LIU, C-C., J. L. FREHN and A. D. LAPORTA. Liver and brown fat mitochondrial response to cold in hibernators and nonhibernators. J. Appl. Physiol. 27: 83-89, 1969.

21. MANASEK, F. J., S. J. ADELSTEIN, and C. P. LYMAN. The effects of hibernation on the in vitro synthesis of DNA by hamster lymphoid tissue. J. Cellular Comp. Physiol. 65: 319-324, 1965.

22. SHUG, A. L., S. FERGUSON, E. SHRAGO, and R. F. BURLINGTON. Changes in respiratory control and cytochromes in liver mitochondria during hibernation. Biochem. Biophys. Acta. 226: 309-312, 1971.

23. SMALLEY, R. L., and R. L. DRYER. Brown fat in hibernation. In: Mammalian Hibernation III, edited by K. C. Fisher, A. R. Dawe, C. P. Lyman, E. Schonbaum, and F. E. South. Edinburgh: Oliver and Boyd: 1967, p. 325-345.

24. SMITH, R. E., and B. A. HORWITZ. Brown fat and thermogenesis. Physiol. Rev. 49: 330-425, 1969.

25. SOUTH, F. E., and W. A. HOUSE. Energy metabolism in hibernation In: Mammalian Hibernation III, edited by K. C. Fisher, A. R. Dawe, C. P. Lyman, E. Schonbaum, and F. E. South. Edinburgh, Oliver and Boyd, 1969, p. 305-324.

26. TASHIMA, L. S., S. J. ADELSTEIN, and C. P. LYMAN. Radioglucose utilization by active, hibernating, and arousing ground squirrels. Amer. J. Physiol. 218: 303-309, 1970.

27. TWENTE, J. W., and J. A. TWENTE. Concentrations of D-glucose in the blood of Citellus lateralis after known intervals of hibernating periods. J. Mammol. 48: 381-386, 1967.

28. WHITTEN, B. K., M. A. POSIVIATA, and W. D. BOWERS. Seasonal changes in hepatic ribosome aggregation and protein synthesis in the hibernator. Physiologist 13: 339, 1970.

29. WHITTEN, B. K., and G. J. KLAIN. Protein metabolism in hepatic tissue of hibernating and arousing ground squirrels. Amer. J. Physiol. 214: 1360-1362, 1968.

30. WHITTEN, B. K., and G. J. KLAIN. NADP specific dehydrogenases and hepatic lipogenesis in the hibernator. Comp. Biochem. Physiol. 29: 1099-1104, 1969.

31. WHITTEN, B. K., L. E. SCHRADER, R. L. HUSTON, and G. R. HONALD. Hepatic polyribosomes and protein synthesis: seasonal changes in a hibernator. Intern. J. Biochem., 1971. In press.

32. WHITTEN, B. K., and R. F. BURLINGTON. Amino acid catabolism during arousal from hibernation. (Abstract) Federation Proc. 30: 483, 1971.

33. WILLIS, J. S. Cold adaptation of activities of tissues of hibernating mammals. In: Mammalian Hibernation III, edited by K. C. Fisher, A. R. Dawe, C. P. Lyman, E. Schonbaum, and F. E. South. Edinburgh: Oliver and Boyd, 1967, p. 356-381.

34. ZIMNY, M. L., and J. E. MORELAND. Mitochondrial populations and succinic dehydrogenase in the heart of a hibernator. Can. J. Physiol. Pharmacol. 46: 911-913, 1968.

EFFECTS OF LOW TEMPERATURE ON COLD SENSITIVE ENZYMES FROM MAMMALIAN TISSUES

ROBERT E. BEYER

Laboratory of Chemical Biology
Department of Zoology
The University of Michigan
Ann Arbor, Michigan 48104

INTRODUCTION

Although I have had some interest in the problem of the mechanism of cold acclimation in the laboratory rat (2,6,36), I have devoted little thought to the problems of hibernation. I have always assumed, in my naivete, that certain key metabolic reactions were subject to Q_{10} considerations of the effect of temperature on chemical reactions and that at low temperature, i.e. those approaching 0°C, substrate became limiting but was produced in sufficient quantity to enable the hibernator to maintain core temperature slightly above ambient. The reality of participation in this Symposium has prompted a deeper consideration of this problem for me and has forced the question: What information is there available from biochemical studies of enzyme regulation which might lead to a possible novel approach to biochemical regulation during hibernation? Studies of poikilotherms have already dealt with various energetic parameters (37,39,49) and I could see no advantage in attempting to deal with such an approach here. Another interesting and valuable approach has been the examination of temperature effects on isozymes (23) which Dr. Somero will examine in considerable detail. My own interest in the role of protein coupling factors in oxidative phosphorylation (3,4) has kept me abreast of developments in the area of low temperature inactivation of enzymes since the enzyme complex most probably responsible for catalyzing the terminal reactions of oxidative phosphorylation (44) has been shown to be cold-labile (46), a characteristic which has been studied in some detail (45). Since approximately twenty enzymes, from various organisms and participating in a variety of metabolic pathways, have been shown to be reversibly cold-labile, the major thrust of this paper will deal with several aspects of cold inactivation of enzymes and the possible relevance of this phenomenon to animals whose tissues are subjected to low temperature during hibernation.

COLD-LABILE ENZYMES - GENERAL

Cold-inactivation of enzymes was first reported in some detail from Racker's laboratory (44,46) in 1960 and since that time has been observed with a number of enzymes (Table 1).

ADENOSINETRIPHOSPHATASE, COUPLING FACTOR 1

In the first of a long series of publications on coupling factors of mitochondrial oxidative phosphorylation (46) by Racker and his colleagues it was observed that during the purification of a soluble ATPase from beef heart mitochondria the enzyme became cold-labile. Twenty minutes of incubation at 0° lowered ATPase activity to approximately half of its maximal activity. Of particular interest was the observation that partial reactivation of ATPase activity could be obtained by incubation of the cold-inactivated enzyme at room temperature. The ability of the coupling factor to increase the ATP synthetic capacity of poorly phosphorylating submitochondrial particles was also shown to be cold-labile (44). Somewhat different results have been reported for a coupling factor from beef heart mitochondria by Sanadi and his co-workers (1). The two coupling factors, F1 and factor A, are similar in that they both catalyze the hydrolysis of ATP, increase the phosphorylative capacity, the ^{32}P-ATP exchange rate, and the rate of reversed electron flow

TABLE 1. Cold-labile enzymes (14).

Enzyme	Source	Reference
Urease	Jack Bean	24
Glutamate dehydrogenase	Neurospora crassa	15
Adenosine triphosphatase, F1	Beef heart mitochondria	45
Carbamyl phosphate synthetase	Frog liver	48
D(-) β-Hydroxybutyrate dehydrogenase	R. rubrum	56
Glutamate decarboxylase	E. coli	55
Nitrogen-fixing enzyme	C. pasteurianum	13
Arginosuccinase	Beef liver	20
Glycogen phosphorylase	Rabbit muscle	19
Pyruvate carboxylase	Chicken liver mitochondria	53,27
Pyruvate kinase	Baker's yeast	34
Glucose 6-phosphate dehydrogenase	Human erythrocyte	31
17 β-Hydroxysteroid dehydrogenase	Human placenta	28
Prephenate dehydrogenase	Bacillus subtilis	7
Proline-tRNA synthetase	E. coli	42
Glyceraldehyde 3-phosphate dehydrogenase	Rabbit muscle	8,9,10
Threonine deaminase	S. typhimurium	18
Threonine deaminase	R. rubrum	25
Homoserine dehydrogenase	R. rubrum	12
Coupling factor 1	Spinach chloroplasts	35
Lactic streptococcus proteinase	Strep. lactis	11
Coupling factor A	Beef heart mitochondria	1
Acetyl-CoA carboxylase	Rat liver	41
Tryptophanase	E. coli	38

of submitochondrial particles, and display cold-lability of ATPase catalysis. However, although ATPase activity of Factor A is sensitive to cold, cold treatment does not destroy its ability to increase the phosphorylative activity, the ability to stimulate ^{32}P-ATP exchange, or reversed electron flow of submitochondrial particles (1). Penefski and Warner (45) have studied the mechanism of cold inactivation of F1 in some detail. The intact, active enzyme preparation has a single boundary in the analytical ultracentrifuge consisting of 11.9S particles of 284,000 molecular weight. Cold inactivation produces 9.1S and 3.5S particles while prolonged cold treatment results in 3.5S subunits which are not capable of reactivation or physical reassociation. Penefsky and Warner (45) suggest that cold inactivation consists of at least two steps which are depicted in Scheme 1. Cold treatment results in an equilibrim mixture

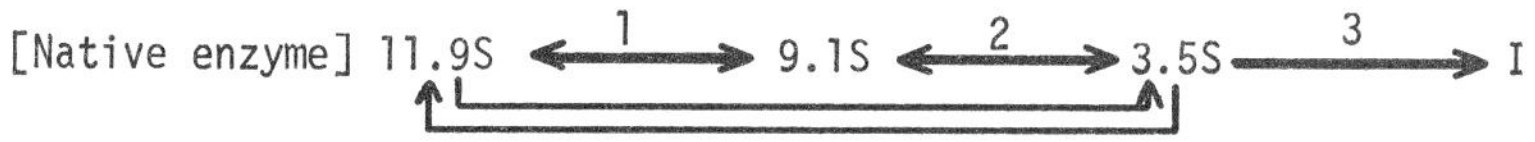

$$[\text{Native enzyme}]\ 11.9S \xrightleftharpoons{1} 9.1S \xrightleftharpoons{2} 3.5S \xrightarrow{3} I$$

SCHEME 1. Subunit interrelations of coupling factor 1 (45).

of 11.9S, 9.1S and 3.5S components capable of transformation to the native 11.9S enzyme upon incubation at room temperature. Upon prolonged incubation in the cold, irreversible changes appear to occur in the 3.5S component. From the effects of ionic strength and pH on dissociation the authors suggest that ionic, as well as hydrophobic, bonding is involved in the maintenance of the quarternary structure and function of the native enzyme. It is of interest that chemical modification of F1 with iodine transforms the native 11.9S component completely to 3.5S components (43). Iodinated F1 activities are similar to those displayed by prolonged cold treated enzyme, i.e. ATPase activity is absent while ATP synthetic, reversed electron flow and ^{32}P-ATP exchange activities of the test submitochondrial particles are enhanced by the iodine-treated F1. The view of F1 subunit composition presented above may be simplistic, however, in view of the recent suggestion (54) that the homogeneous F1 complex contains four components with molecular weights of 8, 12.5, 25 and 55 x 10^3. Further work is obviously required on this interesting and important system.

Since the mitochondrial phosphorylative system not only serves the cell as the major contributor of ATP, but also regulates the rate of electron transfer in the presence of excess oxidizable substrate, and thus the rate of heat production, it is of importance that we examine the possible relevance of the temperature regulated quarternary structure of F1 (and Factor A). A current concept of the reactions leading to the synthesis of cellular ATP differentiates between the roles of endogenous, bound mitochondrial adenine nucleotide and exogenous, cytosolic adenine nucleotide (5). The initial convalently bound phosphorylated compound is considered to be ATP bound to the inner mitochondrial membrane (Scheme 2), the energy for the formation of the bond between membrane-bound ADP and inorganic phosphate being derived, by a still unknown mechanism, from oxidation-reduction energy released during electron transfer in the same inner mitochondrial membrane. Since the newly synthesized ATP, presumably catalyzed by F1 or Factor A, is bound to the inner mitochondrial membrane it is not available in that form to do extra-mitochondrial work. In order for the terminal phosphate of bound ATP to be transferred to exogenous ADP the soluble ADP must first pass through the outer mitochondrial membrane, presumed to be freely permeable to adenine nucleotide, then through the inner mitochondrial membrane where an enzymic exchange (Scheme 3) between bound ATP and soluble ADP occurs. The inner mitochondrial membrane is not freely permeable to ADP and requires transport

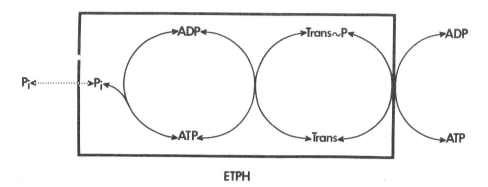

SCHEME 2. High phosphate potential transfer between mitochondrial and cytosolic adenine nucleotides (5).

across the membrane by a specific translocator (32). Once inside the inner mitochondrial (matrix) space the soluble ADP is phosphorylated, possibly by the enzyme phosphoryl transferase (5). The studies of Heldt (21) and of Kemp et al. (30) indicate that the rate limiting step in the overall process of exogenous ATP synthesis at low temperature is the ADP translocator. The effect of low temperature on the rate of uncoupler stimulated ATPase, the ^{32}P-ATP and ADP-ATP exchanges, and the rate of substrate oxidation catalyzed by intact mitochondria in the presence of excess ADP all show discontinuous Arrhenius plots with a break at approximately 17°C. These Arrhenius plots, together with the observation (30) that the rate of phosphorylation of endogenous ADP exceeds the rate of phosphorylation of exogenous ADP at all temperatures between 0° and 23°C, suggest that the rate limiting step of oxidative phosphorylation between 0° and 17°C resides at the adenine nucleotide translocator.

Unfortunately, information about the adenine nucleotide translocator shares the same paucity as most other components of membrane mitochondrial

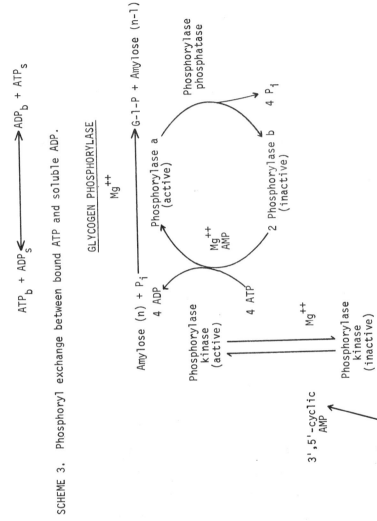

SCHEME 3. Phosphoryl exchange between bound ATP and soluble ADP.

SCHEME 4. Regulation of glycogen phosphorylase.

systems in that it has not been isolated and studied in detail, although a start has been made to identify the carrier sites (59). Thus, essentially nothing is known about the regulatory interrelationships between the ATP synthetase (or F1) system and the adenine nucleotide translocator system. Since it has been shown (47) that F1 is an allotrophic effector (i.e. has an effect on inner mitochondrial membrane function in addition to and separate from its catalytic function) the possibility exists that in intact mitochondrial systems at low temperature the rate limiting reaction in oxidative phosphorylation, the adenine nucleotide translocator, is allotrophically regulated by the subunit status of F1, the enzyme responsible for the terminal step in the synthesis of endogenous ATP.

GLYCOGEN PHOSPHORYLASE

To this point, we have dealt with control of metabolism at a metabolic level which involves the terminal sequence of reactions of the cell, the transfer of electrons from high electron transfer potential substrates to oxygen, catalyzed by the mitochondrial electron transfer chain. One of the major energetic reserves of the cell is glycogen and the enzyme which catalyzes the production of a glycolytic substrate, glucose-1-phosphate, from glycogen is glycogen phosphorylase, an enzyme which is under an elaborate set of controls (cf. 17). The activation of phosphorylase is shown in Scheme 4. Epinephrine effects the synthesis of 3',5'-cyclic AMP catalyzed by adenyl cyclase. Cyclic AMP activates the formation of active from inactive phosphorylase kinase which in turn is able to transform two dimer subunits of inactive phosphorylase _b_ to the active phosphorylase _a_ tetramer by the transfer of the terminal phosphate of four ATP molecules to four serine residues, one on each of the four subunits of the enzyme. Each phosphorylase a subunit contains, in addition to an active catalytic site and phosphorylated serine residue, bound pyridoxal-5'-phosphate and an allosteric site. AMP and glucose-6-phosphate compete for the allosteric site, AMP acting as an allosteric effector and glucose-6-phosphate as an allosteric inhibitor. Active phosphorylase a may be converted to the inactive dimer by the action of phosphorylase phosphatase which dephosphorylates a phosphorylserine residue on each of the four subunits.

The first indication that phosphorylase might be cold sensitive was the observation (58) that glycogen phosphorylase a was unstable to dialysis at high ionic strength and at 3°-4°C. Upon further investigation of the inactivation of phosphorylase at low temperature it was found (19) that the inactive form of the enzyme, phosphorylase b, lost approximately 50% of its activity after 1.5 hours at 0°C and that the dimer was considerably more sensitive to cold than the active tetramer form of the enzyme. As with the case of the ATPase-F1, cold inactivated phosphorylase b is reversible by warming (Table 2). The rate of reactivation of the enzyme is proportional to the enzyme concentration and is more efficient in the presence of pyridoxal phosphate than in its absence. The effect of pyridoxal phosphate on the reactivation extent, as well as its essentiality for the maintenance of the native conformation of the enzyme (26), suggested a role for this coenzyme in the reactivation phenomenon. Kinetic analysis, however, indicated that the loss of enzymic activity could be more directly related to a conformational alteration of the protein structure than to loss of pyridoxal phosphate (19). Analytical ultracentrifugation date of the native, cold-inactivated, and reactivated cold-inactivated enzyme support this view since a heavier component is formed during inactivation and the pattern reverts to the normal, single protein species upon reactivation.

TABLE 2. Reversal of cold inactivation of phosphorylase b. (Data from ref. 19.)

Time at 20° after 4 hours at 0° C	% Original activity	
	No additions	+ Pyridoxal-p
0	14	
120	76	101

An Arrhenius plot of log enzyme activity vs the reciprocal of temperature showed a distinct discontinuity at 13°C. The activation energy below 13°C was calculated to be 46,000 cal as contrasted to 17,000 cal for temperatures above 13°C. The mechanistic basis for such differential activation energies is not known, but the distinctly different activation energies and the abrupt break in the Arrhenius plot clearly indicate two distinctly different populations of enzyme. It is easy to envision that at low temperature a conformational form of phosphorylase b exists which is thermodynamically in an unfavorable form for conversion to catalytically active form. That glycogen phosphorylase in muscle is predominantly in the dimer b form has already been observed (16). Thus, even in the presence of ATP required for phosphorylase b to a conversion, the tissue at low temperature may have an unusual thermodynamic control over the rate of substrate provision for glycolysis and, consequently, many other cellular pathways required for cellular maintenance. What I am suggesting is that at low temperature, in situ, an extremely low concentration of phosphorylase is in the active form and that the controlling factor is the low temperature effect on the tertiary and quaternary structure of the inactive b form imposing an extreme thermodynamic difficulty for the b to a conversion.

GLYCERALDEHYDE 3-PHOSPHATE DEHYDROGENASE

Glyceraldehyde 3-phosphate dehydrogenase catalyzes a key step in glycolysis. In addition to catalyzing the obligatory synthesis of ATP, this step in the glycolytic sequence is unique in being an oxidative reaction and, in addition, comprises approximately 10% of the soluble protein of rabbit skeletal muscle. Although this enzyme has not been considered one of the regulators of glycolysis, recent evidence indicates that this view should be reconsidered and that glyceraldehyde 3-phosphate dehydrogenase may serve as an important control point. The evidence to support this contention (8) is (a) 3',5'-AMP is a potent inhibitor of the enzyme, (b) ATP causes inhibition and dissociation of the enzyme into subunits, and (c) NAD^+ is required for reassociation of the subunits into an active enzyme complex. This enzyme also dissociates reversibly into dimers or monomers in the presence of ATP at low temperature (8). Upon incubation of rabbit skeletal muscle glyceraldehyde 3-phosphate dehydrogenase at low temperatures, enzymic activity is lost (Table 3). Loss of activity at 7° and 0°C was most pronounced during the first hour and was relatively slow thereafter. Enzyme incubated at 12°C or above showed little loss of activity during the first six hours and inactivation was relatively slow subsequently.

Sucrose gradient analysis of the native and ATP-cold inactivated preparations indicate that the 7.4S native tetramer dissociates into 3.1S monomers when incubated in the presence of ATP and at low protein concentration.

An interesting aspect of low temperature inactivation of glyceraldehyde 3-phosphate dehydrogenase is the requirement for ATP (Table 4). The requirement for ATP is absolute; its effect depends upon its concentration and is related to the protein concentration during cold-inactivation. Nevertheless, even at high protein concentration, dissociation of tetramers into inactive dimers occurred at 0°C in the presence of ATP.

TABLE 3. Effect of low temperature on glyceraldehyde 3-phosphate dehydrogenase activity.

Temperature °C	Specific activity, μmoles/min/mg protein
23	50
7	25
0	5

Enzyme, 0.05 mg/ml; ATP, 2 mM; Time of incubation, 60 min. (Data from ref. 8.)

TABLE 4. Effect of ATP on low temperature inactivation of glyceraldehyde 3-phosphate dehydrogenase. (Data from ref. 8.)

ATP, mM during inactivation at 0° C, 4 hr	Spec. act., μmoles/min/mg protein		
	0.1 mg/ml	0.5 mg/ml	2 mg/ml
0	54	54	54
1	18	39	47
10	0	20	36

As is the case of the two previous enzymes discussed above, cold inactivated glyceraldehyde 3-phosphate dehydrogenase may be reactivated by warming. In the case of this glycolytic enzyme, reactivation was extremely rapid (Fig. 1). An interesting aspect of the effect of low temperature and ATP on inactivation of this enzyme is the complete protection against inactivation and dissociation afforded by its coenzyme, NAD^+ (Fig. 1) (9). Since, under aerobic conditions, tissues would be expected to contain a high level of NAD^+ when cellular ATP concentration is low, the relative concentrations of NAD^+ and ATP may act to regulate the subunit relationship of

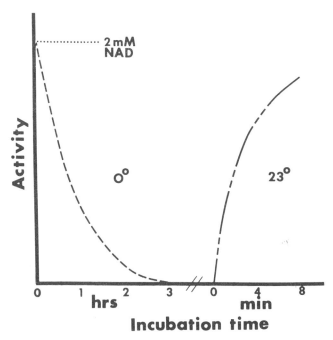

FIG. 1. Temperature dependent inactivation and reactivation of glyceraldehyde 3-phosphate dehydrogenase. (Redrawn from ref. 8.)

glyceraldehyde 3-phosphate dehydrogenase under low temperature conditions and thus provide a regulatory step in glycolysis under such conditions. Another interesting aspect in the consideration of regulation of glycolysis at low temperature by subunit alteration is the finding (9) that several other glycolytic enzymes which are composed of subunits e.g. skeletal muscle aldolase and pyruvate kinase, and heart lactate dehydrogenase, do not show dissociation and inactivation at low temperature.

It would thus appear that glyceraldehyde 3-phosphate dehydrogenase, until recently thought to be a relatively stable 7.4S tetrameric monodisperse, in actuality exists in equilibrium with its 4.4S dimer and 3.2S monomer in aqueous solution. At low temperature the equilibrium is in favor of dissociation while at high protein concentrations and at high temperatures the equilibrium favors the formation of the tetrameric, active form.

PYRUVATE CARBOXYLASE

In addition to the important function of providing high electron transfer potential substrates for oxidative phosphorylation, the Krebs cycle may supply precursors for anabolic pathways. One important Krebs cycle intermediate which may be removed from the mitochondrion and serve as a substrate for amino acid synthesis is oxaloacetate. When oxaloacetate is in excess in the mitochondrion it may be removed; since the reaction which catalyzes the removal of oxaloacetate is reversible this auxiliary pathway may also provide oxaloacetate when its intramitochondrial concentration is low. The enzyme which catalyzes this amphibolic reaction is pyruvate carboxylase which is localized in liver mitochondria of most species and which catalyzes the reversible endergonic carboxylation of pyruvate to form oxaloacetic acid, at the expense of ATP. When the concentration of its allosteric regulator, acetyl-CoA, is low the rate of the forward reaction is very slow. However, in the presence of excess acetyl-CoA, pyruvate carboxylase catalyzes the formation of oxaloacetic acid which then may act as a Krebs cycle acceptor for excess acetyl-CoA, enabling the Krebs cycle to oxidize more acetyl-CoA.

$$\text{Pyruvate} + \text{ATP} + \text{HCO}_3^- \xrightleftharpoons[]{\text{Acetyl-CoA, Mg}^{2+}} \text{OAA} + \text{ADP} + P_i$$

SCHEME 5. Reaction catalyzed by pyruvate carboxylase.

SCHEME 6. Conversion of pyruvate to phosphenolpyruvate in gluconeogenesis.

$$\text{Pyruvate} + CO_2 + ATP \xrightarrow[\text{Acetyl-CoA}]{\text{Pyruvate carboxylase}} OAA + ADP + P_i$$

$$OAA + NADH + H^+ \xrightarrow{\text{M-malate DH}} NAD^+ = \text{malate}$$

$$\text{Malate} + NAD^+ \xrightarrow{\text{C-malate DH}} OAA + NADH + H^+$$

$$OAA + GTP \xrightarrow[\text{Acetyl-CoA}]{\text{PEP carboxykinase}} PEP + CO_2 + GDP$$

Sum: $\text{Pyruvate} + ATP + GTP \longrightarrow PEP + ADP + GDP + P_i$

In addition to being an important reaction regulating the supply of Krebs cycle intermediates for oxidative, energy-yielding reactions, pyruvate carboxylase also plays an important role in gluconeogenesis from amino acids. Oxaloacetic acid produced as a result of pyruvate carboxylase action in the presence of excess acetyl-CoA is converted in the mitochondrion to malate by mitochondrial malate dehydrogenase (M-malate DH), the reducing equivalents being supplied by NADH. Malate may then diffuse out of the mitochondrion into the cytosol and be converted by extramitochondrial malate dehydrogenase (C-malate DH) to oxaloacetic acid. Oxaloacetic acid in the cytosol is converted to the glycolytic intermediate phosphoenolpyruvate in the presence of cytosolic acetyl-CoA, the phosphate group being donated from GTP. Phosphoenolpyruvate may now participate in the usual series of reactions of gluconeogenesis.

In addition to the type of control of pyruvate carboxylase described above the enzyme is inducible. Fasting for 24 to 36 hours result in a 2- to 3-fold increase in the specific activity of the enzyme in chicken liver mitochondria (53).

Pyruvate carboxylase was reported in 1963 to be very labile during its purification (57), a lability which has since been shown to be due to its

TABLE 5. Effect of temperature on activity of pyruvate carboxylase. (Data from ref. 53.)

Temperature, °C	% Inactivation at 150 min
2	69
5	62
10	33
15	9
22	0

inactivation at low temperature (Table 5). The break in the inactivation rate would appear to be in a temperature range similar to other enzymes discussed above with subunit structure, i.e. between 10° and 15°C.

As indicated by the data in Table 6, acetyl-CoA present during cold treatment provided a high degree of protection against inactivation by cold. In addition, the combination of all substrates, cofactor and allosteric effector provided virtually complete protection during four hours of treatment at low temperature. Although the mechanism of protection by these components is not known, the indication of conformational alteration of enzyme structure upon binding of substrates (i.e. induced fit) (33) would suggest that the binding between pyruvate carboxylase and its substrates, metal cofactor and allosteric effector induces a conformational alteration in the enzyme which places the protein in a thermodynamically more stable conformation than in its unbound state. The stabilization of enzymes during storage or purification by the presence of substrate has long been recognized.

As in the case with other cold labile enzymes already discussed, pyruvate carboxylase is capable of reactivation simply by warming. This phenomenon is shown in Figure 2, which is very similar in character to that

TABLE 6. Protection of pyruvate carboxylase against cold inactivation.

	% of original activity	
Present during cold treatment	-acetyl-CoA	+acetyl-CoA
	13.5	62.5
ATP, Mg^{2+}, HCO_3^-	48.5	78.0
Pyruvate, ATP, Mg^{2+}, HCO_3^-	53.5	98.8

Enzyme was incubated at 0° for 4 hours prior to assay. (Data from ref. 53.)

depicted in Figure 1 for glyceraldehyde 3-phosphate dehydrogenase. Following 96 minutes at 2°C only 8% of the original activity of the enzyme remained. Partial and rapid reactivation could be obtained by rewarming the enzyme preparation (63% in the experiment shown in Figure 2) but all attempts to obtain further reactivation, including the addition of acetyl-CoA immediately prior to rewarming and the use of higher temperatures, failed. Evidently a significant degree of irreversible inactivation occurs during cold treatment of pyruvate carboxylase under the in vitro experimental conditions employed.

Subunit structural alteration appears to be a fairly common property of enzymes which are cold-sensitive, pyruvate carboxylase notwithstanding. Ultracentrifugal studies (53) of cold-inactivated preparations indicate that cold-inactivation is accompanied by a transformation of the native tetrameric 15S structure to four 7S subunits. However, these authors claim that "the extent of dissociation is clearly inadequate to account for the loss of activity observed" and they suggest that "the physical changes may be quite complex" (53).

In a later publication from the same laboratory (27) it was reported that the addition of ATP to the cold-inactivated enzyme prior to warming

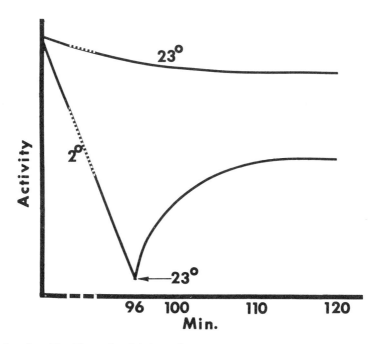

FIG. 2. Reactivation of cold-inactivated pyruvate carboxylase by warming. 92% of the original activity remained after 96 minutes at 2°. (Data from ref. 53.)

resulted in an almost complete degree of reactivation and that under such conditions the kinetics of reactivation are very rapid. Further, the allosteric effector, acetyl-CoA, at the relatively low concentrations required for catalytic activity (ca. 30 µM), provides almost complete protection against cold inactivation. Similarly, one of the products of the reaction, oxaloacetate, completely protects the enzyme from cold inactivation. This would appear to have regulatory significance since the substrate pyruvate, which was formerly thought to provide some protection from cold-inactivation (53), has been found to actually accelerate the inactivation process in the absence of ATP (27).

COLD SENSITIVE ENZYMES

On the basis of observations reported herein, as well as others not included, Irias et al. (27) propose a tentative scheme to describe the conformational status and activity of pyruvate carboxylase with regard to

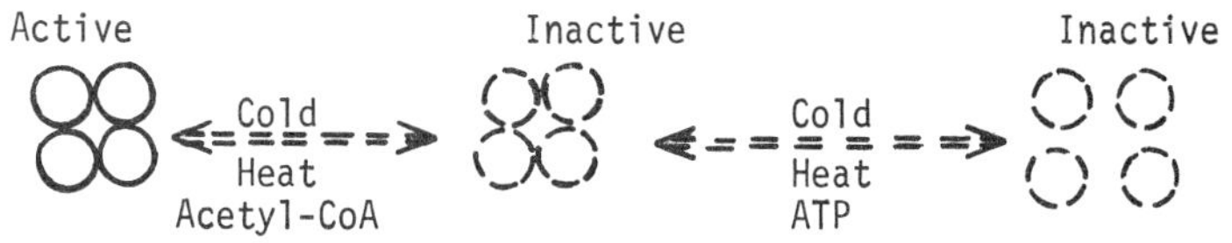

```
    Active                    Inactive                   Inactive
     ⬡⬡    <--Cold-->          ⬡⬡      <--Cold-->         ○ ○
     ⬡⬡       Heat             ⬡⬡         Heat            ○ ○
          Acetyl-CoA                       ATP
```

SCHEME 7. Inactivation of pyruvate carboxylase. (Modified from 27.)

cold inactivation (Scheme 7). On exposure to cold the native, active tetramer is converted into an inactive tetramer which, upon further cold treatment, dissociates into four inactive monomers. It should be noted that the reactions in Scheme 7 are reversible. In addition, the presence of acetyl-CoA and ATP would favor equilibrium toward the left, thus favoring an active species of enzyme. Of particular interest in Scheme 7 is the postulation of an inactive tetramer. Since pyruvate carboxylase, as well as mitochondrial ATPase (F1), is membrane-bound in vivo the likelihood of such subunit complexes dissociating into monomers in vivo, even at low temperatures, is not great. The existence of a reversible reaction, controlled by low temperature as one parameter, between active and inactive tetramers, involving conformational alterations from one state to another, is thus attractive.

ARGINOSUCCINASE

The urea cycle functions in ureotelic organisms to provide a means whereby ammonia released from amino acids may be excreted in the form of a non-toxic end product, urea. The urea cycle sequence of reactions is catalyzed by enzymes contained in liver mitochondria which provides a compartmentation between the reactions of amino acid catabolism and the urea cycle. The overall reaction of the urea cycle is shown in Scheme 8 in which it should be recognized that the two molecules of ammonia are derived from amino acid α-amino groups.

$$2\ NH_3 + CO_2 + 3\ ATP + 2\ H_2O \longrightarrow Urea + 2\ ADP + 2\ P_i + AMP + PP_i$$

SCHEME 8. Overall equation of the urea cycle.

In the urea cycle two amino groups, derived from α-amino acids by deamination, enter the cycle and, with a molecule of carbon dioxide, form arginine from ornithine. In ureotelic animals large amounts of the enzyme arginase catalyze the hydrolysis of arginine to form urea and regenerate ornithine. In the cycle one of the compounds formed is arginosuccinate which is cleaved by arginosuccinase to form arginine and fumaric acid (Scheme 9). The product fumarate may participate in the Krebs cycle while

$$\text{Arginosuccinic acid} \xrightarrow{\text{Arginosuccinase}} \text{Arginine + Fumarate}$$

SCHEME 9. Reaction catalyzed by arginosuccinase.

arginine is converted to urea and ornithine by arginase. There exist several points of control within the sequence of reactions which could possibly limit the rate of the cycle. The level of NAD^+ required for the deamination of glutamic acid could regulate the rate of ammonia formation. The level ATP could conceivably regulate the formation of carbamyl phosphate required for the formation of citrulline from ornithine. In addition, inactivation of a urea cycle enzyme at low temperature could conceivably regulate the rate of the cycle at low temperature. The one enzyme which, to date, has been identified as cold-labile is arginosuccinase (20).

The data in Table 7 indicate that the loss of arginosuccinase activity at 0°C is slow in relation to the series of enzymes discussed above. In addition, little or no cold inactivation was observed, even at 22 hours,

TABLE 7. Loss of arginosuccinase activity at 0°C. (Data from ref. 20.)

Hrs. at 0°C	Buffer, pH 7.5	% Original activity
3	Tris	65
6	Tris	50
22	Tris	26
22	Phosphate	83

when phosphate buffer was substituted for Tris buffer. In a later publication, the rate of inactivation at 0°C was shown to be considerably more rapid in imidazole buffer (52). The presence of the substrate, arginosuccinate, protects the enzyme from low temperature inactivation when present before exposure to low temperature or when added at any time during cold treatment (Fig. 3). The authors suggest the reasonable explanation that the binding of substrate, in association with the active site of the enzyme, induces a conformational alteration in the vicinity of the active site which results in a stabilization of the protein structure (52).

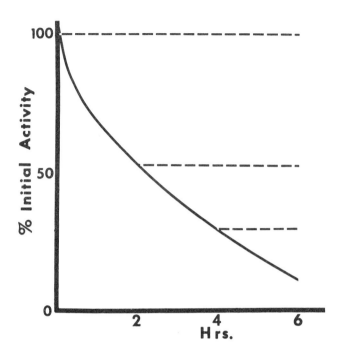

FIG. 3. Effect of substrate on cold-inactivation of arginosuccinase. Substrate added to approximately 2 mM when indicated by broken line. (Data from 52).

By sucrose density gradient centrifugation analysis it was demonstrated that during cold inactivation the native dimer (9.3S) was converted to an inactive monomer (5.6S). Although this point does not appear to have been studied in any detail, Havir et al. (20) report that the rate of inactivation was independent of the stage of purity. The possible significance of this will be discussed later.

As was the case with the other cold-sensitive enzymes discussed above, arginosuccinase activity is restored by warming, with the reestablishment of the original dimer structure. However, the rate of reactivation of arginosuccinase was extremely rapid, full reactivation of a 50% inactivated preparation (18 hour at 0°C being accomplished in 6 minutes [20]).

An analysis of kinetic data with regard to association and dissociation of arginosuccinase has led to the proposal (52) of three forms of the enzyme in equilibrium with one another as in Scheme 10 where A_2 represents

$$A_2 \underset{k_3}{\overset{k_4}{\rightleftarrows}} A_2^{in} \underset{k_1}{\overset{k_2}{\rightleftarrows}} 2A$$

SCHEME 10. Kinetic equilibria between active and inactive subunits of arginosuccinase. (Modified after ref. 52.)

enzymatically active dimer and A_2^{in} represents an assumed enzymatically inactive dimer. At temperatures above 10°C k_1 is greater than k_2 while at lower temperatures $k_1 < k_2$. Low temperature would also favor k_4. The kinetic data also indicate that A_2 and A_2^{in} are in very rapid equilibrium with one another. An interesting aspect of the kinetic data is that approach to equilibrium with samples moderately inactivated (60%) by cold showed an acceleration of reassociation which continued as well as approach to equilibrium permitted. Following extensive inactivation (> 80%) acceleration during assay was very great during the first few minutes and then slowed as equilibrium was approached. This phenomenon is not fully understood but Schulze et al. (52) suggest that the change from fast to slow rates of reassociation is too great to be a result of approach to equilibrium. Analysis of the curves depicting this phenomenon suggests that only k_3, i.e. the rate of conversion of inactive dimer to active dimer, is of first order, and that the governing rate after this initial conversion is k_1 which is not first order. Thus, the change in kinetic order may be the result of critical changes in the concentration of the intermediate, A_2^{in} (52). This consideration may have significance for the urea cycle in tissues at low temperature since any alteration in the concentration of the enzyme (i.e. A_2) would alter the concentration of the intermediate form A_2^{in}. Since the concentration of this enzyme is reduced during dietary change (51) a larger proprtion may be in the A_2^{in} and 2A forms during this, and perhaps other, physiological conditions and thus reduce enzymic capacity.

ACETYL-CoA CARBOXYLASE

The reaction catalyzed by acetyl-CoA carboxylase is an important step in the biosynthesis of fatty acids. The reaction cycle occurs in two steps, viz:

$$CO_2 + ATP + \text{biotin-enzyme} \rightleftharpoons \text{carboxybiotin-enzyme} + ADP + P_i$$

$$\text{Carboxybiotin-enzyme} + \text{acetyl-CoA} \rightleftharpoons \text{malonyl-CoA} + \text{biotin-enzyme}$$

The malonyl group is transferred to acyl carrier protein (ACP) by malonyl transacylase and participates as a substrate, after decarboxylation, in the synthesis of long chain fatty acids. As is the case with most of the cold-sensitive enzymes described above, cold-sensitive acetyl-CoA carboxylase from rat-liver consists of subunit structure containing four monomers each of approximately 100,000 molecular weight. The enzyme is allosterically regulated by the positive effectors citrate, isocitrate and α-ketoglutarate. The regulators transform the enzyme to the active polymeric form.

When rat-liver acetyl-CoA carboxylase is activated by citrate and incubated at 0°C approximately half of its activity is lost in 30 minutes (41). Inactivation is accompanied by conversion of the enzyme to smaller subunits, and incubation of the inactivated enzyme at 25°C results in relatively complete and rapid restoration of activity and polymeric structure. Similarly to those cold-sensitive mammalian enzymes for which data are available on the rate of inactivation at various temperatures, the break point for acetyl-CoA carboxylase from rat-liver occurs between 7° and 15°C.

The great regulatory nature, as well as its key position in fatty acid synthesis, make acetyl-CoA carboxylase attractive in terms of speculating on its regulatory role in tissues at low temperature. This aspect is discussed below.

DISCUSSION

For the purpose of discussing the phenomenon of cold sensitivity in proteins it would indeed simplify matters if the details of inactivation, protection, reactivation, and the modification of these phenomena, were similar for the entire group. Unfortunately, such is not the case, and it is therefore not possible to generalize about the class of proteins which possess this common trait. In addition, although a particular enzyme from one organism may be endowed with the property of cold sensitivity, proteins possessing the capability of catalyzing the same reaction from other organisms have not been shown to be cold sensitive. The rates of inactivation may differ widely as is the case of 17β-hydroxysteroid dehydrogenase (28) which is rapid and pyruvate carboxylase (27,53) which is slow. Although the rate of cold-inactivation of pyruvate kinase from Baker's yeast was very slow, it could be increased about 1000-fold by fructose 1,6-diphosphate, its allosteric activator (34). On the other hand, the rate of inactivation by cold of glyceraldehyde 3-phosphate dehydrogenase was enhanced by its allosteric inhibitor, ATP (8). Still another dissimilar effect was seen with isoleucine, the allosteric inhibitor of threonine deaminase from Rhodospirillum rubrum, which protected the enzyme from inactivation (14,18, 25).

One property which the cold sensitive enzymes appear to have in common is the rapid, initial reversible phase of inactivation followed by a slower and irreversible stage, the irreversibility of the latter stage generally being proportional to the time of cold treatment. An exception to this latter generality is glutamate decarboxylase from E. coli which appears capable of reactivation after long periods of cold incubation (55).

Although I have indicated that enzymic inactivation is accompanied by subunit dissociation, even this is not invarient. Pyruvate carboxylase appears to undergo cold-inactivation by virtue of a conformational alteration prior to falling apart into subunits (27) and inactivation of 17β-hydroxy-steroid dehydrogenase coincides with the formation of higher aggregate forms of the enzyme (28).

One feature which all of the cold-sensitive enzymes may share is an absolute dependence on hydrophobic bonding to ensure proper conformation for enzymic activity. The ability of glycerol to protect cold-sensitive proteins, together with the general property of sensitivity to low temperature, has prompted the suggestion that sensitivity to cold involves protein-water interactions. Specifically, it would appear that cold sensitivity of proteins involves a weakening of hydrophobic bonds within the protein. "Hydrophobic bonding" (the discussion of hydrophobic bonding which follows is, in some degree, paraphrased from the thesis by Feldberg [14]) is generally understood as the tendency of nonpolar groups to approach one another when in an aqueous environment. Kauzmann (29) and Nemethy and Scheraga (40) have attempted to explain hydrophobic bonding. They suggest that water molecules arrange themselves around nonpolar groups when such groups are in an aqueous phase. When two nonpolar groups approach each other, the number of water molecules associated with each nonpolar group decreases. This results in a decrease in the degree of order of the water molecules and therefore an increase in entropy. The increase in entropy necessitates a negative free energy change for the pairing of the nonpolar groups as necessitated by the thermodynamic expression:

$$\Delta F = \Delta H - T\Delta S$$

This expression explains why cold sensitivity may be explained by the rupturing of hydrophobic bonds in proteins. As the temperature of the protein solution is lowered, the quantity of the $-T\Delta S$ term will decrease and the free energy of formation of the hydrophobic bond will decrease (50).

Any discussion of the details of hydrophobic bonding in proteins will, of necessity, depend on the nature of the structure of water. Since the structure of water is far from being a settled question, and since the structure of water will be the subject of a lecture in this symposium, I shall not attempt a deeper analysis of this question here. However, it has been pointed out (22) that even the term hydrophobic bond may be a misnomer. The problem with the term "bond" is that the two nonpolar groups do not share electrons, nor are the dispersion forces great enough to be a significant factor in the free energy of association. It would appear, therefore, that interacting polar groups are not formally bonded and the attractive forces lie in the rearrangement of water molecules and their properties. The important point, however, with respect to hydrophobic interactions in proteins is that they are strong enough to be the major force maintaining the characteristic structure of proteins. Since the Second Law of Thermodynamics assures us that the free energy of association will become more negative with a decrease in temperature, the decrease in temperature results in an increase in the entropy of the system, namely the water molecules associated with the hydrophobic groups. Thus, it would appear reasonable to conclude that the forces which determine the correct conformation and subunit association of cold sensitive enzymes are hydrophobic in nature. It should not be concluded, however, that these forces are exclusively hydrophobic, since some of the cold sensitive enzymes lose activity more rapidly in high salt indicating an electrostatic component. Consistant with this suggestion is the pH dependence of the rate of cold-induced inactivation of, for example, threonine deaminase which also suggests a role for ionic or hydrogen bonding in the maintainence of the native protein structure (14). As pointed out by Feldberg (14), "it would indeed be surprising if ionic bonds were not involved in the maintainence of protein structure. It is likely that cold sensitivity does not reflect any unusual or unique forces, but rather is a phenomenon found in proteins in which ionic interactions are not the major ones in the native conformation."

It would appear that most cold sensitive enzymes undergo a conformational alteration to an inactive state prior to dissociation into subunits. The most conclusive evidence for this statement is the work on arginosuccinase (20) which is conformationally stabilized during cold exposure by the presence of the substrates arginosuccinate or arginine. In addition, the increase in free -SH groups during reversible cold inactivation indicates a conformational alteration. The binding of the exposed -SH groups by -SH reagents, resulting from conformational transition to the inactive state, inhibits reactivation. That the binding of the -SH groups does not interfere with active site function is indicated by the non-essentiality of the -SH groups for catalytic activity under normal functional conditions.

The last point which should be discussed is the possible relevance of cold inactivation of enzymes to the regulation of metabolism. As indicated in Figure 4, six enzymes which operate in mammalian cells display cold sensitivity. Although at least fifteen additional enzymes have been reported to be inactivated by cold treatment (cf. Table 1), I have considered only those which might be relevant to hibernating mammals whose tissues would be slightly above freezing temperatures during hibernation. This is not to say that other mammalian enzymes operating in the metabolic scheme shown in Figure 4 are not similarly affected by cold exposure, since it is possible that many enzymes which have been thought of as unstable are actually cold sensitive and have not been studied from this point of view as yet.

Consideration of the sites in metabolic pathways containing cold sensitive enzymes reveals that several key processes may possibly be controlled by such enzymes at low temperature. Two important enzymes which operate in the breakdown of carbohydrates in mammalian systems, glycogen phosphorylase and glyceraldehyde 3-phosphate dehydrogenase, are inactivated in the cold. Pyruvate carboxylase, which may act either as a control point for the Krebs cycle or as a carbon supplier for gluconeogenesis, is also cold labile. Acetyl-CoA carboxylase, which is an essential initial step in fatty acid synthesis, is also inactivated in the cold. Arginosuccinase, which functions in the urea cycle, also fits into this category. Lastly, the enzyme which appears to catalyze the synthesis of ATP coupled to the electron transfer chain and which might be expected to participate in the

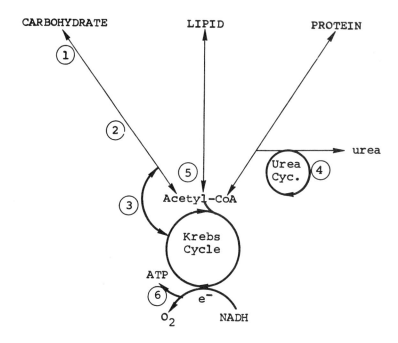

FIG. 4. Cold sensitive points in metabolic pathways of mammalian cells.
1 = glycogen phosphorylase, 2 = glyceraldehyde 3-phosphate dehydrogenase,
3 = pyruvate carboxylase, 4 = argino-succinase, 5 = acetyl-CoA carboxylase,
6 = ATPase (F1).

control of cellular ATP levels, is also inactivated by cold. It would thus appear that many of the important metabolic pathways of the cell contain at lease one step which is catalyzed by an enzyme which, when isolated, is cold labile. The question which should properly be asked is whether it is reasonable to believe that reversible cold inactivation of enzymes can be seriously entertained as a regulatory mechanism in the metabolism of cells at low temperature. Since I have not been able to find studies in the literature directed to this question, the answer, for the present, must remain mute. The question can be entertained on a conceptual basis, however. With one exception, the cold labile enzymes appear to require some degree of purification before they display cold susceptability. From this it might be argued that the phenomenon, since it is only observed in preparations at some stage of purification, has no relevance to metabolic control under cold conditions. On the other hand, since the phenomenon, in this context, has not been studied, we really do not know if this is the case. I would like to suggest the possibility that, since most of the cold labile enzymes do not necessitate dissociation of subunits for inactivation but do appear to be in equilibrium with an intact, but conformationally altered, inactive form, cold treatment may effect hydrophobic interactions of the in situ enzymes to alter equilibrium toward the inactive intermediate. I am suggesting, therefore, that cold may not necessarily drive equilibrium to subunit dissociation in vivo, but may alter the conformation sufficiently, as in allosteric inhibition, to render the pathways in question inoperative, or operative at low levels.

Although it is not possible with our current level of information to provide a definitive answer to the possibility of a role for cold inactivation of enzymes in intact mammalian systems at low temperature, I hope that this review will direct attention to such a possibility and will act to stimulate research in this direction.

ACKNOWLEDGMENTS

Research in the author's laboratory on cold acclimation and enzyme regulation is supported by grants from the National Institute of Arthritis and Metabolic Diseases (AM 10056) and the National Science Foundation

(GB-13496). Permission from Dr. Ross S. Feldberg to quote from his unpublished doctoral thesis is gratefully acknowledged.

REFERENCES

1. ANDREOLI, T. E., K. -W. LAM, and D. R. SANADI. Studies on oxidative phosphorylation. X. A coupling enzyme which activates reversed electron transfer. J. Biol. Chem. 240: 2644-2653, 1965.

2. BEYER, R. E. Regulation of energy metabolism during acclimation of the laboratory rat to a cold environment. Federation Proc. 22: 874-880, 1963.

3. BEYER, R. E. Reconstitution of oxidative phosphorylation in submitochondrial particles by a soluble protein, phosphoryl transferase. Arch. Biochem. Biophys. 123: 41-54, 1968.

4. BEYER, R. E. The isolation and reactions of a phosphorylated form of phosphoryl transferase from beef heart mitochondria. Arch. Biochem. Biophys. 125: 884-894, 1968.

5. BEYER, R. E. The interaction between phosphoryl transferase and submitochondrial particles. Canadian J. Biochem. 46: 677-683, 1968.

6. BEYER, R. E. Biochemical aspects of acclimation to a cold environment. In: Biochemical Responses to Environmental Stress, edited by I. Bernstein. New York: Plenum Publ. Co., 1971, p. 94-118.

7. COATS, J. H., and E. W. NESTER. Regulation reversal mutation: characteristics of end product-activated mutants of Bacillus subtilis. J. Biol. Chem. 242: 4948-4955, 1967.

8. CONSTANTINIDES, S. M., and W. C. DEAL. Reversible dissociation of tetrameric rabbit muscle glyceraldehyde 3-phosphate dehydrogenase into dimers or monomers by adenosine triphosphate. J. Biol. Chem. 244: 5695-5702, 1969.

9. CONSTANTINIDES, S. M., and W. C. DEAL. Reversible dissociation of tetrameric 7.4S rabbit muscle glyceraldehyde 3-phosphate dehydrogenase into 4.4S dimers by ammonium sulfate and into 3.2S monomers by KCl. J. Biol. Chem. 245: 246-253, 1970.

10. CORI, G. T., M. W. SLEIN, and C. F. CORI. Crystalline D-glyceraldehyde 3-phosphate dehydrogenase from rabbit muscle. J. Biol. Chem. 173: 605-618, 1948.

11. COWMAN, R. A., H. E. SWAISGOOD, and M. L. SPECK. Proteinase enzyme system of lactic streptococci. II. Role of membrane proteinase in cellular function. J. Bacteriol. 94: 942-948, 1967.

12. DATTA, P., and H. GEST. Homoserine dehydrogenase of Rhodospirillum rubrum. Purification, properties, and feedback control of activity. J. Biol. Chem. 240: 3023-3033, 1965.

13. DUA, R. D., and R. H. BURRIS. Stability of nitrogen-fixing enzymes and the reactivation of a cold labile enzyme. Proc. Natl. Acad. Sci. U.S. 50: 169-181, 1963.

14. FELDBERG, R. S. The purification and properties of L-threonine deaminase from Rhodospirillum rubrum. (Ph.D. Thesis.) Univ. of Michigan, Ann Arbor, Mich., 1970.

15. FICHAM, J. R. S. A modified glutamic acid dehydrogenase as a result of gene mutation in Neurospora crassa. Biochem. J. 65: 721-728, 1957.

16. FISCHER, E. H., and E. G. KREBS. The isolation and crystallization of rabbit skeletal muscle phosphorylase b. J. Biol. Chem. 231: 65-71, 1958.

17. FISCHER, E. H., and E. G. KREBS. Relationship of structure to function of muscle phosphorylase. Federation Proc. 25: 1511-1520, 1966.

18. FREUNDLICH, M., and H. E. UMBARGER. The effects of analogues of threonine and of isoleucine on the properties of threonine deaminase. Cold Spring Harbor Symp. Quant. Biol. 28: 505-511, 1963.

19. GRAVES, D. J., R. W. SEALOCK, and J. H. WANG. Cold inactivation of glycogen phosphorylase. Biochemistry 4: 290-296, 1965.

20. HAVIR, E. A., H. TAMIR, S. RATNER, and R. C. WARNER. Biosynthesis of urea. XI. Preparation and properties of crystalline argininosuccinase. J. Biol. Chem. 240: 3079-3088, 1965.

21. HELDT, H. W. The effect of low temperature and of atractyloside on the reaction of endogenous and exogenous adenine nucleotides. In: Mitochondrial Structure and Compartmentation, edited by E. Quagliariello, S. Papa, E. C. Slater, and J. M. Tager. Bari, Italy: Adriatica Editrice, 1967, p. 260-267.

22. HILDEBRAND, J. H. A criticism of the term "hydrophobic bond." J. Phys. Chem. 72: 1841-1842, 1968.

23. HOCHACHKA, P. W., and G. N. SOMERO. The adaptation of enzymes to temperature. Comp. Biochem. Physiol. 27: 659-668, 1968.

24. HOFSTEE, B. H. J. The activation of urease. J. Gen. Physiol. 32: 339-349, 1949.

25. HUGHES, M., C. BRENNEMAN, and H. GEST. Feedback sensitivity of threonine deaminases in two species of photosynthetic bacteria. J. Bacteriol. 88: 1201-1202, 1964.

26. ILLINGWORTH, B., H. S. JANSZ, D. H. BROWN, and C. F. CORI. Observations on the function of pyridoxal-5-phosphate in phosphorylase. Proc. Natl. Acad. Sci. U.S. 44: 1180-1191, 1958.

27. IRIAS, J. J., M. R. OLMSTED, and M. F. UTTER. Pyruvate carboxylase. Reversible inactivation by cold. Biochemistry 8: 5136-5148, 1969.

28. JARABEK, J., A. E. SEEDS, and P. TALALAY. Reversible cold inactivation of a 17β-hydroxysteroid dehydrogenase of human placenta: protective effect of glycerol. Biochemistry 5: 1269-1279, 1966.

29. KAUZMANN, W. Some factors in the interpretation of protein denaturation. Adv. Protein Chem. 14: 1-63, 1959.

30. KEMP, A., G. S. P. GROOT, and H. J. REITSMA. Oxidative phosphorylation as a function of temperature. Biochem. Biophys. Acta 180: 28-34, 1969.

31. KIRKMAN, H. N., and E. M. HENDRICKSON. Glucose 6-phosphate dehydrogenase from human erythrocytes. II. Subactive states of the enzyme from normal persons. J. Biol. Chem. 237: 2371-2376, 1962.

32. KLINGENBERG, M., and E. PFAFF. Structural and functional compartmentation in mitochondria. In: Regulation of Metabolic Processes in Mitochondria, edited by J. M. Tager, S. Papa, E. Quagliariello, and E. C. Slater. Amsterdam: Elsevier Publ. Co., 1966, p. 180-201.

33. KOSHLAND, D. E., and K. E. NEET. The catalytic and regulatory properties of enzymes. Ann. Rev. Biochem. 37: 359-410, 1968.

34. KUCZENSKI, R. T., and C. H. SUELTER. Effect of temperature and effectors on the conformations of yeast pyruvate kinase. Biochemistry 9: 939-945, 1970.

35. LIVNE, A., and E. RACKER. Partial resolution of enzymes catalyzing photophosphorylation. V. Interactions of coupling factor 1 from chloroplasts with ribonucleic acid and lipids. J. Biol. Chem. 244: 1332-1338, 1969.

36. LIANIDES, S. P., and R. E. BEYER. Oxidative phosphorylation in liver mitochondria from cold-exposed rats. Amer. J. Physiol. 199: 836-840, 1960.

37. LICHT, P. Thermal adaptation in the enzymes of lizards in relation to preferred body temperatures. In: Molecular Mechanisms of Temperature Adaptation, edited by C. L. Prosser. Washington, D.C.: AAAS Publ., p. 131-145, 1967.

38. MORINO, Y., and E. E. SNELL. The subunit structure of tryptophanase. I. The effect of pyridoxal phosphate on the subunit structure and physical properties of tryptophanase. J. Biol. Chem. 242: 5591-5601, 1967.

39. MUTCHMOR, J. A. Temperature adaptation in insects. In: Molecular Mechanisms of Temperature Adaptation, edited by C. L. Prosser. Washington, D.C.: AAAS Publ., 1967, p. 165-176.

40. NEMETHY, G., and H. A. SCHERAGA. Structure of water and hydrophobic bonding in proteins. II. Model for the thermodynamic properties of aqueous solutions of hydrocarbons. J. Chem. Phys. 36: 3401-3417, 1962.

41. NUMA, S., and E. RINGELMANN. Zur Aufhebung der Citrat-Aktivierung der Acetyl-CoA-Carboxylase durch Kälte. Biochem. Z. 343: 258-268, 1965.

42. PAPAS, T. S., and A. H. MEHLER. Modification of the transfer function of proline transfer ribonucleic acid synthetase by temperature. J. Biol. Chem. 243: 3767-3769, 1968.

43. PENEFSKI, H. S. Partial resolution of the enzymes catalyzing oxidative phosphorylation. XVI. Chemical modification of mitochondrial adenosine triphosphatase. J. Biol. Chem. 242: 5789-5795, 1967.

44. PENEFSKY, H. S., M. E. PULLMAN, A. DATTA, and E. RACKER. Partial resolution of the enzymes catalyzing oxidative phosphorylation. II. Participation of a soluble adenosine triphosphatase in oxidative phosphorylation. J. Biol. Chem. 235: 3330-3336, 1960.

45. PENEFSKY, H. S., and R. C. WARNER. Partial resolution of the enzymes catalyzing oxidative phosphorylation. VI. Studies on the mechanism of cold inactivation of mitochondrial adenosine triphosphatase. J. Biol. Chem. 240: 4694-4702, 1965.

46. PULLMAN, M. E., H. H. PENEFSKY, A. DATTA, and E. RACKER. Partial resolution of the enzymes catalyzing oxidative phosphorylation. I. Purification and properties of soluble, dinitrophenol-stimulated adenosine triphosphatase. J. Biol. Chem. 235: 3322-3329, 1960.

47. RACKER, E. Function and structure of the inner membrane of mitochondria and chloroplasts. In: Membranes of Mitochondria and Chloroplasts edited by E. Racker. New York: Van Nostrand Reinhold Co., 1970, p. 127-171.

48. RAIJMAN, L., and S. GRISOLIA. New aspects of acetylglutamate action. Cold inactivation of frog carbamyl phosphate synthetase. Biochem. Biophys. Res. Commun. 4: 262-265, 1961.

49. READ, K. R. H. Thermostability of proteins in poikilotherms. In: Molecular Mechanisms of Temperature Adaptation, edited by C. L. Prosser. Washington, D.C.: AAAS Publ. 1967, p. 93-106.

50. SCHERAGA, H. A., G. NEMETHY, and I. Z. STEINBERG. The contribution of hydrophobic bonds to the thermal stability of protein conformations. J. Biol. Chem. 237: 2506-2508, 1962.

51. SCHIMKE, R. T. Adaptive characteristics of urea cycle enzymes in the rat. J. Biol. Chem. 237: 459-468, 1962.

52. SCHULZE, I. T., C. J. LUSTY, and S. RATNER. Biosynthesis of urea. XIII. Dissociation-association kinetics and equilibria of argininosuccinase. J. Biol. Chem. 245: 4534-4543, 1970.

53. SCRUTTON, M. C., and M. F. UTTER. Pyruvate carboxylase. III. Some physical and chemical properties of the highly purified enzyme. J. Biol. Chem. 240: 1-9, 1965.

54. SENIOR, A. E., and J. C. BROOKS. Studies on the mitochondrial oligomycin-insensitive ATPase. I. An improved method of purification and the behavior of the enzyme in solutions of various depolymerizing agents. Arch. Biochem. Biophys. 140: 257-266, 1970.

55. SHUKUYA, R., and G. W. SCHWERT. Glutamic acid decarboxylase. III. The inactivation of the enzyme at low temperatures. J. Biol. Chem. 235: 1658-1661, 1960.

56. SHUSTER, C. W., and M. DOUDOROFF. A cold-sensitive D(-)β-hydroxybutyric acid dehydrogenase from Rhodospirillum rubrum. J. Biol. Chem. 237: 603-607, 1962.

57. UTTER, M. F., and D. B. KEECH. Pyruvate carboxylase. I. Nature of the reaction. J. Biol. Chem. 238: 2603-2608, 1963.

58. WANG, J. H., and D. J. GRAVES. Effect of ionic strength on the sedimentation of glycogen phosphorylase a. J. Biol. Chem. 238: 2386-2389, 1963.

59. WEIDEMANN, M. J., H. ERDELT, and M. KLINGENBERG. Adenine nucleotide translocation of mitochondria. Identification of carrier sites. European J. Biochem. 16: 313-335, 1970.

CREDITS

The editors and authors are grateful to the following publishers, organizations and individuals for permission to use materials previously published:

Figure 1: The American Society of Biological Chemists Inc., S. M. Constantinides and W. C. Deal, The Journal of Biological Chemistry 244: 5695-5702, 1969. Copyright The American Society of Biological Chemists Inc., 1969.

Figure 2: The American Society of Biological Chemists Inc.; M. C. Scrutton and M. F. Utter, The Journal of Biological Chemistry 240: 1-9, 1965. Copyright The American Society of Biological Chemists Inc., 1965.

Figure 3: The American Society of Biological Chemists Inc.; I. T. Schulze, C. J. Lusty and S. Ratner, The Journal of Biological Chemistry 245: 4534-4543, 1970. Copyright The American Society of Biological Chemists Inc., 1970.

Table 2: The American Chemical Society; D. J. Graves, R. W. Sealock and J. H. Wang, Biochemistry 4: 290-296, 1965. Copyright The American Chemical Society, 1965.

Table 3: The American Society of Biological Chemists Inc.; S. M. Constantinides and W. C. Deal, The Journal of Biological Chemistry 244: 5695-5702, 1969. Copyright The American Society of Biological Chemists Inc., 1969.

Table 4: The American Society of Biological Chemists Inc.; S. M. Constantinides and W. C. Deal, The Journal of Biological Chemistry 244: 5695-5702, 1969. Copyright The American Society of Biological Chemists Inc., 1969.

Table 5: The American Society of Biological Chemists Inc.; M. C. Scrutton and M. F. Utter, The Journal of Biological Chemistry 240: 1-9, 1965. Copyright the American Society of Biological Chemists Inc., 1965.

Table 6: The American Society of Biological Chemists Inc.; M. C. Scrutton and M. F. Utter, The Journal of Biological Chemistry 240: 1-9, 1965. Copyright The American Society of Biological Chemists Inc., 1965.

Table 7: The American Society of Biological Chemists Inc.,; E. A. Havir, H. Tamir, S. Ratner and R. C. Warner, The Journal of Biological Chemistry 240: 3079-3088, 1965. Copyright The American Society of Biological Chemists Inc., 1965.

Scheme 1: The American Society of Biological Chemists Inc.; H. S. Penefsky and R. C. Warner, The Journal of Biological Chemistry 240: 4694-4702, 1965. Copyright The American Society of Biological Chemists Inc., 1965.

Scheme 2: National Research Council of Canada; R. E. Beyer, Canadian Journal of Biochemistry 46: 677-683, 1968.

Scheme 7: The American Chemical Society; J. J. Irias, M. R. Olmsted and M. F. Utter, Biochemistry 8: 5136-5148, 1969. Copyright The American Chemical Society, 1969.

Scheme 10: The American Society of Biological Chemists Inc.; I. T. Schulze, C. J. Lusty and S. Ratner, The Journal of Biological Chemistry 245: 4534-4543, 1970. Copyright The American Society of Biological Chemists Inc., 1970.

MOLECULAR MECHANISMS OF TEMPERATURE COMPENSATION IN AQUATIC POIKILOTHERMS

GEORGE N. SOMERO

Scripps Institution of Oceanography
University of California-San Diego
La Jolla, California 92037

INTRODUCTION

The effects of temperature on the metabolic functions of poikilotherms have long been a favorite interest of comparative physiologists. The reasons for this interest are perhaps obvious: from the organism's standpoint, temperature is an environmental parameter whose effects are at once extremely important and very difficult to avoid; from the experimenter's point of view, temperature is a relatively easy environmental parameter to manipulate in the laboratory. Hence the large amount of data we find available on temperature effects.

It would be impossible in this short paper to review adequately these many data. Any attempt to discuss metabolic temperature compensation as it occurs in all poikilothermic groups, among which we must logically include such diverse organisms as bacteria, insects, trees and fish, would lead to a bewildering disgorgement of facts, with there being little chance of anything remotely resembling a synthesis or an overview of the field. By necessity, therefore, I shall greatly limit the scope of my coverage and concentrate on a small number of topics which will likely be of some mutual interest to hibernation specialists and to comparative physiologists/biochemists concerned with the responses of poikilotherms to changes in temperature. In particular, I wish to discuss in some detail the biochemical changes which occur in certain fishes as these organisms prepare for winter, for the "biochemical restructurings" which these organisms undergo as they acclimate to reduced temperature may have important parallels in mammals preparing for hibernation.

The basic question to which this paper is addressed might be phrased in the following manner: how do poikilothermic organisms alter their biochemistry to permit continued physiological function at very low temperatures? What, in other words, are the biochemical differences between warm- and cold-adapted species and warm- and cold-acclimated members of the same species?

In approaching these questions I shall restrict my attention to aquatic poikilotherms which, unlike many terrestrial poikilotherms and homeotherms, generally lack behavioral avenues of escape from thermal stress. Aquatic species such as many temperate zone fishes, which normally remain active throughout the year even though their habitat (body) temperatures may vary as much as 20°-30°C, furnish us with some of our best examples of metabolic compensation to temperature. The many studies of metabolic compensation in such eurythermal fishes as the common goldfish (Carassius auratus) and the rainbow trout (Salmo gairdneri) have firmly established that, as a rule, the metabolic rates of warm-acclimated individuals are lower than the rates exhibited by cold-acclimated individuals, when measurements are made at temperatures intermediate between the two acclimation temperatures (9,22). And, when comparisons are made of the metabolic rates of different fish species evolutionally adapted to widely different temperatures, a similar pattern of metabolic compensation is observed (34,38). The biochemical mechanisms which effect these seasonal and evolutionary metabolic compensations shall receive the major emphasis of this paper.

In addition to these compensatory adjustments in the overall metabolic rate, recent work has revealed that important qualitative changes in metabolism may also occur in response to temperature changes. Thus not only does the quantity of metabolism depend on temperature, but in addition the relative contributions which different metabolic pathways make to metabolism may vary under different thermal regimes (9,10,14,15,16,18,23,34). This second aspect of metabolic compensation, namely metabolic "reorganization", shall be considered after discussing the more classical phenomenon of metabolic rate compensation.

MECHANISMS OF METABOLIC RATE COMPENSATION.

Changes in enzyme concentration. Since metabolism is a closely integrated set of enzymic reactions, the basic problem encountered by poikilotherms in metabolically compensating to changes in temperature is one of overcoming the effects of temperature on a vast array of individual enzymic reactions. There would seem to be three basic mechanisms by which poikilotherms could maintain relatively temperature independent rates of enzymic, i.e., metabolic, function: (a) the concentrations of enzymes in the cells could be varied in compensatory manners; (b) enzymes having compensatory differences in catalytic efficiency could be produced at different temperatures; and (c) the composition of the "catalytic environment" (12) could be altered to promote rate stability, e.g., through variations in the concentrations of enzyme activators and inhibitors.

Of these three possible mechanisms, the first may seem the most probable. Certainly the increased enzymic activity which may follow dietary changes, hormonal stimulation, etc. is frequently due to increases in enzyme titre (31). However appealing this "quantitative" hypothesis may be in the case of metabolic temperature compensation, it has yet to be rigorously tested.

I know of no case in which accurate estimates of enzyme concentration, as opposed to enzymic activity, have been made in studies of temperature compensation. Thus, even though there are numerous examples of changes in enzymic activity during temperature acclimation (see 18 for a review of these data), in no case has it been unambiguously shown that these activity changes were the result of changes in enzyme concentration. Until someone does take the great pains necessary to measure enzyme concentrations in the tissues of warm- and cold-acclimated (or -adapted) poikilotherms, preferably by using modern antibody techniques (31), we must regard the "quantitative" hypothesis as unverified, albeit highly attractive.

Changes in the catalytic efficiencies of enzymes.

(a) Temperature-dependent changes in enzyme-substrate affinity. An alternative to producing more enzyme molecules at low temperatures is the synthesis of new enzyme variants which are better catalysts at low temperatures than the enzyme variants present in warm-adapted or warm-acclimated individuals. This hypothesis, which we might term the "qualitative" hypothesis, can be broadened to include other classes of molecules, e.g., phospholipids and structural proteins; we shall discuss certain of these other molecular changes later.

The "qualitative" hypothesis would seem to have much to recommend it. For example, from the standpoint of economy in energy metabolism, it would seem preferable to synthesize a relatively small number of "good" enzyme molecules rather than a large number of "poor" molecules. And, as we shall discuss later, the substitution of new enzyme variants may be essential to preserve the regulatory functions, as well as the catalytic capacities, of enzyme systems.

In seeking to compare the suitability of different variants of a given enzyme for function at low temperatures one is initially confronted with the problem of selecting biologically meaningful criteria for making his comparisons. This is no mean feat, and much effort has been wasted on studies where poor criteria have been chosen. For example, there have been numerous studies of the effects of temperature on the maximal velocities (Vmax's) of enzymic reactions, and the results of these studies have yielded few insights into the biochemical mechanisms of temperature compensation. One major shortcoming of these studies lies in the fact that substrate concentrations in vivo seldom rise to Vmax (saturating) levels. Thus the temperature dependence of the Vmax of an enzymic reaction may be a highly misleading index of what is actually occurring in the cell.

A more meaningful approach to the study of temperature effects on enzymic activity is suggested by current models of enzyme regulation (2). These theories stress the overwhelming importance of enzyme-substrate affinities in governing rates of enzymic activity in the cell. Because cellular concentrations of substrate are normally of the same order of magnitude as the Michaelis constants (Km's) of the enzymes, slight changes in enzyme-substrate affinity, which is normally defined as the reciprocal of the apparent Michaelis constant of substrate for enzymes exhibiting hyperbolic substrate saturation curves, can lead to large changes in the rates of catalysis. It is now appreciated that most modulators of enzymic activity exert their influence by changing enzyme-substrate affinity rather than the Vmax of the reaction. Thus positive modulators promote increases in enzyme-substrate affinity, i.e., decreases in the apparent Km of substrate, while negative modulators exert the opposite effect. Because of the importance of enzyme-substrate affinity in governing rates of catalysis in homeothermic systems at constant temperature, we began a detailed investigation of the effects of temperature on this parameter for enzymes from aquatic poikilotherms.

For all poikilothermic enzymes we have studied, a characteristic relationship between temperature and enzyme-substrate affinity has been observed: over most of the range of temperatures the poikilotherm would normally encounter in its habitat, decreases in temperature promote increases in enzyme-substrate affinity (decreases in the apparent Km of substrate). Thus, to use the terminology of enzyme regulation theory, temperature decreases over the physiological temperature range affect the enzymes of poikilotherms in a manner analogous to positive modulators.

An important effect of this inverse relationship between temperatures and enzyme-substrate affinity is a major reduction in the Q_{10} values of poikilothermic enzyme reactions at the low concentrations of substrate present _in vivo_. In marked contrast to the high Q_{10} values (approximately 2-3) which are characteristic of the Vmax of an enzymic reaction, the Q_{10} values we have observed at physiological substrate concentrations at times approach unity (32). Thus, for at least some enzymic reactions of poikilotherms, the temperature-enzyme-substrate affinity relationship enables the organism to maintain essentially temperature-independent rates of catalysis.

The finding that the Q_{10} characteristic of an enzymic reaction is strongly dependent on the substrate concentration indicates that biologically meaningful Q_{10} values for enzymic reactions can result only if physiological substrate concentrations are used in the assay system. In addition, the dependency of Q_{10} on substrate concentration suggests that the temperature dependency of an enzymic process will vary as the physiological substrate concentration fluctuates, e.g., during vigorous exercise (18,32). Thus it is possible that basal metabolism may be characterized by different Q_{10} values from those associated with active metabolism.

Although the inverse relationship between temperature and enzyme-substrate affinity holds over the greater portion of the organism's physiological temperature range, quite marked decreases in enzyme-substrate affinity are frequently observed at temperatures near, or just below, the lower limit of the thermal tolerance range (4,32). Thus, below a certain critical temperature, further decreases in temperature can act as strong negative modulators of enzymic activity.

We first noted these low temperature Km increases in studies of glycolytic enzymes of rainbow trout which were acclimated to summer temperatures (12-18°C). For example, in the case of summer trout pyruvate kinase a rapid decrease in enzyme-substrate affinity occurred at temperatures below 10-12°C (35). This finding presented a paradox. Although an enzyme with this type of temperature-Km relationship would pose no problems for trout in summer, when temperatures would likely remain above 10°C, in the winter, when water temperatures would drop to 4°C and lower, an enzyme with this type of thermal property would function very poorly. To determine how rainbow trout "solved" this problem we isolated pyruvate kinase from winter (4°C) trout and compared its kinetic properties with those of the summer enzyme (36). These results are shown in Figure 1. Each pyruvate kinase variant exhibits a similar minimal value for the apparent Km of phosphoenol pyruvate (PEP). However, the temperatures at which these minimal apparent Km values occur are markedly different. The winter or "cold" pyruvate kinase continues to bind PEP well at temperatures near the winter minimum. The summer or "warm" enzyme functions well, by this criterion, only down to temperatures near the summer thermal minimum.

We have found similar "warm" and "cold" enzyme variants in several other rainbow trout enzyme systems, including brain acetylcholinesterase (4), liver citrate synthase (17) and isocitrate dehydrogenase (Thomas Moon,

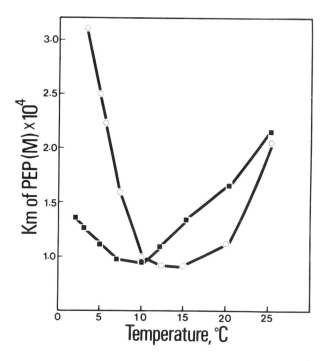

FIG. 1. The influence of temperature on the apparent Michaelis constant (Km) of phosphoenol pyruvate (PEP) for pyruvate kinase enzymes from warm- and cold-acclimated trout. ■ : "cold" pyruvate kinase; ○ : "warm" pyruvate kinase (36).

personal communication), and muscle phosphofructokinase (36). In every case the "cold" variant continues to bind substrate well at temperatures approximating the winter minimum, whereas the "warm" variants exhibit rapid increases in the apparent Km of substrate at temperatures below 10-15°C. Conversely, the "cold" variants frequently exhibit low enzyme-substrate affinity over portions of the summer temperature range. It should be noted that these differences between "warm" and "cold" enzyme variants of this single species have a parallel among enzymes from differently-adapted fish species. For tropical fishes, enzyme-substrate affinity may decrease sharply at temperatures much below 25°C; for Antarctic fishes adapted to -2°C, enzyme-substrate affinity remains high at temperatures of 0°C and below (32,36).

The full disadvantages of the "warm" trout enzyme variants for function under winter temperature conditions extend to the catalytic and the regulatory functions of the enzymes. It should be obvious that, in terms of catalytic capacity per se, the "cold" enzymes are at a great advantage in winter due to their much higher abilities to bind substrate molecules. However, the major disadvantages of the "warm" enzymes at low temperatures would seem to reside in the failure of the "warm" enzymes to function well in regulatory capacities, i.e., in their abilities to maintain control of the rate of catalysis as the concentrations of substrate and the temperature vary.

Because the affinity for substrate of the "warm" enzymes is rapidly decreasing below approximately 10°C, fluctuations in temperature over the winter temperature range will cause enormous Q_{10} values at physiological substrate concentrations. These large Q_{10} values, which result from simultaneous decreases in the kinetic energy of the reactants and enzyme-substrate affinity, may be of sufficient magnitude to inactivate effectively the enzymic <u>reaction</u> at low temperatures, even though no damage is done to the enzyme molecules themselves (32). Q_{10} effects of this sort could be instrumental in establishing the thermal tolerance limits of poikilotherms (32).

Finally, the "warm" enzymes function poorly at winter temperatures with respect to another vital regulatory capacity, namely the capacity of the enzyme to vary the rate of catalysis in response to changes in substrate concentration. Because the apparent Km values of the "warm" enzymes are so high at winter temperatures, rates of catalysis will be relatively insensitive to changes in substrate concentration. This inability to vary the rates of catalysis, i.e., metabolism, as substrate levels change is obviously to the organism's disadvantage. Furthermore, it is conceivable that substrate concentrations could increase under such circumstances to the extent that uncontrolled reactions between substrates and other molecules in the cell could occur, following the laws of mass action rather than the confines of controlled enzymic catalysis (3).

In summary, it is apparent that simple alterations in enzyme concentration cannot fully effect metabolic compensation to temperature in organisms like the rainbow trout. The preservation of metabolic integrity dictates that new enzyme variants, with proper catalytic _and_ regulatory properties, be substituted for the enzyme variants already present in the cell.

(b) Changes in activation energies and substrate turnover numbers. Another parameter which has been used to compare the relative catalytic efficiencies of different enzymes is activation energy (E_a). Most introductory biochemistry textbooks begin their treatment of enzyme kinetics with a statement to the effect that "the role of enzymes is to reduce the activation energies of reactions to the extent that the reactions can occur at body temperatures." Thus it seems reasonable to state that the greater the reduction in E_a which the enzyme can effect, the "better" will the enzyme perform as a catalyst. This argument has been extended to the topic of temperature compensation and, in this context, it takes the following form: if enzymes from cold-adapted (-acclimated) organisms are adapted to function particularly well at low temperatures, relative to enzymes from warm-adapted (-acclimated) organisms, then the cold-adapted enzymes should be more efficient in lowering activation energies than the homologous enzymes from warm-adapted organisms.

This hypothesis has been part of the literature of comparative biochemistry for decades (for a review of this topic, cf. 18), and numerous comparative studies of activation energies of reactions catalyzed by different variants of the same enzyme have been conducted. However, no correlation between adaptation temperature (or acclimation temperature) and activation energy has emerged (4,18). For the reasons discussed below, this state of affairs is less than surprising, and it would seem advisable now to discard the activation energy hypothesis as it pertains to temperature compensation.

The shortcomings of the activation energy hypothesis are the result of an important oversimplification in its treatment of reaction kinetics. Modern theories of reaction kinetics (20) stress that the key rate-determining parameter is the free energy of activation ($\Delta G^{\ddagger}$) and not the energy of activation, which is incorporated into the enthalpy term of the equation, $\Delta G^{\ddagger} = \Delta H^{\ddagger} - T\Delta S^{\ddagger}$. In short, modern theories stress that both the activation energy and the activation entropy can make important contributions to determination of reaction velocity. Unfortunately, most earlier studies of activation energies erroneously neglected the entropy term. The errors which can result from this neglect are well illustrated by the results of a comparative study of the catalytic properties of the heart and muscle lactate dehydrogenase (LDH) isozymes of the rabbit (39). On the basis of E_a differences between the heart- and muscle-LDH reactions, one would predict that the muscle isozyme would have a 3300-fold higher substrate turnover number than the heart isozyme. However, when the isozymes were purified and substrate turnover numbers were determined, the muscle isozyme was found to have a substrate turnover number only 3-4 times that of the heart isozyme. The basis of this large discrepancy between the predicted and actual values of the substrate turnover numbers is a large difference in the activation entropies of the two reactions. Thus the high activation energy of the heart-LDH reaction was accompanied by a high entropy of activation. Quite obviously, high values for both of these parameters will have mutually cancelling effects on the reaction velocity (substrate turnover number).

Activation energy is therefore an unreliable criterion for comparing the catalytic efficiencies of different variants of an enzyme. Ideally, such comparisons should be made as directly as possible, i.e., by measuring substrate turnover numbers. This type of study has been conducted for a small number of enzymes from differently-adapted organisms, and what data are available do suggest that relatively cold-adapted organisms possess enzymes with higher substrate turnover numbers than those of more warm-adapted species. For example, glycogen phosphorylase of lobster muscle has a six-fold higher substrate turnover number than the rabbit muscle enzyme, at an assay temperature of 0°C (1). Similarly, the substrate turnover numbers of cod and lobster D-glyceraldehyde-3-phosphate dehydrogenase are considerably higher, at low temperatures, than that of the rabbit enzyme (7). It therefore seems probable that, for at least some enzymes, evolutionary adaptation to low temperatures is marked by increases in the substrate turnover numbers. The importance of this type of enzymic adaptation in seasonal acclimation is unknown. Studies of the turnover numbers of "warm" and "cold" trout enzyme variants are sorely needed.

Before leaving this topic I wish to raise a question which, in a logical sense, will complete our treatment of this issue: since mammals evolved from poikilothermic ancestors which were well-adapted to temperatures below 37°C, is it correct to ask if poikilotherms have evolved more efficient enzymes than those of mammals? Is it not equally, or even more, reasonable to ask whether mammals have evolved enzymes having lower substrate turnover numbers than poikilothermic enzymes?

(c) Mechanisms for producing new enzyme variants. Up to this point in our discussion no mention has been made of the mechanisms by which new enzyme variants might be produced. Certainly our biases, which are shaped in large part by models of bacterial enzyme induction, lead us to predict that the new enzyme variants found in temperature acclimation are the result of the transcription of new messenger RNA species, followed by the translation of the latter messages into new proteins. While this type of mechanism is likely involved in effecting the enzyme substitutions we have discussed, albeit proof of this point is lacking, there is an alternative mechanism for generating kinetically distinct enzyme variants which merits our consideration.

This alternative mechanism involves direct temperature-mediated interconversions of enzyme variants. In this system a single protein, in terms of primary structure, will exist in kinetically distinct forms at different temperatures as a result of temperature-dependent changes in the "higher" levels of protein structure. Temperature could affect one or more of the following "higher" structures: (a) the tertiary conformation; (b) the aggregation of subunits (quarternary structure); or (c) the binding of the protein to other cellular structures.

This type of mechanism for generating kinetically distinct enzyme variants does appear to occur in certain cases. For example, pyruvate kinase from leg muscle of the Alaskan king crab (<u>Paralithodes</u> <u>camtschatica</u>) exists in two kinetically distinct forms, albeit only a single pyruvate kinase protein appears to be present (33). As the temperature of the organism (enzyme) changes over the physiological temperature range, this single protein undergoes an interconversion from one form to another (Fig. 2). At low temperatures almost all of the pyruvate kinase activity is due to the "cold" variant, which exhibits classical hyperbolic kinetics. As the temperature is raised, an increasingly large fraction of the total pyruvate kinase activity is due to the "warm" variant, which is characterized by sigmoidal kinetics. The net effect of this interconversion phenomenon is the formation of two pyruvate kinase variants with kinetic properties (Fig. 3) strikingly analogous to those exhibited by the "warm" and "cold" rainbow trout pyruvate kinases (Fig. 1). The important distinction between

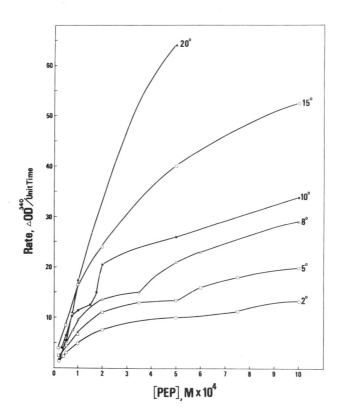

FIG. 2. The effects of temperature and substrate (phosphoenol pyruvate [PEP]) concentration on the activity of king crab leg muscle pyruvate kinase (33).

these two systems is that, in the case of king crab pyruvate kinase, the "warm" and "cold" variants are produced nearly instantaneously in response to temperature changes, whereas in rainbow trout the enzyme variant changes appear to require days or weeks (32).

It is important to note that, in the king crab pyruvate kinase interconversion system, temperature changes mimic the effect of enzyme modulators in a second important way. Thus, temperature decreases over the physiological temperature range promote the conversion of a sigmoidal enzyme to a hyperbolic enzyme, a function frequently served by positive enzyme modulators.

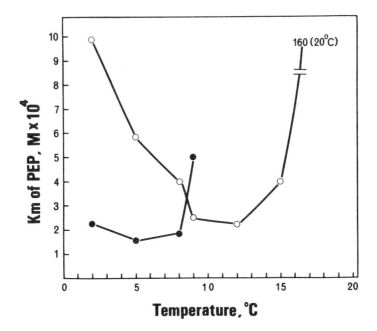

FIG. 3. The influence of temperature on enzyme-substrate affinity for the two variants of pyruvate kinase present in Alaskan king crab leg muscle. ● : "cold" pyruvate kinase; ○ : "warm" pyruvate kinase. (after 33).

(d) Changes in the "catalytic environment". We have so far considered two mechanisms whereby enzymic activity can be modulated to compensate for changes in temperature, namely by changing the quantities or the qualities of enzymes. A third potential compensatory mechanism does not involve changes in the enzyme molecules per se, but rather in the milieu in which the enzymes function. Such alterations in the "catalytic environment" could be of major importance in the temperature compensation process.

One of the potentially most important of these "environmental" factors is pH. Although mammals closely regulate pH, most poikilotherms appear to regulate the ratio of H^+ to OH^-, rather than pH (26). Since the ionization constant of water decreases as temperature decreases, the pH of a poikilothermic organism will tend to increase at low temperatures if the $H^+:OH^-$ ratio is held nearly constant. This variation in pH with temperature could have profound consequences for poikilothermic metabolic control if key enzymes have sharp pH dependencies in the physiological pH range. In at lease one case, citrate synthase of the rainbow trout (17), such a pH dependency exists, and the organism appears to have "taken advantage of" the temperature-pH relationship. Increases in pH through the physiological pH range activate citrate synthase. Thus the effects of temperature decreases on enzymic activity are partially counteracted by the temperature-pH response. The combined effects of this response and the increase in enzyme-substrate affinity which also occurs as temperature decreases yield highly temperature-independent rates of citrate synthase activity (17).

A second group of "environmental" factors likely of importance in governing the metabolic rates of poikilotherms is the diverse array of inorganic ions present in the cells and body fluids. The concentrations of such important ions as Mg^{++}, K^+ and Ca^{++} are known to vary significantly during temperature acclimation in some poikilotherms (13,19,27). Since these and other ions are known to affect the activities of important regulatory enzymes, it is very likely that ionic changes could have important influences on the extent of metabolic compensation to temperature.

In the context of this discussion of the "catalytic environment" it is worth mentioning two studies which illustrate the importance of changes in the intra- and extracellular environments, even though these particular studies did not determine the chemical factors which were responsible for the observed effects. In studies of the in vitro metabolism of tissues from warm- and cold-acclimated fish, Precht (24,25) discovered that the addition of blood serum from cold-acclimated fish to the manometer flasks containing the tissues led to significant increases in oxygen consumption; "warm" fish serum did not have this effect on metabolism.

A clear example of the importance of changes in the intracellular environment in governing the behavior of proteins is found in the case of hemoglobins from differently acclimated bullheads (11). Temperature acclimation of the fish led to adaptive changes in the oxygen binding properties of the hemoglobins. These changes appeared to be due to alterations in the composition of the erythrocyte cytosol, rather than in the hemoglobin proteins per se.

Finally, the most important "environmental" factor for enzymes may be the cellular membranes. Since many enzymes are at least loosely associated with membranes, and because most enzymes are dependent on membrane controlled transport processes for their supply of substrates, cofactors, ions, etc., it is to be expected that the preservation of membrane integrity will be a major "goal" of temperature compensation. Indeed, recent studies of the membrane compositions of differently-acclimated fishes have illustrated that profound alterations in lipid composition occur during temperature acclimation

(6,21,29,30). As in the case of trout enzyme variants, the lipid changes are marked by the appearance of new molecular variants which are particularly well suited for function under the new thermal regime. In general, these lipid changes appear to be qualitative rather than quantitative. Thus, for example, the total phospholipid content of brain membranes may remain constant while the saturation of the phospholipids is significantly altered (29).

The biological importance of these lipid changes during thermal acclimation has recently been revealed in a series of elegant studies of microbial membranes of closely controlled lipid composition (6,37). Cultures of Mycoplasma laidlawii were grown in the presence of lipids of varying degrees of saturation. The lipid composition of the culture medium was precisely reflected in the resulting lipid composition of the organisms' cell membranes. Differential scanning calorimetry was used to observe thermal phase transitions between "crystal" and "liquid crystal" states of the membranes and the purified membrane lipids. The following important facts emerged from these experiments: (a) the phase transition temperatures of the membranes in situ, of the isolated membranes, and of the purified membrane lipids were the same for a given saturation state; and (b) when the growth temperature was lower than the phase transition temperature, osmotic imbalance occurred in the cells; thus, for transport processes to occur properly, the hydrocarbon chains of the lipids had to be in a "liquid crystalline" rather than a "crystalline" state. These careful studies thus confirmed the earlier hypotheses raised by the comparative biochemists who studied membrane changes during acclimation (6,21,29,30).

While the role of membrane "restructuring" in maintaining transport functions is now clear, the influence which qualitative changes in membrane composition may have on membrane-bound enzymes is still obscure. However, there is one study which provides at least circumstantial evidence that membrane lipid changes may affect the activities of membrane-bound enzymes. Caldwell (5) observed significant changes in the activities of several mitochondrial membrane-bound respiratory enzymes during thermal acclimation. He found no evidence that the concentrations of enzymes had changed and, therefore, he suggested that the lipid composition changes which were observed in these same membranes (6) might have been responsible for the altered enzymic activities. Unfortunately, it is not known if different variants of these respiratory enzymes were present in warm- and cold-acclimated fish.

METABOLIC "REORGANIZATION": CHANGES IN THE RELATIVE ACTIVITIES OF METABOLIC PATHWAYS AT DIFFERENT TEMPERATURES

We have earlier mentioned the fact that the types of metabolic activity, as well as the overall quantity of metabolic flow, are influenced by temperature. Temperature-dependent changes in the relative activities of metabolic pathways may occur over two distinct time-courses; each of these shall be treated in turn.

<u>Immediate changes in the relative activities of metabolic pathways</u>. Since not all metabolic pathways are likely to be characterized by identical Q_{10} values, one would predict that sudden temperature changes would be accompanied by immediate alterations in the relative activities of some metabolic pathways. Indeed, such changes appear to be very common. For example, in homogenates of lungfish and electric eel livers, the metabolism of glucose-6-C^{14} can be markedly independent of temperature, while the oxidation of glucose labelled in the 1-carbon exhibits "classical" Q_{10} values of approximately two (15). As a result of these different temperature responses, the pentose shunt assumes a larger role in glucose oxidation at

elevated temperatures. Similar differences in the temperature coefficients of metabolic pathways of poikilotherms have been reported by other authors (8).

The molecular basis of this immediate change in relative pathway participations likely resides in differences in the temperature-dependent enzyme-substrate affinity relationships of enzymes competing for common substrates. Thus should a temperature decrease promote a more rapid increase in enzyme-substrate affinity for one enzyme competing for a common substrate, relative to the change in affinity occurring for the "competitor" enzyme, then the relative abilities of the two enzymes to compete for the common substrate will change, as will the relative activities of the two pathways.

<u>Changes in metabolic pathway participation during acclimation</u>. The earliest examples of changes in the relative participation of different metabolic pathways at different temperatures came from studies of glucose metabolism in differently-acclimated fish (9,10,16,23). In all cases it was observed that cold acclimation promoted a relative increase in the activity of the pentose shunt, while glycolysis was relatively reduced in importance. More recently, other metabolic "reorganizations" have been discovered; these are discussed in detail in recent reviews (14,18,36) and will not be elaborated at this time.

Even though the occurrence of metabolic pathway "reorganization" is now an established fact, the significance of these changes remains somewhat obscure. In part, this uncertainty about the role of metabolic "reorganization" in the temperature compensation process is due to the fact that the time-course of temperature acclimation is poorly understood. Thus it is not known whether certain of the metabolic "reorganizations" which have been reported in acclimation studies are merely transitory changes which serve to bring about a new biochemical steady state (35,36). For example, increased pentose shunt activity is normally associated with demands for large quantities of NADPH, a cofactor of vital importance in reductive biosyntheses. It therefore seems likely that increased pentose shunt activity might be important during the biochemical "restructuring" of the cell, e.g., during the synthesis of new membrane lipids. Following this biochemical "restructuring" the activity of the shunt might be reduced. If this is indeed the role of the pentose shunt during acclimation, then one would predict that (a) the activity of the shunt would exhibit transitory increases during acclimation and return to a more basal level once "restructuring" is complete, and (b) both warm- and cold-acclimation would involve these transitory increases in pentose shunt activity. The importance of this and other pathway participation changes in acclimation could be resolved largely by carefully charting the time-course of pathway changes during acclimation (36).

CONCLUSIONS AND SPECULATIONS

The major point I have tried to make in the preceding discussion is that a wide variety of quantitative and qualitative biochemical mechanisms are available to poikilotherms as they metabolically compensate to changes in temperature. In particular, I have tried to stress that the production of new molecular variants is a key factor in successful temperature compensation. Undoubtedly the list of such qualitative biochemical changes will increase as more studies of temperature compensation are performed. It will be particularly interesting to learn whether changes in so-called "structural" proteins, e.g., ribosomal and membrane proteins, occur.

The capabilities of adult organisms to extensively "restructure" themselves biochemically on a seasonal basis raise a great number of intriguing questions for a wide range of biologists. For example, what environmental factors trigger the seasonal acclimation changes? Are temperature and photoperiod important? Are hormones involved in mediating the acclimation changes? Does temperature act directly on the cells, e.g., at the levels of transcription and/or translation? How rapidly is biochemical "restructuring" effected? And, lastly, do similar "restructurings" occur in hibernators as they prepare for the onset of winter? It is hoped that questions such as these may serve as fruitful guides for further research into the intricacies of temperature compensation phenomena.

REFERENCES

1. ASSAF, S. A., and D. J. GRAVES. Structural and catalytic properties of lobster muscle glycogen phosphorylase. J. Biol. Chem. 244: 5544-5555, 1969.

2. ATKINSON, D. E. Regulation of enzyme activity Ann. Rev. Biochem. 35: 85-124, 1966.

3. ATKINSON, D. E. Limitation of metabolite concentrations and the conservation of solvent capacity in the living cell. Current Topics Cellular Regul. 1: 29-43, 1969.

4. BALDWIN, J., and P. W. HOCHACHKA. Functional significance of isoenzymes in thermal acclimatization: acetylcholinesterase from trout brain. Biochem. J. 116: 883-887, 1970.

5. CALDWELL, R. S. Thermal compensation of respiratory enzymes in tissues of the goldfish (Carassius auratus L.). Comp. Biochem. Physiol. 31: 79-93, 1969.

6. CALDWELL, R. S., and F. J. VERNBERG. The influence of acclimation temperature on the lipid composition of fish gill mitochondria. Comp. Biochem. Physiol. 34: 179-191, 1970.

7. COWEY, C. B. Comparative studies on the activity of D-glyceraldehyde-3-phosphate dehydrogenase from cold and warm-blooded animals with reference to temperature. Comp. Biochem. Physiol. 23: 969-976, 1967.

8. DEAN, J. M. The metabolism of tissues of thermally acclimated trout (Salmo gairdneri). Comp. Biochem. Physiol. 29: 185-196, 1969.

9. EKBERG, D. R. Respiration in tissues of goldfish adapted to high and low temperatures. Biol. Bull. 114: 308-316, 1958.

10. EKBERG, D. R. Anaerobic and aerobic metabolism in gills of the crucian carp adapted to high and low temperatures. Comp. Biochem. Physiol. 5: 123-128, 1962.

11. GRIGG, G. C. Temperature-induced changes in the oxygen equilibrium curve of the blood of the brown bullhead Ictalurus nebulosus. Comp. Biochem. Physiol. 28: 1203-1223, 1969.

12. GRISOLIA, S. The catalytic environment and its biological implications. Physiol. Rev. 44: 657-712, 1964.

13. HICKMAN, C. P., R. A. McNABB, J. S. NELSON, E. D. VAN BREEMAN, and D. COMFORT. Effect of cold acclimation on electrolyte distribution in rainbow trout (Salmo gairdnerii). Can. J. Zool. 42: 577-597, 1964.

14. HOCHACHKA, P. W. Organization of metabolism during temperature compensation. In: Molecular Mechanisms of Temperature Adaptation, edited by C. L. Prosser. Washington, D.C.: American Association for the Advance of Science, 1967, p. 177-203.

15. HOCHACHKA, P. W. Action of temperature on branch points in glucose and acetate metabolism. Comp. Biochem. Physiol. 25: 107-118, 1968.

16. HOCHACHKA, P. W., and F. R. HAYES. The effect of temperature acclimation on pathways of glucose metabolism in the trout. Can. J. Zool. 40: 261-270, 1962.

17. HOCHACHKA, P. W., and J. K. LEWIS. Enzyme variants in thermal acclimation: trout liver citrate synthases. J. Biol. Chem. 245: 6567-6573, 1970.

18. HOCHACHKA, P. W., and G. N. SOMERO. Biochemical adaptation to the environment. In: Fish Physiology, edited by W. S. Hoar and D. J. Randall. New York: Academic Press, 1971, vol. VI. In press.

19. HOUSTON, A. H., and J. A. MADDEN. Environmental temperature and plasma electrolyte regulation in the carp, Cyprinus carpio. Nature 217: 969-970, 1968.

20. JOHNSON, F. H., H. EYRING, and M. J. POLISSAR. The Kinetic Basis of Molecular Biology. New York: John Wiley & Sons, 1954.

21. JOHNSTON, P. V., and B. I. ROOTS. Brain fatty acids and temperature acclimation. Comp. Biochem. Physiol. 11: 303-309, 1964.

22. KANUNGO, M. S., and C. L. PROSSER. Physiological and biochemical adaptation of goldfish to cold and warm temperatures. I. Standard and active oxygen consumptions of cold- and warm-acclimated goldfish at various temperatures. J. Cellular Comp. Physiol. 54: 259-263, 1959.

23. KANUNGO, M. S., and C. L. PROSSER. Physiological and biochemical adaptation of goldfish to cold and warm temperatures. II. Oxygen consumption of liver homogenate; oxygen consumption and oxidative phosphorylation of liver mitochondria. J. Cellular Comp. Physiol. 54: 265-273, 1959.

24. PRECHT, H. Über die Bedeutung des Blutes für die Temperaturadaptation von Fischen. Zool. Jahrb. abt. Allg. Zool. Physiol. 71: 313-328, 1964.

25. PRECHT, H. Ergänzende Versuche zur Bedeutung des Blutes für die Temperaturadaptation bei Fischen. Zool. Anz. 175: 301-310, 1965.

26. RAHN, H. Gas transport from the external environment to the cell. In: Ciba Foundation Symposium on Development of the Lung, edited by A. V. S. De Reuck and R. Porter. Boston: Little, Brown & Company, 1967, p. 3-23.

27. RAO, K. P. Physiology of acclimation to low temperature in poikilotherms. Science 137: 682-683, 1962.

28. REINERT, J. C. and J. M. Steim. Calorimetric detection of a membrane-lipid phase transition in living cells. Science 168: 1580-1582, 1970.

29. ROOTS, B. I. Phospholipids of goldfish (Carassius auratus L.) brain: the influence of environmental temperature. Comp. Biochem. Physiol. 25: 457-466.

30. ROOTS, B. I., and P. V. JOHNSTON. Plasmalogens of the nervous system and environmental temperature. Comp. Biochem. Physiol. 26: 553-560, 1968.

31. SCHIMKE, R. T. Roles of synthesis and degradation in regulation of enzyme levels in mammalian tissues. Current Topics Cellular Regul. 1: 77-124, 1969.

32. SOMERO, G. N. Enzymic mechanisms of temperature compensation: immediate and evolutionary effects of temperature on enzymes of aquatic poikilotherms. Amer. Natur. 103: 517-530, 1969.

33. SOMERO, G. N. Pyruvate kinase variants of the Alaskan king-crab: evidence for a temperature-dependent interconversion between two forms having distinct and adaptive kinetic properties. Biochem. J. 114: 237-241, 1969.

34. SOMERO, G. N., A. C. GIESE, and D. E. WOHLSCHLAG. Cold adaptation of the Antarctic fish Trematomus bernacchii. Comp. Biochem. Physiol. 26: 223-233, 1968.

35. SOMERO, G. N., and P. W. HOCHACHKA. The effect of temperature on catalytic and regulatory functions of pyruvate kinases of the rainbow trout and the Antarctic fish Trematomus bernacchii. Biochem. J. 110: 395-400, 1968.

36. SOMERO, G. N., and P. W. HOCHACHKA. Biochemical adaptation to the environment. Amer. Zool., 1971. In press.

37. STEIM, J. M., M. E. TOURTELLOTTE, J. C. REINERT, R. N. McELHANEY, and R. L. RADER. Calorimetric evidence for the liquid-crystalline state of lipids in a biomembrane. Proc. Nat. Acad. Sci. U.S. 63: 104-109, 1969.

38. WOHLSCHLAG, D. E. Respiratory metabolism and ecological characteristics of some fishes in McMurdo Sound, Antarctica. In: Biology of the Antarctic Seas, edited by M. O. Lee. Washington, D.C.: American Geophysical Union, 1964, Vol. I, 33-62.

39. WROBLEWSKI, F., and K. F. GREGORY. Lactic dehydrogenase isozymes and their distribution in normal tissues and plasma and in disease states. Ann. New York Acad. Sci. 94: 912-932, 1961.

CREDITS

The editors and authors are grateful to the following publishers, organizations and individuals for permission to use materials previously published:

Figure 2: The Biochemical Journal; G. N. Somero, The Biochemical Journal 114: 237-241, 1969.

Figure 3: The Biochemical Journal; G. N. Somero, The Biochemical Journal 114: 237-241, 1969.

STRUCTURED WATER
IN BIOLOGICAL SYSTEMS

JOHN H. BLAND

University of Vermont College of Medicine
Department of Medicine
Rheumatology Unit
Burlington, Vermont 05401

INTRODUCTION

Water is the most usual, unusual of liquids. It is the only substance on the planet that exists in all three of its forms, liquid, solid and gas, in common, every day experience. It was a mistake to study water first among the liquids for it is so anomalous in physical behavior as to be quite unlike typical liquid. In fact, gases and solids have effective, satisfactory laws and they are appropriately law abiding while lawless liquids often refuse to conform to predictions, particularly water.

Water has attracted little attention as a factor in the phenomenom of natural hibernation; however, it obviously must play a major role, seemingly hiding in plain sight (12). Virtually all reactions in biological systems, fast or slow, must go on in water solutions. From the days of Arrhenius and van't Hoff, we have known that rates of chemical reactions are a function of temperature. Hibernation is associated with a fall in body temperature, preceded by a fall in environmental temperature, and both events result in changes in the structure of water and, hence, changes must occur in the types as well as the rates of most, if not all, chemical reactions in the body.

Some ancient and still puzzling facts about water, its universal biological significance as a solvent, the now known temperature related structuring of water, and the recent evidence for the existence of water as a high polymer (super, poly or anomalous water), led me to select the topic, structured water in biological systems, for review.

SOME UNUSUAL PROPERTIES OF WATER

Hypotheses about the structure of water date at least to Roentgen in 1892 who proposed that water must have ice-like molecular aggregates even though a liquid (23). Bernal and Fowler (1) in 1933 interpreted their X-ray diffraction data on liquid water as indicating a 4 to 6 molecule polymer. There is excellent supporting evidence (3,17) that hydrogen bonds in liquid water form an extensive three dimensional network, the structure being short lived (10^{-10} to 10^{-11} sec.) and constantly changing. This structure may form in several ways from units of nearly tetrahedral symmetry. The nine now known modifications of ice are all stable under specific temperature and pressure ranges (16). The water molecules of the compounds called clathrate hydrates or clathrate ices form hydrogen bonded network structure in which the molecules are tetrahedrally coordinated - or very nearly so. Such structural networks are less dense than ice and contain large polyhedral cavities which are occupied by "guest" molecules (15), notably anesthetic agents.

Water's heat capacity is remarkable, i.e., water can absorb much heat without raising its own temperature. Equal volumes of sand and water receiving the same amount of heat in unit time will show an increase in temperature of the sand 5 times greater than that of the water. This heat capacity can be described as the ability to store heat energy with relatively little atomic and molecular vibration, agitation and bond stretching.

The specific heat of water (amount of heat necessary to raise the temperature of water 1°C) is also remarkable. The amount of heat required to raise the temperature of water 1°C decreases from 0°C to 35°C at which point it increases steadily to the boiling point. This observation coincides nicely with the fact that a transition toward less structuring of water occurs between 20°C and 40°C, rising temperature, and the reverse, more structuring with falling temperature, 40°C to 20°C (11). The high latent heat of water is responsible for such favorite things as skiing, ice cream and Turkish baths.

The surface tension of water is higher than all common liquids, with mercury the only exception, suggesting much structuring and having broad implications in biological systems. Everybody knows ice floats but it is still not clear why it expands and loses density between 4°C and 0°C; all other liquids contract as they form solids but ice increases in volume 1/11th of its volume as a liquid. Thus, water is one of the few substances that is less dense as a solid than as a liquid. If this were not so, life on this planet could not exist, for all the lakes, rivers and oceans would freeze from the bottom up immobilizing vast amounts of the world's water.

Substances of similar molecular structure behave very differently than water, e.g. H_2S, H_2Te and H_2Se. H_2Te has a molecular weight of 129, freezes at -51°C and boils at -4°C; H_2Se has a molecular weight of 80, freezes at -64°C and boils at -42°C; H_2S with a molecular weight of 34, freezes at -82°C and boils at -18°C; but water with lowest molecular weight, 18, freezes at 0°C and boils at 100°C, paradoxical and unusual physical properties for a liquid.

Certain substrates make water crystallize to ice at quite warm temperature, therefore, freezing temperature is not the only mechanism for ice formation. Pipelines of methane gas may become obstructed by ice at warm or even hot temperatures. Certain plants are "winter hardy", and do not freeze until the environmental temperature is -30°C to -40°C. Corn has been shown to sustain "frost" damage though the temperature was no lower than 4°C;

presumably damage occurs as water crystallizes on macromolecular proteins and glycosaminoglycans (protein-polysaccharides) of plant connective tissue. Trees, under certain fairly common weather conditions, make known and unknown physiological changes enabling them to withstand freezing temperatures, -5°C to -10°C, without damage to leaves, twigs or buds (22) - and continue photosynthesis; the tissue water is highly structured but does not freeze. When trees have completed their physiologic changes in preparation for winter (tree dormancy), though their tissues contain water, they are able to withstand dry, cold winter air, frozen ground and inability to obtain water at environmental temperatures of -50°C or greater (22). There is no data on water behavior in dormant trees but it clearly does not form ice in the conventional sense. As winter chilling proceeds, there is a change in the types of enzymes, their number and their form (isoenzymes) present in the buds. In the elm, there are at least two temperature sensitive enzyme systems involved, one active at 5°C and the other near 25°C. Subtle changes in water structure, temperature mediated, or produced by the hydration of ions and polyions may affect biological activity. Lettuce seed germination, production of heat by bacteria, and transport of ^{22}Na in corn seedlings were all accelerated by putting them in an environment of clays which changed from sol to gel (thixotropic), thus changing the water structure (19). Biological activity was greater when the clay was in the sol state.

Ordering of water structure by ions and polyions has long been known. In a solution of sodium chloride, the cations, positively charged sodium, has water molecules as a hydration shell about it with their negative ends (oxygen) pointed inward; the chloride anion, negatively charged, has oriented all the water molecules about it with their positive ends pointed inward (hydrogen). Thus the ions are in solution, apart, and their recombination is prevented. Hydration shells of water molecules are packed about positive and negative charges on macromolecular proteins, carbohydrate polymers, and other intra- and extracellular substances that have charges. The water of hydration of collagen was shown by Dehl (7) to have a high state of mobility at temperatures well below the freezing point of ordinary water, suggesting that water structured by the protein behaves quite differently in response to temperature. Berendsen and Migchelsen showed previously that even when the water content was high, the water of hydration remained unfrozen, even at -30°C.

In 1949 P. W. Bridgman demonstrated that when water was subjected to pressure, between 0°C and 30°C, its viscosity decreased, i.e., its fluidity increased. S. F. Feates and D. J. G. Ives in 1956 noted that this increase in fludiity on pressure decreased and disappeared between 30°C and 40°C. In 1966 R. A. Horne and D. S. Johnson proved that the maximal effect of pressure on decreasing viscosity of water occurred below 15°C. These research results have been interpreted as indicating that water must contain collapsible spaces, expanded structures, small "icebergs" that could be crushed on pressure. These presumably disappear above 40°C since there is no longer a fall in viscosity on application of pressure above that temperature. Though perhaps oversimplified, it appears that water is increasingly structured from about 60°C to 0°C in some very unusual way.

Life is moist and if it does not maintain a certain moistness, it disappears. All living things require about the same concentration of water. A healthy adult human's average body weight is about 55 to 65 per cent water; the same is true of a giraffe, a rainbow trout or a dandelion. A grossly obvious fact is that in any biological system most of the water does not flow. It is what was called "bound water" and currently there is a question of whether it is structured or ice-like - or a true water polymer, perhaps mixed with ordinary water and some in various ice-like forms but "ice" at 37°C. A 70 kgm man has 60% or 42 kgm of his body weight made up of water; 23 kgm in his cells and 19 kgm extracellularly. However, of this 42 liters of water relatively little has flow properties, most is in cells and tissues immobilized in a poorly understood way. About 1.6-2.0 liters of water - as plasma, aqueous humor, spinal fluid and 100 ml to 500 ml of urine - is in the form of usual water, but the remainder does not flow. For instance, a piece of skin, cleaned of its fat and compressed up to 200 atmospheres yields very little water; when "ashed" it is found that the skin is 50% to 60% water, recoverable by ashing. Could this extracellular, extra-vascular water be electrostatically bound to connective tissue polyelectro-lyte macromolecules and intracellular protein polyelectrolyte?

IONIC HYDRATION SHELLS

In 1957 Frank and Wen (13) proposed and proved that hydrated ions may possess two different types of hydration shells, called A regions and B regions. Erlander (10,11) showed that ions could not have more than two types of hydration shell, an inner A region and an outer B region. Both sodium and fluoride ions have a charge on their surface which results in only one hydration shell, an A region. Ions with only a B region have a charge per unit surface area less than or equal to that on potassium and chloride ions while those having two shells, A and B, have a charge on their outer surface at least equal to that on a lithium ion.

The A region differs from the B region in that its water molecules are tightly bound, it cannot be easily penetrated by solutes, and the very strong attraction causes a greater separation of the charges on the water molecules. A regions can augment solubility of nitrogen compounds, such as ammonia, and decrease water solubility of acids with hydroxyl groups such as carboxylic, sulfuric or phosphoric acids. Erlander (11) points out that a lithium salt

in water solution results in very highly organized water molecules in the
A regions, and that the A regions themselves interact with hydrogen bonds
to increase and enhance the surrounding water structure. Thus, salts
forming A regions so organize and restrict movement of water molecules that
heat is produced.

Salts which produce B regions only are quite different and their
properties are correlated with the mobility of the water molecules. The
hydrated water molecules of the B regions can be more mobile than those of
pure water and such an ion can break hydrogen bonds in polymers and between
water molecules. The B region molecules, highly mobile, can break up the
structure of water by successful competition for the hydrogens in the bonds.
Thus, if both ions of a salt have B regions, the addition of such a salt
will cause a reduction in the viscosity of the water, and the increase in
fluidity in the B region and in the surrounding water makes the water colder.

STRUCTURAL MODELS FOR WATER

There are now three basic structural models for liquid water. First,
that of Schiffer and Hornig, the continuum model, describing water as a
hydrogen bonded network, with continuous distribution of bond energies and
nearly symmetrical geometries (25). It assumes that hydrogen bonds bend
but do not break. This model does not account for the properties of water
in the transition zone from 15°C to 40°C.

Second model is that of Frank et al. (13) and Nemethy and Scheraga (21)
the flickering cluster or mixture model, visualizing liquid water as in two
or more states in equilibrium; one being clusters of about 50 water molecules,
fairly ordered, hydrogen bonded, iceberg like, tetrahedrally coordinated,
termed the "bulky phase" (cluster); the other phase is nonhydrogen bonded,
disordered molecules, called the dense phase. The flickering cluster is
regarded as short lived (10^{-10} to 10^{-11} sec), continually exchanging

molecules with the adjacent unstructured phase (20). Again the model doesn't explain water behavior in the transition zone between 15°C and 40°C and the minimum volume at 4°C with the 11% expansion between 4°C and 0°C when water freezes is not well correlated with either the continuum or the mixture model.

The third model, described by Kamb (16), proposes that liquid water may be a mixture of three cluster types, resembling ice I, ice II and ice III (9). Varying ice structures are produced by applying pressure at various temperatures to ice I or normal ice. Water molecules are hydrogen bonded, each with 4 others to result in kinked, hexagonal "chair" structures, which are stacked into columns, which also are hydrogen bonded as neighboring columns. In ice I, the hydrogen bonds are random; ice II, a denser ice, the columns are alternated in that one hexagon is slightly flattened and the next one slightly puckered; hydrogen positions are specific, not random. The increased density is not due to any change in the structure of the "chairs" but to a change in the way the columns of "chairs" are stacked one against the other (11). In ice II any "chair" of one column will be midway between any two "chairs" of a neighboring column, with a relative but slight twist. Ice III is a further modification, less stable than ice I or ice II. This model (Fig. 1) the cluster-aggregate model, seems to explain reasonably the properties of water, though there remain real objections to it.

A fourth model, that proposed by Erlander, is unique and provides explanation for the properties of water, in biological systems as well as elsewhere. Only one type of crystal structure, ice I, is proposed and the clusters of water molecules are quite stable, while the cluster-aggregates are not. It could be appropriately termed the cluster-aggregate model.

In the cluster-aggregate model the clusters are regarded as quite stable while the aggregates of clusters are not. As temperature drops from

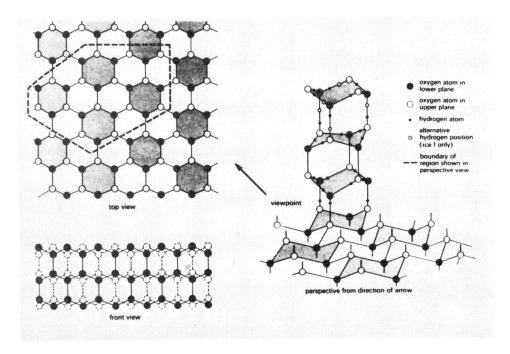

FIG. 1. Normal ice has a structure in the form of "chairs" formed from kinked hexagons superimposed in stacks. Hydrogen atoms are located 10.1 micrometres from center of the nearest oxygen atom and 16.5 micrometres from the adjacent oxygen. In ice I a hydrogen can adopt either of its two possible positions (shown dotted in the perspective view) but in the high pressure ice II type structure the positions are fixed. Denser ices are made by rearranging the hexagonal stacks. Ordinary ice floats because empty (and waterfilled) pockets are formed in it as a result of faults and other errors produced by the interaction of adjacent cluster aggregates of different sizes and shapes. (From Prof. S. R. Erlander, ref. 11).

100°C to 60°C, there is an increase in the size and number of water clusters but no aggregation of clusters. At 60°C aggregates begin to form, made up of different sizes of clusters. Pockets or spaces form within as well as between the aggregates, due to imperfections in fitting together. At 40°C and below there are increasing numbers of pockets; between 40°C and 15°C, the transition zone, both clusters and aggregates form rapidly; below 15°C, the principle mechanism is formation of large cluster-aggregates and at 4°C to 0°C the structure is that of ice I. The transition from water to ice occurs with difficulty since the cluster-aggregates of varying shapes and sizes, must each find surfaces on all neighboring aggregates to which they can become "comfortably" attached. This final aggregation between 4°C and 0°C results in the 11% increase in volume. The spaces and pockets provide explanation for the faults and imperfections in ice I. Erlander (11) points out that rapidly cooled hot water freezes faster than cold water cooled at the same rate. Many cluster-aggregates in cold water cannot readily attach to adjacent aggregates until they first reform, while completely disorganized monomer water at 100°C permits formation of cluster-aggregates of the proper structure, and incorrect, poor fit aggregates are broken down or eliminated to reform in correct structure.

A final explanation for the events occurring when water is heated from 0°C (ice) to 4°C is as follows: as heat is applied two mechanisms operate, (a) the cluster-aggregates break up contracting the volume; (b) expansion occurs due to increasing thermal motion of clusters and cluster-aggregates. From 0°C to 4°C the first mechanism is dominant while at 4°C and above the density falls as thermal motion predominates. The water loses viscosity (becomes more fluid) as pressure prevents formation of cluster-aggregates. "Specific heat has its least value at 35°C since at this temperature, the empty pockets between aggregates have almost all been eliminated but pockets formed by thermal motion of clusters and (monomer) free water molecules have not appeared in great numbers." (11).

ANOMALOUS WATER

Anomalous water, polywater, super water and water II are names given to a strange substance that appears when water condenses in tiny, fused quartz and glass capillary tubes in a partial vacuum. It was first described in 1962 by a Russian chemist, N. N. Fedyakin, who, in a series of papers, with

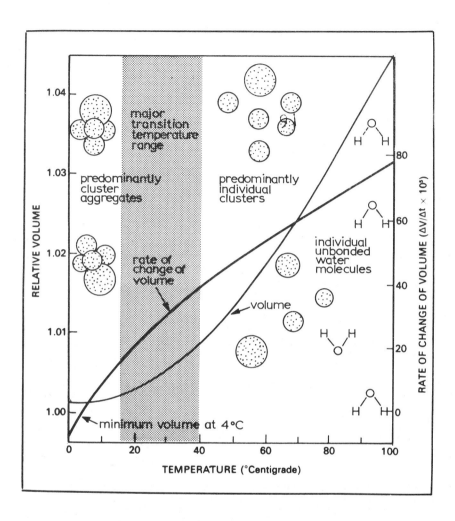

FIG. 2. Structure of water changes as its temperature passes through the range 15-40°C. As temperature is increased above 0°C the volume first decreases slightly as a result of the loss of empty pocket holes in the disappearing cluster aggregates. Thus the volume of a given mass of water is a minimum at 4°C. This volume can be further reduced by applying pressure, which causes the aggregates to separate into individual clusters thus freeing the entrapped pocket spaces between them. Further heating causes volume to rise. Rate of increase of volume with temperature falls away, as shown by the curvature of the graph, until about 60°C is reached. Above this no further water cluster aggregates are formed. (From Prof. S. R. Erlander, ref. 11).

B. V. Derjaguin reported on preparation of this water condensate and described a number of its properties (8). Lippincott et al. (18) confirmed and extended these observations and used the term polywater to describe the substance, believing it was a true polymer of water. There followed a flurry of published papers, some supporting, others denying the existence of "polywater", attributing the properties of the substance to contaminants. The controversy continues but at the 44th National Colloid Symposium at Lehigh University, Bethlehem, Pennsylvania, June 1970, several reports supported the negative side with rather strong evidence. These have since been published (6,24).

In condensing ordinary water in very fine quartz capillary tubes at low pressure, the water molecules presumably become very closely packed together by their interaction with the crystal lattice of the surface hydrogen bond and become polymeric. The substance is 15 times more viscous and 40% denser than ordinary water. It seems a dense, very stable polymer of water with a molecular weight of 72; 4 times that of water (8). It is stable up to 500°C and freezes only with great difficulty, forming a glass-like solid at -40°C.

Polywater calls to mind the concept of bound water in biological systems, about which little is said nowadays - the great bulk of cellular and extracellular water, bound in and out of cells, structured, unfrozen and unable to function as a solvent. Plants do not freeze at 0°C as does a pond; winter hardy plants and many trees do not freeze at -30°C, though they are 50% to 60% water by weight. Theoretically, polywater or water II could be an important constituent of living systems, intra and extra cellularly, making it much more common than previously considered. Ice-like, structured water could be polywater; however, at present this seems quite unlikely. Solvent properties have been little explored but polywater density and high viscosity would be expected to decrease markedly the diffusion of solutes, as well as oxygen and carbon dioxide in solution. Thus cells surrounded by polywater or having polywater intracellularly (perhaps in and about specific structures, nuclei, mitochondria, lysosomes, endoplasmic reticulum, or a Golgi apparatus) would have very limited metabolic activity including transport, enzymatic synthesis and reactions, protein and carbohydrate polymer synthesis. As gross speculation one can relate polywater to the phenomenon of hibernation, i.e., if polywater could be formed in an animal system preceding the hibernating state, the limited metabolic activity expected would be consistent with that observed in true hibernation. At the present writing, however, scientists are finding polywater more and more difficult to swallow.

STRUCTURED WATER IN WHOLE TISSUE

Several recent studies support the concept that cell water and extracellular water at surfaces, or water hydrating polyelectrolyte macromolecules of connective tissue (collagen, glycoproteins and glycosaminoglycans) are normally more like ice than liquid water, belying the classical picture of the cell as a set of bags within a membranous bag containing an aqueous solution of small molecules, ions and proteins. In fact, there has long been a minority group which regard the cell as a very organized, non-liquid, solid phase made up of macromolecules embedded in a matrix of semi-crystalline to ice-like water. (Gilbert N. Ling, Freeman W. Cope, Albert Szent-Gyorgi, A. S. Troshin and Ross A. Gortner). Their view is that cells should be treated as semiconducting crystals. Experimental evidence cited below is the first hard data supporting this point of view.

Using nuclear magnetic resonance (NMR) technique Cope (4,5) showed that the NMR spectra absorption peaks in muscle and brain tissue D_2O (heavy water) were broader than those of free water with the same concentration of D_2O;

tissue D_2O had a shorter spin lattice relaxation time than that of D_2O in free liquid water. The differences were not due to paramagnetic ions or magnetic inhomogeneities. The tissue water had a significantly greater degree of crystallinity than liquid water. A second fraction of tissue water with even shorter relaxation time and more structure was shown. These two fractions of tissue water may well represent many subfractions with varying magnitudes of crystallinity. The only tenable hypothesis consistent with the observed data is that cell water is more like a solid than a liquid.

NMR is an excellent means of probing the structure of water since the NMR spectra of water's protons (hydrogen nuclei) are influenced by electrical and magnetic fields in the surrounding water lattice. These fields finally reflect the long range order or structure of the water lattice.

Hazlewood et al. (14), also using NMR, examined absorption peaks in muscle water and free water, finding that the peaks in muscle water were broader than those in free water; also they noted that muscle water produced two NMR signals, one broader than the other. This, like Cope's study, suggests that tissue water exists in at least two different phases, both more crystalline than free water.

A further experiment was to examine NMR spectra of muscle water both before and after heat denaturation of tissue proteins. They found that spectral line broadening was clearly due to interaction between water molecules and cellular protein, i.e., the broad peaks narrowed after denaturation. Such interaction (presumed structuring of the water by the polyelectrolyte protein) restricts the motional freedom of water molecules causing a more ordered system than occurs in free water.

The Cope-Ling hypothesis proposes that intracellular ions are complexed to polyionic sites in cellular macromolecules (structural and otherwise) and the maintenance of equilibrium of ionic concentrations depends on the relative affinities the different ions have for complexing. Thus, ion pumps are unnecessary and no consumption of energy is needed to maintain equilibrium concentrations. Intracellular Na+ and K+ are located on anionic sites on intracellular polyelectrolyte macromolecules, and are capable of moving from one site to another through the ice-like water lattice in which they are only slightly soluble.

In such a model the cell surface is not regarded as a membrane but an interface or junction between two dissimilar materials, the membrane acting like a semi-conductor junction. Ling's long standing theory that the adsorption of cell water to cell proteins produces ice-like formation in the water is supported by the above quoted studies.

There now seems no question that water molecules do interact strongly with individual ions as well as polyions in such macromolecules as globular protein, collagen, glycosaminoglycans and other polyelectrolytes. The OSO_3^- and COO^- groups of the A regions are capable of restricting the motion of the deuterium atom of heavy water, D_2O. There is no good evidence that DNA or RNA have structured water about their charged phosphate groups.

POSSIBLE IMPLICATIONS OF STRUCTURED WATER TO HIBERNATION

The following comments are, of necessity, highly speculative. The relationship between the structuring of water and temperature suggests that in the range 5°C to 20°C, when water structure is that of cluster-aggregates, that biological activity must be altered by that factor alone in the hibernator. One could reasonably speculate that the animal will manifest changes in kinds of enzymes produced and number of enzymes as well as their form (iso-enzymes) in response to temperature fall and its effect on water structure. Cold inactivation of enzymes may occur as well as possible enzyme induction by low temperature. Hydration shells of the connective tissue structural macromolecules in tendon, skin, ligament, fascia and soft tissue may be altered, perhaps increased by temperature decline. The apparent stiffness of the tissues, muscles and joints of the hibernating animal may be in part a function of the structuring of water (2). Mitochondrial or lysosomal functions may be regulated secondarily to a drop in body temperature as a part of the organism's (bacteria, insects, trees, fish, mammals) biochemical restructuring in preparation for winter.

SUMMARY

The ordinary and extraordinary physical properties of water have been reviewed, both as water exists free and in its interaction with individual ions and polyions. Water-ion interaction and the three principle models for structured water were considered. Evidence for "polywater" in biological systems was reviewed and recent studies were presented supporting the view that water in living systems is crystalline and highly structured. Possible relationship of structured water to the phenomenon of hibernation was briefly considered.

REFERENCES

1. BERNAL, J. D., and R. H. FOWLER. A theory of water and ionic solution, with particular reference to hydrogen and hydroxyl ions. J. Chem. Phys. 1: 515-548, 1933.

2. BLAND, J. H. Theoretical mechanism of production of the symptom stiffness. Federation Proc. 28: 1073-1079, 1969.

3. CONWAY, B. E. Electrolyte solutions: solvation and structural aspects. Ann. Rev. Phys. Chem. 11: 481-528, 1966.

4. COPE, F. W. NMR evidence for complexing of Na^+ in muscle, kidney, and brain, and by actomyosin. The relation of cellular complexing of Na^+ to water structure and to transport kinetics. J. Gen. Physiol., 50: 1353-1375, 1957.

5. COPE, F. W. Nuclear magnetic resonance evidence using deuterated water for structured water in muscle and brain. Biophys. J. 9: 303-319, 1969.

6. DAVIS, R. E., D. L. ROUSSEAU, and R. D. BOARD. "Polywater": evidence from electron spectroscopy for chemical analysis (ESCA) of a complex salt mixture. Science 171: 167-170, 1971.

7. DEHL, R. E. Collagen: mobile water content of frozen fibers. Science 170: 738, 1970.

8. DERYAGIN, B. V., and N. V. CHURAEV. Do we know water? (Russian) Priroda 4: 16-22, 1968.

9. EISENBERG, D., and W. KAUZMANN. The Structure and Properties of Water. New York: Oxford Univ. Press, 1969, p. 71.

10. ERLANDER, S. R. The structure of water and its relationship to hydrocarbon-water interactions. J. Macromol. Sci.—Chem. A2: 595-621, 1968.

11. ERLANDER, S. R. The structure of water. Sci. J. 5A: 60-65, 1969.

12. FISHER, K. C., and J. F. MANERY. Water and electrolyte metabolism in heterotherms. In: Mammalian Hibernation III, edited by K.C. Fisher, A.R. Dawe, C.P. Lyman, E. Schönbaum, and F.E. South. Edinburgh: Oliver and Boyd, 1967, p. 235-279.

13. FRANK, H. S., and WEN, W.-Y. III. Ion-solvent interaction. Structural aspects of ion-solvent interaction in aqueous solutions. A suggested picture of water structure. Disc. Farraday Soc. 24: 133-140, 1957.

14. HAZLEWOOD, C. F., B. L. NICHOLS, and N. F. CHAMBERLAIN. Evidence for the existence of a minimum of two phases of ordered water in skeletal muscle. Nature 222: 747-753, 1969.

15. JEFFREY, G. A., and R. K. MCMULLAN. The clathrate hydrate. Prog. Inorg. Chem. 8: 43-108, 1967.

16. KAMB, B. Ice polymorphism and the structure of water. In: *Structural Chemistry and Molecular Biology*, edited by A. Rich, and N. Davidson. San Francisco: W.H. Freeman and Co., 1968, p. 507-542.

17. KAVANAU, J. L. *Water and Solute Water Interactions*. San Francisco: Holden-Day, 1964.

18. LIPPINCOTT, E. R., R. R. STROMBERG, W. H. GRANT, and G. L. CESSAC. Polywater. *Science* 164: 1482-1487, 1969.

19. LOW, P. F., B. G. DAVEY, K. W. LEE, and D. E. BAKER. Clay sols versus clay gels: biological activities compared. *Science* 161: 897, 1968.

20. NARTEN, A. H., and H. A. LEVY. Observed diffraction pattern and proposed models of liquid water. *Science* 165: 447-454, 1969.

21. NEMETHY, G., and H. A. SCHERAGA. Structure of water and hydrophobic bonding in proteins. IV. The thermodynamic properties of liquid deuterium oxide. *J. Chem. Phys.* 41: 680-689, 1964.

22. PERRY, T. O. Dormancy of trees in winter. *Science* 171: 29-36, 1971.

23. ROENTGEN, W. K. *Ann. Phys. Chim.* (Weid) 45: 91, 1892.

24. ROUSSEAU, D. L. "Polywater" and sweat: similarities between the infrared spectra. *Science* 171: 170-172, 1971.

25. SCHIFFER, J., and D. F. HORNIG. Vibrational dynamics in liquid water: a new interpretation of the infrared spectrum of the liquid. *J. Chem. Phys.* 49: 4150-4160, 1968.

CREDITS

The editors and authors are grateful to the following publishers, organizations and individuals for permission to use materials previously published:

Figure 1: Syndication International IPC Services Ltd. and Transworld Feature Syndicate Inc.; S. R. Erlander, *The Science Journal* 5A: 60-65, 1969. Copyright Syndication International IPC Services Ltd., 1969.

Figure 2: Syndication International IPC Services Ltd. and Transworld Feature Syndicate Inc.; S. R. Erlander, *The Science Journal* 5A: 60-65, 1969. Copyright Syndication International IPC Services Ltd., 1969.

A GUIDE FOR FUTURE STUDIES
OF LOW TEMPERATURE METABOLIC FUNCTION

As the temperature of cells and tissues is gradually lowered, three successive, and to a certain extent overlapping, changes in functional capacity are observed. The first of these is a progressive depression of the functions which are characteristic of the particular cells or tissues under investigation. The second is either a distortion or complete loss of these functions and the third is actual cellular or tissue damage. Depressed functions, within limits, are usually reversible while dysfunction or loss of function depends upon the duration of low temperature exposure. Cellular damage, on the other hand, usually ensues from excessive periods of low temperature inactivity and eventually leads to a loss of viability. Depressed function or loss of function, therefore, does not necessarily imply cellular damage. Thus, if cells or tissues upon rewarming again demonstrate their usual functional capabilities, then damage at low temperature did not occur. Myocardial contractility, for example, may be completely lost at 5° C, yet if this state is maintained for only a short period of time, contractility upon rewarming will be unimpaired.

Cellular sensitivity to low temperature varies considerably from species to species, particularly between hibernators and non-hibernators. The hypothermic Arctic ground squirrel heart, for example, has been shown (9) to maintain in vitro contractility at temperatures down to 0° C., while the hypothermic rabbit heart loses contractility at a temperature of about 10° C. (Fig. 1). Other similar differences in low temperature tissue function have been recognized for many years. In fact, as long ago as 1881 Horvath (22) suggested that the tolerance to low temperature could be used to distinguish hibernating from non-hibernating species. His suggestion has been amply confirmed over the intervening years. Thus, hypothermic hibernators maintain

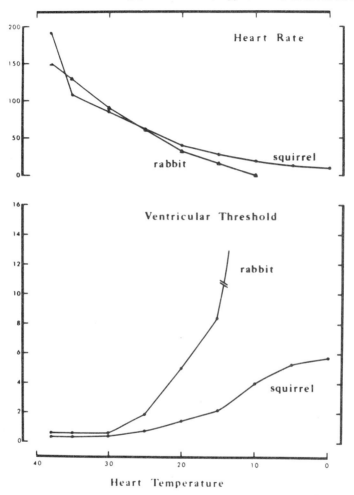

FIG. 1. Effects of temperature on the mean rate of diastolic ventricular threshold of isolated, perfused rabbit and Arctic ground squirrel hearts. Ordinate units are beats per minute for heart rate and milliamps for threshold stimulus strength. All electrical activity in the rabbit heart ceased at a mean temperature of 14.1 ± 2.03° C (S.D.). Ground squirrel hearts were contracting at a rate of $\overline{2}$ - 10 beats per minute at 0° C. The rabbit ventricles were completely refractory to a diastolic stimulus of 15 ma at a temperature of 10° C or lower. The squirrel ventricles exhibited an average threshold of 5.7 ma at 0° C. (Redrawn from ref. 9.)

various neural functions including spontaneous cortical activity (6,29), peripheral neuron excitability (7,46), autonomic activity (31,33), and neuromuscular transmission (46,47) at far lower temperatures than hypothermic non-hibernating species. They also show enhanced low temperature tolerance in such diverse functions as ventilation (22,28), skeletal muscle contraction (45,46) and possibly renal tubular functions (23).

Cellular sensitivity to low temperature also varies considerably from tissue to tissue. Accordingly, skin cells readily tolerate wide temperature fluctuations whereas liver cells lose most of their functional capabilities at temperatures of 18 to 20° C. Skin cells, furthermore, maintain their integrity for prolonged periods at temperatures near freezing, e.g., the ears of cold exposed rats have been shown to maintain cellular integrity over a period of days or weeks. The hypothermic liver, on the other hand, shows evidence of cellular degeneration in a matter of hours (6).

It is now reasonably well established that metabolic factors are in large measure responsible for the differences in low temperature cellular function not only between hibernating and non-hibernating species but also between tissues of the same species. It is also reasonably well established that most, if not all, of the functional changes observed at low temperature are endergonic. That is, they are associated with a utilization of metabolic energy. The hypothermic liver, for instance, exhibits marked reductions in bile formation (6,24,49), bromphenoltetrabrompthalein disulfate (BSP) extraction (6), pentobarbital removal and detoxification (42), morphine conjugation (19) epinephrine removal and degradation (15), sulfanilamide acetylation and excretion (25) and biliary atropine excretion (26). In the kidney, mild to moderate hypothermia also causes a decrease in glomerular filtration rate and an increase in urine flow (4,20,27,38), the latter effect being attributable to a decrease in tubular water reabsorption (20,27).

Tubular transport of glucose (27), sodium and potassium (20) and chloride (14) are similarly depressed during hypothermia.

Since most endergonic processes of the body are coupled to the release of energy from ATP, one might hypothesize that limitations in ATP stores are responsible for the functional decrements which are observed at low temperature. Indeed, support for such a hypothesis was obtained in comparative studies of *in vitro* contractility loss by hypothermic rabbit and ground squirrel hearts (9). As shown in Figure 2, loss of contractility is associated with a sizeable conversion of ATP to ADP in the rabbit heart while retention of contractility in the ground squirrel heart is associated with the maintenance of normal ATP levels. Not all decrements in cellular function, however, are this simply explained. Thus, loss of sodium pump activity may occur at low temperature in spite of normal or near normal intracellular ATP levels. In other words, ATP was available but not apparently used, for sodium transport.

At the molecular level we know that the transition phase of increased structuring of water occurs between 40° C and 20° C; maximal and predominant formation of large and varying sized clusters of aggregates of water molecules occur from 15° C to 4° C. In this latter range one might expect a critical temperature below which dysfunction of cells and tissues occurs and from which energy requirement to maintain a given biochemical function cannot be met. Such an event might be due secondarily to the structuring, ice-likeness of both intracellular and extracellular water, predominantly the former.

The rather obvious conclusion to be drawn from the foregoing discussion is that diverse factors may control or influence the various functions of cells and tissues subjected to low temperature exposure. In our investigations of low temperature metabolic function, therefore, it would seem only reasonable to design metabolic experiments within the framework of rather general hypothesis.

LOW TEMPERATURE METABOLIC FUNCTION 103

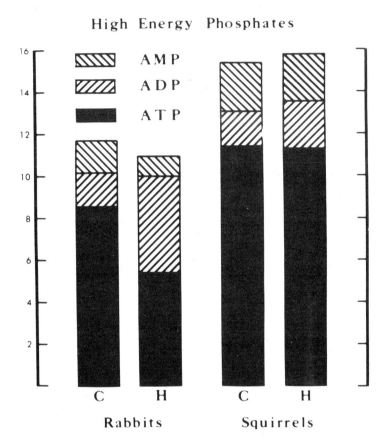

FIG. 2. Adenine nucleotide levels of normal (37° C) and hypothermic (15° C) Arctic ground squirrel ventricles. Ordinate units are μmoles per kg wet weight. Significant differences ($P \leq 0.05$) were: the loss of ATP and the gain of ADP in the hypothermic rabbit heart and the differences in total ATP and adenine nucleotide levels of rabbit and squirrel ventricles at both normo- and hypothermic temperatures. [Redrawn from ref. 9.]

We might, for example, state that dysfunction appears at some critical temperature at which the energy requirement to support a given function exceeds that which is available. This hypothesis, as illustrated in Figure 3, makes no attempt to pinpoint the exact site of the metabolic defect. It could be at any stage of the energy transfer system from substrate degradation to its release from the terminal bonds of ATP or other, similar storage sites. According to this hypothesis, functional defects could result from one or more of the following factors: (a) reduced substrate availability, (b) product inhibition, (c) decreases or losses of enzyme activity, (d) impaired or distorted patterns of electron transport, (e) defects in energy transfer to the terminal bonds of ATP, (f) restrictions in the transfer of ADP or ATP between the intra- and extramitochondrial space, (g) defects in the transfer of energy from terminal ATP bonds or similar storage sites to accomplish the particular cellular function under consideration and (h) effects of water structuring below 20° C on intracellular protein synthesis, ion transport, localized pH, osmolarity, etc.

Investigation of the metabolic factors underlying cellular dysfunction at low temperature is, therefore, far from a simple matter. Usually a wide variety of procedures must be employed before the nature of the metabolic defect or defects is fully understood. In general, these procedures fall into two categories: those associated with studies of intact cells, either *in vivo* or *in vitro*, and those associated with studies of subcellular preparations, which may range from crude homogenates to highly purified enzyme systems. Most of these procedures will provide valuable information if the experiments are properly conducted and interpreted. Conversely, each will produce erroneous or misleading information if improperly conducted. In the paragraphs which follow some of the more promising procedures for

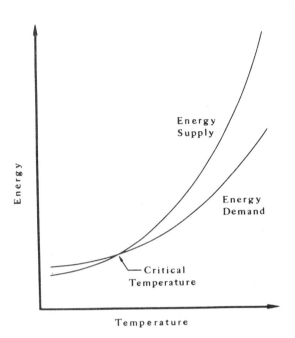

FIG. 3. Schematic representation showing the effects of temperature on the relationship of cellular energy requirements to cellular energy supply or availability. The critical temperature refers to that point where the energy required to support a given function exceeds that which is available. At this point dysfunction commences. A number of factors (see text) could be responsible for reduced energy supply.

studying metabolic function at low temperature, as well as the limitations of these procedures and the precautions that must be taken with them, are outlined.

INTACT ANIMALS STUDIES

Studies with intact animals serve three purposes: First, they allow overall investigation of metabolic and other functions under the usual neurohumoral, cardiopulmonary and other controls that exist under a particular set of experimental conditions, e.g., hypothermia. Second, they allow study of the interactions of various tissue and organ systems; for example, the relationship of liver glycogenolysis on muscle glucose utilization. And third, they permit identification of particular organs, tissues or cells which are affected by a given experimental condition. The latter purpose is particularly important since it permits us to relate cellular and biochemical changes which occur in vitro to changes which occur in the intact organ in vivo. It is most important, in other words, to establish whether or not a given functional change observed in vitro has any meaning in the "real world," that is, in vivo.

Perhaps the most important single prerequisite to successful in vivo studies is to minimize the effect of the experimental procedures per se on the results which are obtained. This can never be completely accomplished, but modern techniques allow us to avoid such functionally distorting procedures as anesthesia, restraint, etc. Excellent procedures are thus available for obtaining blood or other fluid samples, assessing hemodynamic functions, and measuring or modifying bioelectrical activity, in the intact, unanesthetized and unrestrained animal. Recent technological advances in physiology and biochemistry have produced an abundance of miniature transducers and other equipment as well as microsurgical and microchemical techniques which are readily applicable to the intact animal.

Metabolic studies in the intact animal can involve either the entire body or can be limited to one or more organs. In either case the investigator must be thoroughly familiar with the techniques he is using and the factors which are critical to their proper application if he is to obtain meaningful data. One still sees, for instance, reports in which isotope-labeled substrates are used to assess some aspect of metabolic function and the investigator has failed to measure the pool size of the particular metabolite under study. Or, equally objectionable, he has injected an isotope-labeled substrate intraperitoneally and has ignored potential changes in the diffusion barrier between the peritoneal cavity and the blood. Data obtained from such studies are most difficult to interpret and are oftentimes worthless. In vivo studies of organ metabolism can be equally meaningless if all of the critical variables are not measured or controlled. Arteriovenous differences in the level of a given metabolite are nearly impossible to evaluate if the blood flow through the organ is not measured. In addition, abnormal conditions, such as excessively high or low substrate levels, perfusion pressures, etc., can so distort the physiology of an organ, that the observed functions bear little resemblance to those occurring in the unencumbered animal.

Promising specific studies might include measurements of urinary excretion of hydroxyproline, hexoseamine and uronic acid in the hibernator, before, during and post-hibernation could provide data on relative turnover rates of the connective tissue macromolecules, collagen and the glycosaminoglycans. Reliable and simple methods are available. These substances have been successfully used to measure growth rates, growth arrest, intense organ catabolism (normal post partum uterus), magnitude of connective tissue dissolution in certain clinical states (Paget's disease, hyperparathyroidism, massive burn, extensive destructive malignant disease). Interpretation of the data would require taking into account the atrophying consequence of immobilization, the fall in body temperature, the effect in protein and carbohydrate polymer synthesis. Such data could have implications in space travel. Synthesis and degradation of the macromolecular matrix of the supporting system, (bone, ligament, tendon, fascia, skin, cartilage, etc.)

seems even more important than the electrolyte changes, i.e., calcium and phosphorus turnover studies in astronauts, to which most attention has been paid.

ISOLATED ORGAN STUDIES

Isolated organ systems offer new approaches to the study of low temperaturn metabolic function. A principal advantage of these techniques lies in the capability of the investigator to control experimental parameters precisely while the effects of extraorgan factors which are commonly encountered in vivo (hormone effects) can be eliminated. Also, these systems can be designed to approximate certain in vivo conditions quite closely, yet factors such as temperature, substrate level or ion concentration can be varied at will. Of course, interpretation and extrapolation of results obtained with these systems must be guarded but in the hands of the careful worker, such techniques offer approaches to problems which heretofore have been very difficult to solve.

A number of investigators have utilized isolated organs to better understand the effects of low temperature on physiological function in hibernating species. South (46) used a nerve-diaphragm preparation to demonstrate a functional adaptation to low temperature in the hamster. More recently, cardiac function has been studied at low temperatures utilizing isolated atria or whole hearts from a variety of hibernating mammals (30,32,34,41,43). Musacchia and Barr (39) employed an everted intestinal sac technique to assess in vitro active transport of glucose in tissues from cold acclimated hamsters.

There is, however, a paucity of data regarding metabolic changes which occur at low temperature. Covino and Hannon (9) brought attention to the use of isolated organs by comparing energy metabolism in perfused hearts from rabbits and ground squirrels at low temperatures. Further studies of substrate utilization by normothermic and hypothermic perfused rat and ground squirrel hearts (11) offer an example of the type of approach which will be useful in future studies. Factors which control energy metabolism can be elucidated with isolated organ systems. For example, using the methods of Weissler, et al. (50) isolated perfused ground squirrel hearts were used to study the effects of hypoxia on the factors which control glycolytic flux. In this study, physiological function was correlated with levels of adenine nucleotides, glycolytic intermediates and other biochemical parameters. Earlier work by Morgan, et al. (36,37) provides a prime example of the value of isolated perfused heart preparations in studies of myocardial metabolic function.

The original Langendorff perfused heart preparation has undergone considerable modification and sophistication in recent years. Crass and his coworkers (10), have developed an isolated heart preparation which has the properties of pressure development and substrate concentrations characteristic of the *in vivo* state. Techniques described by Feinberg, et al. (13) provide simultaneous measurement of oxygen consumption, energy utilized in terms of force, and the rate of force development at fixed cardiac dimensions (isovolumic) in the isolated rabbit heart. Similar sophistication of methods of using isolated rat hearts has been developed by Gamble, et al. (16). Other workers have utilized parts of the heart (e.g., papillary muscle) to investigate the relationship between physiological function and metabolism (40).

Aside from the heart, other isolated organ systems have recently been successfully employed in metabolic studies. Gilboe, et al. (17,18) have made significant advances in the use of isolated dog brain preparations.

Similar techniques have been developed for the rat brain (1,48). Bowman (5) has successfully utilized an isolated perfused rat kidney to study gluconeogenesis. Since the development of a viable isolated perfused liver system (35), many investigators have used the preparation for metabolic studies. For example, Exton and Park (12), Soling, et al. (44) and Williamson, et al., (51) have conducted extensive studies of hepatic gluconeogenesis in the isolated organ.

The cellular metabolic events associated with the transition from one physiological state to another (e.g., entrance into and arousal from hibernation) have not been extensively studied. Furthermore, the mechanisms which control these events deserve consideration. In this regard, studies by Mendler et al., which employ perfused rat brains _in vivo_ offer exciting possibilities for further experimentation. Mendler has also utilized a similar _in vivo_ perfusion system to study energy metabolism in hypothermic rat hearts (personal communication). The use of these techniques will no doubt promote in depth understanding of the cellular events associated with hibernation.

CELL CULTURE STUDIES

Cell culture preparations, although rarely used in low temperature studies are likely to yield some basic and important information. Skin fibroblasts and liver cell cultures are suggested, utilizing cells from a true hibernator as experimental and cells from a comparably sized animal as a control, (such as the thirteen-lined ground squirrel and a nonhibernating chipmunk). One might, for example, run the cultures in tandem and after 8 to 10 transfers, grow monolayer cultures at 37° C, 25° C, 15° C, and 5° C to determine if differences existed, either macro- or microscopically, in the capacity for growth and division as it relates to temperature. Micro methods are available to measure total protein, hydroxyproline and fractionated studies of the glycosaminoglycans in cell cultures (3). Thus, synthetic function could be correlated with temperature in the cells of the hibernator and the non-hibernator. In this regard it would be of interest to compare skin fibroblasts and liver cells since the latter are known to be cold sensitive.

SUBCELLULAR STUDIES

A number of biochemical mechanisms can be proposed to account for the abilities of hibernating mammals to maintain viable rates of energy metabolism at low temperatures which prove lethal for non-hibernating mammals. It is the intent of the following paragraphs to suggest several such mechanisms and, further, to propose in rather general terms the experimental approaches necessary to determine whether any or all of these hypothetical mechanisms, do, indeed, occur in hibernants.

In line with the discussion of low temperature acclimation by aquatic poikilotherms we can propose three basic means by which metabolic, i.e., enzymic, activity could be increased at low temperatures to compensate for the reduced kinetic energy of the system. First, enzyme concentrations could be increased. Secondly, enzymes with higher catalytic efficiencies could be synthesized. And, thirdly, the composition of the intracellular environment within which most enzymes function could be altered in compensatory manners, that is, through increased concentrations of enzyme activators at low temperature.

In approaching these hypothetical mechanisms experimentally, a multi-faceted attack would appear necessary. It would first seem desirable to measure total activity for a given enzyme in crude, i.e., unfractionated and undialyzed, homogenates of tissues. At this point in the analysis, saturating substrate concentrations should be used since enzymic activity in crude homogenates is apt to be low. The precise levels of substrate which are saturating at different assay temperatures should, of course, be determined with care.

At this point, it may be wise to initiate kinetic experiments, using dialyzed homogenates, to determine whether enzyme variants are present. In aquatic poikilotherms kinetically distinct enzyme variants may be synthesized during seasonal acclimation. In all cases the "winter" and "summer" variants of a given enzyme are characterized by different temperature-versus-apparent K_m responses; the total activities of the enzymes may be the same in "winter" and "summer" tissues. Thus, gross activity measurements alone may give no hint about the presence of enzyme variants. The simplest way to check for kinetic variants of the sort found in some aquatic poikilotherms is to estimate apparent K_m values at different temperatures. If kinetic variants are detected then electrophoretic resolution of the enzyme systems is desirable. Finally, as a word of warming, different variants of an enzyme may have identical activation energy characteristics and, therefore, this parameter is likely of limited usefulness in determining whether or not "warm" or "cold" enzyme variants are present.

To return to the experimental approaches which should be followed to determine whether enzyme quantities and/or the enzymes' intracellular environments are changing, the following rationale may be useful: If it is found that activities in crude homogenates differ, e.g., between the hibernator and non-hibernator, then one should examine the effect of exhaustive dialysis on the two homogenates. If the differences in enzymic activity are due to changes in the composition of the cytosol, then it is very likely that dialysis will abolish the activity differences. However, care must be taken to ascertain that the enzymes in the two homogenates are not differentially labile during dialysis.

If the enzymic activity difference persists after dialysis, then there is likely more enzyme in the more active tissue and/or the enzyme in the more active system is a "better" catalyst. If enzyme variants have been found in kinetic and/or electrophoretic analysis, as discussed above, then the second possibility likely provides at least a partial explanation of the observed rate differences. If enzyme variants do occur, the substrate turnover numbers of the enzymes should be determined to gauge the absolute rates at which the enzyme variants can promote catalysis.

If no enzyme variants appear to be present in the system, then enzyme concentration differences likely occur. To accurately estimate enzyme concentrations, complex immunochemical procedures are required. However, crude estimates of enzyme concentration can be obtained by measuring the intensity of staining (activity or protein) in electrophoretic media, e.g., acrylimide gel.

Finally, one vital aspect of poikilothermic temperature acclimation is the "restructuring" of cellular membranes. A detailed study of the lipid compositions of membranes (neural, mitochondrial, etc.) from the hibernating and non-hibernating life stages of a hibernator would be a most important addition to our knowledge.

In Dr. Beyer's paper, which forms part of the formal symposium, an attempt was made to identify specific key enzymic reactions which, being cold labile, might operate as regulators of metabolism in tissues at low temperature. Since, with one possible exception, all of the mammalian enzymes which have been studied in this respect have been relatively purified systems, the question arises as to whether such enzymes are inactivated by cold treatment in the intact cell. Perhaps the simplest, if not most direct, approach to this question would be to test cold-lability at more and more crude stages in the purification of the particular enzyme in question. If, for example, such enzymes still displayed cold-sensitivity at the crude homogenate stage, one would be justified in pursuing the question further. It might then be extremely interesting to work with higher levels of cellular and tissue organization. Intact isolated cells from some tissues may be prepared by treatment with colagenase and enzyme systems assayed at various temperatures by sensitive polarographic fluorescent or isotope methods. Similar methods might be applied to tissue slices and, perhaps most rewardingly, to perfused organs where rapid freezing techniques might be applied.

STUDIES OF STRUCTURED WATER - MEMBRANE RELATIONSHIPS

Conceptually it seems reasonable to regard the cell membrane as well as intracellular membranes (microsome, lysomes, pinocytotic vacuoles, nuclear and nucleolar membranes, mitochondrial membranes, ribosomes, centrosomes and endoplasmic reticulum) as having structured water over their surfaces, with diminishing magnitudes of structuring or ice-likeness at some distance from the membrane surfaces. Wherever there are charged macromolecules and membrane surfaces, water will behave very differently than in a free liquid state, forming very stable ice-like hydration sheets. Such layering of structured water must alter the properties of the membranes for transport as well as diffusion. Water molecules themselves are not free to move and ions could not be transported through such "dense" water. Bernal (2) proposed that such hydration of ice-like layers are 6-8 molecules thick or 10Å to 20Å with less dense structuring occurring in water beyond 10Å to 20Å from the membrane surface. Since the cell inclusion bodies and their membranes are so very close together within the cell, it is expected that there may be overlapping areas of these ice-like interfaces and all parts of the cell are near or actually are in continuity with phase boundaries.

Membranes can no longer be regarded as serving a gate-keeper function only. Since water structure is a function of temperature (as well as ions and polyions in the water) the hibernating animal with body temperature of 5° C to 20° C may be reasonably expected to show both cellular and intracellular membrane coatings of ice-like structured water and extracellular water structuring - all in greater magnitude than would occur in the non-hibernating state at 37° C body temperature.

These factors are expected to influence enzyme synthesis and intracellular enzyme reactions, protein synthesis, substrate concentrations, pH regulation, osmolarity, etc. This theory could be investigated using nuclear magnetic resonance technique and heavy water (D_2O) on the tissues of the hibernator, before and during induced hibernation as a means of getting some measurement of magnitudes of water structuring in tissues prior to and during the hibernating state (8,21).

Symposium Committee on Metabolism
in Hibernation and Hypothermia

John P. Hannon, Chairman
Robert E. Beyer
Roy F. Burlington
George N. Somero
John H. Bland

REFERENCES

1. ANDJUS, R. K. Some mechanisms of mammalian tolerance to low body temperatures. In: S.E.B. Symposium XXIII. Dormancy and Survival. Cambridge, Eng.: The University Press, 1969, p. 351-394.

2. BERNAL, J. D. The structure of water and its biological implications. In: S.E.B. Symposium XIX. The State of Movement of Water in Living Organisms, edited by G. E. Fogg. Cambridge, Eng.: The University Press, 1965, p. 17-32.

3. BLAND, J. H., G. O'CONNOR, and W. I. SCHAFFER. Fibrodysplasis ossificans progressiva: A study of collagen and glycosaminoglycans in tissue culture. Arthritis Rheumat 1971. In press.

4. BLATTEIS, C. H. and S. M. HORVATH. Renal, cardiovascular and respiratory responses and their interrelations during hypothermia. Amer. J. Physiol. 192: 357-363, 1958.

5. BOWMAN, R. H. Gluconeogenesis in the isolated perfused rat kidney. J. Biol. Chem. 245: 1604-1612, 1970.

6. BRAUER, R. W., R. J. HOLLOWAY, J. S. KREBS, G. F. LEONG, and H. W. CARROL. The liver in hypothermia. Ann. N.Y. Acad. Sci. 80: 395-423, 1959.

7. CHATFIELD, P. O., A. F. BATTISTA, C. P. LYMAN, and J. P. GARCIA. Effects of cooling on nerve conduction in a hibernator (golden hamster) and non-hibernator (albino rat). Amer. J. Physiol. 155: 179-185, 1948.

8. COPE, F. W. Nuclear magnetic resonance evidence using deuterated water for structured water in muscle and brain. Biophys. J. 9: 303-319, 1969.

9. COVINO, B. G., and J. P. HANNON. Myocardial metabolic and electrical properties of rabbits and ground squirrels at low temperatures. Amer. J. Physiol. 197: 494-498, 1959.

10. CRASS, M. F., E. S. McCASKILL, and J. C. SHIPP. Glucose-free fatty acid interactions in the working heart. J. Appl. Physiol. 29: 87-91, 1970.

11. ERASMUS, B., and D. W. RENNIE. Substrate uptake and utilization in normothermic and hypothermic perfused rat and ground squirrel hearts. Physiologist 10: 163, 1967.

12. EXTON, J. H., and C. R. PARK. Control of gluconeogenesis in liver. I. General features of gluconeogenesis in the perfused livers of rats. J. Biol. Chem. 242: 2622-2636, 1967.

13. FEINBERG, H., E. BOYD, and G. TANZINI. Mechanical performance and oxygen utilization of the isovolumic rabbit heart. Amer. J. Physiol. 215: 132-139, 1968.

14. FISHER, E. R., C. COPLAND and B. FISHER. Correlation of ultra structure and function following hypothermic preservation of canine kidneys. Lab. Invest. 17: 99-119, 1967.

15. FUHRMAN, F. A., J. M. CRISMON, G. F. FUHRMAN, and J. FIELD. The effect of temperature on the inactivation of epinephrine in vivo and in vitro. J. Pharmacol. Exptl. Therap. 80: 323-334, 1944.

16. GAMBLE, W. J., P. A. CONN, A. E. KUMAR, R. PLENGE, and R. G. MONROE. Myocardial oxygen consumption of blood perfused, isolated, supported, rat heart. Amer. J. Physiol. 219: 604-612, 1970.

17. GILBOE, D. D., W. W. COTANCH, M. B. GLOVER, and V. A. LEVIN. Changes in electrolytes, pH, and pressure of blood perfusing isolated dog brain. Amer. J. Physiol. 212: 589-594, 1967.

18. GILBOE, D. D., M. B. GLOVER, and W. W. COTANCH. Blood filtration and its effect on glucose metabolism by the isolated dog brain. Amer. J. Physiol. 213: 11-15, 1967.

19. GRAY, I., R. R. RUECKERT, and R. R. RINK. Effect of hypothermia on metabolism and drug detoxification in the isolated perfused rabbit liver. In: The Physiology of Induced Hypothermia. Proceedings of a Symposium, edited by R. D. Dripps. Washington, D.C.: National Academic of Science-National Research Council Publ. 451, 1956, p. 226-234.

20. HARVEY, R. B. Effect of temperature on function of isolated dog kidney. Amer. J. Physiol. 197: 181-186, 1959.

21. HAZELWOOD, C. F., B. L. NICHOLS, and N. F. CHAMBERLAIN. Evidence for the existence of a minimum of two phases of ordered water in skeletal muscle. Nature 222: 747-750, 1969.

22. HORVATH, A. Zur Abkühlung der Warmblütter. Arch. Ges. Physiol. 12: 278-282, 1876.

23. KALLEN, F. C., and H. A. KANTHOR. Urine production in the hibernating bat. In: Mammalian Hibernation III edited by K. C. Fisher, A. R. Dawe, C. P. Lyman, E. Schöenbaum, and F. E. South. Edinburgh: Oliver and Boyd, 1967, p. 280-304.

24. KALLOW. W. Choleresis induced by synthetic drugs. 4. Poikilothermic study of the anesthetized rat. Arch. Exptl. Pathol. Pharmacol. 210: 336-345, 1950.

25. KALSER, S. C., E. J. KELVINGTON, and M. M. RANDOLPH. Drug metabolism in hypothermia. Uptake, metabolism and excretion of S^{35}-sulfanilamide by the isolated, perfused rat liver. J. Pharmacol. Exptl. Therap. 159: 389-398, 1968.

26. KALSER, S. C., E. J. KELVINGTON, M. M. RANDOLPH, and D. M. SANTOMENA. Drug metabolism in hypothermia. II. C^{14}-atropine uptake, metabolism and excretion by the isolated, perfused rat liver. J. Pharmacol. Exptl. Therap. 147: 260-269, 1965.

27. KANTER, G. S. Renal clearance of glucose in hypothermic dogs. Amer. J. Physiol. 196: 866-872, 1959.

28. KAYSER, C., and A. MALAN. Central nervous system and hibernation. Experientia 19: 441-451, 1963.

29. KAYSER, C., F. ROHMER, and C. HIEBEL. L'E.C.G. de l'hibernant. Lethargie et reveil spontane du spermophile. Rev. Neurol. 84: 570-578, 1951.

30. LYMAN, C. P. The effect of low temperature on the isolated hearts of Citellus lecuras and C. mohavensis. J. Mammal. 45: 122-126, 1964.

31. LYMAN, C. P., and R. C. O'BRIEN. Autonomic control of circulation during the hibernating cycle in ground squirrels. J. Physiol. 168: 477-499, 1963.

32. MARSHALL, J. M., and J. S. WILLIS. The effects of temperature on the membrane potentials in atria of the ground squirrel, Citellus tridecem-lineatus. J. Physiol. 164: 64-76, 1962.

33. MASSOPUST, L. C., L. R. WOLIN, and J. MEDER. Spontaneous electrical activity of the brain in hibernators and non-hibernators during hypothermia. Exptl. Neurol. 12: 25-32, 1965.

34. MICHAEL, C. R., and M. MENAKER. The effect of temperature on the isolated heart of the bat, Myotis lucifugus. J. Cellular and Comp. Physiol. 62: 355-358, 1963.

35. MILLER, L. L., C. G. BLY, M. L. WATSON, and W. F. BALE. The dominant role of the liver in plasma protein synthesis. A direct study of the isolated perfused rat liver with the aid of lysine-E-C^{14}. J. Exptl. Med. 94: 431-453, 1951.

36. MORGAN, H. E., E. CADENAS, D. M. REGEN, and C. R. PARK. Regulation of glucose uptake in muscle. II. Rate-limiting steps and effects of insulin and anoxia in heart muscle from diabetic rats. J. Biol. Chem. 236: 262-268, 1961.

37. MORGAN, H. E., and A. PARMEGGIANI. Regulation of glycogenolysis in muscle. II. Control of glycogen phosphorylase reaction in isolated perfused heart. J. Biol. Chem. 239: 2435-2439, 1964.

38. MOYER, J. H., G. MORRIS, and M. E. DEBAKEY. Hypothermia: I. Effect on renal hemodynamics and on excretion of water and electrolytes. Ann. Surg. 145: 26-40, 1957.

39. MUSACCHIA, X. J., and R. E. BARR. Cold exposure and intestinal absorption in the hamster. Federation Proc. 28: 969-973, 1969.

40. POOL, P. E., B. M. CHAMDLER, J. F. SPANN, E. H. SONNENBLICK, and E. BRAUNWALD. Mechanochemistry of cardiac muscle. IV. Utilization of high energy phosphates in experimental heart failure in cats. Circulation Res. 24: 313-320, 1969.

41. POUPA, O., J. PROCHAZKA, and V. PELOUCH. V. Effect of catecholamines on resistance of the myocardium to anoxia and on the heart glycogen concentration. Physiol. Bohemoslov. 17: 36-42, 1968.

42. RAVENTOS, J. The influence of room temperature on the action of barbiturates. J Pharmacol. Exptl. Therap. 64: 355-363, 1938.

43. SENTURIA, J. B., S. STEWART, and M. MENAKER. Rate-temperature relationships in the isolated hearts of ground squirrels. Comp. Biochem. Physiol. 33: 43-50, 1970.

44. SÖLING, H. D., B. WILLMS, D. FRIEDRICHS, and J. KLEINEKE. Regulation of gluconeogenesis by fatty acid oxidation in isolated perfused livers of non-starved rats. European J. Biochem. 4: 364-372, 1968.

45. SOUTH, F. E. Some metabolic specializations in tissues of hibernating mammals. In: Mammalian Hibernation, Proceedings of the First International Symposium on Natural Mammalian Hibernation, edited by C. P. Lyman and A. R. Dawe. Cambridge, Mass.: Mus. Comp. Zool. Harvard Coll., 1960, p. 209-232.

46. SOUTH, F. E. Phrenic nerve-diaphragm preparations in relation to temperature and hibernation. Amer. J. Physiol. 200: 565-571, 1961.

47. TAIT, J. The heart of hibernating animals. Amer. J. Physiol. 59: 467, 1922.

48. THOMPSON, A. M., R. C. ROBERTSON, and A. A. BAUER. A rat head-perfusion technique developed for the study of brain uptake of materials. J. Appl. Physiol. 24: 407-411, 1968.

49. TURNER, M. D., and F. ALICAN. Successful 20-hour storage of the canine liver by continuous hypothermic perfusion. Cryobiology 6: 293-301, 1970.

50. WEISSLER, A. M., F. A. KRUGER, M. BABA, D. G. SCARPELLI, R. F. LEIGHTON, and J. K. GALLIMORE. Role of anaerobic metabolism in the preservation of functional capacity and structure of anoxic myocardium. J. Clin. Invest. 47: 403-416, 1968.

51. WILLIAMSON, J. R., R. A. KREISBERG, and P. W. FELTS. Mechanism for the stimulation of gluconeogenesis by fatty acids in perfused rat liver. Proc. Natl. Acad. Sci. U.S. 56: 247-254, 1966.

CREDITS

The editors and authors are grateful to the following publishers, organizations and individuals for permission to use materials previously published:

Figure 1: The American Physiological Society; B. G. Covino and J. P. Hannon, The American Journal of Physiology 197: 494-498, 1959.

Figure 2: The American Physiological Society; B. G. Covino and J. P. Hannon, The American Journal of Physiology 197: 494-498, 1959.

MEMBRANES AND ELECTROLYTE PHENOMENA

THE SIGNIFICANCE AND ANALYSIS OF MEMBRANE FUNCTION IN HIBERNATION

J. S. WILLIS, L. S. T. FANG and RACHEL F. FOSTER

Department of Physiology and Biophysics
University of Illinois
Urbana, Illinois 61801

INTRODUCTION

In the literature of temperature acclimation of non-homeothermic organisms two classes of adaptation originally suggested by Precht (36) have been widely accepted: capacity adaptation and resistance adaptation.

In their original conception capacity adaptation was intended to apply to organisms making adjustments in order to carry on a perfectly normal life. Resistance adaptation was meant to apply to ability to survive under extreme duress. There are obviously intergradations between these extremes. A less extreme capacity adaptation might be one in which, say, fish experiencing a change of temperature could continue to swim and hunt but could not reproduce, whereas a modest form of resistance adaptation would be one in which life activities might be curtailed but normal physiological function maintained. Adaptation to low temperature found in mammalian hibernation would seem to be of this last type.

Analysis of mechanisms in a case of a resistance adaptation is perhaps somewhat easier than in a case of capacity adaptation, because in the former non-essential activities will frequently be attenuated and the problem will reduce to being one of examining a finite and manageable number of essential activities. In the 1965 symposium Willis (47) suggested that the mechanisms for maintaining normal ionic gradients across cell membranes were perhaps the single most important class of cellular activities to be considered in understanding cold resistance of mammalian hibernators. This suggestion was based upon two propositions. The first was that besides membrane transport only two other energy requiring activities had to be retained - muscle contraction and (in so far as it is a different thing) heat production. The second proposition was that apparent adaptations of these "other two" activities might themselves simply depend upon continued membrane activity at low temperature. Thus, for all we know, failure of muscle contraction at low temperature in non-hibernators could merely be due to depolarization of the membrane. The idea was also put forward that specialized heat production by brown fat (or any other cell) might be linked to increased ion transport as a result of increased permeability due to sympathetic stimulation. Apart from these three, then, all other activities could simply be slowed down to a minimum, assuming that lowering temperature causes no great imbalance between synthetic and degradative function.

Although the suggestion relating to the supremacy of membrane function could legitimately have been construed as a challenge, nothing has transpired since 1965 to force a retreat from that position. On the contrary several developments have occurred to strengthen it. Two separate groups have adduced evidence indicating that increased ion transport may indeed have a thermogenic role in brown fat. Girardier, et al. (16) have found that catecholamines cause decrease in membrane potential and an increase in K efflux from brown fat cells of rats. Horwitz, Smith and their co-workers have made similar observations on an _in situ_ preparation from rats and have also found that the Na,K-ATPase activity of brown fat is stimulated by norepinephrine (23,24).

We have also recently seen the elegant demonstration by Whitten et al. (43) that protein synthesis in liver of ground squirrels is specifically shut down during hibernation, thus eliminating an important competitor for metabolic energy.

The problem of the subcellular nature of cold adaptation of contractility in hibernators, though seemingly far more amenable to a reductionist approach than membrane transport, still, to our knowledge, has not been touched.

So after five years we may still regard adaptation of membrane activity with respect to ions to be a sufficiently relevant subject to warrant this sub-symposium.

I should like to divide my own discussion of this subject into three parts - the first will be re-examination of the extent of retention of ion balance in hibernating mammals and of the significance thereof to the organism, the second will be a brief review of progress in analysing the mechanism of cold adaptation in membranes, and the third will be consideration of importance of acclimation of membrane function to cold as opposed to intrinsic adaptation.

EXTENT AND SIGNIFICANCE OF RETENTION OF ION BALANCE
IN TISSUES OF HIBERNATORS

At the time of the 1965 symposium only two studies had been done on the state of tissues in hibernators with respect to monovalent cations and the results of both were fairly comforting. No decline of K or rise in Na could be seen in the tissues studied which was specifically related to low body temperature of hibernation (9,10,44). Since that time we have re-examined this question using more types of tissue and also looking at longer periods of hibernation, and we find that at least in hamsters and ground squirrels the picture is not quite so simple (52). Some tissues such as brain, cardiac muscle and diaphragm do indeed retain K in hibernation. Others, such as red blood cells and leg skeletal muscle, lose K during hibernation, although the loss from red cells is slow, and one must wait for more than five days to find significant loss in ground squirrels. Still other tissues such as kidney and liver of ground squirrel gain a considerable fraction of K during hibernation. In general the tissues which lose K reaccumulate it upon rewarming, and those which gain it in hibernation lose it immediately upon rewarming (52). Thus, while the tissues of hibernators are certainly adapted to resist loss of K, not all tissues retain the original steady state but some gain what the others lose. Apparently the uptake by kidney and perhaps liver does offset the loss from other cells and, for a time at least, prevent accumulation in the milieu interieur, since there is no compelling evidence for a rise in serum K during hibernation (52).

We have suggested, however, that the uncompensated loss of K from some tissues, leading perhaps eventually to an extracellular accumulation, may be an important factor which, because of partial membrane depolarization leads to rise in sensitivity (33,41) during each bout of hibernation and to the eventual spontaneous arousal of the organism. Viewed in this light the intervals of periodic rewarming could be regarded as an opportunity to restore a normal distribution of K. In support of a similar hypothesis Fisher and Mrosovsky have demonstrated that injections of a solution of KCl were more effective than injections of a solution of NaCl in initiating arousal of 13-lined ground squirrels (13).

Aside from the usual presumptions that normal ionic distribution is necessary to prevent conduction block due to depolarization and to prevent swelling, one wonders whether there are not other implications of failure of ion transport which might have significance with respect to long term survival of cells. It is with respect to its role in the lives of cells that our knowledge of alkali cation transport has seen the greatest expansion in the decade of the 1960's. An intracellular environment high in K and low in Na is now known to be necessary for the initiating stages of protein synthesis (31,32). The ability of cells such as renal tubule, intestinal mucosa and skeletal muscle to accumulate sugars or amino acids against concentration gradients depends upon a normal Na-K gradient (7,29,38). The uptake of Ca by cells is often in exchange for Na (cf. 2) and therefore is influenced by cellular Na concentration.

With respect to some of these roles the situation faced upon rewarming by a Na-loaded, K-depleted cell could be more serious than that in the cold. Thus, while protein synthesis might not be essential during the hypothermia of hibernation, the delay that would be incurred while Na was extruded and K reaccumulated before protein synthesis could proceed at a normal rate might be critical. Again, renal tubular cells may not depend upon active accumulation of sugars for their own metabolism while cold, but the inability to reabsorb glucose from the urine upon rewarming because of low cell K and high Na might be very detrimental.

Other cellular functions of Na-K transport are continually coming to light and one may confidentally expect that this list will be further expanded. The early finding of DeDuve and Wattiaux that lysosomal fragility was increased by anaerobiosis (cf. 8) raises the question in an already sensitized mind of whether lysosomal integrity might not also depend upon a normal K-Na concentration within cells and whether chilling of tissues might not also promote lysosomal rupture. So far as we can discover these possibilities seem not to have been explored.

From these considerations the question naturally arises whether the tissues of hibernators do in fact possess a superior capacity to survive periods of low temperature and, if so, whether this is related to their known ability to retain high K (and low Na) concentrations. A convenient model system for exploring this question is that introduced by the interesting experiments of Segal and co-workers at Children's Hospital in Philadelphia. These workers have shown that when slices of kidney cortex from humans and rats are stored at 4°C they retain a capacity for active accumulation of amino acid and sugars for about 4 to 5 days after which this capacity is progressively lost (30,37).

As a first step toward a study of the question of the effect of cold on cell survival, we have determined K contents of slices of kidneys from hibernators (hamsters and ground squirrels) and from non-hibernating rodents (rats and guinea pigs) stored at 5°C for up to a week or eight days under conditions essentially the same as those used by Segal and Company. We find that the superior capability of kidney slices from hibernators to retain K at this low temperature exists not merely for an hour or two as previously reported (4,45,46) but persists for many days of storage (Fig. 1).

[In these experiments the slices from ground squirrels showed a characteristic but inconvenient amount of variation. About half of them (8 out of 17) either gained K in the first day or lost an amount less than that of one standard deviation from the mean (about 45 µEq/g dry wt). The remaining half lost significantly greater amounts of K in the first 24 hours of storage. For further comparisons, therefore, it has proved useful to distinguish between a "high" group (those which did not lose K) and a "low" group (those which did lose a large amount of K). The comparison between these two selected groups can be seen in Figure 2 and it should be noted that even the "low" group lost less K than the slices from guinea pigs.]

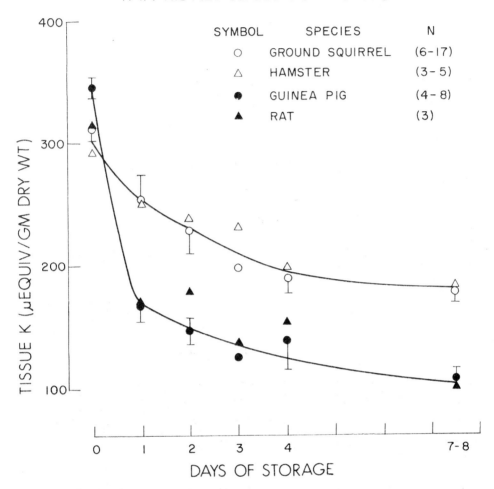

FIG. 1. K content of slices of renal cortex stored at 5°C in bicarbonate-buffered medium with 140 mM Na, 4.0 mM K, 1.5 mM Ca, 10 mM glucose. Standard errors of the mean are indicated.

Since the criterion of survival of cell function used by Segal et al., namely active uptake or aminoisobutyric acid and α-methylglucoside, depends

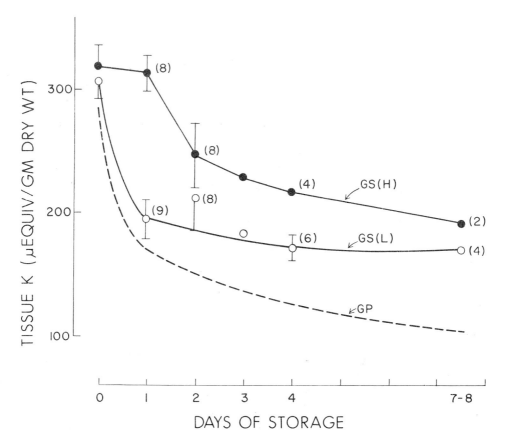

FIG. 2. K content of slices of renal cortex of ground squirrels. The data for ground squirrels shown in Fig. 1 have been sub-divided into two groups - (1) those cases in which there was little or no loss of K during the first 24 hours (GS-H) and (2) those cases in which loss of K was greater than 45 µeq/g dry wt. during first 24 hours (GS-L). The dotted line shows the loss from kidney slices from guinea pig (Fig. 1).

upon the existence of Na-K gradients (29,37), we felt a useful criterion for functional integrity in our experiments would be the extent to which K may be accumulated at some higher temperature. In Table 1 are data which show that after four days of storage the slices of kidney from five ground squirrels (all of which happened to belong to the "low" group) were able to retain a K content somewhat greater than those of rats and guinea pigs when

TABLE 1. K content of slices of renal cortex stored at 5°C and then incubated at a higher temperature. Slices designated "day 0" were only exposed to low temperature during the brief period required to make slices.

Condition of Reincubation	Group	N	K content (µeq/g dry wt.)		
			Day 0	Day 1	Day 4
1. 38°C 1 hour	Ground Squirrel (low group)	(5)	275	217	191
	Guinea Pig	(3)	270	182	155
	Rat	(3)	283	192	137
2. 20°C 3 hour	Ground Squirrel (high group)	(4)	—	321	—
	Ground Squirrel (low group)	(2)	—	323	—
	Guinea Pig	(3)	—	211	—

placed at 38°C for one hour. There was no difference between the three species before storage and only marginal difference after the first day of storage. It is well known, however, that for kidney slices of both hibernators and non-hibernators temperatures in the range of 15° to 25°C are optimal for K uptake and not 38°C (34,46,48). Hence, the uptake at the intermediate temperature of 20°C is also of interest. As shown in Table 2, the higher steady state K at this temperature in ground squirrels as compared with guinea pigs (both "high" and "low") was unmistakable even after only one day of storage.

In these preliminary studies we have also been interested in the effect of rewarming as a prelude to exploring the possibility of in vitro acclimation. We have found so far that, on the average, slices from four ground squirrels of the "high" group respond quite differently to rewarming from those of two "low" group ground squirrels and two guinea pigs (Fig. 3). In those slices from squirrels which have not lost K in the first day, a period of 3 hrs at 30°C causes no average increase in K content but does seem to prevent the sharp loss of K that characteristically occurs on the second day. In those slices either from ground squirrel or from guinea pig, which have lost K, rewarming does lead to uptake of K, but when the slices are replaced in the cold most of this reaccumulated K is lost again.

In the past the main consideration has been how the continued functioning of cells at low temperature enables the organism to survive hypothermia. Only secondarily, if at all, has the question been raised of whether a sojourn at low temperature poses a direct threat to the cells themselves in terms of their ability to resume activities at high temperature. This question is important, however, for a better understanding of hibernation and of the problems faced by tissues of non-hibernators exposed to hypothermia either *in situ* or in isolation. These faltering first steps to studying survival of cell function in stored hibernator tissue may yet be carrying us toward a fascinating new territory.

ANALYSIS OF MECHANISMS OF COLD ADAPTATION OF MEMBRANE FUNCTION

Our approach to the localization of the site of adaptation in hibernators, or the site of lesion in non-hibernators, is outlined in Table 2. The answer to the first question, that there is a retention of high K *in vivo* during hibernation and *in vitro* at low temperature was demonstrated by the studies referred to in the first section of this paper.

Given that there is some kind of adaptation, the question then becomes whether this fact implies a relatively greater rate of transport or a slower rate of leak at low temperature. Even at the 1965 Symposium it could be reported that in kidney slices of hamsters both active transport and passive permeability appeared to be involved (46,47). But the relative proportion of the two and the identification of changes in passive permeability to specific ions has been demonstrated more graphically and quantitatively

TABLE 2. Reductionist analysis of nature of cold adaptation of retention of K in hibernator tissues.

Question	Answer	References
1. Is K gradient maintained		
a) In vivo during hibernation?	Yes, generally, but loss occurs in some, gain in others.	9,10,27,44
b) In vitro at low temperature	Yes, for kidney and red blood cell, not for awake hamster brain.	4,19,20,27,45,46
2. Does (1) require adaptation of		
a) Active transport?	Yes, for kidney and rbc.	28,46
b) Passive permeability?	Yes, for kidney and rbc.	28,46
3. Does (2a) require adaptation of		
a) Metabolism?	No necessarily in kidney and rbc.	48,53
b) Mechanism of transport (e.g. Na,K-ATPase)?	Yes, definitely in kidney, probably in brain and rbc.	5,6,19,20,21, 40,50,51,53
4. Does adaptation of Na,K-ATPase (3b) involve		
a) K-dependent dephosphorylation?	Not strongly in rat vs. hamster.	12
b) Na-dependent phosphorylation?	Very possibly in rat vs. hamster.	
c) Lipid moiety?	No information.	
d) Protein moiety?	No information	

by Kimzey's unidirectional flux studies in red blood cells (26). He showed for example, that between 38° and 5°C the proportion of total unidirectional influx of K which is active (i.e. ouabain inhibited) remains constant in ground squirrel red blood cells and becomes essentially nil in

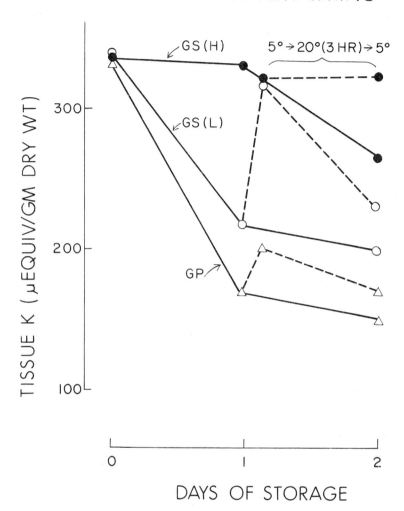

FIG. 3. Effects of brief interval of rewarming on K content of stored kidney slices. Results of rewarming are illustrated for three groups: kidney slices from two guinea pigs (triangles and GP) kidney slices from two ground squirrels of the "low K" group (open circles and GS-L) and kidney slices from four ground squirrels from the "high K" group (closed circles and GS-H). After 24 hours of storage at 5°C, some of the slices were rewarmed to 20°C for 3 hours and then some of these slices were returned to 5°C for a further 21 hours.

guinea pigs at 5°. At the same time the passive fluxes (Na influx, K efflux and ouabain-insensitive K influx and Na efflux) were all more steeply decreased by lowering temperature in ground squirrel red blood cells than in those of guinea pig. Thus, although red blood cells do suffer a net loss of K during hibernation, there can be no doubt that specific adaptations exist which keep this loss to a minimum and that these adaptations involve generally the same activities as in kidney cells.

The next level of discrimination, that is, whether the maintenance of active transport depends upon a greater metabolic rate or simply a more cold resistant pump, is more difficult to resolve. A study on kidney slices which compared the effect of lowering temperature on Na-sensitive or ouabain-sensitive respiration with changes in K uptake led to the conclusion that in two species of non-hibernating rodents the transport-linked respiration was not limited (46). There may, however, have been several pitfalls hidden in the assumptions of that study. Two of the more disturbing of these are that the "basal" metabolic rate of cells measured may depend on the method of inhibition of transport and that considerable intra-tissue transport of K may occur without being reflected in net uptake (49).

In human red blood cells, however, Wood and Beutler have found that those cells stored under conditions allowing retention of high ATP concentrations lost K and gained Na to the same extent as cells with low ATP concentrations (53). They concluded that the inhibition of metabolism was not the cause of failure of transport at low temperature, but rather the inhibition of the transport mechanism itself (53).

Furthermore, if the activity of Na,K-ATPase is acceptable as an index of transport capability, it is safe to state that whether or not low metabolism might be limiting in non-hibernators, ion transport is definitely quite temperature sensitive in non-hibernators and measureably less so in hibernators. Cold persistent activity of Na,K-ATPase has been demonstrated in the kidney cortex of hamsters and in brain of hibernating hedgehogs and hibernating hamsters (6,12,19,20,50,51). These results are in marked contrast to many reports of unusual cold sensitivity of Na,K-ATPase in tissues of non-hibernating mammals (5,21,40,51,53).

Beyond this level further analysis of the mechanisms of ion transport become rather uncertain because of the still comparatively primitive state of understanding of the mechanism of transport generally. It may, however, be possible to find handles for subdividing transport by considering details related to energy conversion on the one hand and translocation of ions on the other.

According to the prevailing conventional wisdom (not universally accepted) the hydrolysis of ATP in the transport mechanism involves the transfer of phosphate to a protein intermediate and then the hydrolysis of this phospho-protein (11,36). The initial phosphorylation of the protein is stimulated by Na and Mg and the dephosphorylation is stimulated by K and inhibited by ouabain. Between these two partial reactions there is thought to be a third intermediate step, inferred from various kinetic studies and involving perhaps a conformational change of the protein (11,35).

The most advanced flux studies have correspondingly indicated a separability between Na extrusion and K uptake in the cycle of coupled Na,K transport. This conclusion is based upon the existence of ouabain-sensitive Na:Na exchange in red blood cells and squid axon and on ouabain-sensitive K efflux in red blood cells (3,14,15,18). Workers in both the ion flux and the ATPase areas have pointed out the mutually corroborative aspects of their separate findings (cf. 17).

Thus, if one accepts the "sequential" view of ion translocation, one can envision an "early" phase beginning with Na within the cell attaching to a site and by so doing stimulating a transphosphorylation of phosphate from ATP to a transport protein. This step would then be followed by translocation of the Na ion to the outside and the conversion of the protein to a K specific form leading to the "later steps" of K translocation into the cell and dephosphorylation of the protein.

It would seem then, that one ought to be able to ask the question of whether the Na dependent "early" stages or the "later" K dependent stages of the cycle are more strongly affected by cold. Conceivably the effects of cold on, for example, Na:Na exchange, might shed some light on this, but such experiments in red blood cells at low temperature would be extremely difficult.

A more promising approach in the short run is offered by the ATPase aspect of transport. In our laboratory, Fang has developed two lines to test this point. The first was to measure the effect of temperature on K dependent phosphatase activity. This activity is found abundantly in preparations also rich in Na,K-ATPase (1,53). While exhibiting a specificity for a variety of phosphates (paranitrophenol phosphate, acetylphosphate, etc.), it resembles the Na,K-ATPase in being stimulated by K and inhibited by ouabain. Some, though by no means all, workers feel that it may reflect the terminal step of activity of the Na,K-ATPase. Fang found that in preparations from kidneys of rats, reducing temperature had less effect on the K-phosphatase than on the net Na,K-ATPase (12). More importantly, there was less difference between rat and hamster preparations in the K-phosphatase activities than in the Na,K-ATPase activities at 5°C.

The second approach has been to use the phosphorylation of protein from a kidney preparation rich in Na,K-ATPase activity. The procedure requires that ATP, labelled with P^{32} in the gamma position, be incubated at low temperature with the enzyme. At the end of exposure the protein is precipitated, washed and the P^{32} counted. Rate of loss of label from protein under various conditions can serve as a measure of dephosphorylation.

In the particular experiment to be described, adapted from a procedure of Post's (35), both the formation and the breakdown of the phosphorylated intermediate was observed at 0°C. To do this a reaction mixture was prepared containing the enzyme, ATP^{32}, Na and K, but no Mg. Except for Mg, these are all the components required for a complete reaction, and the ATP^{32}, presumably becomes bound to the enzyme. The reaction is initiated by the addition of Mg and a 100-fold excess of unlabelled ATP (Fig. 4). The figure thus shows the time course of labelling and loss of label from the enzyme during a single cycle of phosphorylation and dephosphorylation.

The rate of decline of labelled protein is strongly dependent upon K, the rate being very slow in the absence of K. It may be observed that there is little if any difference in the rate of this fall between preparations from hamsters and from ground squirrels.

The results on the K-dependent phosphatase and on the dephosphorylation of phosphoenzyme thus tend toward the same conclusion: the temperature sensitivity does not lie in the K dependent part of the cycle. By elimination, therefore, one is forced to consider the earlier stage of the reaction sequence. That the difference may indeed lie here is suggested by the result in Figure 4 which shows that when the pulse reaction was released by adding Mg and ATP, the phosphoenzyme in the hamster preparation peaked at a higher level than in the rat preparation. Since the last step was going on at essentially the same rate in the two species, this result could mean that the initial phosphorylation steps were slower in the rat. The difficulty of interpreting this result, however, is that we do not know whether this difference between the species is related, as we should

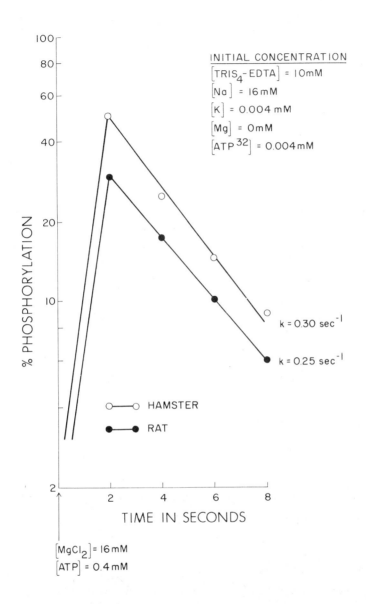

FIG. 4. Labelling of protein in preparations rich in Na,K-ATPase made from kidney cortex of hamsters and rats. The zero-baseline was determined by labelling of protein in a Na-free medium with 16 mM-K. The value for 100 per cent labelling was determined in a medium with 16 mM Na, no K and no non-radioactive ATP. For other details see text.

like to believe, to the very low temperature or whether it is merely fortuitous. The impossibility of studying these rapid reactions at higher temperatures at which rates of overall reaction are similar between the two species make this a difficult question to resolve.

For the moment this is as far as we can carry the molecular analysis of cold adaptation, but another type of meaningful question would be whether it is the protein or the essential lipid component of this enzyme system which confers less cold sensitivity on the hibernators. The hopefulness of resolving this question is suggested by several studies in which it has been possible to deactivate Na,K-ATPase by treatment with phospholipase and to reactivate it by restoration of natural, mixed phospholipid components or purified phospholipids such as phosphotidylserine (22,24,cf,42).

ACCLIMATION OF MEMBRANE FUNCTION

By "acclimation" we mean in the present context an improved capability for functioning at low temperature which occurs either as a result of the organism's preparing to enter hibernation or of its actually entering hibernation. By and large those who have worked on describing adaptation of cellular activity have been content to use the tissues of hibernators that were not actually in hibernation. Usually the tissues of the awake ground squirrel, hamster, or whatever are indeed sufficiently cold resistant to serve as useful objects of study. Nevertheless, it has frequently been observed that actual hibernation does confer an improved performance at low temperature (as in the neuromuscular junction and in resistance to cold swelling of skeletal muscle [39,47]). With regard specifically to ion transport, Willis' initial observations on net uptake of K in hamster kidney, showed that a higher steady state K concentration was achieved at 5°C in hibernating hamster kidneys than in those of awake hamster (46), and a subsequent study revealed that rate of K uptake was twice as great in the hibernating hamster (Haber and Willis, unpublished observation). Curiously, this derivative phenomenon, which carries with it wider implications of cellular and organismic control than can be perceived in merely invariant intrinsic adaptation, has hardly been scratched by any experimental investigation.

That acclimation may be of paramount importance to the organism is sufficiently demonstrated by the studies of Goldman (19,20) who found that the brain slices of awake hamsters revealed no better capacity for ion transport at low temperatures than those of rat and that the Na,K-ATPase was also not less inhibited by cold. Only in hibernating hamster was the Na,K-ATPase of brain more active specifically at low temperature. Similarly, the cold resistance of Na,K-ATPase of brain in hedgehog reported by Bowler and Duncan was also present only in preparations from hibernating individuals (5).

Fang has investigated the difference in Na,K-ATPase of kidney between hamsters that had and had not hibernated (12). Here, too he found that like K transport in that tissue the specific activity of the enzyme was twice as great in the hibernating hamsters as in the awake hamsters. Unlike the enzyme from brain, however, the Na,K-ATPase activity of kidney was greater at every assay temperature, not just at low temperature. Thus, in the kidney, the acclimation is "translational" whereas in the brain it is "rotational."

The next consideration is whether these changes occur prior to hibernation or after hibernation has begun and the tissues themselves have therefore been exposed to cold. In Syrian hamsters, which undergo a long wait in the cold before beginning to hibernate, it seems clear that the changes occur during the "waiting" period. Thus, Goldman found that for the

Na,K-ATPase of brain the "hibernating pattern" of cold resistance was present in hamsters which had been cold exposed for several weeks, but which had not yet begun to hibernate (19,20). For the Na,K-ATPase of kidney, Fang observed that the specific activity rose linearly with time after the animals were placed in the cold and reached a plateau at the same time as the onset of hibernation (12).

Thus, in hamsters these changes would not seem to require direct exposure of the tissues to cold and may depend upon organismic control mechanisms. It is still subject to question whether these observed changes are indeed related to hibernation or, instead, to ordinary cold acclimation such as occurs in other homeotherms. An indication that militates against the latter possibility, however, is the observation of Goldman that in three cold-exposed hamsters, which had failed to hibernate several weeks after their companions had all begun to do so, the enzyme showed the "awake" pattern (20).

While such changes may occur during homeothermic cold exposure in hamsters there is still the interesting possibility that direct cold exposure of the tissues may also stimulate an improvement of cold resistance. We cannot help wondering whether such an acclimation in this more general poikilothermic sense may not be occurring during the short bouts of hibernation characteristic of early season hibernation in the ground squirrel. If so, perhaps, as implied earlier, the *in vitro* storage method may provide a useful model for analysing such changes. In any case, as more is learned about the fundamental mechanisms of cold adaptation the processes of controlling temporal changes in the prevalence of those mechanisms will warrant greater attention.

ACKNOWLEDGMENT

Unpublished research described in this paper was supported in part by a grant from the National Institutes of Health, GM 11494.

REFERENCES

1. BADER, H., and A. K. SEN. (K^+)-dependent acyl phosphate as part of the ($Na^+ + K^+$)-dependent ATPase of cell membranes. Biochem. Biophys. Acta 118: 116-123, 1966.

2. BAKER, P. F., M. P. BLAUSTEIN, A. L. HODGKIN, and R. A. STEINHARDT. The influence of calcium on sodium efflux in squid axons. J. Physiol. 200: 431-458, 1969.

3. BAKER, P. F., M. P. BLAUSTEIN, R. D. KEYNES, J. MANIL, T. I. SHAW, and R. A. STEINHARDT. The ouabain-sensitive fluxes of sodium and potassium in squid giant axons. J. Physiol. 200: 459-496, 1969.

4. BIDET, R., M. M. LECLERC, and C. KAYSER. Effet de la température sur le degré d'hydration et la teneur en Na et K de coupes de cortex rénal de rat et de hamster. (Cricetus cricetus). Arch. Sci. Physiol. 19: 247-257, 1965.

5. BOWLER, K., and C. J. DUNCAN. The effect of temperature on the Mg^{2+}-dependent and Na^+, K^+-ATPases of a rat brain microsomal preparation. Comp. Biochem. Physiol. 24: 1043-1054, 1968.

6. BOWLER, K., and C. J. DUNCAN. The temperature characteristics of brain microsomal ATPases of the hedgehog: changes associated with hibernation. Physiol. Zool. 42: 211-219, 1969.

7. CRANE, R. K. Na^+-dependent transport in the intestine and other animal tissues. Federation Proc. 24: 1000-1006, 1965.

8. DEDUVE, C. Lysosomes and chemotherapy. In: Biological Approaches to Cancer Chemotherapy, edited by R. J. C. Harris. New York: Academic Press, 1961, p. 101-112.

9. ELIASSEN, E., and H. LEIVESTAD. Sodium and potassium content in the muscles of hibernating animals. Nature, 192: 459-460, 1961.

10. ELIASSEN, E., and H. LEIVESTAD. The effect of hibernating hypothermia on the potassium and sodium content of the muscles in the hedgehog, Erinaceus europaeus L. Arb. Univ. Bergen 1962 Mat. Naturv. Serie No. 5: 1-15, 1961.

11. FAHN, S., G. J. KOVAL, and R. W. ALBERS. Na-K-activated ATPase of Electrophorus electric organ. I. An associated sodium-activated transphosphorylation. J. Biol. Chem. 241: 1882-1889, 1966.

12. FANG, L. S. T., and J. S. WILLIS. Further analysis of cold adaptation of Na-K ATPase of hibernating mammals. Physiologist 13: 193, 1970.

13. FISHER, K. C., and N. MROSOVSKY. Effectiveness of KCl and NaCl injections in arousing 13-lined ground squirrels from hibernation. Can. J. Zool. 48: 595-596, 1970.

14. GARRAHAN, P. J., and I. M. GLYNN. The behaviour of the sodium pump in red cells in the absence of external potassium. J. Physiol. 192: 159-174, 1967.

15. GARRAHAN, P. J., and I. M. GLYNN. Factors affecting the relative magnitudes of the sodium:potassium and sodium:sodium exchanges catalysed by the sodium pump. J. Physiol. 192: 189-216, 1967.

16. GIRADIER, L., J. SEYDOUX, and T. CLAUSEN. Membrane potential of brown adipose tissue.. A suggested mechanism for the regulation of thermogenesis. J. Gen. Physiol. 52: 925-940, 1968.

17. GLYNN, I. M. Membrane adenosine triphosphatase and cation transport. Brit. Med. Bull. 24: 165-169, 1968.

18. GLYNN, I. M., V. L. LEW, and U. LÜTHI. Reversal of the potassium entry mechanism in red cells, with and without reversal of the entire pump cycle. J. Physiol. 207: 371-391, 1970.

19. GOLDMAN, S. S., and J. S. WILLIS. Acclimation of active cation transport in the central nervous system during hibernation.(ABSTRACT). Federation Proc. 29: 718, 1970.

20. GOLDMAN, S. S. Cold Resistance of Cation Regulation in the Brain During Hibernation. (Ph.D. Thesis). University of Illinois, Urbana, 1970.

21. GRUENER, N., and Y. AVI-DOR. Temperature dependence of activation and inhibition of rat brain adenosine triphosphatase activated by sodium and potassium ions. Biochem. J. 100: 762-767, 1966.

22. HEGYVARY, C., and R. L. POST. Reversible inactivation of (Na-K)-ATPase by removing and restoring phospholipids. In: The Molecular Basis of Membrane Function, edited by D. C. Tosteson. Englewood Cliffs, N. J.: Prentice-Hall, 1969, p. 519-528.

23. HORWITZ, B. A., P. A. HERD, J. M. HOROWITZ, and R. E. SMITH. Brown fat: Effect of norepinephrine on intracellular potential and Na/K-ATPase activity. Physiologist 12: 257, 1969.

24. HORWITZ, B. A., J. M. HOROWITZ, and R. E. SMITH. Norepinephrine-induced depolarization of brown fat cells. Proc. Nat. Acad. Sci. U.S. 64: 113-120, 1969.

25. ISRAEL, Y. Phospholipid activation of (Na + K)-ATPase. In: The Molecular Basis of Membrane Function, edited by D. C. Tosteson. Englewood Cliffs, N. J.: Prentice-Hall, 1969, p. 529-537.

26. KIMZEY, S. L. Cold resistance of cation regulation in erythrocytes of hibernators. (Ph.D. Thesis). University of Illinois, Urbana, Illinois, 1969.

27. KIMZEY, S. L., CHAPMAN, I., and WILLIS, J. S. Maintenance of K ion concentration by erythrocytes during hibernation. (Abstract). Amer. Zool. 6: 528, 1966.

28. KIMZEY, S. L., and WILLIS, J. S. Temperature resistance of the cation transport system in erythrocytes of hibernators. (Abstract). Fed. Proc. 27: 747, 1968.

29. KLEINZELLER, A. Active sugar transport in renal cortex cells: electrolyte requirement. Biochim. Biophys. Acta 211: 277-292, 1970.

30. LOWENSTEIN, L. M., K. HUMMELER, I. SMITH, and S. SEGAL. The effect of storage at 4° on amino acid transport by rat kidney cortex slices. Biochim. Biophys. Acta 150: 416-423, 1968.

31. LUBIN, M. Cell potassium and the regulation of protein synthesis. In: The Cellular Functions of Membrane Transport, edited by J. F. Hoffman. Englewood Cliffs, N. J.: Prentice-Hall, 1964, p. 193-211.

32. LUBIN, M. Intracellular potassium and control of protein synthesis. Federation Proc. 23: 994-1001, 1964.

33. LYMAN, C. P., and R. C. O'BRIEN. Hyperresponsiveness in hibernation. In: S.E.B. Symposium XXIII, Dormancy and Survival. Cambridge, Eng.: The University Press, 1969, p. 489-509.

34. MUDGE, G. H. Studies on potassium accumulation by rabbit kidney slices: effect of metabolic activity. Amer. J. Physiol. 165: 113-127, 1951.

35. POST, R. L., S. KUME, T. TOBIN, B. ORCUTT, and A. K. SEN. Flexibility of an active center in sodium-plus-potassium adenosine triphosphatase. J. Gen. Physiol. 54: 306-326, 1969.

36. PRECHT, H. Concepts of the temperature adaptation of unchanging reaction systems of cold blooded animals. In: Physiological Adaption, edited by C. L. Prosser. Washington, D. C.: Amer. Physiol. Soc., 1958, p. 50-77.

37. SEGAL, S., I. SMITH, M. GENEL, and P. HOLTZAPPLE. Tubular transport by human kidney stored at 4°C. Nature 222: 387-389, 1969.

38. SCHULTZ, S. G., P. F. CURRAN, R. Z. CHEZ, and R. P. FUISZ. Alanine and sodium fluxes across mucosal border of rabbit ileum. J. Gen. Physiol. 50: 1241-1260, 1967.

39. SOUTH, F. E. Phrenic nerve-diaphragm preparations in relation to temperature and hibernation. Amer. J. Physiol. 200: 565-571, 1961.

40. SWANSON, P. D. Temperature dependence of sodium ion activation of the cerebral microsomal adenosine triphosphatase. J. Neurochem. 13: 229-236, 1966.

41. TWENTE, J. W., and J. A. TWENTE. Progressive irritability of hibernating Citellus lateralis. Comp. Biochem. Physiol. 25: 467-474, 1968.

42. WHEELER, K. P., and R. WHITTAM. The involvement of phosphatidylserine in adenosine triphosphatase activity of the sodium pump. J. Physiol. 207: 303-328.

43. WHITTEN, B. K., M. A. POSIVIATA, and W. D. BOWERS. Seasonal changes in hepatic ribosome aggregation and protein synthesis in the hibernator. Physiologist 13: 339, 1970.

44. WILLIS, J. S. Potassium and sodium content of tissues of hamsters and ground squirrels during hibernation. Science 146: 546-547, 1964.

45. WILLIS, J. S. Uptake of potassium at low temperatures in kidney cortex slices of hibernating mammals. Nature 204: 691-693, 1964.

46. WILLIS, J. S. Characteristics of ion transport in kidney cortex of mammalian hibernators. J. Gen. Physiol. 49: 1221-1239, 1966.

47. WILLIS, J. S. Cold adaptation of activities of tissues of hibernating mammals. In: Mammalian Hibernation III, edited by K. C. Fisher, A. R. Dawe, C. P. Lyman, E. Schönbaum and F. E. South. Edinburgh: Oliver and Boyd, 1967, p. 356-381.

48. WILLIS, J. S. Cold resistance of kidney cells of mammalian hibernators: cation transport vs. respiration. Amer. J. Physiol. 214: 923-928, 1968.

49. WILLIS, J. S. The interaction of K^+, ouabain and Na^+ on the cation transport and respiration of renal cortical cells of hamsters and ground squirrels. Biochim. Biophys. Acta. 163: 516-530, 1968.

50. WILLIS, J. S., and N. MA. Cold resistance of Na, K stimulated ATPase in kidney cortex of hibernating rodents. Physiologists 10: 348, 1967.

51. WILLIS, J. S., and N. MA LI. Cold resistance of Na-K-ATPase of renal cortex of the hamster, a hibernating mammal. Amer. J. Physiol. 217: 321-326, 1969.

52. WILLIS, J. S., S. S. GOLDMAN, and R. F. FOSTER. Tissue K concentration in relation to the role of the kidney in hibernation and the cause of periodic arousal. Comp. Biochem. Physiol. In press.

53. WOOD, L., and E. BEUTLER. Temperature dependence of sodium-potassium activated erythrocyte adenosine triphosphatase. J. Lab. Clin. Med. 70: 287-294, 1967.

54. YOSHIDA, H., K. NAGAI, T. OHASI, and Y. NAKAGAWA. K^+-dependent phosphatase activity observed in the presence of both adenosine triphosphate and Na^+. Biochim. Biophys. Acta 171: 178-185, 1969.

THE Na^+,K^+-ATPase TRANSPORT ENZYME SYSTEM

ARNOLD SCHWARTZ

Division of Myocardial Biology
Baylor College of Medicine
Houston, Texas 77025

INTRODUCTION

The cell membrane probably was a very early evolutionary development. The primitive organism required a means by which it could separate itself from the primeval sea environment, in order to preserve and perpetuate its existence. The further differentiation of the cell membrane into chemically active subunits was required in order for the monovalent cations, sodium and potassium, to be separated with potassium predominately occupying the intracellular environment and sodium remaining in the extracellular milieu. It is well known that potassium is required for a multitude of reductive synthetic processes, as well as for the maintenance or establishment of the normal resting membrane potential.

In simplified terms, it is probable that all excitation processes involve an initial depolarization event. The resting membrane potential is thought to be due to the low permeability of the membrane to potassium, which effects a diffusion gradient or potential. It is thought that a sudden alteration of permeability to sodium results in an influx of very small amounts of sodium, which constitute the rising phase of the action potential or sodium current. By some unknown mechanism, the permeability to sodium suddenly diminishes, while the permeability to potassium increases so that small amounts of potassium diffuse from the outside of the cell across the membrane to the outside, constituting the phase of repolarization. During recovery the two cations, sodium and potassium, are moved against concentration and electrical gradients to the original loci, a phase requiring energy presumably in the form of adenosine triphosphate (ATP). A generalized cell with cation localization is depicted in Figure 1 and an action potential from an excitable cell is shown in Figure 2, with the various phases listed as 0-4.

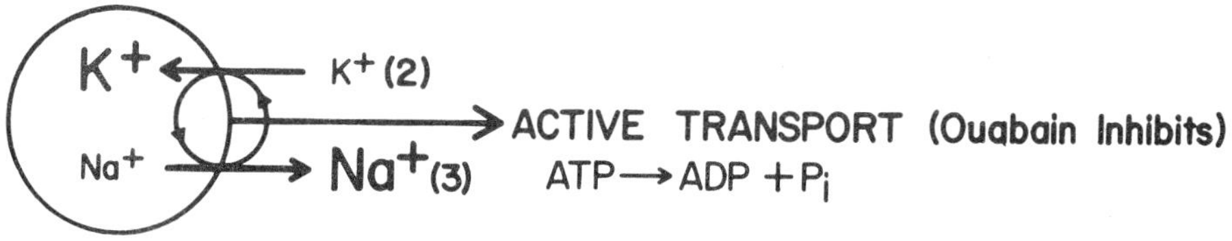

FIG. 1. Proposed stoichiometry of active monovalent cation transport.

In 1953 Schatzmann (16) found that an active cardiac glycoside (a digitalis drug causing an increased force of contraction of heart and the drug of choice in treating congestive heart failure) specifically inhibited the active movements of sodium and potassium across the red blood cell membrane (Fig. 1). Active transport can be defined as that process which results in the translocation of substances across membranous barriers at the expense of energy derived from cellular metabolism. It was shown a few years later by Skou (20) that an adenosine triphosphatase (ATPase) could be identified in a microsomal fraction derived from crab nerve homogenates, that specifically required magnesium, sodium and potassium for maximal hydrolytic activity, and was later shown to be specifically inhibited by active cardiac glycosides. Subsequently, numerous investigators discovered the presence of this Na,K-ATPase in almost every mammalian and non-mammalian cell studied that carried on active transport of sodium and potassium. While the exact cellular localization of this particulate enzyme is still not completely known, the Na,K-ATPase is thought to be primarily associated with the cell or plasma membrane. It is quite clear, at least in the red blood cell ghost, that this enzyme system is, in fact, an intregal part of the cell membrane. Using the latter experimental model,

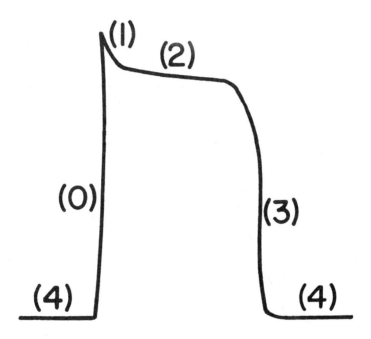

FIG. 2. An idealized action potential. The phases are: (0) rising phase of action potential, sodium current, (1) overshoot, (2) plateau, (3) phase of repolarization, (4) resting membrane potential.

Post (17) and other investigators ascertained an interesting stoichiometry between Na^+, K^+ and ATP. For each three sodium ions that were transported in one direction, two potassium ions were transported in the opposite direction at the expense of one high energy phosphate bond. This stoichiometry reinforces the complexity of the Na,K-ATPase and suggests some type of cooperative ligand-site effect, although other kinetic explanations are also possible and have been advanced.

PREPARATION OF THE Na,K-ATPase

In general, the Na,K-ATPase can be isolated from tissues by means of homogenization in sucrose or saline containing media at neutral pH. Very active preparations can be obtained by the use of various salt treatments, density gradient centrifugation and detergent additions. In general, however, a membranous sediment is obtained which can be resuspended in an appropriate buffered medium. The membrane fraction can also be solubilized by means of Lubrol (13,18) or other types of detergent treatment, yielding a final preparation that is essentially clear after ultracentrifugation. It is entirely possible, however, that even the clear preparations consist of extremely minute membranous sites. The preparations usually are stored in small tubes kept at a very low temperature, and can remain active depending upon the source for several months or longer.

ASSAY OF ENZYME ACTIVITY

By definition, the Na,K-ATPase is that enzymatic activity measured in the presence of Mg^{++}, Na^+ and K^+ and completely inhibited by ouabain, a water-soluble cardiac glycoside, or other active cardiac glycoside principle. One can prepare an enzyme that is essentially free of Mg^{++}-ATPase (ouabain-insensitive) activity.

Enzyme activity can be measured in several ways, one by colorimetric assay depending simply on the release of inorganic phosphate from added

ATP, and another, employing the linked enzyme reaction sequence as follows:

$$ATP \xrightarrow{Mg^{++}, Na^+, K^+} ADP + \text{inorganic phosphate}$$

$$\text{Phosphoenolpyruvate} + ADP \xrightarrow{\text{(pyruvate kinase)}} \text{pyruvate} + ATP$$

$$\text{Pyruvate} + NADH \xrightarrow{\text{(lactate dehydrogenase)}} NAD^+ + \text{lactate}$$

It is clear that, if measured amounts of NADH are added to a cuvette with all other ingredients added in excess, with the exception of the enzyme, each mole of NADH oxidized to NAD is stoichiometrically equivalent to one mole of ATP hydrolyzed to ADP. The latter procedure for measurement has a number of distinctive advantages which include the removal of inhibitory ADP, maintenance of a steady-state level of ATP and the ability to follow the chemical reaction spectrophotometrically from minute to minute. A sample spectrophotometric trace is depicted in Figure 3.

CHARACTERISTICS

Using the above system of recording and other, better known assays as well, Na,K-ATPases isolated from heart, brain, kidney, and other mammalian tissues show the following general approximate characteristics, depending upon the source:

K_m for ATP	=	0.3 mM
K_m for Na^+	=	15.0 mM
K_m for K^+	=	1.0 mM
K_m for Mg^{++}	=	0.3 mM
Optimal pH	=	7.2 - 7.4

Probably the most significant feature of the enzyme activity is its ability to interact directly and specifically with active cardiac

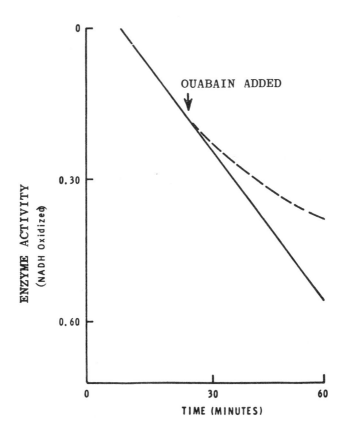

FIG. 3. A spectrophotometric recording of Na,K-ATPase activity. (μmoles NADH oxidized at 340 mμ.)

glycosides. Ouabain, in extremely low concentrations, for example, causes a complete inhibition of the Na,K-stimulated portion of the ATPase. This inhibition can be most conveniently observed using linked enzyme assay system (see Fig. 3). It is of interest that the inhibition induced by ouabain is curvilinear with time and is temperature-dependent. Consequently, any calculation of a half-maximal inhibitory value must take these factors into consideration. In general, using standard assay conditions, approximately 5×10^{-7} M ouabain induces half-maximal inhibition of Na,K-ATPase enzyme system isolated from all organs. This, however, does vary with species; in particular, the half-maximal inhibitory concentration of ouabain for cardiac Na,K-ATPase isolated from rat, a species which is remarkably insensitive to cardiac glycosides, is approximately 10^{-4} M. This drug-enzyme interaction may explain the well known insensitivity of the rat to cardiac glycosides. On the other hand, Na,K-ATPase isolated from pig hearts is remarkably sensitive to cardiac glycosides with half-maximal inhibitory concentrations in the range of 10^{-11} M and threshold concentrations around 10^{-12} M. These concentrations are well within the therapeutic range, that is, the amount of drug required to produce a positive inotropic effect in the failing human heart. Accordingly, it is attractive to suggest that the Na,K-ATPase may be an important pharmacological receptor for this class of drugs. A direct relationship has been found between the onset of the positive inotropic effect and the enzyme inhibition (5).

Another important characteristic of the cardiac glycoside-induced inhibition is the fact that potassium specifically diminishes the action of the inhibitor. That is, in the presence of increasing amounts of potassium in the medium more inhibitor is required to produce the same degree of inhibition. A kinetic examination of this event, however, revealed a non-competitive interaction (cf. Fig. 4), suggesting that potassium and ouabain do not act at the same site on the membrane. Further examination revealed that both sodium and potassium are required in a specific ratio for the inhibitor to effect its characteristic action. The

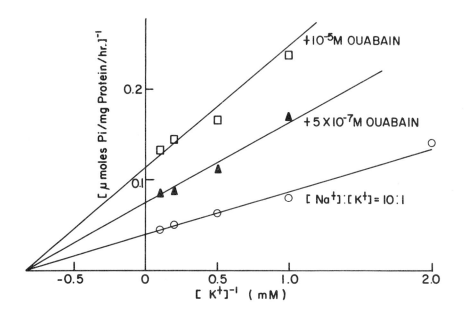

FIG. 4. Influence of K^+ on ouabain-induced inhibition of Na^+, K^+-ATPase. Na^+, K^+-ATPase activity was determined at various K^+ concentrations with a constant Na^+/K^+ ratio of 10 in the presence and absence of 5×10^{-7} M and 10^{-6} M ouabain; $1/v$ was plotted against $1/K^+$ (=$1/Na^+ \times 10$) (11).

relation between the Ki and the Na^+/K^+ ratio may be expressed as follows:

$$Ki = 3.5 \times 10^{-6} \times (Na^+/K^+)^{-0.82}$$

This relationship is seen graphically in Figure 5 (11).

These characteristics of ouabain-enzyme interaction suggested that the drug might be inhibiting the enzyme activity in a manner not dissimilar from an allosteric inhibitor (8). To examine this further, we undertook a series of experiments employing H^3-digoxin and H^3-ouabain. The results clearly indicated that the drug bound quite specifically to the Na,K-ATPase

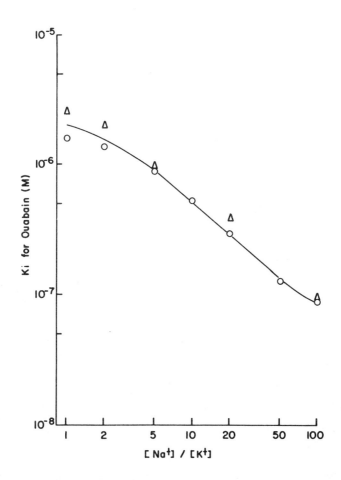

FIG. 5. Relationship between half-maximal inhibitory concentration of ouabain on Na,K-ATPase and the ratio of Na^+/K^+. Ki obtained was plotted against Na^+/K^+ ratios on a full logarithmic scale. The symbols, O and Δ, represent different enzyme preparations (11).

enzyme system maximally in the presence of ATP, Mg^{++} and Na^+. Addition of potassium strikingly diminished the binding. These initial findings were consistent with the following two-stage mechanism:

$$\text{ATP} + \text{Membrane} \underset{K^+}{\overset{Na^+}{\rightleftharpoons}} \text{Membrane-P} + \text{ADP}$$

$$\text{Membrane-P} \longrightarrow \text{Membrane} + \text{Pi}$$

It appeared that the cardiac glycosides were binding to the enzyme perhaps at the intermediary phosphoenzyme state. This reaction scheme was formulated from experiments involving the use of ATP^{32} and studies of ADP-ATP exchange reactions (1). Further studies revealed a number of interesting facets. Maximum finding of the drug to the Na,K-ATPase occurred in the presence of Mg^{++} and inorganic phosphate and in the absence of any energy donor (15). Other experiments indicated that inorganic phosphate could incorporate into the enzyme in the presence of Mg^{++}, but only after the enzyme had been treated with ouabain. In other words, a phosphoenzyme could be formed in the presence of ouabain (9,19). Experiments by a number of investigators revealed that the phosphoenzyme "intermediate" formed in the presence of Mg^{++} + inorganic phosphate + ouabain or in the presence of Mg^{++} + Na^+ + ATP were chromatographically identical (6). However, very recent information from our laboratory suggests that the phosphorylated enzyme formed in the presence of Mg^{++} + inorganic phosphate is probably different from that formed in the presence of Mg^{++} + Na^+ + ATP, a fact which further complicates the picture (2). More recently, we have also found that H^3-ouabain or H^3-digoxin specifically binds to the enzyme in the presence of Mg^{++} + acetate or Mg^{++} + sulfate. These data strongly suggest that the drug can interact with the enzyme in the complete absence of the phosphate precursor, that is, the drug binds to either a phospho- or dephosphoenzyme.

It should be apparent from the above that the two-stage hypothesis of enzyme action requires revision and, in fact, most investigators in this field now believe that the Na,K-ATPase probably consists of a number of steps as follows:

$$ATP + Membrane \xrightarrow{Na^+} Membrane_1\text{-}P + ADP$$

$$Membrane_1\text{-}P \longrightarrow Membrane_2\text{-}P \longrightarrow Membrane_n\text{-}P$$

$$Membrane_n\text{-}P \longrightarrow Membrane + Pi$$

The phosphorylated membrane may represent different energy states, although this has not definitively been shown. The role of the socalled phosphorylated intermediate in the Na,K-ATPase reaction is probably at the same level of understanding that it is in the field of oxidative phosphorylation.

In this complex scheme, digitalis would interact with one of the intermediary stages, either phospho or dephospho, causing a stabilization of this particular state and probably preventing potassium from reacting as follows (3):

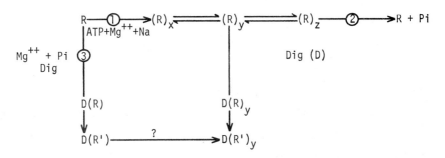

R = digitalis receptor and $(R)_x$, $(R)_y$ and $(R)_z$ are various conformations of the receptor, Dig or D = digitalis (ouabain). Under appropriate conditions (Mg^{++}, Na^+, K^+ and ATP ①-②), the enzyme cycle is complete and passes through an indeterminant number of intermediates. $(R)_y$ is the primary conformation within the turnover cycle that can interact with ouabain. Potassium induces a completion of the enzyme cycle (Na^+, K^+-ATPase)②, thus temporarily reducing the steady-state "concentration" of the conformation $(R)_y$, which interacts with the drug to form $D(R)_y$; hence, the time-dependency of ouabain binding and enzyme inhibition in the presence of potassium. $D(R)_y$ would be relatively unstable, with the drug loosely bound. $D(R)_y$ converts to a more stable form to which the drug is more tightly bound $D(R')_y$.

The insensitivity of the rat to cardiac glycosides can be explained by the insensitivity of the Na,K-ATPase to these drugs. Furthermore, we have found that the cardiac glycoside-enzyme complex from this species is easily dissociated even at 0° C by just one washing, whereas similar complexes obtained from other species are remarkably stable under these conditions (4).

H^3-ouabain or H^3-digoxin, when reacted with the enzyme from glycoside-sensitive species, yield a tightly bound complex. This complex cannot be dissociated at low temperatures, but is easily dissociated at 37° or 45° C with restoration of enzyme activity. Curiously enough, monovalent cations, particularly potassium, stabilize or inhibit the dissociation of the complex at all temperatures.

The above kinetic and binding data strongly suggests that the cardiac glycosides are allosteric-type inhibitors and further that the Na,K-ATPase may be a conformational or structurally-flexible enzyme. To test this hypothesis, we carried out two types of experiments. Firstly, we examined the fluorescent properties of 1-anilino-8-naphthalene-sulfonic acid (ANS) and 2-p-toluidinyl-naphthalene-6- sulfonic acid (TNS) bound to the Na,K-ATPases in the presence and absence of various ligands and ouabain. Secondly, we examined circular dichroism spectra (10) of Na,K-ATPases in the presence and absence of various ligands and ouabain. As can be seen in Figures 6 and 7, ouabain causes specific changes which may be interpreted as conformational in nature.

Some studies have been carried out attempting to elucidate the nature of one of the intermediary stages or steps in the reaction sequence. It has been suggested that a γ-glutamyl phosphate peptide may represent an important phosphorylated intermediate (7).

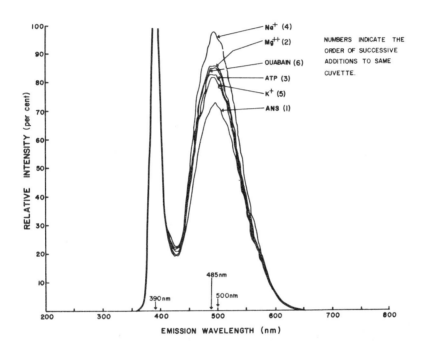

FIG. 6. Fluorescence spectra of ANS bound to synaptic membranes (Specific activities = 800-1000 nmoles Pi/mg/min). Numbers in parentheses denote order of successive additions to the same cuvette. Final concentrations were 100 μg/ml protein, 30 μM ANS, 4 mM $MgCl_2$, 0.2 mM Tris-ATP, 40 mM NaCl, 10 mM KCl, and 0.1 mM ouabain. The data are representative of numerous experiments employing different preparations, and are statistically significant. Temperature, 37°C. The ouabain effect is shown in this figure as a partial reversal; in numerous other experiments, ouabain induced a complete reversal of the K^+ effect (14).

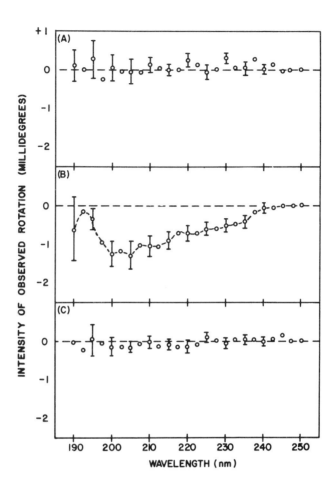

FIG. 7. Difference (circular dichroism) spectra of Lubrol-WX-treated Na,K-ATPase preparations. The means ± standard error are the differences between rotation values in the presence and absence of ouabain. Panel A represents results obtained in the presence of 1 mM Tris-Cl, (pH 7.4) ± 10 μM ouabain (n = 6). Panel B was obtained in the presence of 1 mM Tris-Cl (pH 7.4) 1 mM $MgCl_2$, and 0.5 mM $Tris-PO_4$ ± 10 μM ouabain (n = 10), and Panel C presents results obtained under the same conditions as in Panel B, except that a heat-inactivated (boiling water bath for 10 min) preparation was used (n = 5). n = number of samples scanned. True spectra were estimated from recordings (cf. Fig. 2) by taking the mean of the noise amplitude at every 2.5 nm (10).

CONCLUSIONS

The Na,K-ATPase enzyme system is probably present in the cell membrane and functions as a translocator of the monovalent cations sodium and potassium. In addition, this important enzyme may be the pharmacological receptor for cardiac glycosides. Kinetic, labelled-cardiac glycoside experiments, and spectroscopic studies suggest that the enzyme consists of a series of intermediary stages or steps and that conformational changes lead to appropriate movements of ions.

ACKNOWLEDGMENTS

Sincere appreciation is extended to Dr. Julius C. Allen for his critical assistance with the manuscript.

This study was supported by USPHS grants HE 05435 and HE 07906.

This work was done during the tenure of a Career Research and Development Award (K_3 HE 11, 875).

REFERENCES

1. ALBERS, R. W. Biochemical aspects of active transport. Ann. Rev. Biochem. 36: 727-750, 1967.

2. ALLEN, J. C., R. A. HARRIS, and A. SCHWARTZ. The nature of the transport ATPase-digitalis complex. I. Formation and reversibility in the presence and absence of a phosphorylated enzyme. Biochem. Biophys. Res. Commun. In press.

3. ALLEN, J. C., G. E. LINDENMAYER, and A. SCHWARTZ. An allosteric explanation for ouabain induced time dependent inhibition of sodium, potassium-adenosine triphosphatase. Arch. Biochem. Biophys. 141: 322-328, 1970.

4. ALLEN, J. C., and A. SCHWARTZ. A possible biochemical explanation for the insensitivity of the rat to cardiac glycosides. J. Pharmacol. Exptl. Therap. 168: 42-46, 1969.

5. BESCH, H. R., J. C. ALLEN, G. GLICK, and A. SCHWARTZ. Correlation between the inotropic action of ouabain and its effects on subcellular enzyme systems from canine myocardium. J. Pharmacol. Exptl. Therap. 171: 1-12, 1970.

6. CHIGNELL, C. F., and E. TITUS. Identification of components of (Na^+ K^+) (adenosine triphosphatase) by double isotopic labeling and electrophoresis. Proc. Natl. Acad. Sci. U.S. 64: 324-329, 1969.

7. KAHLENBERG, A., P. R. GALSWORTHY, and L. E. HOKIN. Studies on the characterization of the sodium-potassium transport adenosine triphosphatase. II. Characterization of the acyl phosphate intermediate as an L-glutamyl-α-phosphate residue. Arch. Biochem. Biophys. 126: 331-342, 1968.

8. KOSHLAND, D. E. Conformation changes at the active site during enzyme action. Federation Proc. 23: 719-726, 1964.

9. LINDENMAYER, G. E., A. H. LAUGHTER, and A. SCHWARTZ. Incorporation of inorganic phosphate-32 into a Na^+, K^+-ATPase preparation: stimulation by ouabain. Arch. Biochem. Biophys. 127: 187-192, 1968.

10. LINDEMAYER, G. E., and A. SCHWARTZ. Conformational changes induced in Na^+, K^+-ATPase by ouabain through a K^+-sensitive reaction: kinetic and spectroscopic studies. Arch. Biochem. Biophys. 140: 371-378, 1970.

11. MATSUI, H., and A. SCHWARTZ. Kinetic analysis of ouabain-K^+ and Na^+ interaction on a Na^+, K^+-dependent adenosinetriphosphatase from cardiac tissue. Biochem. Biophys. Res. Commun. 25: 147-152, 1966.

12. MATSUI, H., and A. SCHWARTZ. Purification and properties of a highly active ouabain-sensitive Na^+, K^+-dependent adenosinetriphosphatase from cardiac tissue. Biochim. Biophys. Acta 128: 380-390, 1968.

13. MEDZIHRADSKY, F., M. H. KLINE, and L. E. HOKIN. Studies on the characterization of the sodium-potassium transport adenosinetriphosphatase. I. Solubilization, stabilization and estimation of the apparent molecular weight. Arch. Biochem. Biophys. 121: 311-316, 1967.

14. NAGAI, K., G. E. LINDENMAYER, and A. SCHWARTZ. Direct evidence for the conformational nature of the Na^+, K^+-ATPase system: fluorescence and circular dichroism studies. Arch. Biochem. Biophys. 139: 252-254, 1970.

15. SCHWARTZ, A., H. MATSUI, and A. H. LAUGHTER. Tritiated digoxin binding to ($Na^+ + K^+$)-activated adenosine triphosphatase: possible allosteric site. Science 159: 323-325, 1968.

16. SCHATZMANN, H. J. Herzglykoside als Hemmstoffe für den aktiven Kalium- und Natriumtransport durch die Erythrocytenmenbran. Helv. Physiol. Acta 11: 341-354, 1953.

17. SEN, A. K., and R. L. POST. Stoichiometry and localization of adenosine triphosphatase dependent sodium and potassium transport in the erythrocyte. J. Biol. Chem. 239: 345-352, 1964.

18. SHIRACHI, D. Y., A. A. ALLARD, and A. J. TREVOR. Partial purification and ouabain sensitivity of lubrol-extracted sodium-potassium transport adenosine triphosphatase from brain and cardiac tissues. Biochem. Pharmacol. 19: 2893-2906, 1970.

19. SIEGEL, G. J., G. J. KOVAL, and R. W. ALBERS. Sodium-potassium activated adenosine triphosphatase. VI. Characterization of the phosphoprotein formed from orthophosphate in the presence of ouabain. J. Biol. Chem. 244: 3264-3269, 1969.

20. SKOU, J. C. The influence of some cations on the adenosine triphosphatase from peripheral nerve. Biochim. Biophys. Acta 23: 394-401, 1957.

CREDITS

The editors and authors are grateful to the following publishers, organizations and individuals for permission to use materials previously published:

Figure 4: Academic Press, Inc.; H. Matsui and A. Schwartz, Biochemical Biophysical Research Communications 25: 147-152, 1966. Copyright Academic Press, Inc., 1966.

Figure 5: Academic Press, Inc.; H. Matsui and A. Schwartz, Biochemical Biophysical Research Communications 25: 147-152, 1966. Copyright Academic Press, Inc., 1966.

Figure 6: Academic Press, Inc.; K. Nagai, G. E. Lindenmayer and A. Schwartz, Archives of Biochemistry and Biophysics 139: 252-254, 1970. Copyright Academic Press, Inc., 1970.

Figure 7: Academic Press, Inc.; G. E. Lindenmayer and A. Schwartz, Archives of Biochemistry and Biophysics 140: 371-378, 1970. Copyright Academic Press, Inc., 1970.

Scheme, bottom page 159: Academic Press, Inc.; J. C. Allen, G. E. Lindenmayer and A. Schwartz, Archives of Biochemistry and Biophysics 141: 322-328, 1970. Copyright Academic Press, Inc., 1970.

COLD SWELLING AND ENERGY METABOLISM IN THE HYPOTHERMIC BRAIN OF RATS AND DOGS

N. MENDLER, H. J. REULEN AND W. BRENDEL

Department of Cardiac Surgery and
Institute of Experimental Surgery,
University of Munich, Germany

INTRODUCTION

Moderate induced hypothermia in man has been well established as a clinical procedure for cardiac and neurosurgery. The beneficial effects of this treatment have encouraged many attempts to extend hypothermia to extremely low temperatures in order to profit from better protection of tissues against circulatory dysfunction. These studies had one finding in common: with decreasing temperature, the benefits of hypothermia are gradually overcome by the fact that temperature reduction itself becomes lethal to homoiothermic species (33).

From a survey of the experimental literature it was suggested that the brain might be especially vulnerable to cold (10,33). In addition, clinical reports of neurological damage after profound hypothermia for open heart surgery have been frequent (7,12,16). In experiments designed to explore further the effects of low temperature on the brain, cerebral edema was found present in rats cooled to progressive degrees of hypothermia, and a relationship between temperatures, severity of edema and survival of the animals could be demonstrated (Fig. 1). It appeared from these studies, that a cold swelling of the brain was one limiting factor for recovery from temperatures below 10°C (10).

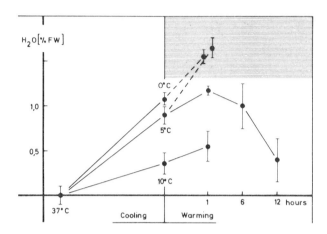

FIG. 1. Increase in water content of the brain in rats cooled to and rewarmed from 10, 5, and 0°C. Shaded area: Animals died during rewarming.

In accordance with observations on isolated tissues, analysis of the edematous brains revealed an intracellular accumulation of sodium and water with a concomitant loss of potassium (23,25,29). The increase of tissue water was shown to be a linear function of sodium uptake - a finding which is also characteristic for brain edema produced by inhibition of active cation transport in normothermia (4,5,18,30,31).

From these results, Reulen and Brendel suggested that a cold-induced imbalance of the actively maintained transcellular cation gradients was the basic mechanism leading to hypothermic brain swelling. Based on the observation that in cooled isolated tissues the activation energy of active transport increases threefold as compared to that of passive diffusion (8,14,17), they concluded that a relative inhibition of active cation fluxes would occur in situ, as well as resulting in edema.

The present report summarizes our efforts to further elucidate the pathophysiology of hypothermic brain edema: the rate-limiting action of cold on ion transport could be thought to be at different steps of the process. Production of energy supplies necessary for transport could be critically reduced or, in spite of sufficient energy resources, their utilization at the site of transport could be impaired. For the brain *in situ*, this alternative might be further complicated by interactions with other pathological changes, e.g. in circulation and respiration.

An approach to these problems was attempted by following simultaneous changes in energy metabolism substrates and electrolyte disturbances in brain tissue during profound hypothermia. Additional information was expected from interference with some parameters of cold swelling by varying the experimental conditions.

METABOLITES AND COLD SWELLING DURING SURFACE COOLING

When profound hypothermia is induced by immersion of artificially ventilated rats in an ice-water bath, 4°C core temperature is reached in one hour (1,29). After this stage of acute cooling all animals have developed a typical cold swelling of the brain, which is more pronounced after three hours at this temperature (Table 1). As in previous studies (10,31), a linear correlation between changes in water and sodium content of the swollen brain can be observed (Figs. 2, 5). The equation of the regression line was calculated to be $H_2O = 0.05\ Na + 78.2$ ($r = 9.0$). At the same time, a marked loss of tissue potassium can be observed.

The metabolic state after acute surface cooling represents a moderate hypoxia (22,35,36): glycolysis has been activated as indicated by a significant loss of glycogen and glucose accompanied by an accumulation of lactate. However, a near normal ATP is still maintained by anaerobic energy production. After an additional three hours of circulatory arrest

TABLE 1: Energy metabolites, water and electrolytes in the rat brain during hypothermia induced by surface cooling in normocapnia and hypoxia-hypercapnia.

	N	ATP	ADP	AMP	Glyc.	Gluc.	La.	H_2O	Na^+	K^+
Normals	14	2.23 ±0.08	0.43 ±0.03	0.39 ±0.03	3.35 ±0.14	3.50 ±0.27	3.35 ±0.09	78.50 ±0.08	238 ±2	484 ±5
Acute cooling to 4°C Normocapnia	9	2.03 ±0.18	0.85* ±0.24	0.66 ±0.25	2.37° ±0.64	1.60*° ±0.64	6.42* ±0.60	78.95° ±0.15	246 ±5	480° ±7
Acute cooling to 4°C Hypercapnia	7	2.47° ±0.13	0.85* ±0.05	0.87* ±0.05	4.49*° ±0.06	4.59*° ±0.41	5.98* ±0.79	78.17*° ±0.12	245* ±2	463*° ±6
3 hours at 4 – 0°C Normocapnia	10	0.32*° ±0.02	0.62*° ±0.04	1.70* ±0.03	0.81* ±0.32	0.40*° ±0.07	12.17*° ±0.75	79.45*° ±0.13	256* ±3	468*° ±4
3 hours at 4 – 0°C Hypercapnia	7	0.60*° ±0.06	0.89*° ±0.04	1.97* ±0.09	0.89* ±0.08	1.35*° ±0.07	13.28*° ±0.14	78.77° ±0.14	257* ±3	448*° ±3

Metabolites in μmol/g fresh weight. Water in g/100 g fresh weight. Electrolytes in mEq/kg dry weight. Values are means ± SE of mean.

*) Significant difference from normal.
°) Significant difference between corresponding normocapnic and hypercapnic groups (Student test, p < 0.02).

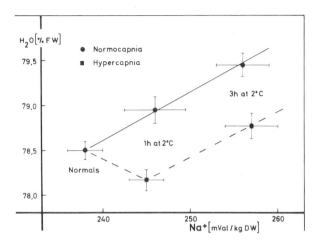

FIG. 2. Correlation between brain water and sodium content during profound hypothermia in rats cooled in normocapnia (ice water) and in hypercapnia (closed-vessel-technique). (mval = mEq.)

in ice water, energy stores have been extensively used up, while severe cold swelling is manifest.

On the basis of these findings, one would be tempted to ascribe the development of cold swelling to a deprivation of energy available for active transport by circulatory depression, as cessation of the ECG activity occurs at a mean colonic temperature of 7.6°C during surface cooling. Thus, it would be a fair assumption to expect edema to be absent if tissue perfusion could be maintained by artificial circulation.

METABOLITES AND COLD SWELLING DURING ARTIFICIAL CIRCULATION

An experimental approach to this hypothesis was made possible by the development of a small pump-oxygenator by means of which rats could be cooled to temperatures below 4°C with fully maintained oxygen supply (23). This technique allows about 50% of normothermic cardiac output (120 ml/kg min)

(9,28) to be extracorporeally circulated (ECC) between right atrium and femoral artery for one hour with permanent survival of the animals. As indicated in Table 2, this procedure does not alter water, electrolyte, and metabolite concentrations in the brains of normothermic control animals. Rapid cooling can be obtained by the heat exchanger incorporated in the artificial circulation: colonic and esophageal temperatures fell below 4°C within 20 - 25 min. Changes in acid-base status were observed as typical for hypothermic perfusion when 100% O_2 is used for oxygenation (23,33): PCO_2 fell to 34 mm Hg after 10 minutes of cooling, pH rose to 7.54.

After acute cooling, brain metabolism has adjusted to a new steady state on a higher energetic level: the ATP/ADP ratio is found elevated from 5.2 to 7.3. Even more pronounced changes are present in the glycolytic substrates, where glucose has increased to a threefold concentration accompanied by a drop in lactate of the same order of magnitude. These findings indicate that utilization of oxygen is not impaired by temperature reduction. Although this apparent elevation of high-energy substrates may be partly due to faster fixation of the tissue by the more rapid freezing in liquid nitrogen of hypothermic animal (3,22,23), an absence of a temperature-induced metabolic energy deficit is, nevertheless, still demonstrated by these findings.

Cold swelling, however, is present after this short time of hypothermia, and after further perfusion for one hour below 4°C it has been fully developed and shows the typical linear correlation between water and sodium changes, the regression curve (Fig. 3) having the equation $H_2O = 0.025\ Na + 77.4$ (r = 0.96). Brain volume after one hour can be calculated (13) as increased by 6%, and an impairment of microcirculation by elevated intracranial pressure must be expected. The metabolic equivalent for such relative tissue hypoxia is found in an increased glycolytic activity (Table 2) as

TABLE 2. Energy metabolites, water and electrolytes in the rat brain during hypothermia induced by extracorporeal circulation (ECC).

	N	ATP	ADP	AMP	Glyc.	Gluc.	La.	H_2O	Na^+	K^+
Normals	6	2.52 ±0.05	0.48 ±0.03	0.32 ±0.06	3.58 ±0.26	1.57 ±0.12	5.03 ±0.33	77.94 ±0.07	204 ±2	428 ±2
ECC-controls 37°C	4	2.38 ±0.07	0.55 ±0.03	0.45 ±0.05	3.44 ±0.42	1.59 ±0.47	4.48 ±0.62	78.07 ±0.10	209 ±4	442 ±4
Acute cooling To 4°C	6	2.63 ±0.08	0.36 ±0.04	0.35 ±0.02	3.66 ±0.11	4.86* ±0.31	1.64* ±0.39	78.37* ±0.21	211 ±10	403 ±23
60 min circulatory Arrest at 4 - 2°C	6	1.05*° ±0.06	1.06*° ±0.07	1.15*° ±0.06	2.31*° ±0.10	1.32*° ±0.12	10.85*° ±0.40	78.75*° ±0.13	238* ±6	458* ±11
60 min ECC At 4 - 2°C	6	2.44° ±0.05	0.50° ±0.04	0.59* ±0.06	4.41° ±0.69	2.39° ±0.23	7.75*° ±0.33	79.24*° ±0.16	248* ±6	465* ±7

Metabolites in μmol/g fresh weight. Water in g/100 g fresh weight. Electrolytes in mEq/kg dry weight. Values are means ± SE of mean.

*) Significant difference from normal.
°) Significant difference between ECC and Circulatory arrest (Student "T" test, $p < 0.02$).

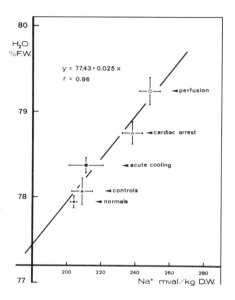

FIG. 3. Correlation between brain water and sodium content in rats cooled to 4°C by extracorporeal circulation and then subjected to cardiac arrest or continuous perfusion for one hour at this temperature. (mval = mEq.)

compared to changes after acute cooling; however, ATP is still maintained at a normal level (22,35,36). If perfusion is discontinued after acute cooling, brain energy metabolites represent a marked hypoxia after one hour at 2 - 4°C. As after acute surface cooling, high energy phosphate breakdown cannot be compensated by anaerobic energy production and a severe deficit is present. However, cold swelling has developed even less than during perfusion at a normal metabolic state.

In contrast to these observations on cold swelling, the development of edema by normothermic metabolic inhibition is closely dependent on the reduction of tissue concentrations of ATP. Reulen et al. (31) showed that in normothermic hypoxia brain edema formation does not start unless ATP values have been below 70% of the normal for prolonged periods of time, and Rossum has presented evidence of a similar dependence of active transport

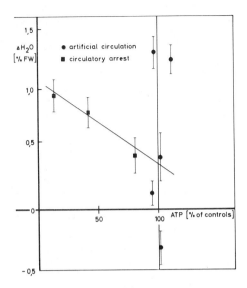

FIG. 4. Increase in brain water compared with loss of ATP in the brain of rats and dogs cooled by immersion with circulatory arrest and by extra-corporeal circulation.

inhibition on relative energy deficits in tumor cells in this symposium.

Obviously, no such coupling of the development of cold swelling to an energy deficit can be found, as graphically demonstrated in Figure 4. It can be seen that during perfusion and in the presence of an unchanged ATP concentration cold swelling develops in an even more pronounced way than during states of reduced energy supply in surface cooling with circulatory arrest, and it becomes evident that the hypothesis stated above cannot be maintained.

On the contrary, one may conclude:

1. Transcellular electrolyte gradients, which are maintained during artificial perfusion but which gradually decrease during circulatory arrest play a more important role in the development of cold swelling than the concentration of high energy substrates.

2. The apparent correlation between ATP-reduction and edema formation presents nothing more than a dependence of changes in both parameters on a common time base.

In addition, a disequilibrium of production and utilization based on major differences in the temperature characteristics of intermediary reactions of aerobic and anaerobic energy metabolism has been discussed as a cause of impaired electrolyte transport in hypothermia (10,25,29). Comparison of the metabolic patterns contained in Tables 1, 2, and 3 with those found under different normoxic and hypoxic conditions in normothermia (22, 30,31,36) does not support this assumption: after cooling under artificial oxygenation normal concentrations of high energy substrates are found. Also, after various degrees of hypoxia in hypothermia the metabolic state does not differ from the picture typical of normothermic hypoxia. Thus, synthesis as well as utilization of energy stores seem to be equally affected by cold and it becomes likely that major differences in the Q_{10} of single steps of energy metabolism are absent <u>in situ</u>.

INFLUENCE OF EXTRACELLULAR SODIUM DEPLETION, INCREASED EXTRACELLULAR OSMOLARITY AND AMIDONICOTINIC ACID (ANA).

On the basis of the results reported so far, the sodium ion represents a leading factor in causing cold swelling, and reduction of its concentration gradient across the cell membrane would be a logical way to evaluate its mechanism further. This was accomplished by artificial perfusion of the isolated head. Details of the method are described elsewhere (31). As indicated in Tables 3 and 4, normothermic perfusion of control animals for one hour leads to no significant deviations from normal values except for a small reduction of sodium in the cortex. Findings after one hour of hypothermic hemoperfusion are in close agreement with changes observed during total body perfusion in the rat: cold swelling develops in a typical

TABLE 3. Energy metabolites in the dog cortex during isolated head perfusion in hypothermia with blood and low-sodium perfusate.

	N	ATP	ADP	AMP	La
Normals	5	1.73	0.71	0.62	2.11
		1.62 - 1.81	0.57 - 0.81	0.42 - 0.78	1.62 - 2.47
Controls	4	1.74	0.75	0.64	2.55
60 min Perfusion at 37°C		1.60 - 1.94	0.58 - 0.91	0.38 - 0.84	1.80 - 3.53
Hypothermia	4	1.99	0.52	0.46	4.46
60 min Perfusion at 4°C		1.82 - 2.25	0.42 - 0.60	0.31 - 0.66	2.31 - 6.09
Hypothermia - low Na^+	4	2.12	0.56	0.58	4.01
60 min Perfusion at 4°C		1.66 - 2.75	0.39 - 0.78	0.46 - 0.77	3.18 - 4.64

Means, highest and lowest values in µmol/g fresh weight.

way, although high energy phosphates are unchanged. Slopes of the sodium-water correlation are also similar ($H_2O = 0.022\ Na + 79.9$).

In the experimental group, heads were cooled by an initial hemoperfusion. After 4°C had been reached, blood was exchanged against an isotonic low-sodium solution, containing 5 mEq/l Na^+, 5 mEq/l K^+, procaine and mannitol, as an osmotic substitute. This perfusate had been proposed by Bretschneider as a membrane-stabilizing cardioplegic (11) and had been shown to protect high energy phosphates of the myocardium against ischemic damages. Serum sodium during perfusion with this medium dropped to 10 mEq/l within 12 min and remained constant at this level for one hour, potassium was found unchanged at a mean value of 4.5 mEq/l.

At the end of the experiments, high energy phosphates were found slightly elevated (Table 3), suggesting an adequate oxygen transport by the hemoglobin-free perfusate below 4°C. As demonstrated graphically in Figure 5,

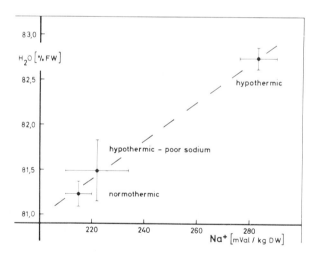

FIG. 5. Prevention of hypothermic brain swelling by extracellular sodium depletion during perfusion of the isolated dog head with a low sodium solution. (mval = mEq.)

TABLE 4. Water, sodium and potassium in cortex and white matter of the dog brain during isolated head perfusion in hypothermia with blood and low-sodium perfusate.

		Cortex			White Matter		
	N	H_2O	Na^+	K^+	H_2O	Na^+	K^+
Normals	5	81.52 ±0.25	241 ± 5	471 ±13	70.04 ±0.34	142 ± 2	242 ±12
Controls	4	81.23 ±0.14	215 ± 5	482 ± 8	69.84 ±0.38	142 ± 4	243 ±11
60 min Perfusion at 37°C							
Hypothermia	4	82.74*° ±0.12	283*° ± 7	440*° ±10	70.17 ±0.39	153*° ± 3	231 ± 7
60 min Perfusion at 4°C							
Hypothermia – poor Na^+	4	81.49° ±0.34	222° ±12	448*° ± 6	69.74 ±0.70	127° ± 5	244 ± 3
60 min Perfusion at 4°C							

Water in g/100 g fresh weight. Electrolytes in mEq/kg dry weight. Mean values ± SE of mean.

*) Significant difference from normals.
°) Significant difference between perfusion with blood and poor-sodium perfusate (Student "T" test, p < 0.02).

cold swelling was absent in both cortex and white matter after perfusion at a reduced extracellular sodium concentration: sodium and water are found within the range of normothermic hemoperfusion, while potassium loss is unaffected as compared to hypothermia during perfusion with blood (Table 3).

Intravenous administration of urea (1.5 g/kg) during the early surface cooling stage in rats leads to an increase of serum osmolarity by 20 - 35 mosmol at 30 minute after injection. Without renal elimination being present, serum osmolarity decreases only slowly within three hours of prolonged hypothermia. Urea does not penetrate readily into the brain cells of the rat as it does in the dog (27). As compared with untreated animals (Fig. 6), the correlation of sodium and water at this stage was found to be modified in a typical way: sodium accumulation had occurred without the usual concomitant rise in water content.

Brain swelling after depression of energy supply by metabolic inhibition in normothermia has been successfully reversed by amidonicotinic acid (ANA), and a protective effect of this substance on energy metabolism was demonstrated (4). In a similar way sodium and water uptake after three hours below 4°C were found to be within the range of changes after one hour in untreated animals (Fig. 6). This reduction of cold swelling may be explained by an improved energy supply to active transport during the cooling process, when hypoxia is otherwise known to contribute considerably to edema formation (31).

The differential modifications of sodium and water movements after extracellular sodium depletion and application of urea and ANA, are accompanied by a similar loss of potassium in all experimental groups compared (Table 5). Thus, these data strongly support the concept of sodium being the leading factor in causing cold swelling - and water uptake occurring secondarily for osmotic reasons. Potassium changes seem to reflect a variety of experimental conditions which will be discussed in a later section.

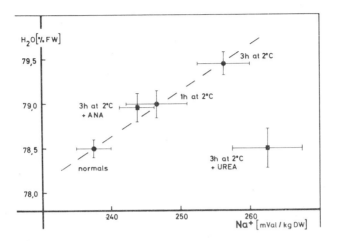

FIG. 6. Modification of hypothermic brain edema in rats cooled by immersion in ice water after intravenous administration of ANA and urea. (mval = mEq.)

TABLE 5. Influence of urea and amidonicotinic acid on cold swelling of the rat brain after 3 hours below 4°C.

	H_2O	Na^+	K^+
Normals	78.50	238	484
	±0.08	± 2	± 5
3 Hrs below 4°C	79.45*	256*	468
	±0.13	± 3	± 4
3 Hrs below 4°C + Urea	78.52*° ±0.22	262° ± 4	462 ± 5
3 Hrs below 4°C + ANA	78.97*° ±0.16	244*° ± 2	465 ± 4

Water in g/100 g fresh weight. Electrolytes in mEq/kg dry weight.
Values are means ± SE of mean.

*) Significant difference between treated and untreated animals.
°) Significant difference between treatments (Student "T" test, p < 0.02)

INFLUENCE OF HYPERCAPNIA

Resistance to profound hypothermia in rodents has been markedly improved by cooling the animals in a hypoxic and hypercapnic environment (2,15,26). It seemed, therefore, of interest to study cold swelling of the brain after cooling by this technique. Striking differences from effects of surface cooling of animals ventilated with air were observed. At the time of arrival at 2°C, potassium loss was found to be five times greater (Fig. 7) than after cooling in ice water, while sodium uptake did not differ in the two groups (Table 2). Together with the overall cation loss of 14 mEq/kg DW, water content had decreased by 0.33% FW, i.e., brain shrinkage had occurred (20). At the same time, high energy phosphates increased even above the normal level, indicating that oxygen demands were sufficiently met during cooling in the low-oxygen atmosphere of the closed vessel. Whether the beneficial action of the high PCO_2 on brain energy stores is due to vasodilation, changes in hemoglobin dissociation characteristics, or direct action on intermediary metabolism (2,33,34,37) cannot be decided under these experimental conditions. With respect to cold swelling, the state of the brain after acute cooling in hypercapnia represents a more favorable situation of the start of prolonged cooling, which is further supported by the finding that these animals survive readily after rewarming (unpublished observations). However, upon maintaining hypothermia for three hours cold swelling develops in a parallel way but at a lower level than after normocapnic cooling (Figs. 2,7).

These results might give some explanation for the beneficial effect of this mode of cooling on the tolerance to prolonged deep hypothermia observed by a number of authors. The amount of tissue swelling associated with and governed by the overall changes in cation composition (20) probably

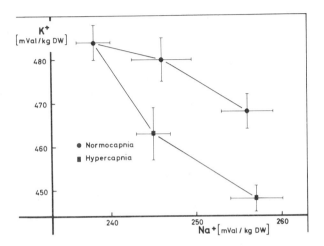

FIG. 7. Potassium loss as a function of sodium uptake in the brain of rats cooled in normocapnia and hypercapnia. (mval = mEq.)

represents the final limitation to survival, as the elevation of intracranial pressure - which is proportional to the amount of edema - "shuts off" the cerebral microcirculation during rewarming (cf. also Fig. 1). The reversibility of ionic disturbances - though possible in principle would thus be prevented by an insufficient cerebral perfusion adding ischemic damage to that of hypothermia. This interpretation is supported further by the observation that rewarming animals after prolonged perfusion in hypothermia leads to a steep decrease in cerebral ATP contents together with a further rise in vascular resistance (unpublished data). The brain, encapsulated in the rigid skull without any possibility to expand, can be assumed to be especially vulnerable to this mechanism.

INTERACTIONS OF COLD SWELLING WITH ACID-BASE AND ENERGY METABOLISM

So far, it has been proposed in this study that a temperature-induced inhibition of active transport presents the basic mechanism of hypothermic brain edema, and that sodium uptake by the cells presents one leading parameter of this phenomenon. This behaviour of the brain *in situ* would thus reflect the generally agreed higher sensitivity to cold of active as compared to passive cation fluxes (14,17). In contrast to standardized conditions in experiments on isolated tissues, the organ *in situ* is subject to influences imposed by changes in the general pathophysiology of the organism in hypothermia. Interactions of some of these influences with the quantitative manifestation of cold swelling are demonstrated in Table 6. It can be seen that the slope of the sodium-water correlation curve is highest in those instances where a considerable energy deficit - represented by a decrease in ATP concentrations - is present. At the same time a marked acidosis is produced under these conditions by inadequate respiration and/or tissue perfusion. Where this acidosis is most pronounced the highest potassium losses from brain tissue can be observed. Alkalosis, however, as present during cooling by total body artificial circulation at low PCO_2 is found associated with an increased tissue potassium. While it is difficult to quantify or separate these interactions of energy and acid-base metabolism, it appears from Table 4 that the manifestation of edema observed under a given experimental condition is a more complex result of a combined action of "pure" temperature, energy deficit and transcellular cation-hydrogen exchange.

COLD SWELLING IN THE HIBERNATOR - SITE OF TEMPERATURE ACTION

Though modified in its extent by multiple parameters, cold swelling of the brain appears to develop in every instance where homiothermic animals are cooled to profound hypothermia, even where energy donators for active transport are available in abundance. Assuming that this energy is accessible at the site of transport the action of cold must be visualized as being in the temperature sensitivity of the enzyme system, mediating sodium and potassium fluxes at the membrane. Findings of Willis (38,39,40) support this explanation. He could show that cation transport in kidney cortex slices is more sensitive to cold than the supporting respiration and recently (41) that activation of the membrane ATPase by sodium is abolished at 15°C in the rat but still present at 5°C in the hamster, a hibernator. In addition, Fang and Willis have shown that the activity of Na,K-ATPase rose steadily in the hamster brain during preparation of the animals for hibernation (for reference see Willis, Fang and Foster, this symposium). Our own finding showing cold swelling of the brain in the European dormouse (Glis glis) to be absent when animals are cooled during hibernation but readily produced in summer animals (Fig. 8) supports the

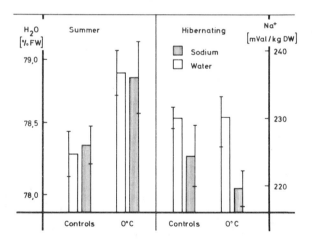

FIG. 8. Sodium and water contents of the brain of Glis glis cooled to 0°C in summer and during hibernation. (Mval = mEq.)

TABLE 6. Slope of the sodium-water correlation, available energy for active transport, acid-base status and brain potassium in different forms of experimental brain edema.

Author (Ref.)	Species	Model	$\frac{H_2O}{Na^+}$	ATP	Acidosis	K^+
Reulen et al. (30)	rat	2.4 - DNP	0.068	↓	+	↓
This study	rat	hypothermia	0.050	↓↓	+	↓
This study	rat	hypothermic + CO_2	0.050	↑	++	↓↓
Reulen et al. (31)	dog	isol. head, hypotens.	0.039	↓	+	↓
Messmer et al. (25)	dog	ECC - hypothermia	0.026	±	-	↑
This study	dog	isol. head, - hypother.	0.022	↑	+	↓
Mendler (23)	rat	ECC - cooling	0.025	±	-	↑

view that an enzyme acclimation unique to the hibernator is responsible for the tolerance of these species to low temperature. Conversely, it seems to be impossible for the homoiothermic non-hibernator to adjust his active transport activity to the demands of profound hypothermia. This fact might well be regarded as a natural barrier limiting the ability of these species to survive extreme cooling.

ACKNOWLEDGMENT

This research was supported by the "Deutsche Forschungsgemeinschaft".

REFERENCES

1. ADOLPH, E. F., and J. GOLDSTEIN. Survival of rats and mice without oxygen in deep hypothermia. J. Appl. Physiol. 14: 599-604, 1959.

2. ANDJUS, R. K., and A. U. SMITH. Reanimation of adult rats from body temperatures between 0 and +2°C. J. Physiol. 128: 446-472, 1955.

3. ANDJUS, R. K. Some mechanisms of mammalian tolerance to low body temperatures. In: S.E.B. Symposium XXIII. Dormancy and Survival Cambridge, Eng.: The University Press, 1969, p. 351-394.

4. BAETHMANN, A., H. J. REULEN, and W. BRENDEL. Die Wirkung des Antimetaboliten 6-Aminonicotinamid (6-ANA) auf Wasser- und Elektrolytgehalt des Rattenhirns und ihre Hemmung durch Nikotinsäure. Z. Ges. Exptl. Med. 146: 226-240, 1968.

5. BAKAY, L. and J. C. LEE. Cerebral Edema. Springfield, Ill.: Charles C. Thomas, 1965.

6. BERGMEYER, H. U. Methods of Enzymatic Analysis. New York: Academic Press, 1965. Translated from the German.

7. BJÖRK, V. O., and G. HULTQUIST. Brain damage in children after deep hypothermia for open heart surgery. Thorax 15: 284-291, 1960.

8. BOWLER, K., and C. J. DUNCAN. The effect of temperature on the Mg^{2+}-dependent and Na^+-K^+ ATPases of a rat brain microsomal preparation. Comp. Biochem. Physiol. 24: 1043-1054, 1968.

9. BULLARD, R. W. Cardiac output of the hypothermic rat. Amer. J. Physiol. 196: 415-419, 1959.

10. BRENDEL, W., C. MUELLER, H. J. REULEN, and K. MESSMER. Elektrolytveränderungen in tiefer Hypothermie. II. Beziehungen zur klinischen und biologischen Überlebenszeit. Arch. Ges. Physiol. 288: 220-239, 1966.

11. BRETSCHNEIDER, H. C. Überlebenszeit und Wiederbelebungszeit des Herzens bei Normo- und Hypothermie. Verh. Deut. Ges. Kreislaufforsch. 30: 11-64, 1964.

12. EGERTON, N., W. S. EGERTON, and J. H. KAY. Neurologic changes following profound hypothermia. Ann. Surg. 157: 366-374, 1963.

13. ELLIOTT, K. A. C., and H. JASPER. Measurement of experimentally induced brain swelling and shrinkage. Amer. J. Physiol. 157: 122, 1949.

14. FRANKENHAUSER, B., and L. MOORE. The effect of temperature on the sodium and potassium permeability changes in myelinated nerve fibres of Xenopus laevis. J. Physiol. 169: 431-437, 1963.

15. GIAJA, J., and R. K. ANDJUS. Sur l'emploi de l'anesthesie hypoxique en physiologie operatoire. Compt. Rend. Acad. Sci., 229: 1170-1172, 1949.

16. GILMAN, S. Cerebral disorders after open-heart operations. New Engl. J. Med. 272: 489-498, 1963.

17. HODGKIN, A. L., and R. D. KEYNES. Active transport of cations in giant axons from Sepia and Loligo. J. Physiol. 128: 28-60, 1955.

18. KOCH, A., J. B. RANCK, and B. L. NEWMAN. Ionic content of neuroglia. Exptl. Neurol. 6: 186-200, 1962.

19. KUFFLER, S. W., and J. G. NICHOLLS. The physiology of neuroglial cells. Ergeb. Physiol. 57: 1-90, 1966.

20. LEAF, A. On the mechanism of fluid exchange of tissues in vitro. Biochem. J. 62: 241-248, 1956.

21. LEE, J. C., and I. BAKAY. Ultrastructural changes in the edematous central nervous system. II. Cold induced edema. Arch. Neurol. 14: 36-49, 1966.

22. LOWRY, O. H., L. V. PASSONNEAU, F. X. HASSELBERGER, and D. W. SCHULTZ. Effect of ischemia on known substrates and cofactors of the glycolytic pathway in brain. J. Biol. Chem. 239: 18-30, 1964.

23. MENDLER, N. Elektrolyt- und Metabolitveränderungen im Gehirn der Ratte nach Kühlung auf 4 - 2°C mit einer Herzlungemaschine. Z. Ges. Exptl. Med. 146: 206-225, 1968.

24. MENDLER, N., W. WEISHAAR, and W. BRENDEL. Eine Herzlungemaschine für Ratten als experimentelles Modell der extrakorporalen Zirkulation. Thoraxchirurgie 17: 534, 1969.

25. MESSMER, K., W. BRENDEL, H. J. REULEN and K. NORDMANN. Elektrolytveränderungen in tiefer Hypothermie. III. Beziehungen zur biologischen Überlebenszeit bei extrakorpalem Kreislauf. Arch. Ges. Physiol. 288: 240-261, 1966.

26. MILLER, F. S., and J. A. MILLER. Cardiac arrest and recovery of contraction in mice cooled with hypoxia-hypercapnia (Abstract). Anat. Rec. 148: 312, 1964.

27. PAPPIUS, H. M. The distribution of water in brain tissue swollen in vitro and in vivo. In: Biology of Neuroglia, edited by E. D. P. de Roberts and R. Carred. New York: Elsevier Pub. Co., 1965.

28. POPOVIC, V. P., and K. M. KENT. 120-day study of cardiac output in unanesthetized rats. Amer. J. Physiol. 207: 767-770, 1964.

29. REULEN, H. J., P. AIGNER, W. BRENDEL, and K. MESSMER. Elektrolytveränderungen in tiefer Hypothermie. I. Die Wirkung akuter Aushukühlung bis 0°C und Wiedererwärmung. Arch. Ges. Physiol. 288: 197-219, 1966.

30. REULEN, H. J., and A. BAETHMANN. Das Dinitrophenol-Odem. Ein Modell zur Pathophysiologie des Hirnödems. Klin. Wochschr. 45: 149-154, 1967.

31. REULEN, H. J., U. STEUDE, W. BRENDEL, C. HILBER, and S. PRUSINER. Energetische Störung des Kationentransports als Ursache des intrazellulären Hirnödems. Acta Neurochir. 22: 129-166, 1970.

32. SCHMAHL, F. W., E. BETZ, E. DETTINGER, and H. J. HOHORST. Energiestoffwechsel der Grosshirnrinde und Electroencephalogramm bei Sauerstoffmangel. Arch. Ges. Physiol. 292: 46-59, 1966.

33. THAUER, R., and W. BRENDEL. Hypothermie. Progr. Surg. 2: 73-271, 1962.

34. THEWS, G. Implications to physiology and pathology of oxygen diffusion at the capillary level. In: Selective Vulnerability of the Brain in Hypoxaemia, edited by J.P. Schade and W.H. McMenemy. Philadelphia: F.A. Davis Co., 1963.

35. THORN, W.H., H. SCHOOL, G. PFLEIDERER, and B. MUELDENER. Stoffwechselvorgänge im Gehirn bei normaler und herabgesetzter Körpertemperatur unter ischämischer und anoxischer Belastung. J. Neurochem. 2: 150-165, 1958.

36. THORN, W., G. PFLEIDERER, R. A. FROWEIN, and I. ROSS. Stoffwechselvorgänge im Gehirn bei akuter Anoxie, akuter Ischämie und in der Erholung. Arch. Ges. Physiol. 261: 334-360, 1955.

37. WHITE, J. C., M. VERLOT, B. SELVERSTONE, and H. K. BEECHER. Changes in brain volume during anaesthesia: the effects of anoxia and hypercapnia. Arch. Surg. 44: 1-21, 1942.

38. WILLIS, J. S. Resistance of tissues of hibernating mammals to cold swelling. J. Physiol. 164: 51-63, 1962.

39. WILLIS, J. S., and N. MA. Cold resistance of Na, K stimulated ATPase in kidney cortex of hibernating rodents. Physiologist 10: 348, 1967.

40. WILLIS, J. S. Cold resistance of kidney cells of mammalian hibernators: cation transport vs. respiration. Amer. J. Physiol. 214: 923-928, 1968.

41. WILLIS, J. S., and N. M. LI. Cold resistance of Na-K-ATPase of renal cortex of the hamster, a hibernating mammal. Amer. J. Physiol. 217: 321-326, 1969.

THE METABOLIC COUPLING OF ION TRANSPORT

G. D. V. VAN ROSSUM

Department of Pharmacology
Temple University School of Medicine
Philadelphia, Pa., 19140

INTRODUCTION

The maintenance of ionic gradients between cells and their surroundings is vital to the functioning of nerve and muscle cells, to the activity of absorptive and secretary organs, and to the balanced activity of the multi-enzyme systems that catalyze cell metabolism. The maintenance of the concentrations is to a large extent dependent on the movement of ions against electrochemical gradients, the energy for which is derived from metabolic sources. This paper will discuss some aspects of the role of the energy-conserving pathways of the cell (i.e., glycolysis and oxidative phosphorylation) as energy-donors for ion transport and, conversely, the way in which the rates of these pathways are geared to the requirements of the transport mechanism. Similar considerations should apply to any of the other energy-requiring reactions of a cell. Therefore, a further important aspect of the energy-coupling is the way in which the various transport and anabolic activities "share" with each other the available cellular energy, especially when the rate of metabolism becomes limiting. A possible approach to this question, using ion transport as a model, will be indicated.

METABOLIC SOURCE OF ENERGY FOR ION TRANSPORT

Respiration and glycolysis. There is little argument, at least for the time being, that the provision of metabolic energy for ion transport is mediated by high-energy compounds that are synthesized during oxidative phosphorylation and glycolysis. In cells that contain mitochondria, oxidative phosphorylation may be expected to support the transport adequately under aerobic conditions. The ability of cells to maintain ion transport by using ATP derived from glycolysis, when oxidative phosphorylation is inactive, varies considerably between tissues, and seems to be broadly related to the potential glycolytic activity of the tissue. Thus, the nearly absolute dependence of some tissues on oxidative phosphorylation is probably largely dictated by their rates of glycolysis, which remain low even after the Pasteur effect has come into play. However, it is interesting to look a little beyond this generalization and to consider the extent to which the total rate of ATP synthesis under anaerobic conditions compares with that in aerobiosis, and then to relate this to the ability to maintain ion transport anaerobically. Such estimates are summarized for a few tissues in Table 1. In ascites tumor cells, the Pasteur effect almost completely compensates for the absence of oxidative phosphorylation when the cells are treated with cyanide; in confirmation of the validity of these estimates, Hempling (17) found equal levels of ATP in ascites cells whether they were actively respiring or whether they were treated with glucose and a respiratory inhibitor, Amytal. As may be expected, Na^+ transport was little affected by the presence of cyanide. However, the values obtained with other tissues indicate two more intriguing conclusions: (a) In the rapidly glycolysing hepatoma 3924A of the Morris series (28), and in seminal vesicles, the total rate of anaerobic ATP production did not have to attain the aerobic level for the transport of Na^+ to continue largely unimpaired, suggesting that much of the ATP from oxidative phosphorylation was in excess of the requirements of transport. A similar result was obtained with liver slices from the 21 day rat fetus. (b) Conversely, the post-natal liver slices illustrate that a given rate of anerobic ATP production does not necessarily lead to commensurate transport activity.

These two points are shown rather more directly in a series of experiments with hepatoma 3924A (unpublished observations). Slices of the tumor were depleted of K^+ and loaded with Na^+ during incubation at 1°C for 90-120 minutes, and transport activity was then measured as the net uptake of K^+ or loss of Na^+ upon restoration of metabolically favorable conditions. A similar procedure was used in many of the experiments described below. In Figure 1b the slices were incubated with cyanide and glucose so that glycolysis was the sole source of ATP. Variations in the K^+ accumulation

TABLE 1. Na^+-transporting activity and calculated rate of ATP production under aerobic and anaerobic conditions. ATP production from oxidative phosphorylation was calculated by assuming $ATP/O_2 = 3$. In the case of ATP synthesis from glycolysis, ATP/lactate was taken to be 1.0 when glucose was the substrate, and 1.5 when endogenous glycogen was the substrate. Estimates of Na^+-transporting activity relate to rate constant for ^{24}Na efflux (ascites cells), maintenance of Na^+ cell content (seminal vesicles), or net extrusion of Na^+ as described for Figure 1 (hepatoma 3924A and rat liver).

Tissue	ATP (μM/h-tissue unit)		Tissue unit	Na^+ transport Anaerobic x100 Aerobic	Reference
	Aerobic	Anaerobic			
Ascites tumor	291.0	269.0	ml cells	85	26
Hepatoma 3924A	1.2	.8	mg dry wt	75	unpublished data
Seminal vesicle mucosa	1.4	.3	mg dry wt	100	6
Rat liver:					
Fetus 17-18 days	6.5	1.3	mg protein	47	31,32
Fetus 21 days	4.1	1.4		74	
Postnatal 1 day	5.0	0.5		0	
Adult	3.1	0.3		0	

are expressed as a percentage of that occurring under aerobic conditions in control slices, and are plotted against the rate of glycolysis. Variations in glycolytic rate were those found inherently in different samples of the tumor. (Similar unpublished results were obtained when the lactate production of rapidly-glycolysing tumor samples were reduced by titration with iodoacetate). There was a nearly linear dependence of the K^+ accumulation on the rate of glycolysis, but transport approached nil when the rate of glycolysis was still substantial - about 20% of maximal. Secondly, in Figure 1a the slices were incubated aerobically and with endogenous substrate only, so that glycolysis was negligible (cf. Table 2); variations in the rate of respiration were induced by different concentrations of cyanide. Plotting the net transport, which was entirely dependent on respiration, against the rate of respiration shows that the latter had to be inhibited substantially before the transport was affected; thus, as suggested above, a considerable fraction (in this case, upwards of 50%) of the energy released by respiration was indeed in excess of the requirements of transport in this tissue. Further, comparing the calculated rates of ATP production, we see that the rate of oxidative phosphorylation at the Q_{O_2} critical for maintenance of maximal K^+ transport, was 800-900 mM/kg fat-free dry weight/h (ffDW). This was very similar to the rate of ATP production from anaerobic glycolysis that was just sufficient to maintain the anaerobic transport at the aerobic level (100% on the ordinate of Fig. 1b). Thus, ATP was used with equal efficiency for the maintenance of transport, whether it was synthesized during glycolysis or oxidative phosphorylation.

The type of relation between transport and respiratory activity seen in Figure 1a was the first observed with liver slices (unpublished observations), in which, despite endogenous glycogen stores, are almost entirely dependent on respiration for the maintenance of net active

METABOLIC COUPLING OF ION TRANSPORT

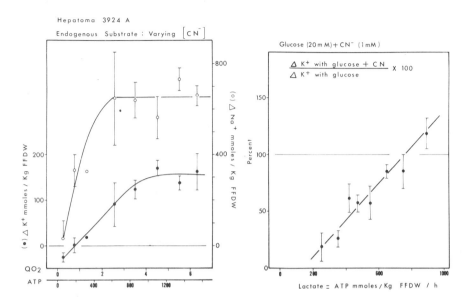

FIG. 1. Relation of net ion movements to (a) respiration and (b) anaerobic glycolysis in slices of hepatoma 3924A. Slices were incubated in the Warburg manometric apparatus for 90-120 min at 1°C, and then for 70 min at 38°C, in an oxygenated, phosphate-buffered Ringer's solution. In (a) the incubation vessels contained various concentrations of cyanide. In (b) all vessels contained 20 mM glucose + 1 mM cyanide, except for control slices that were incubated with glucose alone in order to determine the aerobic level of K^+ accumulation. The net gain of K^+ or loss of Na^+ (ΔK^+ or ΔNa^+) during the incubation at 38°C is plotted against (a) the rate of O_2 uptake (μl O_2/mg ffDW/h) or (b) the rate of lactate appearance in the incubation medium. Rates of ATP production in (a) are calculated from the Q_{O_2} values by assuming P/O = 3; they are expressed as μmoles ATP/kg ffDW/h.

movements of Na^+ and K^+. A series of experiments in which Amytal was used to vary the respiratory rate is illustrated in Figure 2 (cf. Fig. 4 for comparable experiments with cyanide). The rate of respiration had to be reduced below 6 μl O_2/mg protein/h before the total net transport was affected; and in other experiments the same was found to be true of the initial rate of the net transport. At lower rates of respiration the Na^+ extrusion fell off sharply and was not significant at an O_2 uptake of 2.5

µl/mg/h. The O_2 uptake persisting at this point was insensitive to inhibitors and was probably due to microsomal mixed-function oxidases (44); it may therefore be subtracted from the total respiration in order to get an estimate of the mitochondrial respiration which is alone coupled to oxidative phosphorylation. Then, at the critical point for onset of transport inhibition, the mitochondrial respiration was about 6.0 - 2.5 = 3.5 µl/kg protein/h. In contrast to the Na^+ extrusion, the ATP content of the slices fell almost linearly with the rate of respiration until it reached a steady, minimal value of about 0.5 mM/kg protein - a level presumably maintained by anaerobic glycolysis. Apparently, the total mitochondrial O_2 uptake was concerned in maintenance of tissue ATP levels, but much of it was in excess of that required to produce high-energy compounds for the support of transport.

These results suggest: (a) that the transport mechanism of liver slices is considerably protected against reduction of metabolic activity, at least when this is due to a specific decline of energy conserving reactions; (b) if one assumes that at least most of the total mitochondrial respiration is usefully employed to provide energy for metabolic processes, then transport can compete successfully with many other processes for the available high-energy compounds when the amount of the latter is limited.

High-energy intermediates of oxidative phosphorylation. That adenosine triphosphate can act as an energy donor for membrane transport has been thoroughly established, especially by experiments of the type in which ATP was directly introduced into various cell preparations (50). Several authors have considered the suggestion, originally made by Slater (42), that intramitochondrial high-energy intermediates of oxidative phosphorylation might also provide a direct source of energy for biochemical processes. Use of inhibitors of oxidative phosphorylation such as oligomycin allows this to be tested experimentally and several functions of isolated mitochondria, including accumulation of Ca^{2+} and K^+, are recognized to derive energy directly from the intermediates (43). The intermediates can be formed either: (a) from respiratory activity or (b) from extramitochondrial ATP derived from glycolysis in the intact cell) by reversal of part of the phosphorylation system (Fig. 3).

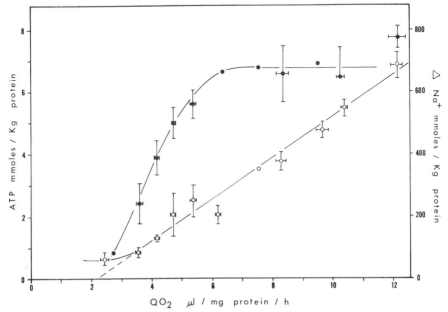

FIG. 2. Relation of Na^+ transport and ATP content to the varying rates of respiration induced by different concentrations of Amytal in rat-liver slices. The methods were as described in Figure 1a. At the end of the incubation period, the slices were put into 10% perchloric acid/40% ethanol at -20°C, homogenized and centrifuged. Protein was determined on the precipitate by a biuret method, while Na^+ and ATP were analyzed in the supernatant (the latter by an enzymic method).

In case (a), utilization of the intermediates to support a process is indicated if the process shows some insensitivity to inhibition by oligomycin, in the absence of a significant production of ATP from substrate-level phosphorylations. In work with intact cells, several aerobic processes involving ion transport in brain, kidney, ascites tumor, thyroid and gastric mucosa are completely inhibited by oligomycin, or other inhibitors of oxidative phosphorylation, and are therefore dependent on ATP (17,27,39,45,46,49). On the other hand, ion transport in crab muscle (2) and rat-liver slices (33) shows much resistance to oligomycin, and intermediates of oxidative phosphorylation were suggested to act as energy donors for the transport mechanisms. In neither of these papers was ATP measured, so that it was not possible to be sure that substrate-level ATP formation was reduced sufficiently to exclude it as a source of energy. The considerable insensitivity of transport

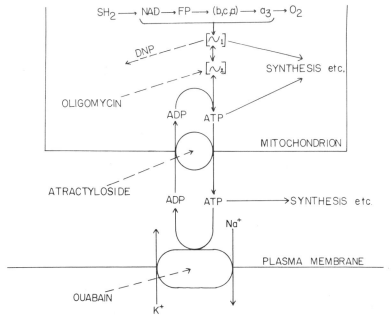

FIG. 3. Schematic representation of the relation between ion transport at the plasma membrane and intramitochondrial energy-conserving reactions.

in liver slices to reduced ATP levels (Fig. 2) indicate this to be a serious deficiency in the evidence. The early results of a current attempt to fill this gap are shown in Figure 4b, where the varying Q_{O_2} values on the abscissa were obtained by varying oligomycin concentrations between 1.25 and 125 µM. For comparison, Figure 4a shows results obtained with cyanide which were qualitatively similar to those with Amytal (Fig. 2). Oligomycin only caused a partial (40%) inhibition of net Na^+ extrusion. The inhibition of transport was first seen in slices with ATP contents below 2.7 mM/kg protein, 2.8 mM/kg in the presence of Amytal and 1.6 mM/kg in the presence of cyanide. However, the minimal ATP level seen in the presence of oligomycin was 2.1 mM/kg; this is well above the lowest levels with cyanide or Amytal (0.7 and 0.6 mM/kg), and a comparable level in the presence of these inhibitors is associated with the persistence of considerable net transport of Na^+. Therefore, the tentative conclusion is that the insensitivity to oligomycin of aerobic transport in the adult liver may be a consequence of the continued occurrence of low, but adequate, production of ATP from substrate-level phosphorylation rather than to derivation of energy from high-energy intermediates.

Incidentally, various aspects of Figure 4b indicate that oligomycin is acting predominantly as an inhibitor of oxidative phosphorylation in the slices, rather than as an inhibitor of the Na,K-ATPase of the membranes (contrast ref. 49). For example, with oligomycin as inhibitor, the slope of the line relating ATP decrease to decrease of O_2 uptake is almost double that seen with the two respiratory inhibitors, cyanide and Amytal; on the other hand, ouabain (1 mM) had no significant effect on slice ATP contents (Fig. 7).

In the case of transport postulated to be supported by intramitochondrial intermediates that are formed from extramitochondrial ATP, the necessary evidence is that the transport process should be inhibited by oligomycin despite the presence of normally adequate concentrations of ATP. Uncoupling agents (e.g. 2,4-dinitrophenol) should also inhibit because they activate the mitochondrial ATPase, and oligomycin should not relieve this inhibition by uncouplers (43). Reasonably complete evidence on these lines has been

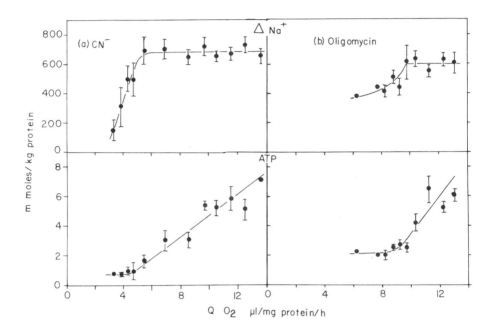

FIG. 4. Relation of Na$^+$ transport and ATP content to the varying rate of respiration induced by different concentrations of (a) cyanide and (b) oligomycin in rat-liver slices. Details as for Figures 1 and 2. In (b) all media contained 85 mM ethanol (the solvent for oligomycin), except for control slices which showed no difference from slices incubated with ethanol but without oligomycin.

provided for a few cases of transport, namely in fetal liver (33) and ascites cells (17) and also, incidentally, for protein synthesis in lymphocytes (18). However, in other cases such as Na$^+$ transport in turtle bladder (7) and the beating of heart cells (16), DNP inhibited the process and lowered ATP levels, but oligomycin did not inhibit. This would seem to indicate that both processes were exclusively dependent on ATP for their energy. However, in the turtle bladder DNP inhibited transport, although the ATP content dropped by only 25%, and on this basis it was postulated (7) than an extramitochondrial high-energy intermediate that is sensitive to DNP, but is not part of the oxidative phosphorylation sequence, was the source of energy for Na^+ transport. In this case, it clearly would be of interest to determine the level of ATP associated with maintenance of normal transport activity in the bladder, along the lines described above, as one way of testing the independence of turtle bladder Na^+ transport of ATP.

In general, conclusive evidence for the direct utilization of intramitochondrial high-energy intermediates as energy donors for transport is sparse, and the best documented cases at the moment seem to be those in which the intermediates are formed from ATP by the reversal pathway.

Cell Ca^{2+} transport. Most work on the metabolic coupling of ion transport has been done with the Na^+ and K^+ transporting activities of plasma membranes. Recently however, some attention has been paid to the metabolism-dependent extrusion of Ca^{2+} across the plasma membranes of some cells. In the case of red cells (40) and slices of liver and hepatoma 3924A (37 and unpublished data) this transport seems to occur by a mechanism which requires high-energy compounds and which is independent of the Na,K-transporting system. However, with cardiac muscle (30) and squid axon (3) the Ca^{2+} extrusion takes place by a hetero-exchange diffusion for Na^+. Sodium ions diffuse down their electrochemical gradient into the cells, and this movement is coupled to an outward movement of Ca^{2+}. Thus the energy for the outward Ca^{2+} movement is, in these cases, indirectly obtained from the concentration gradient set up by the activity of the Na^+ "pump".

CONTROL OF ENERGY METABOLISM BY TRANSPORT

Glycolysis. When cells are transporting ions by a mechanism deriving energy from glycolysis, inhibition of transport by ouabain is accompanied by 15-30% decline of the rate of glycolysis. This has been shown for red

TABLE 2. Effects of ouabain on K^+ accumulation, respiration and lactate production by slices of hepatoma 3924A. For details of procedure, see Figure 1. n = 8-12. Results of van Rossum et al. (unpublished).

Conditions	Net K^+ uptake (mM/kg ffDW)	O_2 uptake (μl/mg ffDW/h)	Lactate production (mM/kg ffDW)
Aerobic:			
Endogenous substrate	154 ± 24	5.5 ± 0.3	-3.0 ± 3.9
+ Ouabain (1 mM)	-53 ± 14	4.0 ± 0.3	6.8 ± 5.5
Cyanide (2 mM):			
Glucose (20 mM)	86 ± 18		470 ± 29
+ Ouabain (1 mM)	-61 ± 8		304 ± 25

cells (12,48), frog muscle (11), and ascites cells (15); and a similar result with hepatoma 3924A is shown in Table 2, where ouabain inhibited transport of K^+ completely in the presence of glucose plus cyanide, and reduced lactate production by 30%. It is believed that alterations in the levels of ATP and ADP (arising from the changing activity of the transport mechanisms) mediate alterations in the rate of glycolysis and thus permit a control of the energy-providing system by the processes utilizing ATP. From measurements of the changes in levels of glycolytic intermediates when red cell transport is stimulated, the rate-limiting step seems to be due to the requirement of ADP as substrate for 3-phosphoglycerate kinase (12). However, in the electric organ of *Electrophorus*, glycogenolysis and phosphofructokinase seem to be the rate-limiting steps of glycolysis that are activated by the onset of ion transport (8).

Respiration. In a similar way, the rate of respiration of tissues which are dependent on oxidative phosphorylation for the support of ion transport, is partially controlled by the rate of transport. This principle has been established by two types of experiment. In some cases the rate of the respiration so controlled has been equated to the energy-requirement of the transport system, but in each type of experiment there are considerations which suggest that such a comparison must be made with caution.

(a) The rate of respiration of actively transporting tissue samples is compared with that of samples incubated with ouabain, or in potassium free or sodium free medium (1,13,24,29,47). In Table 2, for example, ouabain reduced the rate of respiration in slices of hepatoma 3924A by 30%. The snag with this approach is that the non-transporting cells inevitably have a lowered K^+ content, and this can itself adversely affect the respiration independently of an effect on transport. For example, the inhibition of the respiration of various tissues (4,29) including rat liver (38) by ouabain is partially relieved by increasing the K^+ concentration of the medium (Table 3) although the transport was not activated. In this table the rate of the respiratory fraction actually controlled by the transport mechanism is maximally 0.8 µl/mg/h and not 2.4 µl/mg/h. Other experiments suggest that the direct effect of K^+ on respiration (independently of its effect on transport) is related to an activation of the citric acid cycle (38).

(b) The second type of experiment is applied in tissues where some form of stimulation (e.g. electrical, hormonal) causes an increase of transporting activity above the resting level (1,5,29,45). This stimulation is accompanied by an increased rate of respiration which is inhibited by ouabain. This method avoids large changes in cellular cation composition, but may be subject to the possibility (as may experiments of type a) that the stimulated transport activity withdraws ATP from other reactions, especially in view of the apparently high affinity of the transport process for ATP (Figs. 2 and 4). The stimulated respiration would then underestimate the

TABLE 3. Effect of ouabain on the respiration of rat-liver slices in media of varying K^+ and Na^+ concentrations. Slices were incubated from 90 min at 1°C and 70 min at 38°C in phosphate-buffered media in which varying amounts of K^+ were substituted for Na^+. n = 8.

Medium [Na^+] (mM)	[K^+]	Q_{O_2} (μl/mg ffDW/h) Control	+ Ouabain (0.75mM)	Δ
161.5	5	9.2 ± 0.5	6.8 ± 0.4	-2.4
141.5	25	9.1 ± 0.7	7.7 ± 0.6	-1.4
91.5	75	10.1 ± 0.6	9.3 ± 0.5	-0.8

respiratory requirement of transport. Some indication of this has been obtained with acid secretion by gastric mucosa, where under certain conditions onset of acid secretion was accompanied by negligible changes of O_2 uptake (39).

Whittam (47) related such findings to the observations of Chance & Williams (10) on respiratory control in isolated mitochondria, suggesting that ADP released at the transporting site enters the mitochondria and so influences the oxidative phosphorylation activity. In support of this, it has been shown that the O_2 uptake of homogenates, or reconstituted mixtures of mitochondria and microsomes, is partly related to extramitochondrial ATPase activities (4,23) although there was no direct demonstration of the passage of adenine nucleotides across the mitochondrial membranes. Important though such work is, one would also like to obtain relevant evidence directly from intact cells. The following experiments represent two possible approaches.

<u>Measurement of redox levels of electron carriers</u>. Alterations in the rate of respiration of isolated mitochondria due to the presence (state 3) or absence (state 4) of ADP, are accompanied by characteristic changes in

the percent reduction of electron carriers of the respiratory chain (10). Such changes can be followed in preparations of whole tissues by observing changes of light absorption (for cytochromes) or fluorescence (for nicotinamide nucleotides) (9,19). Figure 5 illustrates an experiment on fluorescence

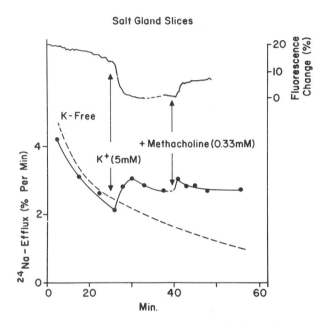

FIG. 5. ^{24}Na efflux, and fluorescence of NAD(P)H in slices of herring-gull salt gland. Slices were incubated for 120-180 min at 1°C in K$^+$-free, phosphate-buffered Ringer's solution containing ^{24}Na. Two slices were then placed in the slice-holder of the fluorometer described by Chance et al. (38) and were washed with a continuous stream of the oxygenated medium at 30°C without ^{24}Na. Effluents were collected at intervals and their radioactivity determined. The control slice (dashed line) was washed in K$^+$-free medium throughout; the experimental slice was treated as indicated. The fluorescence trace represents the difference between the fluorescence of the two slices; a downward movement of the trace represents a decline of fluorescence in the experimental slice, indicative of an oxidation of NAD(P)H.

changes in a slice of avian salt gland. Addition of K$^+$ stimulated efflux of ^{24}Na, oxidized NAD(P)H and reduced cytochromes c and a (not shown). These changes of the electron carriers are characteristic of a transition towards state 3; they were reversed by ouabain in other experiments suggesting a return towards state 4 upon inhibition of transport. It is not to be expected that the extreme states 3 and 4 obtained in isolated mitochondria will be achieved in whole liver cells because of the continual utilization of high energy compounds for a wide variety of reactions. Inhibition or activation of ion transport will only move the metabolic state of the intracellular mitochondria somewhat closer to one or other of the extremes. Results consistent with similar transitions between states 3 and 4 upon initiation or inhibition of ion transport have been obtained with frog muscle (when treated with insulin or lactate [21]) and rat-liver slices (35). The results obtained in such experiments with whole cells are not always as simple to interpret as are comparable experiments with isolated mitochondria, as indeed one may anticipate from the fact that the cell contains many more interacting processes. In particular, addition of K^+ to salt gland or liver slices often caused either little change or even a reduction of NAD(P). These are, of course, experiments of type a (see above) and the varying response of NAD(P) can be accounted for if one considers that K^+ appears to activate the substrate-level oxidation by the citric acid cycle, in addition to inducing state 3 in the respiratory chain.

Experiments of type b can also be done with the salt gland slices (34,36), since the organ is stimulated to secrete Na^+, with increased O_2 uptake, by cholinergic agents (5). In Figure 5, methacholine stimulated Na^+ efflux somewhat and this was accompanied by a **reduction** of NAD(P) (Figs. 5,6). There was also a reduction of cytochromes c and a, compatible with a transition towards state 3, which was reversed by ouabain (Fig. 6). However, the reduction of NAD(P) induced by K^+ again implies an activation of substrate-level oxidations which was not directly due to the transport mechanism since it was not reversed by ouabain.

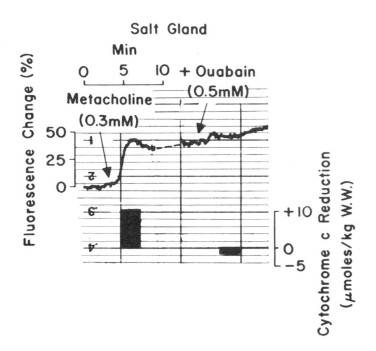

FIG. 6. Effects of methacholine and ouabain on the oxido-reduction level of NAD(P)H and cytochrome c, in slices of herring-gull salt gland. Details as for Figure 5, except that Ringer's containing 5 mM K^+ was used throughout. The cytochrome c values indicated by the blocks, were determined by placing the slice-holder in a split-beam spectrophotometer temporarily, and determining the different spectra between experimental and control slices at 500-630 nm. The height of the blocks represents the change of reduced cytochrome c content compared to the initial level before addition of methacholine. Changes of cytochrome a were qualitatively similar to those of c (34).

In general, these experiments on the electron carriers support Whittam's view that alterations in transport activity at the plasma membrane are reflected in transitions of the mitochondria between states 3 and 4. But additional factors at the substrate level also affect respiration, apparently by a mechanism which is only incidentally related to the occurrence of transport itself.

Atractyloside. A likely tool for examining the importance of the actual transfer of adenine nucleotides between mitochondria and the cytosol is atractyloside, which specifically inhibits the ADP-ATP exchange mechanism of the inner mitochondrial membrane (22). The few studies so far reported have shown variable effects of atractyloside on Na^+ and K^+ transport processes in crab muscle, gastric mucosa and thyroid tissue (2,39,46). In preliminary experiments with liver slices, the following points have been seen (Fig. 7): (a) At 50-100 μM atractyloside, respiration was somewhat inhibited but the slice ATP content was raised. (b) At higher concentrations, respiration was reduced by 40% and ATP was still at the control levels, in contrast to the substantial fall of ATP content seen when a comparable inhibition of respiration was caused by respiratory inhibitors or oligomycin (Figs. 2 and 4). (c) Despite the normal ATP level, Na^+ extrusion was slightly but significantly reduced. These results are consistent with a partial reduction in the entry of ADP into the mitochondrial matrix and with some limitation of the access of ATP to the transport mechanism - possibly by retention within the mitochondria. However, Na^+ extrusion continued at 80% of the control level. The effect of atractyloside on the K^+-induced stimulation of respiration in liver slices is shown in Figure 8. At 0.2 mM, atractyloside greatly reduced the effect of K^+ on respiration, but transport was unaffected and the residual stimulation of respiration by K^+ was roughly equal to the normal, transport-coupled respiration. These effects clearly require further study, but the present indication is that while atractyloside affects mitochondrial functions in the cells, the interaction between oxidative phosphorylation and membrane transport is relatively insensitive to the ADP-ATP transport system of the mitochondria. Thus, there must be a

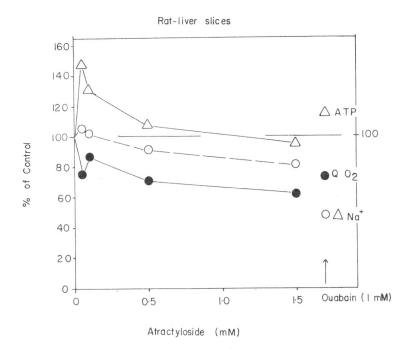

FIG. 7. Effect of different concentrations of atractyloside on Na^+ extrusion, O_2 uptake and final ATP content, of rat-liver slices. Details of methods as for Figures 1 and 2. For comparison, results obtained with ouabain (1 mM) in the same experiments are included on the extreme right where the meaning of the symbols is also indicated. Each point is the mean of 4 observations.

substantial amount of atractyloside-insensitive nucleotide transfer into and out of the mitochondria, probably permitted by the competitive nature of the inhibition of the translocase by atractyloside (22). It is worth noting that Kemp (20) has shown a considerable efflux of ATP can occur from liver mitochondria incubated with atractyloside, provided that phosphate (1 mM or more) is present in the extra mitochondrial medium. However, no entry of ADP occurred in the presence of atractyloside.

Whatever the precise mechanism of this coupling, the fact that the respiratory rate is partly linked to the ion-transporting activity clearly

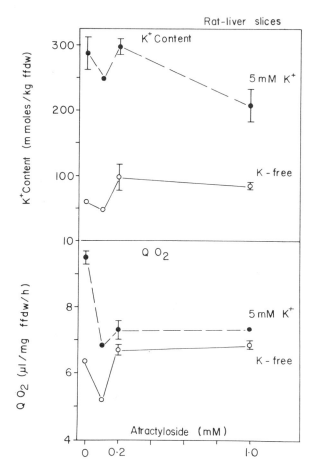

FIG. 8. Effect of atractyloside on the respiration and K^+ content of rat-liver slices incubated with Ringer's solutions containing either no K^+ or 5 mM K^+. Details as for Figure 1, except that the tissue K^+ values shown represent the K^+ contents after the incubation at 38°C was finished. At the onset of this incubation, the slices of K^+-free medium contained 59 mM K^+/kg ffDW, and those in the 5 mM K^+ medium contained 75 mM/kg.

provides a means by which energy provision in the cells can be geared to the energy requirements. An interesting suggestion recently made (14) is that a rapid rate of alkali-cation transport (in response to a hormonally-induced increase of cell-membrane permeability) and of the respiration coupled to it may be a mechanism for thermogenesis in brown fat cells.

Energy requirements for ion transport. The expenditure of respiratory energy necessary to support the active transport of Na^+ can be considered at three levels. This will be exemplified by results with liver cells (Table 4).

a) The minimal respiratory requirement would be that which would provide sufficient energy to fulfill the thermodynamic requirements of the osmotic and electrical work done in moving ions across the plasma membrane at the observed rate. The O_2 uptake thus required can be calculated (making several assumptions) to be either 0.8 or 1.2 µl O_2/mg ffDW/h, depending on whether one assumes a transport mechanism which requires an obligatory 1:1 coupling of Na^+ to K^+ movements, or one which involves solely passive movements of K^+ secondarily to the Na^+ transport. In fact, there is evidence for a substantial degree of coupling in liver cells (13,35).

b) A second definition would take into account the likelihood that the energy of hydrolysis of ATP is not entirely converted into transport work, i.e. that the "pump" is not 100% efficient. From results with frog muscle, Dydynska and Harris (11) calculated that the transport work actually performed corresponds to about 5 kcal/M of ATP used. If the same holds in liver, the transport mechanism would work at an efficiency of 50%, and the O_2 uptake actually required to fulfill the energy requirements of the observed rate of Na^+ transport would be 1.5 - 2.3 µl/mg-h. This value can be compared to the O_2 uptake which is directly controlled by the transporting activity. This is in the range 0.7 - 1.2 µl/mg-h (38) and is thus somewhat less than the O_2 uptake calculated to be required by the coupled Na-K pump (1.5 µl/mg-h). Taken at their face value these numbers would imply that Na^+ transport draws part of its energy from high-energy sources that would otherwise be utilized by other processes.

TABLE 4. Estimates of the O_2 consumption required to support Na^+ extrusion by slices of rat liver. The energy requirement is calculated from either

$$\text{equation (1)} \quad E = QRT \cdot \ln \frac{[Na]_o[K]_i}{[Na]_i[K]_o} \quad \text{or,}$$

$$\text{equation (2)} \quad E = QRT \cdot \ln \frac{[Na]_o}{[Na]_i} + E_mF$$

where E = rate of energy production required,
Q = the initial rate of net Na^+ extrusion at 38°C = 0.815 M/kg ffDW/h, and
E_m (for the membrane potential in liver slices = -35mV.) (25).
In converting the energy requirement into the equivalent O_2 uptake, the following values were assumed; P/O = 3 and
ΔG' (for the terminal phosphate bond of ATP under the conditions pertaining within the liver slices) = -11 kcal/M (38).

	Energy required Kcal/kg ffDW/h		Equivalent O_2 uptake μl/mg ffDW/h	
	Eqn. 1	Eqn. 2	Eqn. 1	Eqn. 2
Calculated energy requirement				
Minimum required for transport work.	2.0	3.2	0.8	1.2
Required if ATP ⟶ transport is 50% efficient.	4.0	6.4	1.5	2.3
Observed O_2 uptakes				
O_2 "coupled" to transport.			0.7 to	1.2
Minimal O_2 required to maintain transport in presence of respiratory inhibitors.			2.4 to	2.9

c) Conversely, one also has to consider the possibility that there are some reactions which have a higher affinity for high-energy compounds than the transport system. Therefore, a third definition of the energy requirement would be the minimal respiratory activity required to maintain transport in the whole cell in the presence of other reactions competing for high-energy compounds. The critical rate of mitochondrial respiration within the slices, at which transport just starts to be inhibited in experiments of the type seen in Figures 2 and 4a, would seem to be an estimate of this. In six series of experiments, done with different substrates and inhibitors, the critical Q_{O_2} was in the range 2.4 - 2.9 µl/mg ffDW/h. This is higher than the rate calculated to be required in paragraph b, and seems to suggest that in addition to transport other energy-requiring processes continue at these low rates of respiration. Another indirect indication of the occurrence of systems with a greater affinity for ATP is provided by the observation that in adult liver slices (Table 1) and in hepatoma 3924A (Fig. 1), a substantial rate of anaerobic glycolysis was unaccompanied by a commensurate ion-transport. Thus, the O_2 uptake required to maintain transport in the living cell must include that required to maintain the other reactions which are apparently less sensitive to a paucity of energy.

Obviously, many of the numberical values I have just quoted are to be taken cum grana salis. But the reasoning is valid and awaits the availability of more precise, quantitative results in order that more definite conclusions may be drawn. The finding of a critical rate of respiration above which ion transport is not inhibited suggests that a similar situation may hold for other energy-requiring processes of the liver cell, and that their respective critical respiratory rates may differ. For example, Scholz et al. (41) have noted different responses of gluconeogenesis and urea synthesis to titration of perfused liver with Amytal. There is thus the possibility of establishing a "pecking" order of cellular functions, which would be determined by such biochemical factors as the relative apparent K_m's for ATP and the concentrations of the enzymes, and which might well have a survival value under conditions in which energy production is limited.

REFERENCES

1. BAKER, P. F., and C. M. CONNELLY. Some properties of the external activation site of the sodium pump in crab nerve. J. Physiol. 185: 270-297, 1966.

2. BITTAR, E. E. Effect of inhibitors and uncouplers on the Na pump of the Maia muscle fibre. J. Physiol. 187: 81-103, 1966.

3. BLAUSTEIN, M. P., and A. L. HODGKIN. The effect of cyanide on calcium efflux in squid axons. J. Physiol. 198: 46P-48P, 1968.

4. BLOND, D. M., and R. WHITTAM. The regulation of kidney respiration by sodium and potassium ions. Biochem. J. 92: 158-167, 1964.

5. BORUT, A., and K. SCHMIDT-NIELSEN. Respiration of avian salt-secreting gland in tissue slice experiments. Amer. J. Physiol. 204: 573-581, 1963.

6. BREUER, H., and R. WHITTAM. Ion movements of seminal vesicle mucosa. J. Physiol. 135: 213-225, 1957.

7. BRICKER, N. S., and S. KLAHR. Effects of dinitrophenol and oligomycin on the coupling between anaerobic metabolism and anaerobic sodium transport by the isolated turtle bladder. J. Gen. Physiol. 49: 483-499, 1966.

8. CHANCE, B. Biochemical studies of transitions from rest to activity. In: Sleep and Altered States of Consciousness. XIV, edited by S. S. Kety, E. V. Evarts, and H. L. Williams. Baltimore: Williams & Wilkins Co., 1967, p. 48-63.

9. CHANCE, B., P. COHEN, F. JOBSIS, and B. SCHOENER. Intracellular oxidation-reduction states in vivo. Science 137: 499-508, 1962.

10. CHANCE, B., and G. R. WILLIAMS. Respiratory enzymes in oxidative phosphorylation. III. The steady state. J. Biol. Chem. 217: 409-427, 1955.

11. DYDYNSKA, M., and E. J. HARRIS. Consumption of high-energy phosphates during active sodium and potassium interchange in frog muscle. J. Physiol. 182: 92-109, 1966.

12. ECKEL, R. E., S. R. RIZZO, H. LODISH, and A. B. BERGGREN. Potassium transport and control of glycolysis in human erythrocytes. Amer. J. Physiol. 210: 737-743, 1966.

13. ELSHOVE, A., and G. D. V. VAN ROSSUM. Net movements of sodium and potassium, and their relation to respiration, in slices of rat liver incubated in vitro. J. Physiol. 168: 531-553, 1963.

14. GIRARDIER, L., J. SEYDOUX, and T. CLAUSEN. Membrane potential of brown adipose tissue. A suggested mechanism for the regulation of thermogenesis. J. Gen. Physiol. 52: 925-940, 1968.

15. GORDON, E. E., and M. DE HARTOG. Valinomycin-stimulated glycolysis in Ehrlich ascites tumor cells. Biochim. Biophys. Acta 162: 220-229, 1968.

16. HARARY, I., and E. C. SLATER. Studies in vitro on single beating heart cells. VIII. The effect of oligomycin, dinitrophenol and ouabain on the beating rate. Biochim. Biophys. Acta 99: 227-233, 1965.

17. HEMPLING, H. G. Sources of energy for the transport of potassium and sodium across the membrane of the Ehrlich mouse ascites tumor cell. Biochim. Biophys. Acta 112: 503-518, 1966.

18. JARETT, L., and D. M. KIPNIS. Differential response of protein synthesis in Ehrlich ascites tumor cells and normal thymocytes to 2,4-dinitrophenol and oligomycin. Nature 216: 714-715, 1967.

19. JOBSIS, F. F. Spectrophotometric studies on intact muscle. I. Components of the respiratory chain. J. Gen. Physiol. 46: 905-928, 1963.

20. KEMP, A. Onderzoekingen Betreffende Gefosforyleerde Verbindingen in Mitochondrien en hun Relatie Tot de ATP Synthese. (Ph.D. Thesis). University of Amsterdam, Amsterdam, 1968.

21. KERNAN, R. P. Spectroscopic studies of frog muscle during sodium uptake and excretion. J. Physiol. 169: 862-878, 1963.

22. KLINGENBERG, M., and E. PFAFF. Structural and functional compartmentation in mitochondria. In: Regulation of Metabolic Processes in Mitochondria, edited by J. M. Tager, S. Papa, E. Quagliariello, and E.C. Slater. New York: Elsevier, 1966, p. 180-201.

23. LANDON, E. J. Interaction of mammalian kidney membrane and mitochondria in vitro. Biochim. Biophys. Acta 143: 518-521, 1967.

24. LEVINSON, C., and H. G. HEMPLING. The role of ion transport in the regulation of respiration in the Ehrlich mouse ascites-tumor cell. Biochim. Biophys. Acta 135: 306-318, 1967.

25. LI, C.-L., and H. MCILWAIN. Maintenance of resting membrane potentials in slices of mammalian cerebral cortex and other tissues in vitro. J. Physiol. 139: 178-190, 1957.

26. MAIZELS, M., M. REMINGTON, and R. TRUSCOE. Metabolism and sodium transfer of mouse ascites tumor cells. J. Physiol. 140: 80-93, 1958.

27. MINAKAMI, S., K. KAKINUMA, and H. YOSHIKAWA. The control of respiration in brain slices. Biochim. Biophys. Acta 78: 808-811, 1963.

28. MORRIS, H. P., and B. P. WAGNER. Induction and transplantation of rat hepatomas with different growth rate (including "minimal deviation" hepatomas). In: Methods in Cancer Research IV, edited by H. Busch. New York: Academic Press, 1968, p. 125-152.

29. RANG, H. P., and J. M. RITCHIE. The dependence on external cations of the oxygen consumption of mammalian non-myelinated fibres at rest and during activity. J. Physiol. 196: 163-181, 1968.

30. REUTER, H., and SEITZ, N. The dependence of calcium efflux from cardiac muscle on temperature and external ion composition. J. Physiol. 195: 451-470, 1968.

31. VAN ROSSUM, G.D.V. Net sodium and potassium movements in liver slices prepared from rats of different foetal and post-natal ages. Biochim. Biophys. Acta 74: 1-14, 1963.

32. VAN ROSSUM, G.D.V. Respiration and glycolysis in liver slices prepared from rats of different foetal and post-natal ages. Biochim. Biophys. Acta 74: 15-23, 1963.

33. VAN ROSSUM, G.D.V. The effect of oligomycin on net movements of sodium and potassium in mammalian cells in vitro. Biochim. Biophys. Acta 82: 556-571, 1964.

34. VAN ROSSUM, G.D.V. Measurements of respiratory pigments and sodium efflux in slices of avian salt-gland. Biochem. Biophys. Res. Commun. 15: 540-545, 1964.

35. VAN ROSSUM, G.D.V. Effects of potassium, ouabain and valinomycin on the efflux of $^{24}Na^+$ and pyridine nucleotides of rat-liver slices. Biochim. Biophys. Acta 122: 323-332, 1966.

36. VAN ROSSUM, G.D.V. Relation of the oxidoreduction level of electron carriers to ion transport in slices of avian salt gland. Biochim. Biophys. Acta 153: 124-131, 1968.

37. VAN ROSSUM, G.D.V. Net movements of calcium and magnesium in slices of rat liver. J. Gen. Physiol. 55: 18-32, 1970.

38. VAN ROSSUM, G.D.V. On the coupling of respiration to cation transport in slices of rat liver. Biochim. Biophys. Acta 205: 7-17, 1970.

39. SACHS, G., R. H. COLLIER, R. L. SHOEMAKER, and B. I. HIRSCHOWITZ. The energy source for gastric H^+ secretion. Biochim. Biophys. Acta 162: 210-219, 1968.

40. SCHATZMANN, H. J. ATP-dependent Ca^{++}- extrusion from human red cells. Experientia 22: 364-365, 1966.

41. SCHOLZ, R., E. SCHWARZ, and T. BUCHER. Barbiturate und energie liefernder Stoffwechsel in der hamoglobinfrei durchstromten Leber der Ratte. Z. Klin. Chem. 4: 179-189, 1966.

42. SLATER, E. C. Mechanism of phosphorylation in the respiratory chain. Nature 172: 975-978, 1953.

43. SLATER, E. C., and H. F. TER WELLE. Applications of oligomycin and related inhibitors in bioenergetics. In: 20 Colloquium der Gesellschaft fur Biologische Chemie. Berlin: Springer-Verlag, 1970, p. 258-278.

44. THURMAN, R. G., and R. SCHOLZ. Mixed function oxidation in perfused rat liver. The effect of aminopyrine on oxygen uptake. European J. Biochem. 10: 459-467, 1969.

45. TOBIN, R. B., and H. MCILWAIN. The effect of oligomycin on the respiration and glycolysis of electrically-stimulated brain slices. Biochim. Biophys. Acta 105: 191-192, 1965.

46. TYLER, D. D., J. GONZE, F. LAMY, and J. E. DUMONT. Influence of mitochondrial inhibitors on the respiration and energy-dependent uptake of iodide by thyroid slices. Biochem. J. 106: 123-133, 1968.

47. WHITTAM, R. The dependence of the respiration of brain cortex on active cation transport. Biochem. J. 82: 205-212, 1962.

48. WHITTAM, R., M. E. AGER, and J. S. WILEY. Control of lactate production by membrane adenosine triphosphatase activity in human erythrocytes. Nature 202: 1111-1112, 1964.

49. WHITTAM, R., K. P. WHEELER, and A. BLAKE. Oligomycin and active transport reactions in cell membranes. Nature 203: 720-724, 1964.

50. WHITTAM, R., and K. P. WHEELER. Transport across cell membranes. Ann. Rev. Physiol. 32: 21-60, 1970.

CREDITS

The editors and authors are grateful to the following publishers, organizations and individuals for permission to use materials previously published:

Figure 6: Academic Press, Inc.; G. D. V. van Rossum, Biochemical Biophysical Research Communications 15: 540-545, 1964. Copyright Academic Press, Inc., 1964.

Table 4: Elsevier Publishing Company, Amsterdam, The Netherlands; G. D. V. van Rossum, Biochimica Biophysica Acta 205: 409-413, 1954.

TEMPERATURE ADAPTATION OF TRANSPORT PROPERTIES IN POIKILOTHERMS

M. W. SMITH

Department of Physiology
A.R.C. Institute of Animal Physiology
Babraham, Cambridge, Great Britain

There is much to be gained in the general field of membrane transport from drawing analogies and making comparisons between tissues and between species. The toad bladder has, for instance, served as a useful model in showing how aldosterone exerts its effect on sodium transport in other epithelial tissues (6,9,23) and the identification of different amino acid transport systems in Ehrlich cells (5,21) has helped in the analysis of other carrier-mediated transport processes in the intestinal epithelium (7,17,19). Similarly, the comparison of cation transport systems in the tissues of animals showing a high or low resistance to changes of temperature has been valuable in showing how animals might maintain their cellular ionic gradients during hibernation (33). The Na,K-activated adenosine triphosphatase (Na,K-ATPase) of homogenates prepared from the hamster kidney, for example, operates at temperatures approaching 0° C (34) while other preparations of Na,K-ATPase from non-hibernating animals do not (3,12,31,34). This enzyme system, which is generally held responsible for the active transfer of Na and K across cell membranes (24), also operates at low temperatures when prepared from cold-adapted goldfish (26) and frogs (2). There is then some reason to press the view that a study of transport properties in cold-adapted poikilotherms will give information relevant to hibernation. The ability of some poikilotherms to change their transport systems to suit different environmental temperatures (28) also has its parallel in studies of hibernation. The K concentration of hibernating hamster kidneys, for instance, is greater than that of the awake hamster (32) and the cold-resistance of the Na,K-ATPase prepared from hedgehog brain is said to increase during hibernation (4). Further examples of how preparation for hibernation affects transport are given by Willis (34).

There is, however, one major difference which should not be forgotten when making such comparisons. The adapting poikilotherm changes its transport properties in order to pursue a normal existence at different body temperatures whereas the hibernator does not. One might suppose that the cold-adapted poikilotherm will show, possibly in an exaggerated form, peculiarities of transport similar to those found in the hibernating animal. However it is quite possible that a qualitatively different response will be needed if the animal is to survive and remain active at low body temperatures. This is particularly likely to occur in tissues which perform more than one function. Take as one example the intestinal absorption of nutrients. The cessation of feeding during hibernation will allow both active and passive movements of molecules to be inhibited in the intestinal mucosa, if this increases its ability to maintain normal ionic gradients, and there is some evidence that amino acid transport shuts off in this way in the tortoise intestine during hibernation (11). A similar type of regulation in the cold-adapted poikilotherm would be less likely to succeed, for here ways must be found to continue amino acid absorption at low temperatures and at the same time maintain something approaching normal ionic gradients within the intestinal epithelium. Provided these reservations are taken into account, however, there is still reason to believe that knowledge gained from the study of transport in poikilotherms will help in understanding more about hibernation.

Future reference to adapting transport properties will be confined mainly to those concerned with amino acid movement across the goldfish intestine; how these can be modified through changes in environmental temperature and how a model membrane system can be constructed to mimic certain of the changes seen in the biological tissue. Transport studies were carried out by incubating everted intestinal sacs *in vitro* in the presence of different amino acids. The amounts of amino acids transferred by the intestine were determined by Moore and Stein (18) analysis of solutions taken from the serosal side of sacs at the end of incubation. The technique for preparation and use of everted sacs has been described in detail elsewhere (25). Preliminary experiments showed that amino acid transport took place into the serosal solution at a constant rate following an initial 20 minute lag period (27). The experiments involved switching cold-adapted fish to warm water and observing the changes which take place in amino acid transport. Some mental acrobatics are therefore needed to imagine what would have happened if the change had been in the opposite direction.

The transport of three amino acids across intestines taken from cold-adapted fish, before and 20 days after placing them in water at 30° C, is shown in Table 1. The net transfer of all three amino acids fell but the fall was only significant for valine. The extent to which methionine and phenylalanine were concentrated in the serosal solutions did not change on adaptation but that for valine did. Subsequent work showed the fall in net transport to be partly due to a fall in water and sodium transport which also occurred when cold fish were adapted to warm water. In view of this the ability of the intestine to concentrate amino acids at the serosal surface was chosen to judge whether selective changes in transport had taken place. An "adaptation index" was calculated for individual amino acids by dividing the final concentration reached in the serosal compartment

TABLE 1. Amino acid transport by everted sacs of goldfish intestine. Intestines from 16° C-adapted fish were used before and after placing fish in water at 30° C.

Amino acid	Serosal transfer μmoles/2h/g intestine		t	P
	Day 0	Day 20		
Valine	4.6 ± 0.8	1.8 ± 0.4	2.9	<0.05
Methionine	5.9 ± 1.1	3.3 ± 0.1	2.0	=0.10
Phenylalanine	5.7 ± 1.3	3.2 ± 0.2	1.6	>0.10
	Serosal concentration mM			
Valine	2.5 ± 0.2	1.7 ± 0.1	3.5	<0.02
Methionine	3.1 ± 0.2	2.7 ± 0.1	1.3	>0.20
Phenylalanine	3.0 ± 0.3	2.6 ± 0.1	0.9	>0.40

Initial concentration 0.5 mM. Means ± S.E.

after incubation with warm-adapted intestines, by that reached after transport across cold-adapted intestines. An adaptation index of 1.0 showed that the transport properties being considered had remained constant. Experiments were then undertaken to determine a whole range of adaptation indices for different amino acids. The results are summarized in Figure 1. Fish had been adapted to four different temperatures and the intestinal sacs incubated at 30°C on all occasions. The amino acids divided into two groups. The transport of the largest group (Pro, Ser, Val, Ala, Arg, Lys, Ile and Thr) changed on adaptation while that of the smaller group (His, Phe, Met, Tyr and Leu) did not. Tyrosine and leucine might be considered to span the gap between these two groups but they were thought, on balance, to be more closely associated with the smaller group of amino acids. Adaptation might be due to the intestine failing at high temperatures to concentrate those amino acids whose transport was poor under all conditions. That this

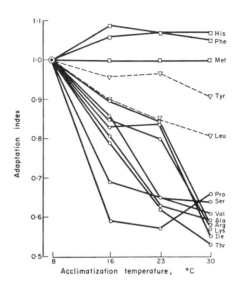

FIG. 1. Effect of temperature acclimatization on the ability of goldfish intestine to concentrate naturally occurring amino acids at its serosal surface. Amino acids were used at an initial concentration of 0.5 mM.

was not so is shown in Table 2. The final serosal concentrations for histidine (1.86 mM) and threonine (1.80 mM) were virtually identical, yet the transport of histidine remained unchanged on raising the environmental temperature while that for threonine fell by half. The same was true for the more actively concentrated amino acids.

The final concentration reached by an amino acid in the serosal solution bathing the intestine is determined both by active transport into the mucosal cell and passive transfer out. The efflux of amino acids from the mucosal cell may be by passive diffusion or by some carrier-mediated process (20). Any of these processes could change on adaptation. However, there

TABLE 2. Serosal concentrations of amino acids obtained *in vitro* using intestines taken from goldfish adapted to different temperatures.

Amino acid	Final serosal concentration (mM)	
	8° C adapted fish	30° C adapted fish
Threonine	1.80	0.97
Serine	2.36	1.52
Proline	1.50	0.99
Alanine	2.04	1.19
Valine	2.91	1.76
Methionine	2.72	2.72
Isoleucine	2.36	1.30
Leucine	3.13	2.52
Tyrosine	2.34	2.12
Phenylalanine	2.47	2.60
Histidine	1.86	1.99
Lysine	2.66	1.52
Arginine	2.73	1.58

was in the literature no reference to carrier systems for amino acids having the specificity shown in the present experiments. A search was therefore made for ways in which the passive transfer of amino acids could be regulated.

Amino acids can be separated chemically by virtue of their differential solubility in organic solvents (8) and by their binding characteristics to ion-exchange resin (18). Each of these properties was examined in turn to see if it could provide an answer to the changing pattern of amino acid transport shown by the goldfish intestine. The order of elution of amino acids from Dowex 50, a polysulphonic resin, is shown in Table 3. The amino acids whose transport properties remained unchanged by adaptation formed a compact group between the less tightly bound neutral amino acids and the basic amino acids. Isoleucine intruded into this grouping. A similar, though less complete, grouping based on ion-binding characteristics is shown by different glass electrodes in Figure 2. The numbers along the abscissa refer to different glasses, those on the left having a higher surface charge density those on the right. Glycine and alanine can be

TABLE 3. The relation between adaptation of transport and ion-binding properties of different amino acids.

Order of elution from Dowex 50	Adaptation index	Classification of amino acid
Threonine	0.54	Adaptor
Serine	0.64	Adaptor
Proline	0.66	Adaptor
Glycine	0.54	Adaptor
Alanine	0.58	Adaptor
Valine	0.60	Adaptor
Methionine	1.00	Non-adaptor
Isoleucine	0.55	Adaptor
Leucine	0.81	Non-adaptor
Tyrosine	0.91	Non-adaptor
Phenylalanine	1.05	Non-adaptor
Histidine	1.07	Non-adaptor
Lysine	0.57	Adaptor
Arginine	0.58	Adaptor

distinguished from phenylalanine and histidine by these glasses provided the surface charge density is sufficiently high. This grouping of amino acids is the same as that seen in transport experiments and one might predict that the cold-adapted membranes had somewhere a high concentration of fixed negative charges. Unfortunately no difference could be found between the zeta potentials of membranes isolated from the intestinal epithelium of goldfish adapted to different temperatures. This does not invalidate the idea that fixed negative charges in mucosal membranes might affect transport, for the technique used to measure zeta potentials is itself subject to error; but it does mean that this aspect of possible control cannot be studied further with the techniques at present available.

The separation of amino acids by two-dimensional chromatography is shown in Figure 3. The area formed by joining the RF values of amino acids whose transport was not regulated contained none of the amino acids whose

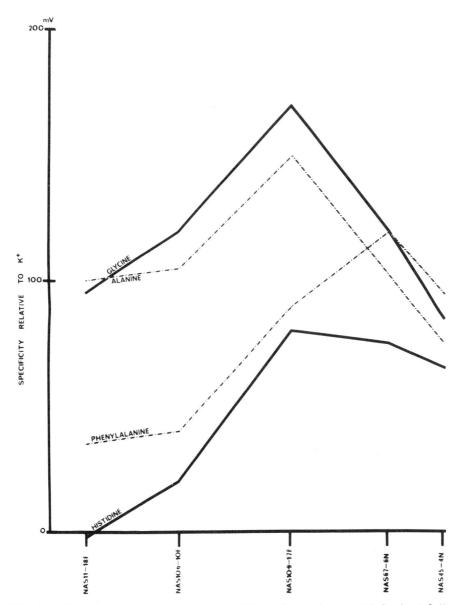

FIG. 2. Electrode responses to amino acids. Electrode potentials in 0.1 N solution of amino acids are plotted as a function of the $Na^+ - K^+$ selectivity of a variety of glass electrodes. Adapted from Eisenman (10).

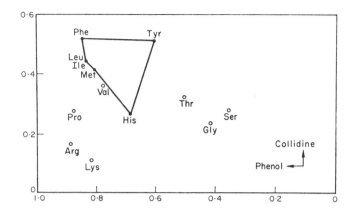

FIG. 3. Two-dimensional chromatography of amino acids in phenol and collidine taken from the paper by Dent (1948). Approximate R_x, values can be deduced from the scales. ●, non-adapting amino acids; ○, adapting amino acids.

transport was regulated. It seemed to be the more lipophilic amino acids whose transport remained unaffected by changes in environmental temperature. An analysis of goldfish intestinal membranes showed that the fatty acid composition of the phospholipids was changing with adaptation (13). The amount of each phospholipid species remained constant but, as is shown in Figure 4, the amounts of docosahexaenoic ($C_{22:6}$) and arachidonic ($C_{20:4}$) fatty acyl groups in the lecithin and phosphatidyl ethanolamine molecules increased four-fold during cold adaptation. The proportion of stearic acid ($C_{18:0}$) decreased, apparently to compensate for the increase in the longer chain, polyunsaturated fatty acids, though it is unlikely that all the $C_{18:0}$ would be found in the β position of the different phospholipids. Further work studying the time course with which changes in transport and in fatty acid composition took place (cf. Fig. 5) suggested that they might be directly connected; both the ratio of $C_{22:6}$ to $C_{18:0}$ fatty acids and the phenylalanine to valine concentration ratio changed to the fully adapted state 36 to 48 hours after raising the environmental temperature of the fish (30).

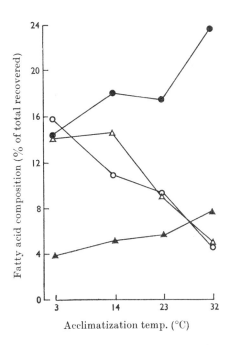

FIG. 4. Percentage recovery of some fatty acids extracted from the intestinal mucosal membranes of goldfish adapted to different temperatures. ●, $C_{18:0}$ fatty acid; △, $C_{20:4}$ fatty acid; ○, $C_{22:6}$ fatty acid; ▲, $C_{20:3}$ fatty acid. The $C_{20:3}$, $C_{20:4}$ and $C_{22:6}$ fatty acids were from a phospholipid fraction and the $C_{18:0}$ fatty acid from a neutral-lipid extract of total lipids.

If the selective changes seen in amino acid transport were connected with passive diffusion across a lipid barrier, one might expect to find similar changes occurring in purely artificial systems. The hydrated phospholipid crystal, described originally by Bangham, Standish and Watkins (1), was chosen to study the kinetics of amino acid diffusion. Lecithins containing acyl groups of varying lengths and degrees of unsaturation were prepared by feeding hens on diets which were rich or poor in unsaturated fatty acids. Lipid extracts from eggs laid subsequently were then chromatographed on silicic acid to separate different lecithins (14). These

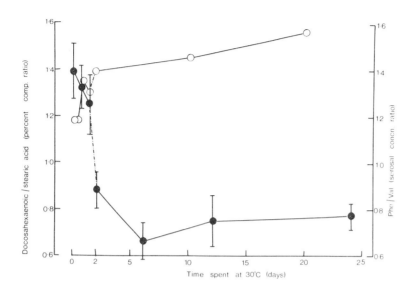

FIG. 5. Time course for changes in fatty acid composition of goldfish mucosal membranes and for changes in amino acid transport following a sudden increase in the environmental temperature of cold-adapted goldfish. ●——●, Docosahexaenoic acid/stearic acid content of intestinal membranes; ○——○, phenylalanine/valine serosal concentration, measured after a 2 hour incubation of everted intestinal sacs. Change of environmental temperature, 16 to 30° C.

were mixed with cholesterol and phosphatidyl serine and aqueous solutions of radioactive amino acids added. The swollen lipids were then agitated mechanically. Free amino acids were removed by gel filtration and the diffusion of amino acids from within the liposomes then measured as described previously (15,16). The results of one series of experiments, using an unsaturated lecithin to form the liposomes, are shown in Figure 6. The efflux was apparently first-order during the first 50 minutes incubation. This is taken to represent the kinetics of amino acid diffusion from the first liposome compartment across the outermost bilayer. At longer

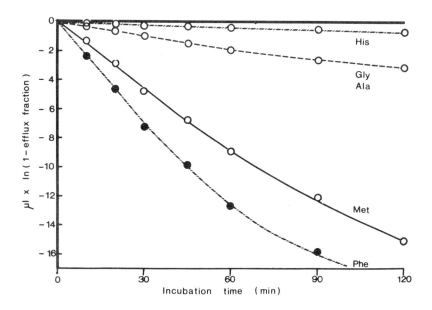

FIG. 6. Amino acid efflux from liposomes. Liposomes were formed from an unsaturated lecithin and incubated at 30° C for 150 minutes in oxygen-free buffer. HIS, histidine; GLY, glycine; ALA, alanine; MET, methionine; PHE, phenylalanine. The initial slope of each line gives the rate constants for efflux in μl/min.

TABLE 4. Amino acid efflux from two preparations of liposomes. The liposomes were prepared using lecithins containing different proportions of $C_{22:6}$ fatty acid in the β-position. L-amino acids were used throughout.

Amino acid	Efflux rate constant (μl/min)		Ratio $^b/_a$
	(a) 5% $C_{22:6}$	(b) 80% $C_{22:6}$	
Phenylalanine	0.13	0.23	1.8
Methionine	0.052	0.16	3.0
Alanine	0.011	0.032	3.0
Glycine	0.011	0.033	3.0
Histidine	0.0036	0.0047	1.3

incubation times the slopes ceased to be linear. Efflux rate constants were calculated from the initial part of each curve. The effect of varying lecithin composition on amino acid efflux is shown in Table 4. Increasing unsaturation raised the rate of efflux of all amino acids tested. Histidine and phenylalanine, however, showed only a small effect compared with glycine and alanine. Subsequent work with phenylalanine showed its efflux to be essentially independent of the type of lecithin used. This selective effect on amino acid diffusion is similar to that seen previously in transport studied across the goldfish intestine. Methionine was an exception to this grouping: its efflux from liposomes responded to the type of lecithin used although its transport across the intestine had remained unaffected during adaptation.

More detailed studies were carried out to test this selectivity. Phenylalanine (intestinal transport unaffected by adaptation) and glycine (intestinal transport changed by adaptation) were used together in a series of liposome preparations containing lecithins having different degrees of unsaturation. The permeability ratio of phenylalanine to glycine was then determined in each case. The results are shown in Figure 7. Virtually the only source of unsaturated long chain fatty acid in these lecithins was $C_{22:6}$ and the permeability ratio fell as the proportion of this fatty acid increased. This effect was entirely due to a selective increase in the rate of glycine diffusion across bilayers made from unsaturated lecithins.

One explanation of how amino acid transport adapts in the goldfish intestine is to say that a change in lipid composition alters the rate at which some transfer process operates. Results with methionine diffusion across lipid bilayers are at variance with this idea. The ion-binding properties of methionine, on the other hand, would lead one to predict what did in fact happen, that the transport of methionine across the

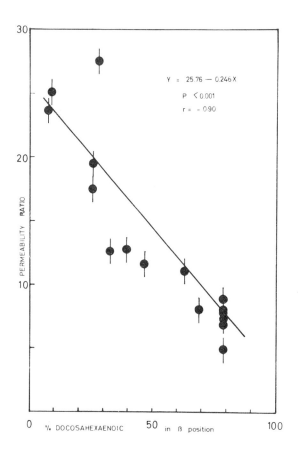

FIG. 7. Relative permeability of liposomes of different lecithin composition to phenylalanine and glycine. The relative permeability is plotted as the ratio of the efflux rate constants phenylalanine to glycine. The lines across each point give the range of results expected assuming a 5% error in the calculation of individual rate constants.

intestine would not be changed by adaptation. But then ion-binding selectivity cannot be used to explain the changes that took place in isoleucine transport. These exceptions to the general pattern cease to be relevant if differential lipid solubilities and ion-binding properties should combine to determine whether the transport of a particular amino acid will be modified. This idea seems in keeping with current theories of electrostatic and hydrophobic interactions for it is very difficult to imagine that one could change one parameter without affecting the other.

The site for regulation of amino acid transport could be the microvillar or lateral/basal membranes of the epithelial cells, but there are reasons for thinking that the microvillar membrane plays little or no part in the regulation process. Direct measurements of short-term influx of phenylalanine and alanine into the mucosa, made using the technique described by Schultz et al. (22), show no change on adaptation of cold fish to warm water (Smith and Ellory, unpublished results). Probably it is the movement of amino acids out of the cell which is being modified. The efflux of amino acids across the lateral and basal membranes of the intestinal epithelium would be down their concentration gradients and control of these downhill movements would be in accord with the idea that it is a passive transfer which is being changed. Selectivity could occur through charge and hydrophobic interactions between the individual amino acid and the mucosal membrane. If the efflux is carrier-mediated, then it has further to be assumed that the amino acid-membrane interaction is, in fact, strong enough to determine the kinetic behaviour of the amino acid-carrier complex.

There is also separate evidence that the passive movements of Na into this tissue can be controlled and that in this case it is the microvillar membrane which is changing. The net transport of Na falls during the change from cold to warm water at a time when the Na,K-ATPase activity is normal. Only much later does the Na,K-ATPase change its characteristics and then it appears to be a change in working efficiency rather than in absolute amount of enzyme present which takes place, for the number of specific ouabain-binding sites does not change with adaptation (29).

Clearly the control of Na transport has to be investigated in greater detail, for there is as yet no positive proof that the supply of ATP remains non-limiting under all conditions. But the evidence at present points to an initial control of leak pathways rather than pump sites, these increasing as the body temperature falls. This is the opposite to what is found for leak processes in mammalian hibernators (32), but it makes sense energetically in the poikilotherm in a tissue which has to continue to perform a transcellular transport function at very low temperatures.

REFERENCES

1. BANGHAM, A. D., M. M. STANDISH, and J. C. WATKINS. Diffusion of univalent ions across the lamellae of swollen phospholipids. $\underline{J.\ Mol.\ Biol.}$ 13: 238-252, 1965.

2. BOWLER, K., and C. J. DUNCAN. The temperature characteristics of the ATPases from a frog brain microsomal preparation. $\underline{Comp.\ Biochem.\ Physiol.}$ 24: 223-227, 1968.

3. BOWLER, K., and C. J. DUNCAN. The effect of temperature on the Mg^{2+}-dependent and $Na^+ - K^+$ ATPases of a rat brain microsomal preparation. $\underline{Comp.\ Biochem.\ Physiol.}$ 24: 1043-1054, 1968.

4. BOWLER, K., and C. J. DUNCAN. The temperature characteristics of brain microsomal ATPases of the hedgehog: changes associated with hibernation. $\underline{Physiol.\ Zool.}$ 42: 211-219, 1969.

5. CHRISTENSEN, H. N. Methods for distinguishing amino acid transport systems of a given cell or tissue. $\underline{Federation\ Proc.}$ 25: 850-853, 1966.

6. CRABBE, J. The mechanism of action of aldosterone. In: $\underline{Progress\ in\ Endocrinology}$, edited by C. Gual and F. J. G. Ebling. Proc. 3rd Intern. Cong. Endocrinol., 1968, p. 41-46.

7. DANIELS, V. G., A. G. DAWSON, H. NEWEY, and D. H. SMYTH. Effect of carbon chain length and amino group position on neutral amino acid transport systems in rat small intestine. $\underline{Biochem.\ Biophys.\ Acta}$, 173: 575-577, 1969.

8. DENT, C. E. A study of the behaviour of some sixty amino-acids and other ninhydrin-reacting substances on phenol-"collidine" filter-paper chromatograms, with notes as to the occurrence of some of them in biological fluids. $\underline{Biochem.\ J.}$ 43: 169-180, 1948.

9. EDELMAN, I. S., R. BOGOROCH, and G. A. PORTER. On the mechanism of action of aldosterone on sodium transport: the role of protein synthesis. $\underline{Proc.\ Natl.\ Acad.\ Sci.\ U.S.}$ 50: 1169-1171, 1963.

10. EISENMAN, G. The electrochemistry of cation-sensitive glass electrodes. In: $\underline{Advances\ in\ Analytic\ Chemistry\ and\ Instrumentation}$, edited by C. M. Reilly, New York: Interscience, 1965, vol. 4, p. 213-269.

11. GILLES-BAILLIEN, M. Permeability characteristics of the intestinal epithelium and hibernation in $\underline{Testudo\ hermanni}$ hermanni Gmelin. $\underline{Arch.\ Intern.\ Physiol.\ Biochem.}$ 78: 327-338, 1970.

12. GRUENER, N. and Y. AVI-DOR. Temperature-dependence of activation and inhibition of rat-brain adenosine triphosphatase activated by sodium and potassium ions. $\underline{Biochem.\ J.}$ 100: 762-767, 1966.

13. KEMP, P. and M. W. SMITH. Effect of temperature acclimatization on the fatty acid composition of goldfish intestinal lipids. $\underline{Biochem.\ J.}$ 117: 9-15, 1970.

14. KLEIN, R. A. The large-scale preparation of unsaturated phosphatidyl choline from egg yolk. Biochem. Biophys. Acta, 219: 496-499, 1970.

15. KLEIN, R. A., M. J. MOORE, and M. W. SMITH. Diffusion of neutral amino acids across lipid bilayers. J. Physiol. 210: 33P-34P, 1970.

16. KLEIN, R. A., M. J. MOORE, and M. W. SMITH. Selective diffusion of amino acids across lipid bilayers. Biochim. Biophys. Acta, 233: 420-433, 1971.

17. MEPHAM, T. B., and M. W. SMITH. Regulation of amino acid transport across intestines of goldfish acclimatized to different environmental temperatures. J. Physiol. 186: 619-631, 1966.

18. MOORE, S., and W. H. STEIN. Procedures for the chromatographic determinations of amino acids on four per cent cross-linked sulfonated polystyrene resins. J. Biol. Chem. 211: 893-906, 1954.

19. MUNCK, B. G. Amino acid transport by the small intestine of the rat. The existence and specificity of the transport mechanism of imino acids and its relation to the transport of glycine. Biochim. Biophys. Acta, 120: 97-103, 1966.

20. MUNCK, B. G., and S. G. SCHULTZ. Lysine transport across isolated rabbit ileum. J. Gen. Physiol. 53: 157-182, 1969.

21. OXENDER, D. L., and H. N. CHRISTENSEN. Distinct mediating systems for the transport of neutral amino acids by the Ehrlich cell. J. Biol. Chem. 238: 3686-3699, 1963.

22. SCHULTZ, S. G., P. F. CURRAN, R. A. CHEZ, and R. E. FUISZ. Alanine and sodium fluxed across mucosal border of rabbit ileum. J. Gen. Physiol. 50: 1241-1260, 1967.

23. SHARP, G. W. G., and A. LEAF. Mechanism of action of aldosterone. Physiol. Rev. 26: 593-633, 1966.

24. SKOU, J. C. Enzymatic basis for active transport of Na^+ and K^+ across cell membrane. Physiol. Rev. 45: 596-617, 1965.

25. SMITH, M. W. Sodium-glucose interactions in the goldfish intestine. J. Physiol. 182: 559-573, 1966.

26. SMITH, M. W. Influence of temperature acclimatization on the temperature dependence and ouabain sensitivity of goldfish intestinal adenosine triphosphatase. Biochem. J. 105: 65-71, 1967.

27. SMITH, M. W. Selective regulation of amino acid transport by the intestine of goldfish (Carassius auratus L.). Comp. Biochem. Physiol. 35: 307-401, 1970.

28. SMITH, M. W., V. E. COLOMBO, and E. A. MUNN. Influence of temperature acclimatization on the ionic activation of goldfish intestinal adenosine triphosphatase. Biochem. J. 107: 691-698, 1968.

29. SMITH, M. W., and J. C. ELLORY. Temperature-induced changes in sodium transport and Na^+/K^+-adenosine triphosphatase activity in the intestine of goldfish (Carassius auratus L.). Comp. Biochem. Physiol. In press.

30. SMITH, M. W., and P. KEMP. Parallel temperature-induced changes in membrane fatty acids and in the transport of amino acids by the intestine of goldfish (Carassius auratus L.). Comp. Biochem. Physiol. In press.

31. SWANSON, P. D. Temperature dependence of sodium ion activation of the cerebral microsomal adenosine triphosphatase. J. Neurochem. 13: 229-236, 1966.

32. WILLIS, J. S. Characteristics of ion transport in kidney cortex of mammalian hibernators. J. Gen. Physiol. 49: 1221-1239, 1966.

33. WILLIS, J. S. Cold adaptation of activities of tissues of hibernating mammals. In: Mammalian Hibernation III, edited by K. C. Fisher, A. R. Dawe, C. P. Lyman, E. Schoenbaum, and F. E. South. Edinburgh: Oliver and Boyd, 1967, p. 356-381.

34. WILLIS, J. S. The significance and analysis of membrane function in hibernation. In: Hibernation-Hypothermia: Perspectives and Challenges, edited by F. E. South, J. P. Hannon, J. S. Willis, E. T. Pengelley, and N. R. Alpert. Amsterdam: Elsevier. In press.

35. WILLIS, J. S., and N. M. LI. Cold resistance of Na-K-ATPase of renal cortex of the hamster, a hibernating mammal. Amer. J. Physiol. 217: 321-326, 1969.

ANALYSIS OF MEMBRANE STRUCTURE

CARLOS GITLER

Department of Biochemistry
Centro de Investigación y de Estudios Avanzados del I.P.N.
Apartado Postal 14-740
Mexico 14, D. F.

INTRODUCTION

Cold adaptation in hibernators must require significant alterations of cell membranes to allow them to continue their functions at temperatures in the vicinity of 5° C (85). Unfortunately, our knowledge of membrane structure and its relation to function is still extremely rudimentary so that any attempt to describe transitions leading to cold adaptation must be postponed until we have a better understanding of the mechanisms which allow normal function.

STATEMENT OF THE PROBLEM

Biological membranes can be considered to be metastable association colloids (5) capable of responding to many different environmental and metabolic conditions. This can be concluded from the fact that it has been shown that the membrane components can be readily dispersed by the use of non ionic and ionic detergents (80,6,24, cf. 8 for summary of some of the evidence) as well as by other treatments such as high concentrations of urea or guanidine or solvents which are not drastic enough to break covalent bonds. Thus there is no evidence for a rigid architecture or matrix held by covalent bonds underlying membranes, and whatever structure is seen under a particular set of conditions is the result of the balance of forces acting on the associated colloidal entities. The problem of membrane structure is then one of defining the alternative quasi-stable states which a set of components can assume under a given set of conditions, as well as the transitions occurring between these various states.

The composition of the different membranes varies widely, although in general the main components are phospholipids, proteins, glycoproteins, glycolipids, water and ions. In eukaryotic cells there is, in addition, the presence of sterols. How these components are held together by non-covalent forces and create a barrier to diffusion but at the same time are capable of such precise functions as are those of solute translocation and impulse conduction is the key problem of membrane structure of which some important features of this structure are just discernible while others are completely suppositional.

INTERACTIONS AMONG THE LIPID COMPONENTS

As early as 70 years ago, with the discovery of mesomorphic or liquid crystalline phases, their possible existence in the lipid components of living systems was the subject of elucubration (7,48). More recent studies (51,76) definitely indicate that phospholipids at the concentrations present in membrane exist as liquid crystalline arrays. The lipids are capable of both thermotropic and lyotropic mesomorphism. As pointed out by Lawrence (47)

> "The action of water (or other solvents) and heat are not dissimilar in regard to the effect of breaking the bonds holding the molecules in their crystalline orientation in one or possibly two directions. In solution this permits the entrance of solvent molecules into the lattice, with a corresponding swelling of the structure."

In living systems we are dealing with water-induced lyotropic mesophases. After swelling, the molecular arrays are capable of maintaining a quasi crystalline order due to the fact that most membrane lipids are amphipathic molecules (23,35), that is, the same molecule contains discrete polar and apolar regions which in the presence of water become hydrophilic and hydrophobic respectively, leading to an orientation of the components with regards to the solvent and to an overall molecular ordered arrangement. As pointed out by Hartley (35) and especially by Luzzati (50), in order to have either micelles or mesophases of amphipathic molecules in water, the hydrocarbon portions of the lipids must exist in a disordered liquid-like state.

The melting of the hydrocarbon portions of the lipids in aqueous dispersions and in membranes have been studied by nuclear magnetic resonance (NMR) (10,14) and differential thermal analysis (DTA) (13,14), as well as by other crystalographic methods (76). The findings indicate that swelling of phospholipids in water occurs only above the melting temperature of the fatty acid chains. Comparison by means of DTA of the endothermic transitions associated with the melting of the paraffin chains of solid phospholipids and in the presence of water, indicate that as mentioned above, water penetration with concommitant disruption of the crystal lattice decreases the melting temperature of the hydrocarbon chains. The transition temperatures for a given phospholipid fall in the presence of water to a new value which now becomes independent of further additions of the solvent (14). A complete disruption of the aggregates by the addition of water does not occur (except at very high dilutions) because the solvation of the polar groups is opposed by the hydrophobic forces resulting from the presence of apolar groups in the phospholipids.

Clearly, any added substance which alters the solvation of the polar groups will change the arrangement of the lipid components and the degree of fluidity of the apolar chains. Addition of counterions to phospholipid dispersions have been shown to result in the liberation of bound water (11). Also, the nature of the cation and anion present in the suspending solution of Nocardia asteroides alters the fluidity of the fatty acid chains and the diffusional loss of cell components (10).

The temperature at which the hydrocarbon chains melt depends also on the degree of unsaturation of the fatty acids of the phospholipids (14). Recent findings of Steim et al. (79) indicate that there is a close parallelism between the temperature at which melting occurs in the phospholipids of Mycoplasma laidlawii, whether these are present in sonic aqueous dispersions or in the membrane. This organism incorporates fatty acids present in the growth media; cells grown in the presence of stearic or oleic acid show DTA transition temperatures of 45° and -20°C respectively. Other studies on this microorganism (64,65) indicate that cells grown in unsaturated, branched-chain and cyclopropane fatty acids are filamentous, while those grown with long-chain saturated acids become coccoid and eventually swell and lyse. This swelling is prevented by 0.3 M sucrose. This represents an excellent example of the requirement for a definite mesomorphic state of the membrane lipids to establish a barrier to diffusion and probably may be generalized to other membrane functions (cf. 10, 19,59,86).

Recently, studies have been made of the apparent microscopic viscosity and order of the hydrocarbon regions of micelles and membranes by the measurements of the degree of fluorescence depolarization of perylene and of 2-methylanthracene dissolved in the apolar regions of detergent micelles (73) and of erythrocyte membranes (Dianoux and Weber, unpublished observations). The values obtained for the detergent micelles are shown in Table 1. In the series of detergents studied, the computed microviscosities at 27° C are all in the range of 17-50 centipoises. The change in microviscosity with temperature was found to follow a simple exponential form with an activation energy (ΔE) in the range of 6.1 - 9.6 Kcal mole^{-1}. The higher ΔE values found for the micelles, as compared to the analogous hydrocarbon solvents, are probably due to strong interactions around the charge edges of the chains, which increase the energy required for crossing the fusion potential barrier between adjacent hydrocarbon chains. The observed values of ΔE are all higher than 6 Kcal mole^{-1}, a value predicted to be the upper limit for linear hydrocarbon liquids (42).

Although added salts affected only slightly the microviscosity values of the hydrocarbon regions of cetyltrimethylammonium bromide (CTABr) micelles, differences were noted in the effect of added sodium bromide (increase in the viscosity) as compared with that of added sodium phosphate (viscosity unchanged up to 2 M salt) which might reflect the capacity of the anion to penetrate the Stern layer of the detergent. In addition, mixed micelles of CTABr with cetyl alcohol or cholesterol and with hexadecyl sulfate were used to test the effect of these additives on the micelle hydrocarbon fluidity. The microviscosity of these mixed micelles was found to increase rapidly with the concentration of the admixed component, and at a molar ratio close to one, microviscosities of several poises were obtained. In this same article (73) Weber derived the equations which allow, from the polarization of fluorescence as a function of wavelength, an assessment of the anisotropy of the solvent containing the fluorescent molecule. The change in the degree of depolarization with wavelength of excitation of perylene in mixed micelles of CTABr: hexadecyl sulfate (1.0:0.85), in glycerol: propyleneglycol (1:1) at 4° C and in propyleneglycol at -14° was found to be very similar. This served as an indication that the interior of the mixed micelles is close to isotropic.

Preliminary measurements (unpublished results, Dianoux and Weber) of the polarization of the fluorescence emission of perylene dissolved in the hydrocarbon regions of hemoglobin-free erythrocyte membranes (HFE membranes), indicate that the freedom of rotation of the probe is greatly reduced and gives viscosities even higher than those in the micelles containing cholesterol. Erythrocyte membranes contain a high cholesterol level; yielding an approximate molar ratio of cholesterol to phospholipids of one. Model experiments have shown that the addition of cholesterol to phospholipids results in an apparent immobilization of the fatty acid chains as judged by NMR and DTA measurements (14).

The results presented above with CTABr micelles show clearly that the addition of cetyl alcohol or cholesterol results in a very marked increase in the viscosity in the micelle interior. Schick and Fowkes (70) have studied the effect of additives on the critical micelle concentration and foam stabilization of ionic detergents and concluded that long-chain alcohols and amides decrease significantly the concentration at which micelle formation occurs and lead to remarkable foam stability. In addition, alcohols have been shown to have a net condensing effect when added to ionic monolayers (21); cholesterol in a similar manner has been shown to condense phospholipid monolayers (3,21). It is likely in the micelles, that the alcohols intercalated between the ionic amphiphiles contribute van der Waals interactions and at the same time shield the charge through ion-dipole interactions and thus decrease charge repulsion. The majority of the studies on membranes, such as those of myelin and of the erythrocyte which contain appreciable amounts of cholesterol, indicate that the lipids exist in the lamellar mesophase. Reports of the existence of other polymorphic forms have not been substantiated (37).

The studies of Steim et al. (79) in Mycoplasma laidlawii also appear to indicate that the membranes of this organism which are devoid of cholesterol exist also in a bilayer or lamellar structure. The experiments involve essentially the comparison, by means of DTA, of the lipids of

Table 1. Microviscosity and fusion activation energy (Δ E) of 2-Methylanthracene (2MA) and Perylene (Per) embedded in various micelles

Material	Concentration Mole/liter	Probe	P X100	(27°C) Centipoise	Δ E Kcal/Mole
LTABr	5×10^{-2}	2MA	5.38	26	7.2
		Per	2.08	17	9.6
MTABr	2×10^{-2}	2MA	5.10	32	7.1
		Per	2.18	21	9.5
CTABr	10^{-2}	2MA	5.73	30	6.2
		Per	2.05	19	9.6
SDBABr	10^{-2}	2MA	3.77	50	6.1
		Per	3.06	37	8.0
n-Dodecane				1.33	2.8
N-Hexadecane				2.98	3.8
White oil U.S.P.				124.00	12.7

From reference (73). LTABr, MTABr and CTABr are the lauryl, myristoyl and cetyltrimethylammonium bromides. SDBABr is stearoyldimethylbenzylammonium bromide.

M. laidlawii in situ and after extraction as an ultrasonic aqueous dispersion. The findings indicate that a major portion (roughly 70%) of the membrane lipids melt in a manner similar to those present in the aqueous dispersion. Since these latter exist in a smectic mesophase or a bilayer form, the conclusion reached by Steim et al. is that some 2/3 of the lipids exist in the form of a bilayer of lamellar phase in the membrane of M. laidlawii.

The presence of hydrophobic interactions leads to the proximity of the lipid molecules and their close packing in turn results in interactions which impart to them new and interesting properties.

Ionic Interactions. The formation of micelles of ionic detergents is accompanied by the neutralization of only a fraction of the micelle charge by adsorbed counterions (54). This leads to the presence of a potential in the micelle surface which has been calculated to be in the order of -85 to -136 millivolts for micelles of sodium dodecyl sulfate and sodium octadecyl sulfate respectively (3). The effect of various univalent counterions on the micelle charge depends not only on their concentration but also on the structure of the ion and how closely it approaches the charged head groups of the micelle-forming amphiphiles (Stern layer). The smaller the distance of closest approach, the greater will be the screening action on the charges of the micelle (54).

In the case of zwitterionic detergents such as N-alkyl-N,N-dimethyl-glycines (N-alkylbetaines) their behavior is equivalent to that of non-ionic detergents, the fraction of gegenions bound to the micelle is zero and the critical micelle concentration is much lower than that of the ionic detergents (54). Beckett and Woodward (2) have proposed that the arrangement of the charges in the micelles of alkylbetaines is such that there is intermolecular neutralization. It is likely that the behavior of phospholipid molecules in micelles, liquid crystals and in membranes should be similar to that of the alkylbetaines mentioned above. That is, there should be extensive intermolecular neutralization of the type shown in Figure 1b. Alternatively, intramolecular neutralization could also occur (Fig. 1a) (72). It is very unlikely that oppositely charged groups can coexist in the presence of neighboring hydrophobic regions since as demonstrated by Packter and Donbrow (60) ion-pair formation is significantly enhanced if the oppositely charged molecules contain apolar regions capable of hydrophobic interactions.

In order to study the formation of ion-pairs between phospholipid molecules, the reactivity of the amino groups of phosphatidyl ethanolamine (PE) and phosphatidyl serine (PS) with trinitrobenzenesulfonate (TNBS)

ANALYSIS OF MEMBRANE STRUCTURE

(31,34,58) has been studied. The overall reaction is shown below:

$$R-NH_3^+ \rightleftharpoons RNH_2 + TNBS^- \longrightarrow R-NH-TNB \quad (a)$$

$$+ \quad\quad\quad +$$
$$+ \quad\quad\quad -$$
$$H \quad\quad\quad HSO_3$$

Any factor which tends to stabilize the amino group in the protonated form will decrease the overall rate. (The reaction might proceed with the initial binding of TNBS⁻ to the surface and the subsequent reaction with the amino group. This possibility is now being explored.) Thus salt formation between the amino group of PE and PS with phosphate groups

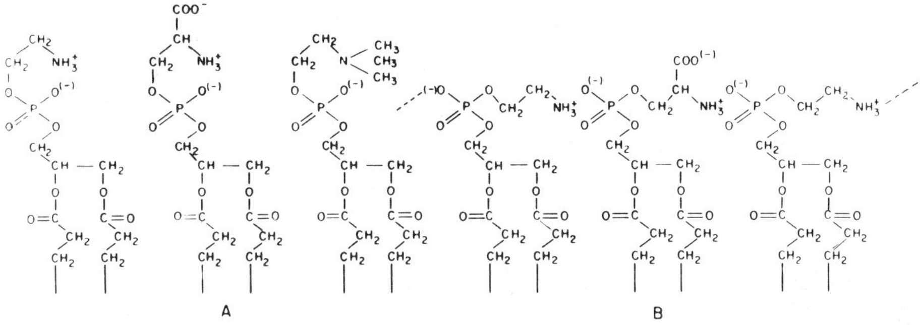

FIG. 1. Diagrammatic representation of intra-(1a) and intermolecular (1b) ion-pairs among phospholipid molecules.

(Figs. 1a, 1b) should markedly lower the reactivity of the amino group with TNBS.

In Figure 2a are shown the rates of reaction of TNBS with aqueous ultrasonic dispersions of PE and PS alone and in the presence of added Triton X-100. In Figure 2b are shown the equivalent rates of reactions of mixtures of PE with phosphatidyl choline (PC). It may be observed that the overall rates of reaction are small for dispersions of PE and almost nil for those of PS; intermediate rates are obtained in the mixtures with PC and a still further enhancement is found in the presence of added Triton X-100. These unpublished results of Gitler and Cid strongly suggest that intra- or intermolecular salt bridges (Fig. 1) stabilize the ammonium form of the amino moiety of PE and PS when present in aqueous dispersions without any other added component. The marked enhancement in the presence of Triton X-100 would appear to favor intermolecular ion-pair formation since in the case of an intramolecular ion-pair intercalation of the nonionic detergent should not have led to a rate enhancement. Specific interactions between the PE and Triton X-100 leading to exposure of the amino group seem unlikely since other nonionic detergents of different formula are equivalent to the Triton X-100. (If as mentioned previously, the reaction proceeds with the initial binding of the TNBS to the surface, the rate enhancement might be due in part to a more favorable binding in the presence of the Triton or PC).

Measurements were made of $k'_{(obs)}$, the apparent pseudo first order rate constant, for the reaction of TNBS with PE and PS in the presence of Triton X-100, as a function of the pH of the suspending solution. The result gave values of $pK_{(app)}$ of 9.2 and 9.1 and $k'_{(obs)}/(NH_2)$ of 6.43 and 1.25 min^{-1} for PE-Triton and PS-Triton respectively. Thus the $pK_{(app)}$ values are essentially equal but the presence of the carboxyl is PS decreases the rate of reaction 5 fold when compared with PE. A surprising finding is the rate enhancement in mixtures of PE and of PS with PC. It seemed likely, a priori, that equivalent intermolecular ion-pairs should have resulted from the interactions of PE-PE and of PS-PS when compared with those of PE-PC and of PS-PC. The actual findings indicate that PC acts nearly as a neutral detergent although a partial stabilization of the NH_3^+ is still present when compared with Triton X-100. This would appear to indicate that part of the charge of the phosphate of the PC is neutralized by intramolecular

ANALYSIS OF MEMBRANE STRUCTURE 249

ion-pair formation (Fig. 1b) leaving very little negative charge to stabilize the NH_3^+ of PE and of PS. Some of these results and the conclusions reached are similar to those of Papahadjopoulos and Weiss (61) who measured reactivity of the amino group of PE and PS with TNBS by determining the overall change in electrophoretic mobilities of the lipid aggregates.

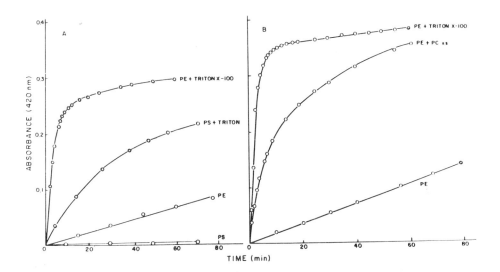

FIG. 2. Rate of reaction of trinitrobenzenesulfonate (TNBS) with aqueous dispersions of phosphatidyl serine (PS), phosphatidyl ethanolamine (PE) with or without added Triton X-100 or phosphatidyl choline (PC). Final TNBS concentration was 1.66 mM; the lipids (10-15 µg of Pi) were suspended in 50 mM sodium phosphate buffer, pH 7.5 and Triton X-100 final concentration was 5 mM. PC/PE ratio in 1b was 10. Temperature 30° C.

Figure 3a shows the rate of reaction of TNBS with hemoglobin-free rabbit erythrocyte ghosts (HFE membranes) with ghosts + Triton X-100, with ghosts + sodium dodecyl sulfate and with lipid-free native ghost proteins + Triton X-100. It is apparent that at short time periods the main reactive amino groups are those of membrane lipids and that a significant enhancement in the rate occurs in the presence of Triton X-100. Addition of sodium dodecyl sulfate decreases the rates essentially to those of the lipid-free membrane proteins in Triton X-100. In Figure 3b are presented for comparison the rates of reaction of TNBS with rabbit erythrocyte total lipids dispersed in water by ultrasound and the enhancement due to the addition of Triton X-100.

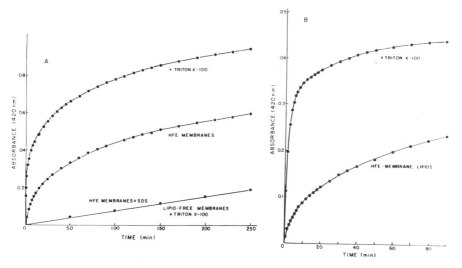

FIG. 3a. Rate of reaction of TNBS with hemoglobin-free rabbit erythrocyte membranes (HFE membranes) and with HFE membrane proteins. The final concentrations of TNBS, Triton X-100 and sodium dodecyl sulfate (SDS) were 1.66 mM, 5 mM and 5 mM respectively. HFE membranes and HFE lipid-free membrane proteins were suspended in 50 mM sodium phosphate buffer, pH 7.5, at a protein concentration of 0.1 mg per ml. Temperature 30° C.

FIG. 3b. Rate of reaction of TNBS with HFE-membrane total lipids. The lipids (12 μg of Pi) dispersed in 50 mM sodium phosphate buffer, pH 7.5. All other conditions are in 3a.

Although the experimental work has not proceeded far enough to give a single explanation for these results, it would appear that the reactivity of the membrane phospholipids is similar to that of PE and PS in the presence of PC, and much faster than that observed for PE or PS alone at pH 7.5. The molar ratios of PC/PE and PC/PS in erythrocyte membranes are roughly 1.25 and 2.0 respectively (84). The mixtures shown in Figure 2b were performed at a PC to PE ratio of about 10. Lower ratios give decreasing rates.

These results indicate that there are marked differences in the ionic interactions between the different phospholipids which must imply different orientations of the polar groups in the surface. Conclusive evidence for the inter- as opposed to intramolecular ion-pairs is as yet not available since binding of the TNBS to the different phospholipid aggregates could significantly alter the overall observed rates.

The most interesting aspect of the above results is the accessibility of the amino groups of the membrane lipids to the TNBS and their nearly equivalent reactivity when compared with the sonic dispersion of the lipids extracted from the same membranes. This argues against extensive ionic lipid-protein interactions as required by the membrane model of Davson and Danielli. Experiments are now in progress to make these results conclusive.

Hydrogen Bonds. Little attention has been given to the presence of stabilizing hydrogen bonds between the lipid membrane components. In general, this is due to the fact that in water, the formation of a solute-solute intermolecular bond is opposed by the formation of hydrogen bonds with the water molecules (20,43). However, shielding of the water as occurs in an apolar environment greatly increases the tendency to form very stable hydrogen bonds (43). For a detailed discussion see Jencks (39). In phospholipids, the fatty acid ester, ether and amide groups contain dipoles which in a medium of low microscopic polarity would have to be stabilized by dipole-dipole interactions. In addition cholesterol in the crystal melts at 148.5° C while cholesteryl chloride melts at 92°. It is clear then that the hydroxyl in position 3 is capable of interactions which lead to a marked stabilization of the crystal lattice. In a similar manner, the glycerol in phosphatidyl glycerol, in cardiolipin and the sugar in glycolipids have available potential hydrogen bond-forming groups.

In liquid crystals we have seen that the various lipid molecules are brought into close proximity so that water is largely excluded or is located mainly in the vicinity of the ionic groups. It is likely, therefore, that the microscopic polarity in the surface is lower than that in the bulk and that this might favor extensive hydrogen bonding.

Actual measurements of the dielectric constant in the surface of micelles and membranes are available. The elegant experiments of Mukerjee and Ray (56) measured the position of the charge transfer band of micelles of N-long chain alkyl pyridine iodides and compared the peak position with that of the charge transfer complex of the same compounds molecularly dispersed in different solvents. Their results indicated a dielectric constant in the surface of the micelle of approximately 36. In studies on the binding of 1-anilino-naphthalene-8-sulfonate (1,8-ANS) to erythrocyte membranes (25), it was observed that the peak position of the fluorescence emission of the bound dye indicated that the 1,8-ANS was present in an environment of polarity equivalent to that of pure ethanol. Since 1,8-ANS is a charged molecule, it is likely to be located very close to the oppositely charged lipid polar groups of the membrane surface. In both of these studies it is clear that the environment in the membrane surface is of significantly lower polarity than that of bulk water, and it is very probable that hydrogen bonds would be quite stable either between the different lipid components or between the lipids and water molecules which would be bound tightly in the surface.

Some evidence is available to indicate that indeed the formation of hydrogen bonds might lead to greater stability of lipid aggregates. Schick and Fowkes (70) found that the greatest foam stability and lowering of the critical micelle concentration occurs when long-chain additives containing groups capable of forming extensive hydrogen bonds are admixed with ionic detergents. Alexander (1) summarized the evidence supporting hydrogen-bond formation in condensed monolayers. Kaplan (40) synthesized N-α-palmitoyl serine as a model compound to mimic some of the properties of phospholipids and found it to form extensive hydrogen bonds in aprotic solvents.

It is important to establish whether the role of hydrogen bonds suggested above is a significant one; if this were the case, some orientation of the components to maximize hydrogen bond formation could be envisaged. In addition, knowledge of the contribution of hydrogen bonds to the stability of the membrane would allow an interpretation of the effect of additives which are capable of interaction through the formation of such bonds. Examples of these include amides, carbamates, local anesthetics and sugars.

INTERACTION AMONG THE PROTEIN COMPONENTS

Studies on membrane protein components have been hampered by their insolubility in aqueous solvents. Singer (74) pointed out some time ago that this insolubility was perhaps an important property of membrane proteins which resulted in their special functional characteristics. He has proposed recently (75), based mainly on the findings of Ito and Sato (38), that membrane proteins consist of discrete regions containing an accumulation of polar groups while others are essentially apolar so that on interaction with lipids and other similarly structured proteins in an aqueous environment, there will be hydrophobic associations between the apolar regions and orientation of the polar regions towards the aqueous environment. Ito and Sato (38) observed that highly purified cytochrome b_5 of liver microsomes obtained following protease treatment (called trypsin b_5) has a molecular weight of about 12,000 and shows no tendency to polymerize in water. On the other hand cytochrome b_5, solubilized with detergents (detergent b_5) without trypsin treatment and purified to an essentially homogeneous state, had a molecular weight of about 25,000 and existed in solution as an oligomer which could be depolymerized in 4.5 M urea. Tryptic digestion of this protein yielded a hemoprotein which was identical with cytochrome b_5 purified from protease treated microsomes. According to the authors:

> "While trypsin b_5 is hydrophilic in nature and shows no tendency for polymerization, detergent b_5 can be kept in the monomeric form only in the presence of urea. In the absence of urea it forms an oligomer probably owing to an intermolecular hydrophobic interaction. Therefore it seems that the nontrypsin b_5 moiety of detergent b_5 confers hydrophobicity on the whole molecule, while the catalytic activity probably resides on the trypsin b_5 moiety. The hydrophobic moiety also seems to be responsible for the firm attachment of cytochrome b_5 to the microsomal membrane. In other words, this moiety appears to play a role of built-in 'structural protein' in this interaction."

Singer (75) has further made an operational definition of integral and perpheral proteins as those which are within the membrane matrix and those that are adsorbed into the membrane respectively. Clearly both can be functional membrane proteins but the first will be difficult to solubilize because of their extensive apolar regions, while the latter may be dissociated from the membrane by mild treatments. The membrane proteins are probably in many cases oligomeric and they may contain hydrophobic and hydrophilic subunits leading to part of the functional oligomer being dispersible in water when separated from the other hydrophobic protomers. Many membranes contain an appreciable fraction of the total proteins in the form of glycoproteins thus in the case of human erythrocyte membranes, 10% of the membrane dry weight is carbohydrate of which the majority is bound covalently to proteins. Since this carbohydrate is responsible for the antigenic properties and the electrokinetic zeta potential, it must be present in the most external region of the cell membrane; these proteins must have, therefore a portion of the molecule designed to anchor them in or on the membrane in such a manner that the sugar moieties are exposed to the external aqueous environment.

Isolation of membrane proteins has in general been possible only with those defined above as "peripheral." That is, those which remain soluble in water after different mild methods which disrupt the membrane. On the other hand extremely drastic procedures have been used to render the majority of the membrane proteins soluble in water. Most of these methods result in proteins which are partially or totally denatured and in the case of membrane enzymes, have lost catalytic activity. In addition, they also lead to the dissociation of oligomeric proteins to subunits or even into constituent polypeptide chains.

An additional problem has been that of removing the membrane lipids without denaturing membrane proteins. One of the best procedures available is that of Hernández (36). Treatment of membranes with phospholipase C converts, in the majority of the cell membranes, some 70% of the phospholipids into the water soluble phosphorylated base and diacylglycerol. Subsequent preparation of an acetone powder results in lipid-free proteins in a native state. Thus phospholipids, insoluble in cold acetone are rendered acetone-soluble by the phospholipase treatment. This procedure has been successfully applied to rabbit erythrocyte membranes (Gitler and Cid, unpublished observations). Attempts to solubilize in the native state the lipid-free proteins from the acetone powder without the aid of nonionic detergents were completely unsuccessful; the use of Triton X-100 resulted in the recovery of the majority of activity of three membrane bound enzymes as a lipid free detergent protein solution. The successful use of Triton X-100 to disperse the membranes (24) and the lack of success in rendering lipid-free membrane proteins soluble in an active state, except by the use of the same detergent, suggested the use of nonionic detergents as solvents to study the characteristics of the membrane proteins. Triton X-100 is a polyoxyethylated branched octyl phenol and thus absorbs light at 280 nm and reacts with the Folin reagents. In order to overcome these difficulties the Triton was replaced by Brij 36T which has approximately the same number of oxyethylene chains (about 10) attached to lauryl alcohol. This detergent shows no absorbance at 280 nm and is as good as or better than Triton in dispersing erythrocyte proteins without loss in enzymatic activity (Gasca-Rivas, Rudy and Gitler, unpublished observations). Figure 4 shows the results obtained from the passage of human erythrocyte membranes dispersed in Brij 36T through a 4% agarose column equilibrated with the detergent (10 mM) in 20 mM Tris, pH 7.4 and 1 mM EDTA. As can be seen from the absorbance at 280 nm some 10 peaks are eluted of which peak V represents some 50% of the total. Peaks V and VI contain the majority of the membrane carbohydrates. Some of the carbohydrates in VI might be due to glycolipids which elute partially in this region. Almost all of the membrane sialic acid elutes in peak V. The position of the elution of acetylcholinesterase is also shown. It appears as a discrete peak even though it is known to be an oligomeric protein. Hence, little dissociation into subunits has occurred. Almost all of the acetylcholinesterase of the intact cell is recovered. The position of the lipids is also shown. It is likely that these peaks are still heterogeneous. However, the very high molecular weights of the membrane proteins (as can be judged from the position of elution of known proteins) indicates that this procedure will allow the study of the oligomeric membrane proteins in a condition resembling that of the intact membrane. Subsequent treatment of each of the peaks by means of chromatographic and electrophoretic procedures will permit the identification of the various components which can then be dissociated further into subunits. Since in the initial stages the use of a nonionic detergent allows the study of enzymic activity and immunological identification of the native proteins (18,24), it seems that this approach is more rational than that of treating the membrane proteins initially with drastic procedures which will yield a large number of polypeptide chains in a denatured form and impossible to identify with respect to their function.

This latter approach has, however, yielded preliminary information to show that membrane proteins might have unusual structural characteristics. Gwynne and Tanford (33) in a preliminary communication have identified in human erythrocyte membranes significant amounts of what appears to be one of the largest single polypeptide chains known. Reported molecular weight based on the elution pattern from a 4% agarose column of total membrane proteins after reduction, alkylation, and denaturation in 6 M guanidine hydrochloride yielded a molecular weight of about 200,000 for this fraction. The only other known polypeptide chain of this length is that of the heavy chain of myosin. It is of interest, in passing, to mention that Ohnishi claimed for some time to have isolated so called contractily proteins for different membranes (57). These results have not been confirmed (46).

On the opposite size scale, Laico et al. (46) have presented evidence which indicates that some 50% of the erythrocyte membrane proteins are so called "miniproteins" of molecular weights in the vicinity of 5000. These glycopeptides or miniproteins are only observed after dissociation of the membrane components with very high concentrations of sodium dodecyl sulfate, mercaptoethanol and γ-aminobutyric acid. They have very high negative charge densities since in electrophoretic mobility they run ahead of the marker dye. The possibility of their arising from fragmentation during preparation is still to be excluded since they seem to have the properties expected for the sialopeptides liberated by trypsin treatment of

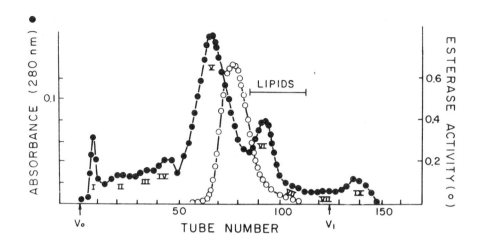

FIG. 4. Elution profile of hemoglobin-free rabbit erythrocytes (4 mg protein), from a 4% agarose column. The membranes in 10 mM Brij 36T and 1 mM EDTA in 20 mM Tris HCl buffer, pH 7.5, were eluted with the same buffer at 25° C.

erythrocyte membranes. More evidence will also be required simply to verify the above results. These two reports are contradictory since in Gwynne and Tanford's experiments no miniproteins were observed even though the procedure used ensures splitting into the component polypeptide chains. Nevertheless, the notion of miniproteins is very appealing in view of their possible functional relation to peptide antibiotics (valinomycin, etc.) which are known to function as artificial ion carriers.

Very little can be said at present about cooperative interactions between subunits. Their possible existence in membrane proteins has been proposed (12) but based only on analogy with the behavior of soluble proteins. One aspect seems of interest in view of the subject matter of this symposium, and that is the dissociation of oligomeric proteins to subunits which occurs on incubation at low temperatures (for a summary, cf. 63). One of the best examples occurring in a membrane protein is that of the ATPase from beef heart mitochondria purified in a soluble form by Racker and coworkers (62) and studied in detail by Penefsky and Warner (63). This enzyme, in addition to its ATPase activity, functions as a factor which couples the esterification of inorganic phosphate to the oxidation of substrates catalyzed by submitochondrial particle. Both activities were found to decrease markedly when the enzyme was incubated for short periods of time at 0° C. The 11.9 S native enzyme (284,000 daltons, about 10 monomers) is transformed at low temperatures to an equilibrium mixture of 11.9 S, 9.15 S and 3.55 S components. The nature of the cold labile forces which are involved in maintaining the native conformation of the ATPase are not known. However, the postulated lability of hydrophobic bonds in proteins was suggested as a possible explanation (63). As reviewed by Penefsky and Warner (63) quite a few other enzymes share this cold sensitivity with the ATPase. These findings might be of interest in view of the report by Melman and Rosenbaum (53) that in winter hibernating bats Mg-activated ATPase was not apparent in the infoldings of the renal tubules unless the animal had been aroused 24 hours before sacrifice.

The results of the enzyme inactivation studies have been obtained with purified proteins. Some factors such as phospholipids are protective against cold inactivation so that the relevance to hibernation adaption might not be as obvious as presented here. It might, however, merit further study especially since the effect of prolonged cold storage has not been determined.

INTERACTIONS OF LIPID AND PROTEINS

No generalizations can as yet be made about the interaction between the membrane lipids and proteins. The studies of Steim et al. (79) indicate that roughly some 70% of the lipids in the membranes of M. laidlawii show hydrophobic interactions with other lipids. The remainder would conceivably be capable of hydrophobic interactions with the membrane proteins. What does seem apparent is that a large fraction of the lipids are not intercalated within the protein chains since this would markedly modify the apparent fatty acid melting temperatures. Evidence from soluble lipoproteins also supports this view (78).

Ionic interactions between acidic lipids and basic proteins such as cytochrome C (17) have been reported although the relevance to membrane structure is not as yet apparent. The results of the reaction of TNBS with membrane amino groups presented earlier appear to indicate the ready accessibility of the amino groups of the lipids which would argue against very extensive stable ionic interactions between the lipid amino groups and the acidic groups of the proteins. These results do not rule out the possibility of salt formation between the carboxyl group of phosphatidyl serine and basic groups of the protein. It should be mentioned that the proteins of the human erythrocyte membrane contain a significant excess of acidic residues (66,4). It is conceivable that lipid-protein interactions could be mediated by divalent cations (8). Specific functional interactions between membrane proteins and phospholipids are also known to exist (32,71) although the fraction of the phospholipids involved is probably rather small since, in the cases where detailed studies are available, the stoichiometry is such that only a few lipid molecules interact per protein molecule (71).

INTERACTIONS OF THE MEMBRANE COMPONENTS WITH WATER

The formation of a liquid crystal of the lipid components will bring into close opposition a series of ionic groups resulting in a high surface charge density which will retain irrotationally bound water molecules; salt formation between the molecules as well as interactions with counterions which can penetrate into the Stern layer would diminish the waters thus held.

The value of 36 as the apparent dielectric constant in the surface of micelles of N-long chain alkyl pyridine iodides found by Mukerjee and Ray could either be due to the proximity of the hydrocarbon chains or alternatively to dielectric saturation of the water molecules due to the very high surface charge density of these micelles (56).

Studies on DTA of phospholipid-water binary mixtures indicate that 1,2-dipalmitoylphosphatidyl choline binds about 20% water (14). Cerbón (11) has shown that phospholipid dispersions bind more water than can be accounted for from that held in the hydration shell of the ionic groups of the lipids. The recent findings of Gent et al. (22) are very interesting in that they show that myelin binds at least 1 g of water per gram of myelin dry weight. The results are based on a comparison of the dielectric relaxation of water molecules in pure water ($\Sigma \alpha = 78.2$) versus that of myelin ($\Sigma \alpha = 59$) at 25° C. The authors conclude that "ice-like" water is present in the myelin since its interaction with the membrane components was found to be much stronger than those present in normal liquid water. The values reported are about 2 to 3 times those determined for pure protein solutions and suggest that water may be considered a membrane structural component (22).

As mentioned previously the presence of a large number of ester, ether and amide bonds near the phospholipid surface could in a low polarity medium result in strong interactions with water molecules. In addition, the high density of sugar residues in some membrane surfaces could also contribute to water immobilization.

PROPERTIES OF THE MEMBRANE SURFACE

Since both phospholipids and proteins are amphipathic molecules (23) their association leads to the formation of so-called amphipathic interfaces (15). These interfaces have some unusual properties which are dependent on the state of aggregation and can change dynamically in the presence of added substances. The location of the interface can be considered to be the limit to which water can penetrate into the membrane and still have the properties of bulk water. This limit will greatly depend on the packing arrangement of the lipids and proteins and will be a sensitive function of counterions and other additives and therefore the properties and location of the interfaces will vary significantly with the conditions to which the membrane is subjected. For example, Figure 5 shows the effect of added cations on the binding of 1,8-ANS to HFE membranes (69). At low ionic concentrations, the number of binding sites is very low and increases linearly as the square root of the salt concentration. Since 1,8-ANS does not fluoresce in water and only does so when it binds to interfaces with a much lower microscopic polarity than that of water, these results indicate that sites of low polarity are being formed in the membrane with the addition of cations. Calcium is some 100 fold more effective than sodium in this regard. The fluorescence of ethidium bromide, a cationic dye which is similar in its sensitivity to the polarity of the binding sites as is 1,8-ANS (26), and in addition shows enhanced fluorescence on binding to amphipathic interfaces of negative charge as compared to those neutral (67), binds very tightly to mitochondrial membranes. Figure 6 shows that the fluorescence is very sensitive to the metabolic state of the mitochondria (Estrada, Rubalcava and Gitler, unpublished results). Conditions which "deenergize" the membrane lower the fluorescence, while the addition of trinucleotides such as ATP enhances the fluorescence. No detailed descriptions of these results are intended here; suffice it to show that the interface where the dye is located is capable of significant modification as a function of additives and of metabolic state. The sensitivity to charge of the ethidium shows that not only the microscopic polarity can be altered but also the surface potential. It is likely that energization of the membrane involves charge separation and that the fluorescent probes can reflect these phenomena (27).

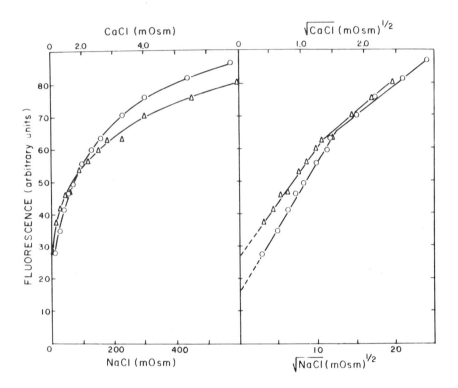

FIG. 5. Binding of 1,8-ANS to HFE membranes as a function of NaCl (O) and $CaCl_2$ (Δ) concentrations (69).

One of the salient properties of amphipathic interfaces is that they allow, in equilibrium with the aqueous environment, a _decrease_ in the microscopic polarity in the presence of an <u>increase</u> in charge density. This is a very unique situation since, for example, apolar solvents have low dielectric constants but do not allow significant solubilization of charges. These properties have been explored in micelle catalysis in analogy with the surface of an enzyme which is also an amphipathic interface. The results have been summarized recently (15) and indicate that remarkable rate modifications can be observed in the presence of the micelles. This is clearly shown by the data in Table 2 which represents a more detailed study (15) of the kinetic parameters presented in previous reports (25). The rate of liberation of p-nitrophenol from p-nitrophenyl esters catalyzed by mixed micelles of CTABr-Nα-myristoylhistidine was compared with the equivalent reaction of the same esters with Nα-acetyl-histadine. Both reactions are essentially similar except that the micelle reaction occurs in the amphipathic interface while that of the N-acetylhistidine takes place in bulk solution. A remarkable rate enhancement was found in the micelle reaction. The last column of Table 2 shows that the rate increase is already apparent with p-nitrophenyl acetate (44 fold). As the chain length of the nitrophenyl ester increases the surface reaction is augmented dramatically while that occurring in the bulk actually decreases. When the chain length of the ester reaches 10 carbons, the ratio of micelle/bulk rates gives values of the order of 10^5. The rate enhancement appears to be due to nonpolar as well as polar stabilization of the transition states. The dual nature of the micelle surface contributes to the catalysis by these surfaces. It is likely that in the membrane surface similar forces will be exerted to stabilize intermolecular complexes between the membrane components or with added solutes.

A study of the behavior of these membrane surfaces has been initiated, as mentioned by means of the changes in fluorescence characteristics of

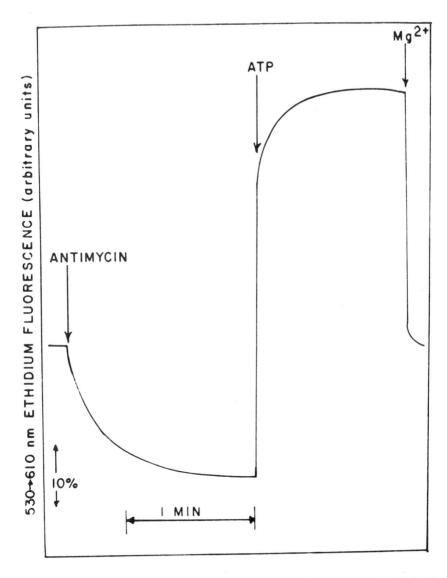

FIG. 6. Fluorescence response of ethidium bromide bound to mitochondria in various metabolic conditions (unpublished work of Estrada, Rubalcava and Gitler).

bound dyes (26,27,67,69 and unpublished observations of Gitler and Rubalcava). It is reasonable to expect that these molecules will be present in different membrane regions as a function of their solubility characteristics. Thus, charged molecules such as 1,8-ANS and ethidium will be located near the polar groups of the phospholipids or in exposed apolar regions of membrane proteins which contain oppositely charged groups. On the other hand, apolar molecules like perylene and 2-methylanthracene which are almost completely insoluble in water, will be located in the essentially apolar regions of the membrane. Evidence to support these contentions comes from studies of polarization of the fluorescence emission of these molecules in detergent micelles. Thus, while the fluorescence emission of perylene in the micelle interior is almost completely depolarized (Table 1) indicating its freedom of rotation in the liquid hydrocarbon environment, that of 1,8-ANS in the same micelles shows a polarization of 12% (27) indicating a significant restriction of the freedom of rotation as would be expected if this molecule were located in the pallisade layer of the detergent micelle.

The use of these probes allows the study of the membrane interface as a function of temperature (41,68,82). Figure 5 shows as an Arrhenius plot the changes in the fluorescence of 1,8-ANS bound to HFE membranes under various conditions and to sonicated egg lecithin. A negative temperature coefficient is observed. In the intact membranes, an initial inflection in the slope occurs at from 10° to 21°C (depending on ANS concentration). A monotonic decrease in fluorescence is then observed which levels off at 44.7° C. Increasing the temperature up to 70° C does not lead to a decrease in the observed fluorescence. The slopes for the linear decrease result in an apparent activation energy of -4 557 cal. The decrease in fluorescence as the temperature is increased is due to a decrease in the quantum yield

Table 2. Kinetic parameters for the reaction of p-nitrophenyl esters with mixed micelles of CTABr-Nα-myristoylhistidine and Nα-acetyhistidine.

p-nitrophenyl ester	Micelles of CTABr-MirHis			N-AcHis	Micelle
	k'_2 (min^{-1}M^{-1})	$1/K_i$ (M^{-1})	$1/K_p$ (M^{-1})	k_2 (min^{-1}M^{-1})	bulk
Acetate	551	61	79	12.6	44
Propionate	1,388	196	184	16.2	86
Pentonoate	4,348	1,220	971	11.6	375
Hexanoate	11,565	2,702	2,778	9.0	1,285
Octanoate	63,938	16,667	-	8.0	7,992
Decanoate	309,000	110,000	-	2.5	1.23x10^5

k'_2 is the apparent second order rate constant for the reaction in the micelle leading to the acylation of MirHis.
$1/K_i$ if the kinetically derived non-productive binding constant of the esters to the CTABr regions of the micelles. $1/K_p$ is the binding constant of the esters to CTABr micelles determined directly.
See Gitler and Ochoa-Solano (15,25) for experimental details.

per bound molecule since the number of 1,8-ANS molecules bound per mg of protein ($\underline{n}$) actually increases as the temperature is raised. Thus, for HFE membranes in 10 mM Tris-HCl, pH 7.4, the values determined at 5°, 26°, and 37° C of $\underline{n}$ are 53, 69 and 78 of $\overline{K}_{D(app)}$ 8.8, 16.6 and 22.6 µM respectively. At the same temperatures, the relative fluorescence intensities per µ mole of ANS bound are 72, 40 and 25 respectively.

The leveling off in the fall in 1,8-ANS fluorescence which occurs in the HFE membranes at 44.7° C seems to be due to the appearance of protein binding sites. Thus on treatment of the membranes with trypsin, this leveling disappears and the membranes now show a behavior similar to that of sonicated egg lecithin. On the other hand, treatment with phospholipase C makes the leveling effect more apparent and it now begins to occur at a much lower temperature (30°C). An actual rise in the fluorescence with temperature is seen which is reminiscent of the findings of Galy and Edelman (cited in 8) who observed such a temperature dependent increase in the fluorescence of 2-p-toluidinyl-naphthalene-6-sulfonate bound to light chains of human globulins. We also have observed a similar effect of temperature on the fluorescence of 1,8 ANS bound to "spectrin." In the case of the membranes, the appearance of protein binding sites does not seem to be due to gross denaturation since the change in fluorescence with temperature are reversible up to 46°. Vanderkooi and Martonosi (82) observed the same fluorescence changes on heating and cooling skeletal muscle microsomal membranes in the presence of 1,8 ANS up to temperature of 70° C. Kasai et al. (41), on the other hand, describe a markedly different behavior of the ANS fluorescence on heating from that which occurs on cooling of electric organ membranes. These authors, however, heated the membranes to 80° prior to cooling.

An interesting aspect of the data shown in Figure 3 is the change in the slope of the linear portions on treatment of the membranes with the enzymes. Thus the slope which for the intact membranes is of -4557 cal (for the linear portion between 8° and 44.7° C) decreases on treatment with either trypsin or phospholipase C to values of -3157 and -3441 cal respectively. The resulting slope is near to that of the sonicated phospholipids.

When the temperature behavior of 1,8-ANS bound to nonionic micelles is studied, it is found that the fluorescence decreases linearly to a point immediately before the cloud point of the detergent and thereafter the slope is increased. If nonionic detergents with different oxyethylene chains are studied, the break in the curve is seen always to occur some 2°-3° C before the known cloud point. It is believed that the solution becomes cloudy at a temperature where the oxyethylene groups become dissolvated. Thus it is very likely that the 1,8-ANS fluorescence falls when the interaction with water is lost. This might indicate that the mobility of water at the membrane surface affects ANS fluroescence and would thus explain the negative temperature coefficients observed when the dye is in the membrane surface (67).

A MEMBRANE MODEL

In an attempt to summarize the evidence presented in the previous sections in a more didactic manner, a model of the membrane is presented. It should be noted that the evidence used comes from different sources such as the erythrocyte, myelin, liver microsomes and bacteria and that in no case is there sufficient evidence available to attempt a generalization of membrane structure. The ideas presented are an extension of those recently detailed by Singer (75) to support his so-called Lipid-Globular Protein Mosaic Model.

Some of the observations on which the model is based are listed below: (a) DTA (79) and X-ray diffraction (19) of membrane lipids indicate that they exist as a lamellar, smectic mesophase. (b) Sufficient lipid is available in the erythrocyte to just barely form a bilayer surrounding the cell (16). (c) Some 2/3 of the membrane lipids in M. laidlawii melt at temperatures which indicate that their apolar interactions are with other lipid molecules and not with proteins. The remaining 1/3 could be interacting with proteins. (d) The amino groups of membrane phospholipids in situ react with TNBS at rates essentially similar to those of the same lipids present in aqueous dispersions. (e) Phospholipase C digests some 70% of membrane phospholipids. That fraction not hydrolysed contains the majority of the membrane phosphatidyl serine. (f) The measured surface tensions of phospholipid bilayers yield values of 1 or less dyn/cm (55). (g) Some membrane proteins appear to be formed of apolar regions which anchor the proteins in the membrane through hydrophobic interactions, while the protein areas containing accumulation of ionic groups are in the aqueous environment. In glycoproteins, the carbohydrate will be exposed to the aqueous environment in the outside of the cell (77). (h) 1,8-ANS appears to bind to boundary regions between lipid and protein (71). This binding is very sensitive to cations, calcium being some 100 fold more effective than sodium.

The derived model is shown in Figure 7. The arguments to present only globular proteins is as follows. Proteins tend to spread at the air-water interface in order to decrease the surface free energy. In bilayers, the surface tension is nil and thus there is no driving force for the spreading of proteins at the bilayer-water interface, (cf. Cecil and Louis 9). Studies of circular dichroism (29,49) and infrared spectroscopy (52,83) support the globular nature of membrane proteins, although it should be noted that assuming 1 mg of protein covers 1 m^2 of surface area at the air-water interface, less than 1% of the membrane protein of the erythrocyte would be required to cover the entire cell surface. The above methods cannot detect such a small fraction of randomly coiled protein among other essentially globular proteins.

In the erythrocyte, membrane proteins have a high net negative charge (4,66) and the lipids would also have a net negative charge; thus the lipid protein interface would be very sensitive to counterions leading to stabilization by the formation of tricomplexes (cf. Bungenberg de Jong 8).

CONCLUSIONS

Some regions of the membrane (depicted in the model in Figure 8 by capital letters) are probably the sites of adaptive changes during exposure of an organism to the cold. Site A, the apolar region of the

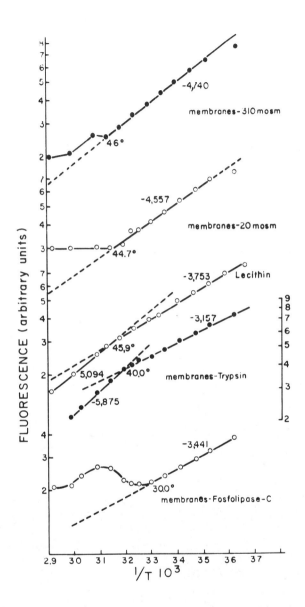

FIG. 7. The effect of temperature of the fluorescence emission of 1,8-ANS (8×10^{-5}M) bound to HFE membranes, HFE membranes treated with trypsin and phospholipase C and to sonicated egg lecithin (68).

lipid bilayer must have an adequately disordered state to function as a diffusion barrier as detailed previously. It can now be studied by fatty acid analysis, DTA and depolarization of the fluorescence of apolar molecules. Region B represents the lipid-water interface while C is the lipid-protein-water interface. Both are probably the loci of ionic effect, changes in structured water, etc. These regions can be studied by the binding of fluorescent probes such as 1,8-ANS ethidium and others, and also by the kinetic analysis of the chemical reactions of specific functional groups. The region of protein-protein interactions (D), is probably the site of allosteric, cooperative phenomena known for other multiple subunit proteins. These might involve effectors such as hormones which could alter enzymatic activities associated with membrane function. These will have to be studied by the procedures developed for soluble enzymes (44). Site E, the most external region of the membrane is involved with recognition phenomena and electrokinetic behavior.

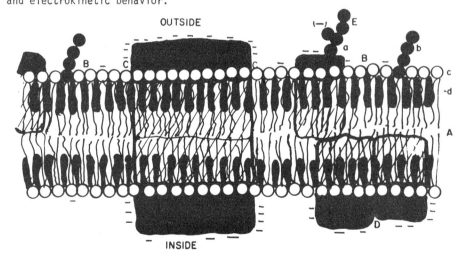

FIG. 8. Schematic membrane model. The components are (a) Scaloproteins, (b) glycolipids, (c) phospholipids, (d) cholesterol. The proteins are shown both spanning the bilayer and intercalated within it. Dark areas contain essentially polar groups, hatched region are essentially apolar.

An attempt has been made to present an integrated, rather than detailed, picture of current knowledge of membrane structure. It is hoped that this might allow the formulation, following future experiments, which will aid in the elucidation of the events leading to cold adaptation.

ACKNOWLEDGMENTS

A significant part of the membrane model presented is based on a similar analysis made available to me prior to publication by Dr. S. J. Singer.

In addition, I would like to thank Drs. M. Montal and B. Rudy for the many discussions which contributed in an important manner to the present text.

REFERENCES

1. ALEXANDER, A. E. The role of hydrogen bonds in condensed monolayers. Proc. Roy. Soc., London, Ser. A 179: 470-485, 1942.

2. BECKETT, A. H., and R. J. WOODWARD. Surface active betaines: N-alkyl-N,N-dimethylglycines and their critical micelle concentrations. J. Pharm. Pharmacol. 15: 442-431, 1963.

3. BERNARD, L. de. Associations moléculaires entre les lipides. II. - Lecithine et cholesterol. Bull. Soc. Chim. Biol. 40: 161-170, 1958.

4. BLUMENFELD, O. O. The proteins of the erythrocyte membrane obtained by solubilization with aqueous pyridine solution. Biochem. Biophys. Res. Commun. 30: 200-205, 1968.

5. BOOIJ, H. L. Association colloids. In: Colloid Science II. Reversible Systems, edited by H. R. Kruyt. New York: Elsevier Pub. Co., 1949, p. 681-722.

6. BRADFORD, H. F., P. D. SWANSON, and D. B. GAMMACK. Constituents of a microsomal fraction from the mammalian brain: their solubilization, especially by detergents. Biochem. J. 92: 247-254, 1964.

7. BRAUNS, R. Flussige Kristalle und Lebewesen. Stuttgart: E. Schweizerhartsche Verlags, 1931. p. 125-202.

8. BUNGENBERG DE JONG, H. G. Complex colloid systems. Colloid Science II Reversible Systems, edited by H. R. Kruyt. New York: Elsevier Pub. Co., 1949, p. 335-432.

9. CECIL, R. and C. F. LOUIS. Protein-hydrocarbon interactions. Interactions of various proteins with decane in the presence of alcohols. Biochem. J. 117: 147-156, 1970.

10. CERBÓN, J. Nuclear magnetic resonance of water in microorganisms. Biochim. Biophys. Acta 88: 444-447, 1964.

11. CERBÓN, J. N.M.R. studies on the water immobilization by lipid systems in vitro and in vivo. Biochim. Biophys. Acta 144: 1-9, 1967.

12. CHANGEUX, J.-P., J. THIERY, Y. TUNG, and C. KITTEL. On the cooperativity of biological membranes. Proc. Natl. Acad. Sci. U.S. 57: 335-341, 1967.

13. CHAPMAN, D., and D. T. COLLIN. Differential thermal analysis of phospholipids. Nature 206: 189, 1965.

14. CHAPMAN, D., and D. F. H. WALLACH. Recent physical studies of phospholipids and natural membranes. In: Biological Membranes, edited by D. Chapman. London: Academic Press, 1968, p. 125-202.

15. CORDES, E. H., and C. GITLER. Reaction kinetics in the presence of micelle forming surfactants. In: Bioorganic Chemistry, edited by T. Keiser and F. Kezdy. Chicago: University of Chicago Press. In press.

16. CORNWELL, D. G., R. E. HEIKKILA, R. S. BAR, and G. L. BIAGI. Blood cell lipids and the plasma membrane. J. Amer. Oil Chem. Soc. 45: 297-304, 1968.

17. DAS, M. L., and F. L. CRANE. Proteolipids. I. Formation of phospholipid-cytochrome c complexes. Biochemistry 3: 696-700, 1964.

18. DEMUS, H., and E. MEHL. Identification of water-insoluble membrane proteins by immunoelectrophoresis in a solubilizing urea-triton solvent. Biochim. Biophys. Acta 211: 148-157, 1970.

19. ENGELMAN, D. M. X-ray diffraction studies of phase transitions in the membrane of Mycoplasma laidlawii. J. Mol. Biol. 47: 115-117, 1970.

20. FRANZEN, J. S., and R. E. STEPHENS. The effect of a dipolar solvent system on interamide hydrogen bonds. Biochemistry 2: 1321-1327, 1963.

21. GAINES, G. L. Insoluble Monolayers at Liquid-Gas Interfaces. New York: Interscience Publishers, 1966.

22. GENT, W. L. G., E. H. GRANT, and S. W. TUCKER. Evidence from dielectric studies for the presence of bound water in myelin. Biopolymers 9: 124-126, 1970.

23. GITLER, C. Propiedades de superficies biológicas. In: Ensayos Bioquimicos, edited by G. Soberón. Mexico: Prensa Medica Mexicana, 1969.

24. GITLER, C., G. MARTINEZ-ZEDILLO, D. MARTINEZ-ROJAS, and G. CHAVEZ-DIAZ. Factors influencing the activity of membrane-bound enzymes in the erythrocyte. Natl. Cancer Res. Inst. Monogr. 27: 153-164, 1967.

25. GITLER, C., and A. OCHOA-SOLANO. Nonpolar contributions to the rate of nucleophilic displacements of p-nitrophenyl esters in micelles. J. Amer. Chem. Soc. 90: 5004-5009, 1968.

26. GITLER, C., B. RUBALCAVA, and A. H. CASWELL. Fluorescence changes of ethidium bromide on binding to erythrocyte and mitochondrial membranes. Biochim. Biophys. Acta 193: 479-481, 1969.

27. GITLER, C., and B. RUBALCAVA. In: Fluorescent monitors for the study of membrane structure and function. Recent Progress in Microbiology. In press.

28. GITLER, C., and B. RUBALCAVA. Interaction of fluorescent probes with hemoglobin-free erythrocyte membranes. In: Structural Probes for the Function of Proteins and Membranes, edited by B. Chance, T. Yonetoni, and C. P. Lee. New York: Academic Press. In press.

29. GLASER, M., H. SIMPKINS, S. J. SINGER, M. SHEETZ and S. I. CHAN. On interactions of lipids and proteins in the red blood cell membrane. Proc. Natl. Acad. Sci. U.S. 65: 721-728, 1969.

30. GODDARD, E. D., O. HARVA, and T. G. JONES. The effect of univalent cations on the critical micelle concentration of sodium dodecyl sulphate. Trans. Faraday Soc. 49: 980-984, 1953.

31. GOLDFARB, A. R. A kinetic study of the reactions of amino acids and peptides with trinitrobenzenesulfonic acid. Biochemistry 5: 2570-2574, 1966.

32. GOTTERER, G. S. Rat liver D-β-hydroxybutyrate dehydrogenase II. Lipid requirement. Biochemistry 6: 2147-2152, 1967.

33. GWYNNE, J. T. and C. TANFORD. A polypeptide chain of very high molecular weight from red blood cell membranes. J. Biol. Chem. 245: 3269-3271, 1970.

34. HABEEB, A. F. S. A. Determination of free amino groups in proteins by trinitrobenzenesulfonic acid. Anal. Biochem. 14: 328-336, 1966.

35. HARTLEY, G. S. Aqueous Solutions of Paraffin Chain Salts. A Study in Micelle Formation. Paris: Hermann et Cie, 1936.

36. HERNÁNDEZ-MONTES, H. Una Glicerol-Ester Hidrolasa en Mitocondrias de Higado de Rata Tratadas con Fosfatidasa C. (M.S. Thesis) Centro de Investigacion y de Estudios Avanzados del I.P.N., México, 1966.

37. HUSSON, F., and V. LUZZATI. Structure of red-cell ghosts and the effect of saponin treatment. Nature 197: 822, 1963.

38. ITO, A., and R. SATO. Purification by means of detergents and properties of cytochrome b_5 from liver microsomes. J. Biol. Chem. 243: 4922-4923, 1968.

39. JENCKS, W. P. Catalysis in Chemistry and Enzymology. New York: McGraw-Hill Series in Advanced Chemistry, 1969.

40. KAPLAN, A. E. Hydrogen bonding properties of N-palmitoyl-L-serine. Studies related to membrane reactions. J. Colloid Interface Sci. 25: 63-70, 1967.

41. KASAI, M., J. P. CHANGEUX, and L. MONNERIE. In vitro interaction of 1-anilino 8-naphthalene sulfonate with excitable membranes isolated from the electric organ of Electrophorus electricus. Biochim. Biophys. Res. Commun. 36: 420-427, 1969.

42. KAUZMAN, W., and H. EYRING. The viscous flow of large molecules. J. Amer. Chem. Soc. 62: 3113-3125, 1940.

43. KLOTZ, I. M., and J. S. FRANZEN. Hydrogen bonds between model peptide groups in solution. J. Amer. Chem. Soc. 84: 3461-3466, 1962.

44. KOSHLAND, D. E., and K. E. NEET. The catalytic and regulatory properties of enzymes. Ann. Rev. Biochem. 37: 359-410, 1968.

45. KURZ, J. L. Effects of micellization on the kinetics of the hydrolysis of monoalkyl sulfates. J. Phys. Chem. 66: 2239-2246, 1962.

46. LAICO, M. T., E. I. RUOSLAHTI, D. S. PAPERMASTER, and W. J. DREYER. Isolation of the fundamental polypeptide subunits of biological membranes. Proc. Natl. Acad. Sci. U.S. 67: 120-127, 1970.

47. LAWRENCE A. S. C. Lyotropic mesomorphism. Trans. Faraday Soc. 29: 1008, 1933.

48. LEHMANN, O. Über fliessende Kristalle. Z. Phys. Chem. 4: 462-472, 1889.

49. LENARD, J., and S. J. SINGER. Structure of membranes: reaction of red blood cell membranes with phospholipase C. Science 159: 738-739, 1968.

50. LUZZATI, V. X-ray diffraction studies of lipid-water systems. In: Biological Membranes, edited by D. Chapman. London: Academic Press, 1968, p. 71-123.

51. LUZZATI, V., and F. HUSSON. The structure of the liquid-crystalline phases of lipid-water systems. J. Cell Biol. 12: 207-219, 1962.

52. MADDY, A. H., and B. R. MALCOM. Protein conformations in the plasma membrane. Science 150: 1616-1618, 1965.

53. MELMAN, A., and R. M. ROSENBAUM. Histochemical correlates for differences in functional activity of kidneys from active and cold-stored summer bats, (Myotis lucifungus). Anat. Rec. 145: 401-411, 1963.

54. MOLYNEAUX, P., C. T. RHODES, and J. SWARBRICK. Thermodynamics of micellization of N-alkyl betaines. Trans. Faraday Soc. 61: 1043-1052, 1965.

55. MORAN, A., and A. ILANI. Surface tension of an artificial bileaflet membrane in comparison to parent lipid solution-water interfacial tension. Chem. Phys. Lipids 4: 169-180, 1970.

56. MUKERJEE, P., and A. RAY. Charge-transfer interactions and the polarity at the surface of micelles of long-chain pyridinium iodides. J. Phys. Chem. 70: 2144-2149, 1966.

57. OHNISHI, T. Extraction of actin- and myosin-like proteins from erythrocyte membrane. J. Biochem., Tokyo, 52: 307-308, 1962.

58. OKUYAMA, T., and K. SATAKE. On the preparation and properties of 2,4,6-trinitrophenyl-amino acids and -peptides. J. Biochem., Tokyo, 47: 454-466, 1960.

59. OVERATH, P., H. U. SCHAIRER, and W. STOFFEL. Correlation of in vivo and in vitro phase transitions of membrane lipids in Escherichia coli. Proc. Natl. Acad. U.S. 67: 606-612, 1970.

60. PACKTER, A., and M. DONBROW. Ion-pair formation in aqueous solutions of organic electrolytes. Proc. Chem. Soc. London, 220-221, 1962. (Note: not issued in volumes.)

61. PAPAHADJOPOULOS, D., and L. WEISS. Amino groups at the surfaces of phospholipid vesicles. Biochim. Biophys. Acta 183: 417-426, 1969.

62. PENEFSKY, M. S., M. E. PULLMAN, A. DATTA, and E. RACKER. Partial resolution of the enzymes catalyzing oxidative phosphorylation. J. Biol. Chem. 235: 3330-3336, 1960.

63. PENEFSKY, H. S., and R. C. WARNER. Partial resolution of the enzymes catalyzing oxidative phosphorylation. VI. Studies on the mechanism of cold inactivation of mitochondrial adenosine triphosphatase. J. Biol. Chem. 240: 4694-4702, 1965.

64. RAZIN, S., B. J. COSENZA, and M. E. TOURTELLOTTE. Variations in Mycoplasma morphology induced by long-chain fatty acids. J. Gen. Microbiol. 42: 139-145, 1966.

65. RAZIN, S., M. E. TOURTELLOTTE, R. N. McELHENY, and J. D. POLLACK. Influence of lipid components of Mycoplasma laidlawii membranes on osmotic fragility of cells. J. Bacteriol. 91: 609-616, 1966.

66. ROSEMBERG, S. A., and G. GUIDOTTI. The protein of human erythrocyte membranes. J. Biol. Chem. 243: 1985-1992, 1968.

67. RUBALCAVA, B. La Interaccion de dos Monitores Fluorescentes con Membranas Biológicas. (Ph.D. Thesis.) Centro de Investigación y de Estudios Avanzados del I. P. N., México, 1970.

68. RUBALCAVA, B., and C. GITLER. Interaction of fluorescent probes with membranes. II. Site of binding of ANS in erythrocyte membranes. Biochemistry. In press.

69. RUBALCAVA, B., D. MARTINEZ de MUÑOZ, and C. GITLER. Interaction of fluorescent probes with membranes. I. Effect of ions on erythrocyte membranes. Biochemistry 8: 2742-2747, 1969.

70. SCHICK, M. J., and F. M. FOWKES. Foam stabilizing additives for synthetic detergents. Interaction of additives and detergents in mixed micelles. J. Phys. Chem. 61: 1062-1068, 1957.

71. SEKUZU, I., P. JURTSHUK, and D. E. GREEN. Studies on the electron transfer system. II. Isolation and characterization of the D-(-)-β-hydroxybutyric apodehydrogenase from beef heart mitochondria. J. Biol. Chem. 238: 975-982, 1963.

72. SHAH, D. O., and J. H. SCHULMAN. Influence of calcium, cholesterol and unsaturation on lecithin monolayers. J. Lipid Res. 8: 215-226, 1967.

73. SHINITZKY, M., A.-C, DIANOUX, C. GITLER, and G. WEBER. Microviscosity and order in the hydrocarbon regions of micelles and membranes determined with fluorescent probes. I. Synthetic micelles. Biochemistry. In press.

74. SINGER, S. J. The properties of proteins in nonaqueous solvents. Adv. Protein Chem. 17: 1-68, 1962.

75. SINGER, S. J. The molecular organization of biological membranes. In: Structure and Function of Biological Membranes, edited by L. I. Rothfield. New York: Academic Press, 1971. In press.

76. SMALL, D. A classification of biological lipids based upon their interactions in aqueous systems. J. Amer. Oil Chem. Soc. 45: 108-117, 1968.

77. STECK, T. L., R. S. WEINSTEIN, J. H. STRAUS, and D. F. H. WALLACH. Inside-out red cell membrane vesicles: preparation and purification. Science 168: 255-257, 1970.

78. STEIM, J. M., O. J. EDNER, and F. G. BARGOOT. Structure of human serum lipoproteins: nuclear magnetic resonance supports a micellar model. Science 162: 909-911, 1968.

79. STEIM, J. M., M. E. TOURTELLOTTE, J. C. REINERT, R. N. McELHANEY, and R. L. RADER. Calorimetric evidence for the liquid-crystalline state of lipids in biomembranes. Proc. Natl. Acad. Sci. U.S. 63: 104-109, 1969.

80. SWANSON, P. D., H. F. BRADFORD, and H. MCILWAIN. Stimulation and solubilization of the sodium ion-activated adenosine triphosphatase of cerebral microsomes by surface-active agents especially polyoxyethylene ethers: action of phospholipases and oneuraminidase. Biochem. J. 92: 235-247, 1964.

81. TRIGGLE, D. J. Some aspects of the role of lipids in lipid-protein interactions and cell membrane structure and function. In: Recent Progress in Surface Science, edited by J. F. Danielli, A. C. Riddiford, and M. D. Rosenberg. New York: Academic Press, 1970, vol. 3, p. 273-290.

82. VANDERKOOI, J., and A. MARTONOSI. Sarcoplasmic reticulum. VIII. Use of 8-anilo-1-naphthalene sulfonate as conformational probe on biologic membranes. Arch. Biochem. Biophys. 133: 153-163, 1969.

83. WALLACH, D. F. H., and P. H. ZAHLER. Infrared spectra of plasma membrane and endoplasmic reticulum of Ehrlich ascites carcinoma. Biochim. Biophys. Acta 150: 186-193, 1968.

84. WAYS, P., and D. J. HANAHAN. Characterization and quantification of red cell lipids in normal man. J. Lipid Res. 5: 318-328, 1964

85. WILLIS, J. S. Cold adaptation of activities of tissues of hibernating mammals. In: *Mammalian Hibernation III*, edited by K. C. Fisher, A. R. Dawe, C. P. Lyman, E. Schonbaum, and F. E. South. Edinburgh: Oliver and Boyd, 1967, p. 356-381.

86. WILSON, G., S. P. ROSE, and C. F. FOX. The effect of membrane lipid unsaturation on glycoside transport. *Biochem. Biophys. Res. Commun.* 38: 617-623, 1970.

CREDITS

The editors and authors are grateful to the following publishers, organizations and individuals for permission to use materials previously published:

Figure 5: American Chemical Society; B. Rubalcava, D. Martinez de Muñoz and C. Gitler, *Biochemistry* 8, 2742-2747, 1969. Copyright American Chemical Society.

PERSPECTIVES ON THE ROLE OF MEMBRANES IN HIBERNATION AND HYPOTHERMIA

INTRODUCTION

In the following discussion we have attempted to do two things. The first aim has been to make statements regarding our view of the current positions of several issues of our topic as they have emerged from common or similar findings amongst the several papers. Our second purpose has been to point to new directions and approaches that may be profitable in the analysis of membrane function in hibernators and failure in hypothermic non-hibernators.

ROLE OF MEMBRANE

Two possible classes of hypothermic injury related to membrane have previously been discussed. One, fairly speculative, envisions direct damage to individual cells. The other is concerned with malfunction occurring because of loss of integrative activity as, for example, in conduction. Changes in hormone secretion would also fall in this latter category. The role of Na^+ and K^+ transport, perhaps mediated by Ca^{2+}, in the activity of certain endocrine glands such as pancreatic islets and thyroid suggests that alteration of ion transport arising from hypothermia could lead to severe disturbance of the metabolism of the whole organism.

A third possible class of membrane-related injury can also be perceived which could occur partly because of the structural and functional organization of certain critical tissues. An example of this class would be the damage following rewarming of the hypothermic brain and heart which results from occlusion of microcirculation which in turn is a secondary consequence of swelling. Studies have shown, too, that in the intestinal and gastric mucosa massive cell death may follow hypothermia, and this may also be related to failure of secretory processes in the mucosal cells rendering them more vulnerable to onslaughts of low pH and proteolytic enzymes.

ROLE OF METABOLISM

A major conclusion we have reached is that, so far as we know, it is not the reduction of metabolism by cold which blocks ion transport in non-hypothermic hibernators. This statement is based upon studies of such intact tissues in vivo as brain and heart, of isolated tissues and cells, and, at the enzymatic level, of Na,K-ATPase. The meager effects on transport of partial metabolic inhibition in liver lends weight to this conclusion because they suggest that at least in that tissue, transport appears to take priority over the majority of other activities in competing for reserves of energy.

This is not to say that the many studies of cold adaptation of energy metabolism in hibernators are totally irrelevant to the problem of survival at low temperature. In some tissues of some species low energy production might limit transport. This might be a particularly good possibility in cells which must carry on multiple activities and with a relatively high work load, such as heart, diaphragm, and skeletal muscle. Then, too, observed adaptations of respiration and glycolysis may represent a "margin of safety" in hibernators. Nevertheless, (to our knowledge) limitation of energy release has not been demonstrated to be a critical problem in hypothermia of any tissue of a non-hibernating mammal. This fact, taken together with the extent and variety of cases for which the converse has been indicated, renders it imperative that for the future the relationship between energy production and utilization by each type of cell of interest be determined.

The above discussion relates entirely to the transport of Na^+ and K^+. The increasing emphasis on the regulator role of Ca^{2+} in living cells indicates the importance of understanding the effects of hypothermia on the metabolic coupling of the transport of this ion, also.

ACCLIMATION

A second major statement is that in the brain of hibernators, acclimation prior to or following the onset of hibernation in the individual organism represents the entire cold adaptation of that tissue. Again, this conclusion is based upon a variety of species and types of preparation extending from intact brains of dormice through slices of brain of hamsters, to the enzymatic analyses of brain of hamsters and hedgehogs. It is important to realize, however, that the brain is very likely an exception in this regard, although a very important one. Thus, earlier demonstrations of cold resistance of whole, perfused hearts, of red blood cells, or kidney cortex, of nerve, and of skeletal muscle were all based upon tissues and organs from non-hibernating ("warm-room awake") individuals of hibernating species. Of course, in some of these cases hibernation may confer an additional resistance on top of that which is apparently present at all times.

The analysis of the nature of this acclimation could follow a program such as that described previously. In addition, however, one must consider whether changes in, say, Na,K-ATPase involve simple increases in enzyme, either by synthesis or deinhibition, or an induction of different isozymes. An important and possible first step may be to determine whether there are increased numbers of pump sites in acclimated tissues by use of ouabain binding to tissues. A related question is whether ouabain binding in the

Na,K-ATPase preparation increases with acclimation or stays the same, yielding higher turn-over numbers for the enzyme.

ANALYSIS OF MEMBRANE FUNCTION

Progress towards understanding the molecular mechanisms of resistance of both structural and enzymatic aspects of membrane activity unfortunately must involve some degree of acceptance of specific models or hypotheses about membrane structure or function which have not been proven or even necessarily widely accepted. Thus, Fang's analysis of ATPase adaptation rests largely upon such a specific model. The approach to various aspects of membrane structure suggested by Gitler is similarly limited. Accepting this loss of innocence, however, as the primeval price of eating from the tree of knowledge, we may envision numerous fruitful lines of endeavor.

Analysis of passive permeability in intact cells of hibernators as a function of temperature has so far been entirely neglected. Even the older techniques of analyzing permeability of organic molecules by hemolysis of red blood cells might help to determine whether greater temperature sensitivity of cation permeability in ground squirrel erythrocytes reflects a more general difference in permeability. In red blood cells and in linear cells, also, it has recently been suggested that a rise in concentration of ionic calcium within the cell leads to an increase in passive K^+ permeability. Greater temperature sensitivity of K^+ permeability observed in guinea pig red blood cells (as compared with those of ground squirrels) might therefore represent a failure of the Ca^{2+} pump.

Depending on type of cell and of the molecule, permeability may be simply diffusive or be by some saturable mechanism. Determination of reflection coefficients to probe molecules in erythrocytes and in epithelial cell systems is a fairly well developed procedure which could yield information on permeability of the diffusive type as a function of temperature. In some epithelial cells passive entry of Na^+ is a saturable process and this can also be examined by careful analysis of the kinetics of Na^+ loading of the cells. The effect of temperature on the interaction of Na^+ and organic molecules in their mutual passive entry into epithelial cells, reticulocytes and muscle is also highly relevant.

Turning now to a more direct consideration of the membrane there is considerable evidence for changes in melting behavior of lipids as a function of changes in relative composition and in degree of saturation of their components. There is also good reason to believe that such alterations influence both the leakiness of membranes and the activity of enzymes bound to membranes.

An important first consideration, then, is whether such changes do occur in, say, brain during preparation for hibernation or whether major differences can be seen between tissues of hibernators and non-hibernators. If such differences are observed, then one must consider how alterations in saturation and molecular composition could be brought about. Two possibilities would be by de novo synthesis of phospholipids or the in situ deacylation and reacylation of phospholipids by enzymes within the membrane. The mechanisms for these processes should be examined in hibernators and non-hibernators with respect to their relative prevalence and their capacity for leading to specific ("adaptive") changes.

(Although similar sophisticated changes in protein structure could also be involved, a corresponding quantitative analysis of amino acid composition would be less likely to be of importance than, say, attempts to determine the effects of temperature on higher orders of protein structure.)

Beyond this, the question resolves to one of analysis of physical changes within membranes as a function of temperature. In addition to the battery of sophisticated physical techniques that can be applied to this purpose the relatively straight forward use of chemical probes described by Gitler offer a wealth of possible approaches. These methods and their application to lipid-lipid, lipid-protein and lipid-ion interactions will not be repeated here. It is worth noting, however, the possibility of exploring protein aggregations by a separation on agarose columns of membranes treated with Brij 36T. This relatively gentle procedure for membrane solvation, which leaves enzymes and proteins intact, may be of special use in detecting changes in aggregation of protein components of Na,K-ATPase.

A bothersome aspect of many studies suggested in the foregoing two paragraphs is the problem of selecting a proper control. For those tissues which exhibit either no acclimation or only minor acclimation, (which may include that most useful cell type, the erythrocyte), comparisons between species may be difficult owing to large non-relevant differences in basic composition. For acclimating tissues, such as brain, comparison of changes within the same species may be possible, but they may still be difficult to manipulate because acclimation may occur over relatively long periods of time. (Possibly reversal of acclimation upon the return to warm conditions would be more approachable). In either case the problem may be somewhat alleviated by the consideration that what one is looking for in each case are the comparisons of elements which vary with temperature.

CYTOPLASMIC ION AND WATER ADSORPTION

The assertion already has been made that it is necessary to adopt, at least provisionally, a particular set of views of membrane function before proceeding to a study of adaptation. This truism applies as well to the fundamental issue of whether the type of membrane activities assumed in the foregoing discussion do, in fact, govern ionic distribution. Alternative mechanisms have been advanced, among which the best known is that of Ling. In its simplest form this hypothesis envisions the cell as a semicrystalline matrix in which most of the water is organized into polarized multilayers. Solutes are either excluded from this matrix and therefore exist at low concentrations within the cell (e.g., Na^+) or are accumulated by being adsorbed to specific sites on both membrane and cytoplasmic molecules (e.g., K^+). According to this hypothesis, the role of metabolism in the resting cell is to maintain a critical concentration of ATP in order to stabilize the matrix, rather than to provide a source of high-energy phosphate bonds for contant turnover. In terms of the general problem of the effect of temperature on membrane function, this hypothesis engenders some of the same questions as the "orthodox" membrane hypothesis, for example, the importance of metabolism and the effect of temperature on protein conformation (but now more particularly, cytoplasmic protein). As newer techniques (such as nuclear magnetic resonance) become more available for estimating the possibility and extent of adsorption of ions and other solutes and the organization of water within the cell, an altertness for their application to studying the effects thereon of temperature and of temperature adaptation will be necessary.

CONCLUSION

The foregoing descriptions of our work and summary of our discussions indicate, in our opinion, that the effects of temperature on the occurrence and control of solute movements between cells and their surroundings, are probably a decisive factor in determining the ability, or otherwise, of organisms to survive at low temperatures. The study of membrane and transport phenomena in hypothermia is clearly of great interest from the standpoints of both general biology and applied medicine. Progress towards mechanisms has so far been meager; many avenues of approach are open and waiting to be traveled by freshly motivated new explorers.

                                            Symposium Committee on Membranes and Electrolyte Phenomena

                                            John S. Willis, Chairman
                                            Leslie S. T. Fang
                                            Carlos Gitler
                                            Nicholas Mendler
                                            Margaret C. Neville
                                            Michael W. Smith
                                            G. D. V. van Rossum

CARDIAC MUSCLE

ULTRASTRUCTURE OF CARDIAC MUSCLE

A comparative review with emphasis on the muscle
fibers of the ventricles

J. R. SOMMER[1], R. L. STEERE[2],

E. A. JOHNSON[3], and P. H. JEWETT[4]

[1]Depts. of Pathology, Physiology and Pediatrics,
Duke University Medical Center and
the Veterans Administration Hospital, Durham, N. C. 27705

and

[2,3,4]Plant Virology Laboratory, Crops Research
Division, Agricultural Research Service
U. S. Department of Agriculture, Beltsville, Md. 20705

INTRODUCTION

It is important to know the structure of cardiac muscle when it is in a state that can reasonably be presumed to be normal. For only then can one begin to attempt a comparison between one set of findings and that of another in search of identifying the abnormal. The definition of normal and abnormal states is especially difficult in ultrastructural enquiries, as opposed to comparable light microscopic ones, because adequate sampling often does not seem practical. Moreover, our understanding of structural alterations produced by preparative procedures is barely contemporary with the light microscope, much less with the electron microscope; the calamity in the case of fixation can aptly be put by suggesting that in commenting on quality of fixation one pronounces Aesthetics. The physiologist is faced with similar problems that, while being peculiar to his field of knowledge and methods of enquiry, are often equally prohibitive to getting precise information. As a result, studies in structure-function correlation are informative only to the extent that the accuracy of the observation from one field of study is matched by that coming from the other, and from this point of view, structure-function correlation on cardiac muscle has fallen behind that of skeletal muscle mainly because cardiac muscle fibers cannot be studied in isolation.

The following is a brief and general review of the ultrastructure of cardiac muscle of several animals taking advantage of the information provided by several kinds of tissue preparation. In complementing one another, not only do these several different preparative procedures combine to give an extended view of cardiac ultrastructure, but they also offer an opportunity to assess the distortions that are peculiar to each.

CONTRACTILE MATERIAL

Cardiac muscle, like skeletal muscle, is striated, the contractile material being divided into identical repeating units, the sarcomeres, the sarcomere being the distance between consecutive Z lines (Plate 1). Additional substriations are generated by the myofilaments, comprising interdigitating thick filaments (myosin) and thin filaments (actin), (Plates 2, 6) which are arranged in such a way as to form bands, the A bands and I bands, respectively (Plate 1). The actin filaments seem to be fastened to the Z lines at one end (14, 34); the free ends slide between the myosin filaments during the contraction-relaxation cycle (25,27). The force responsible for the motion of the filaments is presumed to arise from the cyclic making and breaking of interconnecting sidebridges between the thin and thick filaments (Plates 6, 20). In skeletal muscle the developed force varies with the sarcomere length, that is to say with the extent and kind of overlap of thick and thin filaments and thus, presumable, with the number of side bridges (18). The uncontracted, resting length of the sarcomere of skeletal muscle fibers is 2.0 to 2.2 µ. If the fiber is excited to contract and is prevented from shortening by being held just taut prior to and during tetanic excitation, the developed tetanic tension is maximal in this range of length and declines if the muscle is either allowed to contract to a shorter, or is stretched to greater, length. For example, the tetanic force decreases, linearly, from this maximum to zero as the sarcomere length is increased from 2.2 µ to 3.56 µ. A similar proportionality has not yet been established in cardiac muscle because of the lack of a suitable preparation equivalent to the single skeletal muscle fiber of the frog on which these relationships were determined. Nevertheless, in the absence of evidence to the contrary, it would appear from ultrastructural comparisons that a sliding filament mechanism for contraction, basically identical to that of skeletal muscle, exists in cardiac muscle. That is to say, the basic mechanical properties of cardiac muscle are identical to skeletal muscle in that, for example, an identical relationship exists between developed force and sarcomere length, and apparent differences in the mechanical behavior of populations of cardiac muscle fibers as compared with that of a single skeletal muscle fiber can be explained on the basis of inhomogeneity of fiber geometry and sarcomere length (17). As predicted by the sliding filament theory, H bands occur in cardiac muscle under certain circumstances and, under reverse circumstances, so does the double overlap of actin filaments across the center region (M-line) of the A band (68). The appearance of the contractile material in cardiac muscle of several different animals is quite similar, but M lines are only rarely seen in frog and bird hearts (cf. Plates 1, 6, 18).

CELLS AND BUNDLES

Mammalian cardiac muscle consists of a functional syncitium of small individual cells (4) (10-15 µ in diameter and of uncertain, but short, length) which are attached to one another mainly through the intercalated discs. Skeletal muscle, on the other hand, is composed of many large (100 µ O.D.) and long (centimeters) separate, multinucleated cells which form bundles inserting at either end into an exo- or endoskeleton through a tendon.

In the ventricles of most animals two populations of muscle fibers can be distinguished. One is made up of working fibers which provide the contractile force of the ventricles, and the other is made up of fibers that appear to constitute the conduction system (Purkinje fibers in the broadest sense). The diameter of the individual working and conducting fibers (see below) varies among different animal orders. In all mammals the diameter of the cardiac working fiber is about 10 to 15 µ, in the birds about 6-8 µ, and in the frog about 3 µ. Immature animals have somewhat smaller fiber diameters than those observed in adult animals (28).

Both working and conducting fibers are covered by sarcolemma with a basement membrane, and both kinds of fibers aggregate into bundles. Within the bundles each fiber is surrounded by collagen-containing extracellular space (Plates 2, 3) except where the fibers are closely held together by several kinds of junctional complexes (Plates 6-14). Junctional complexes are discrete structural specializations of the cell surface (11, 55). It is important to note that fibers forming bundles may be loosely assembled or tightly packed (Plates 4, 5) (60). By and large the working fibers of mammals seem to be loosely assembled into bundles, the width of the extracellular spaces between the fibers being in the order of, say, 0.5 µ or more (Plate 4). In frogs and birds several fibers often combine to form small tightly packed bundles in which the intercellular space between individual fibers measures about 30 mµ in width and is devoid of collagen and basement membrane (cf. Plate 5). The geometry of fiber apposition cannot safely be disregarded as trivial when electrophysiologic observations made on bundles of fibers are to be interpreted meaningfully. While this is, perhaps, more readily apparent in the case of the nexus, which is presumably related to low resistance pathways between adjacent cells, other forms of cell apposition are not without consequence to the overall electrical behavior of fiber bundles (see below). The geometry of fiber apposition in different animals, and within different regions of one single heart, needs further clarification.

Mammalian working fibers have transverse tubules (at the level of the Z lines, Plate 1) which are tubular and sometimes branching invaginations of sarcolemma complete with basement membrane (43). The diameters of these tubules are, in general, very much larger than the transverse tubules of skeletal muscle. Transverse tubules have been found to be absent in all Purkinje fibers (60), as well as in the working fibers of several amphibians (61,69), fishes (76), reptiles (35), oppossum (unpublished observations), birds (30,61), immature mammals (29, 50) and of most atrial fibers in mammals (41,60). Generally, the diameter of those working fibers that have transverse tubules is about twice that of working fibers that do not. However, it is premature at this point to conclude that fiber diameter is the determining factor of whether transverse tubules are present or not. For, although Purkinje fibers and most atrial fibers of mammals do not have transverse tubules, in the case of the small mammals the diameter and the amount and location of contractile material within such fibers is close to identical to that seen in the corresponding working fibers which do have transverse tubules. Because of the large diameter (50-100 µ) of skeletal

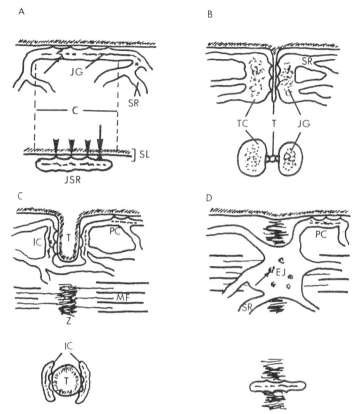

FIG. 1. Schematic drawings in two different planes, of the various kinds of couplings found in striated muscle. A. The coupling (C) consists of sarcolemma (SL), junctional SR (JSR), junctional granules (JG), and junctional processes (60). The junctional processes (arrow heads) have a varied appearance. Most frequently they appear as densities (large arrow) between the cytoplasmic faces of the sarcolemma and junctional SR. The junctional granules in cardiac muscle often appear to fuse to form a membrane (central membrane) (3) which in the case of mammalian cardiac muscle only occasionally extends beyond the region of the coupling (asterisk) while in the hearts of some birds such extensions are the rule. When a coupling is formed with the surface sarcolemma that surrounds the entire cell, the coupling is called peripheral coupling. B. Two couplings in skeletal muscle make up the triad of Porter and Palade (46). The junctional SR (the terminal cisternae, TC) of the triads is dilated, contains granular material (junctional granules, JG), and is attached to a transverse tubule (T) via junctional processes. There are no peripheral couplings in vertebrate skeletal muscle. C. Mammalian cardiac muscle has both interior (IC) and peripheral couplings (PC). When a transverse tubule (T) has two couplings a triad is formed, when only one coupling is present a dyad results. Cardiac fibers of birds and other lower animals and Purkinje fibers have no trans-

(continued)

verse tubules, and thus no interior couplings. However, cardiac fibers of all animals have peripheral couplings, except those of Rana pipiens in which couplings have so far escaped detection. Z = Z line. MF = Myofibrils. D. In the hummingbird and finch, in which the cardiac fibers have no transverse tubules, the junctional SR of peripheral couplings (PC) is extended deep into the interior of the cells where it at least partially envelopes the Z lines and I bands. EJ = extended junctional SR; arrow points to junctional processes which are directed toward Z line material on either side (cf. Plates 18-20).

muscle fibers transverse tubules are needed in skeletal muscle for excitation-contracting coupling (21). In this respect it is not surprising to find that skeletal muscle does not have peripheral couplings (Plates 8, 16, 17 and Fig. 1). The much smaller diameter of mammalian cardiac fibers (10-15 µ) seems to obviate the need for transverse tubules (especially in view of the fact that these fibers do have peripheral couplings, see below). At this time, to us, the function of transverse tubules in cardiac muscle of mammals is not immediately apparent.

JUNCTIONAL COMPLEXES

Junctional complexes, of which there are many different types, are discrete structures at sites where two cells are joined with one another (11,55). The following types occur in cardiac muscle (Plates 8-13):

    a) Macula adherens (the desmosome).

    b) Fascia adherens (the intermediate junction).

    c) Nexus (the gap junction).

Although the f. adherens in frogs, and sometimes in birds, looks different from that seen in mammalian hearts, for the purpose of this review it is grouped with the fascia adherens. Desmosomes and, especially, ff. adherentes combine to form the intercalated disc which traverses the heart muscle fiber transversely. Where it occurs the intercalated disc is in register with Z lines (Plates 1,6) and establishes end-to-end attachments between two or more cells. Within the intercalated disc the ff. adherentes and mm. adherentes display all manner of transition from one form to the other which is, in part, due to varying angles of sectioning, but also depends on previous preparative procedures. Nexuses are often included in the definition of the intercalated disc (72). However, although nexuses can almost always be demonstrated in intercalated discs, nexuses are also found in regions of the cell surface that are not associated with the intercalated disc proper. For this reason we prefer not to include the nexus in the structural definition of the intercalated disc. The nexus which almost always is oriented parallel to the fiber axis (Plates 6,7) consists of the two apposed plasma membranes (unit membranes) of two adjacent cells (8). Between the two unit membranes, however, there seems to be a 2 mμ gap which is accessible to small molecules (e.g. lanthanum) from the extracellular space (48) (Plates 11,12,13).

Recently, attempts have been made to elucidate the tridimensional structure of the nexus further by employing the freeze-etch technique of Steere (70,71) and the accurate topography of the two fracture surfaces previously described (5,35,39) has been clarified by making complimentary replicas of both fracture surfaces (6,62,65). The findings have borne out previous observations in thin sections which demonstrated that there is a hexagonal substructure of some sort within the nexus (31,48,49). The freeze-etch images do show a hexagonal array of pits which on the complimentary surfaces seem to be opposed by particles (Plate 14). In as much as the nexus is presumably a low resistance pathway between cells, several possibilities as to how this may work in cardiac muscle have been discussed (40,61). One of the most crucial questions yet to be answered concerns the precise relationship between the lanthanum space (the regions of the nexus filled by lanthanum) and other structural elements of the nexus.

There has been concern over the fact that gap junctions are totally absent (or, at any rate, so small in area and few in number that they have escaped detection) in different cardiac cells that are known to be resistively coupled (61). It is significant, perhaps, that the frequency of occurrence of nexuses is related to the diameter of the cells studied; the smaller the cells, the smaller and fewer the nexuses (30). These observations are not surprising, however, if one asks just how large an area such nexuses would have to cover in order to electrically couple two small cells as opposed to two large cells. Clearly, to accomplish the feat, nexuses between small cells (cells having a high input resistance) could be much smaller than nexuses that must resistively couple large cells (cells having a lower input resistance). More accurate information on the function of the nexus would come from quantitative estimates of the area covered by nexuses between two cells. For technical reasons this is extraordinarily difficult to do.

SARCOPLASMIC RETICULUM

The sarcoplasmic reticulum (SR) is an intracellular network of tubules that measure between 25 and 60 mμ in diameter (Plates 1,2,6,7,15,16, and Fig. 1) (46). To the extent that it is devoid of attached ribosomes and that it sometimes contains electron-dense material (61), the SR may be viewed as the equivalent of the smooth endoplasmic reticulum of other cells (e.g. liver). One function of the SR in both cardiac and skeletal muscle has been established (10,20,75): The SR harbors an ATP-dependent pump that moves calcium from the cytoplasm into the lumens of the SR network against a considerable gradient. Aside from providing one possible explanation for the phenomenon of relaxation, these observations in combination with recent findings concerning the troponin system (9), have provided data that identify compartments which seem to be capable of sequestrating calcium during the contraction - relaxation cycle. Successive compartmentalization of calcium could provide an explanation for several observed physiologic phenomena, but the precise movements of calcium during the contraction-relaxation cycle remain obscure. Another possible function of the SR, the participation in secretory or transport activity, has recently been suggested for cardiac muscle on solely morphologic grounds (61).

The structure and distribution of the SR in cardiac muscle differs in three significant ways from that in skeletal muscle: (a) The SR in cardiac muscle is widely continuous from one sarcomere to the next (Plates 15,16); (b) close apposition between specialized regions of the SR (the junctional SR of the couplings, Plate 17) in cardiac muscle occurs both at the transverse tubules and at the surface sarcolemma (in vertebrate skeletal muscle only at the transverse tubules); (c) the fine structure of these appositions in cardiac muscle is different from that of presumably homologous structures in skeletal muscle in some respects.

Although the significance of these differences is obscure at the moment, a few possible implications of some of them have previously been discussed (60). Aside from the structural differences between the SR of cardiac muscle on the one hand, and that of skeletal muscle on the other, there are also considerable differences in the appearance of the SR as one looks at the hearts of different animals. But, by and large, such differences are more like variations of a structural continuum, rather than jumps of innovative differentiation.

Generally one finds that the SR is more poorly developed in amphibians (but not necessarily functionally insignificant) than it is in birds and mammals. The most striking anatomical specializations of the SR of both skeletal and cardiac muscle is what we have previously named the coupling (31,60) (Plate 17 and Fig. 1). It consists of a portion of a SR tubule (the so-called junctional SR) which contains junctional granules and which is apposed to sarcolemma by way of junctional processes (16,33,74). The junctional SR of skeletal muscle is a dilated sac, the terminal cisterna of triads and dyads (45), which is terminal to the extent that it appears to end at the transverse tubule to which it attaches. In skeletal muscle two terminal cisternae (junctional SR) oppose each other, but usually do not connect with one another across a transverse tubule. Equivalent structures are present in cardiac cells but the junctional SR does not appear dilated, except in chickens (61), and connections between it and the SR on the opposite side of the transverse tubule are common. Couplings (junctional SR, junctional granules, junctional processes and sarcolemma) in cardiac muscle are not only seen along the invaginated sarcolemma of transverse tubules (interior couplings, Plate 17) but also along the surface sarcolemma (peripheral couplings, Plates 17,18). In many sections of working fibers of mammals and birds one finds one peripheral coupling each on either side of the Z lines that abut sarcolemma. In conducting fibers of larger animals (e.g. ungulates), fewer couplings are apparent than in working fibers of birds and mammals, and they are all peripheral couplings. In frog cardiac muscle (Rana pipiens) couplings, sensu stricto, have escaped detection so far. The function of the couplings is obscure. Circumstantial evidence, however, suggests that couplings are involved in at least one step of excitation-contraction coupling (26).

Simpson et al. (53,54) have recently emphasized the rather consistent, and perhaps functionally significant, association of one tubule of SR with the Z line (Plate 17). They have named it the Z-tubule (54). This tubule seems to occur in cardiac cells of all animals and although it is part of the general SR network (60), it derives its distinction from being anchored in some way to the Z lines. In so doing it often presents as a transversely oriented (with respect to the longitudinal fiber axis) portion of the SR network. This orientation, by way of terminology, may lead to confusion vis-a-vis the so-called transverse tubules. As will be recalled (Plate 1), the latter are invaginations of sarcolemma (the cell surface membrane) and therefore must be clearly distinguished from the other, intracellular, network of tubules namely the sarcoplasmic reticulum. Thus, and while it

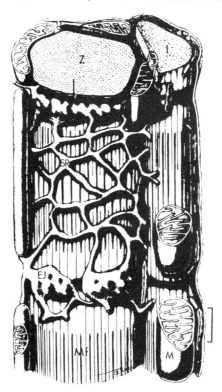

FIG. 2. Drawing of part of a cardiac muscle fiber of the hummingbird (similar to cardiac fibers of the finch and canary). The extended junctional SR (EJ) surrounds the Z lines and I bands (Z, I) and is continuous with the remainder of the SR (SR). The extended junctional SR contains junctional granules (arrow). The myofibrils usually do not show an M line. The fibers have peripheral couplings (brackets). M = Mitochondrion.

is proper to speak of a transversely oriented tubule of the SR, the transverse tubule, sensu stricto, is something else again. The unfortunate zwitter-term "sarcotubular system" can apply equally to either of the two tubular systems or simultaneously to both. The term should not be used.

The sarcoplasmic reticulum in chicken hearts, unlike that in frogs, is similar to that in mammalian cardiac muscle except that (because of the absence of transverse tubules in chicken hearts) all the couplings are peripheral couplings (by definition). In the chicken the junctional SR of these couplings often appears dilated, containing granular dense material within its lumen (61). SR tubules in the interior of such cells sometimes contain similar material. The cardiac fibers of other birds, e.g. of the finch, are similar to that of the chicken. However, the diameter of the cardiac fibers of the finch is somewhat larger than that of the fibers of the chicken and small nexuses, exceedingly rare in young chickens (58), are often present (30). Moreover, by losing contact with the sarcolemma, but without losing its typical morphologic features (junctional granules and junctional processes), the junctional SR of the couplings extends deep into the cytoplasm (extended junctional SR, 30). Often this may be in the form of a belt that is wrapped around the Z-I region of the sarcomeres (Plates 18,19,20). The remainder of the SR tubules which are continuous with the extended junctional SR and which are located between the A bands of adjacent myofibrils, as opposed to being associated with the I bands (Plates 18,19,20), have neither junctional granules nor processes. As a result they appear as essentially smooth empty tubules (Plate 19). The fact that the extended junctional SR occurs in several birds which all have very high heart rates (600 to 1000 per minute), but not in the chicken, a bird that has a much slower heart rate (150 to 300 per minute), raises the possibility that the presence or absence of extended junctional SR may be related to the different rates in birds. Structures similar to the extended junctional SR have been observed very occasionally in dog atrial fibers and in Purkinje fibers of the dog and rabbit (57). Although they often are very closely associated with mitochondria (36, unpublished observations), these profiles have not been demonstrated to be continuous with them, nor has it been established that they are identical with the filamentous mitochondria previously described (22). There is, however, evidence that they are continuous with the SR and that, in fact, they form couplings at the sarcolemma (59). The question of possible continuity between SR and mitochondria has been discussed in the past (13), but has not been resolved as yet.

The important question as to whether or not the SR is open to the extracellular space has been a matter of controversy for some time (13,14,59,60). Evidence purporting to show openings of the SR to the extracellular space (13,64) as opposed to simply showing sarcolemmal invaginations of one sort or another, has been rejected as equivocal (60). The question regarding direct communications between the extracellular space and the SR therefore, remains open. At the moment all one can say is that the possibility of intermittent communications between the SR and the extracellular space cannot be ruled out. In this context it is important to recall recent work (41) in which fusion of coated vesicles with tubes of the SR was shown.

Cytochemical observations have demonstrated that the SR of cardiac muscle has lead phosphate reaction product on its membrane (63). Similar reaction product is seen on the nexus (52,64). Several circumstances suggest that the reaction product at the SR membranes may be related to either the basic or extra ATPase (or both) found *in vitro* in isolated SR membranes, and that the reaction product seen at the nexus may be related to a calcium stimulated ATPase (52,66). Freeze-etch preparations of the SR of dog cardiac muscle show granules along the entire SR (66) (Plate 16).

Corresponding cytochemical studies seem to indicate that reaction product may be deposited around such granules (Plate 16). Aside from suggesting possible sites of enzyme activity, the distribution of reaction product can be viewed as a quasi-intracellular negative stain outlining the reliefs seen in freeze-etch preparations. On the basis of recent comparative studies employing freeze-etching, thin sections and biochemical methods, a relationship between membrane associated particles and a calcium-stimulated ATPase has been suggested (1,2,7). If, in the future, a relationship were to be firmly established between the particles in the SR and the quite similar particles in the nexus, it may turn out to be important that according to our findings (cf. Plates 14,16) the particles in the nexus point toward the gap (i.e. toward the extracellular space) while in the SR membranes the particles point toward the lumen of the SR (in the direction of calcium transport).

CONDUCTING FIBERS

In the early 19th century Purkinje described fibers on the endocardial surface of the hearts of several animals, fibers that microscopically were quite different from those of the working musculature (47). The fibers were subsequently shown to be the most peripheral part of an arborizing network taking its origin from larger bundles that combine in the region of the right atrioventricular valve to form what is now known as the bundle of His. The bundle of His is topographically closely related to the AV or Aschoff-Tawara node in the right atrium.

Electrophysiologic studies have shown that the conduction system programs the sequence of the spread of excitation in the ventricles, presumably to ensure proper expulsion of blood toward the great vessels. Extensive light microscopic studies have been made of the distribution of these fibers (73). Purkinje fibers in a number of animals are easy to recognize with the light microscope; they are much larger in diameter than the working fibers, the contractile material is seen mostly in the periphery of the fibers or is chaotically distributed throughout the cytoplasm and, in a few instances (e. g. dog hearts), the Purkinje fibers contain more glycogen than the working fibers. The application of Purkinje's microscopic and topographic criteria has failed to identify conducting fibers, especially those distal to the bundle branches, in many small mammals (rodents, guinea pigs, rabbits, cats). As a result, fibers that by topography are presumably part of the fiber system described by Purkinje, but that microscopically could not be clearly distinguished from the working fibers in these small animals, were named transitional fibers (73). Electron microscopic studies of these fibers have shown that there are discrete differences between them and the working fibers. The most striking difference is that whereas the working fibers have transverse tubules, these fibers do not (31,60). An additional useful criterion seems to be that the mitochondria of the Purkinje fibers are smaller than those in working fibers and that there is a large population of mitochondria that are elongated (22,31,34), or have elongated processes (12). On transverse sections these processes appear doughnut-shaped. From the point of view of structure-function correlation it should be pointed out that the conduction velocity of an action potential in a fiber without transverse tubules (everything else being equal, e.g. cell diameter, cytoplasmic resistivity and membrane properties) would be greater than that in fibers with them (60). In other animals, such as large mammals (ungulates, man, elephant, whale) and birds (chicken, finch), in which the Purkinje fibers are very large as compared with the working fibers (by a factor of 3 to 10), the comparative sizes of the fibers alone clearly favor faster conduction in the Purkinje fibers (60). In addition to being much larger the Purkinje fibers in large mammals are also very tightly packed, forming bundles in which the individual fibers are separated from one another only by small clefts of approximately 30 mμ in width. From theoretical considerations such an arrangement would tend to further increase conduction velocity since the area of membrane per unit length of fiber that would provide an immediately accessible pathway for excitatory current is effectively diminished (60). In other words, there is a reduction in the area of membrane per unit length which is occupied by that portion which immediately follows the transmembrane potential initial depolarization phase. In this respect the tightly packed bundle would tend to behave, momentarily, as a single fiber of a diameter equal to the overall diameter of the bundle (60). With respect to packing, the Purkinje fibers of the dog seem to occupy an intermediate position between the small and larger mammals (60). In the dog, although all cells contain abundant contractile material and most cells are rather loosely assembled into bundles, one finds several cells that are tightly packed, held together by innumerable desmosomes as is the rule in bundles of Purkinje fibers in the large animals. Similar desmosomal junctions have recently been observed in certain cells within strands of dog Purkinje fibers; they were called transitional cells mainly because transitional potentials were recorded from them (37). These experiments are as difficult to follow anatomically as they are to interpret physiologically, since the recordings made from one cell within a network of electrically connected cells does not necessarily reflect the electrical properties of that single cell. It is not necessary to postulate the existence of a third population of cells having transitional membrane properties to account for transitional electrophysiological behavior in cells lying between two populations that exhibit widely different electrophysiological behavior. Abrupt electrical junction of two populations of cells of different membrane properties would theoretically result in a gradual transition of the electrical behavior of cells as one approaches and crosses the junction.

For the time being it seems appropriate to maintain, as the light microscopists have, that Purkinje fibers have a variable morphology but that in all of them that have been studied the absence of transverse tubules is an invariant feature (60). The validity of defining mammalian Purkinje fibers as those fibers without transverse tubules (60) has been spoiled somewhat by the recent findings that the working fibers of immature mammals during the first few weeks after birth also lack transverse tubules. It is noteworthy, however, that at that time the diameter of such fibers is still very small (6 μ) while adjacent conduction fibers in the left bundle are both larger in diameter, as well as tightly packed (29).

OTHER CELL STRUCTURES

Many of the ultrastructural components seen in other cells are, of course, also found in cardiac cells.

Mitochondria occupy the interfibrillar space (Plate 1) and, though usually not exceeding the length of a sarcomere (Plate 16,17) they may sometimes be very long, extending over several sarcomere lengths (Plate 18). The internal structure of the mitochondria is varied, and peculiar forms of cristae have been observed (56). Some mitochondrial images occasionally may appear to reflect functional states - as seems to be the case in some in vitro experiments with isolated mitochondria (19). However, it is not immediately apparent in what way the in vitro findings apply to observations made in situ.

Golgi complexes are found regularly in the cardiac muscle of all animals studied so far. Lysosomes are present, as are multivesicular bodies, residual bodies, lipofuscin granules (Plate 18), and microtubules (45).

Cardiac muscle contains specific granules of at least two kinds. One kind is found only in atrial fibers of mammals almost exclusively (28,44). The other kind is found in ventricular and atrial fibers in amphibians (42, 69) and to a lesser extent in birds (61). Little is known of the function of the mammalian atrial granules. The dense granules found in amphibians seem to contain catecholamines (51).

ACKNOWLEDGMENTS

Mr. Isaiah Taylor deserves gratitude and admiration for his superb technical assistance in the preparation of the sectioned material and of the photographic reproductions. Mr. James M. Moseley was most helpful in the preparation of freeze-etch replicas.

We are grateful to Drs. D. Robertson and L. Lessin for letting us use their AEI - 6B electron microscopes for some of the pictures illustrated here, and to Mr. D. Powell for the drawing of inset 8 in Plate 14.

Supported in part by NIH grant #12486-01 and Part I fund from the Veterans Administration.

article continued on pages 312–354.

FOR TECHNICAL REASONS IT HAS BEEN NECESSARY TO PLACE THE
ELECTRONMICROGRAPHS ON ALTERNATE RIGHT HAND PAGES

PLATE 1. Longitudinal freeze-etch replica of two adjacent dog heart muscle cells that are connected by an intercalated disc (open arrows). Mitochondria (M) are fractured at various levels and often cristae are exposed. MF = Myofibrils. P = Sarcolemmal vesicle. BV = Blood vessel. Transverse tubules: small arrows. Z grooves (asterisks) are seen in the sarcolemmal surface; transverse tubules originate from these grooves. Shadow direction: Long arrow. (X 6,000) Triangular insets: Stained sections of similar cells. Opposing large arrows indicate Z lines, and define the sarcomere. Opposing arrow heads indicate I band. Small arrow heads point to M lines. Brackets show A band width. E = extracellular space. X 5,000.

PLATE 2. Freeze-etch replica of a cardiac cell fractured transversely.
P = Sarcolemmal vesicle. NC = Nerve cell with axon (A). I = I band.
SR = Sarcoplasmic Reticulum. AB = A band. CO = Collagen. M = Mitochondrion.
Open arrow: Sarcolemma. Note difference in texture between I and X; the texture at X appears similar to the fracture surfaces surrounding collagen fibers (CO) and probably represents an intracellular fixation artifact (cf. Plate 6, X). Long arrow = Shadow direction. X 25,000. <u>Inset</u>: Higher magnification of fractured myosin. Note central depression in myosin filaments. Actin is not very prominent in most cross fractures. X 75,000.

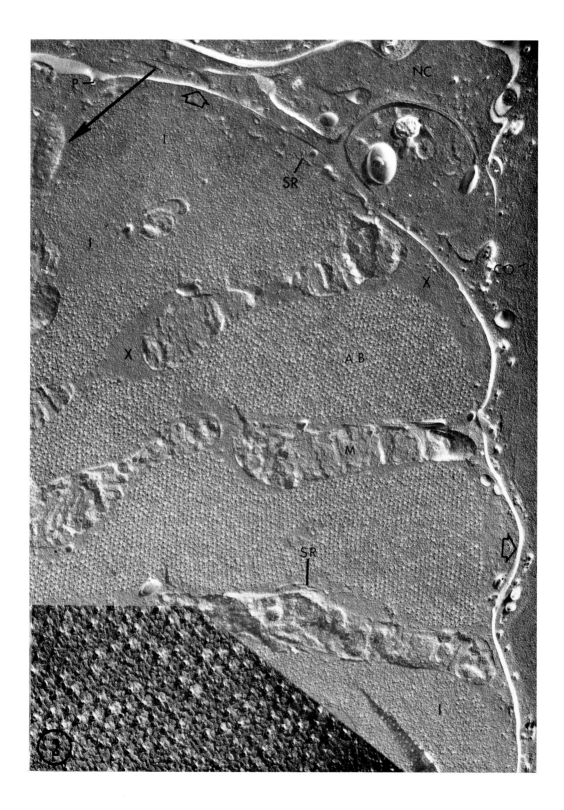

PLATE 3. Freeze-etch preparation of collagen that, perhaps better than any other, gives a visual impression of the potential strength inherent in the collagen "skeleton" that surrounds heart muscle fibers everywhere. Note the typical banding pattern of the collagen fibrils. Arrow: Direction of shadow. The "shadow" is white in this illustration. In all other illustrations of freeze-etch preparations the negatives have been reversed so that the "shadows" are black. For some of us this facilitates interpretations, as our perception of tridimensional geometry may, perhaps, be viewed as having been "programmed" rather consistently for black shadows. X 140,000.

PLATE 4. Transverse section through a bundle of papillary muscle fibers of a 4 week old puppy. The illustration represents loosely assembled fibers having large extracellular spaces between them (E) except where the fibers are joined by junctional complexes (arrows). Most of these junctional complexes include nexuses which are presumed to be low resistance pathways between adjacent cells. BV = Blood vessel. CO = Collagen. Nu = Nucleus. X 4,000.

PLATE 5. Transverse section through a bundle of Purkinje fibers of the chicken (2 fibers, 1,2). These fibers are very similar to the "classic" Purkinje fibers as they are seen especially in the large mammals (ungulates whale, man). These fibers are usually very tightly packed via desmosomes although in the chicken the fibers are often loosely assembled. The contractile material (arrow head) and small mitochondria are chaotically distributed throughout the cytoplasm, which also contains innumerable isolated small fibrils, the nature of which is still in doubt. In the ungulates and other large mammals the contractile material is often concentrated in a peripheral 5 μ zone subjacent to the sarcolemma. The two adjacent cells in Plate 5 are "spotwelded" together over long distances by innumerable desmosomes that are occasionally interspersed by nexuses. Because of the desmosomes the intercellular (extracellular) space is held constant over these long distances at about 30 mμ (cf. inset), which is of considerable physiologic importance (see text). Moreover, it should be realized that the two closely apposed membranes (small arrows) are not visible with the light microscope. For this reason, physiological considerations, as well as experiments, may be based on the erroneous assumption that one is looking at one single fiber (having the diameter as indicated by the two opposing open arrows) while in reality one is dealing with two much smaller cells that are separated by a very narrow intercellular cleft. To complicate matters, a single fiber may separate into several fibers over a short distance (60), a circumstance that would completely escape detection unless serial sections are examined. Nu = Nucleus. CO = Collagen. Large arrow = Sarcolemma. X 4,000. Inset: Higher magnification of two closely apposed sheep Purkinje fibers showing two desmosomes (D) and an intercellular space that measures approximately 25 mμ in width.

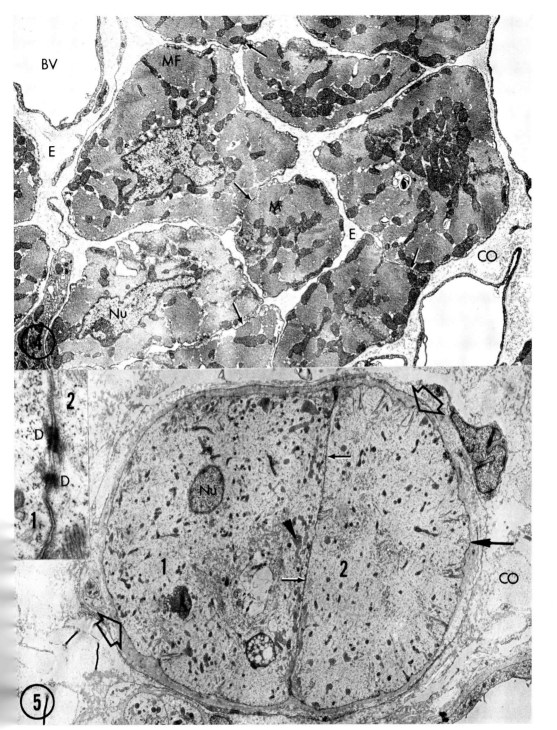

PLATE 6. Longitudinal Epon section through an intercalated disc. The intercalated disc proper shows the typical zig-zag course (ID). Both discs are connected by a nexus (arrow) which runs parallel to the axis of the myofibrils and is surrounded by clear cytoplasm. The disc proper is composed of ff. adherentes interspersed by the more ordered desmosomes which form small patches. The arrow head points to an SR tubule, presumable the Z tubule (cf. discussion). Note side bridges between thick and thin filaments. The brackets indicate a region of the nexus that when further magnified would look as in Plate 12. (cf. Plate 2, X) X 60,000.

PLATE 7. Freeze-etch replica of a region of cardiac muscle similar to the one in Plate 6. The short arrow points to a granular area showing particles associated with the outer face of the inner leaflet of one of the unit membranes forming the nexus. Arrow head points to SR tubules. Shadow direction: Long arrow. X 60,000.

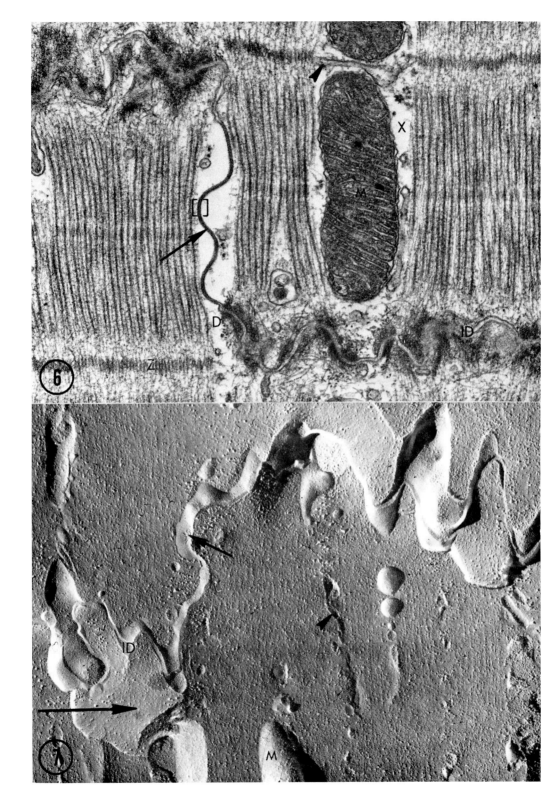

PLATE 8. Transverse Epon section through intercalated disc. Compare I bands (I) and A bands (AB) with similar regions of Plate 2. Open arrows: Couplings. Arrow head: Basement membrane. E = Extracellular space. N = Nexus. D = Desmosome, MY = Myelin figure. P = Sarcolemmal vesicle. M = Mitochondrion. Encircled area is a profile similar to that seen in Plate 11 but of lesser magnification and differently prepared tissue. X 37,000.

PLATE 9. Transverse fracture through an intercalated disc in much the same plane as the section in Plate 8. Arrow heads: Sarcolemma. My = Myelin figures. Plump arrow indicates a tridimensional equivalent of a similar region in Plate 8. Shadow direction: Long arrow. X 34,000.

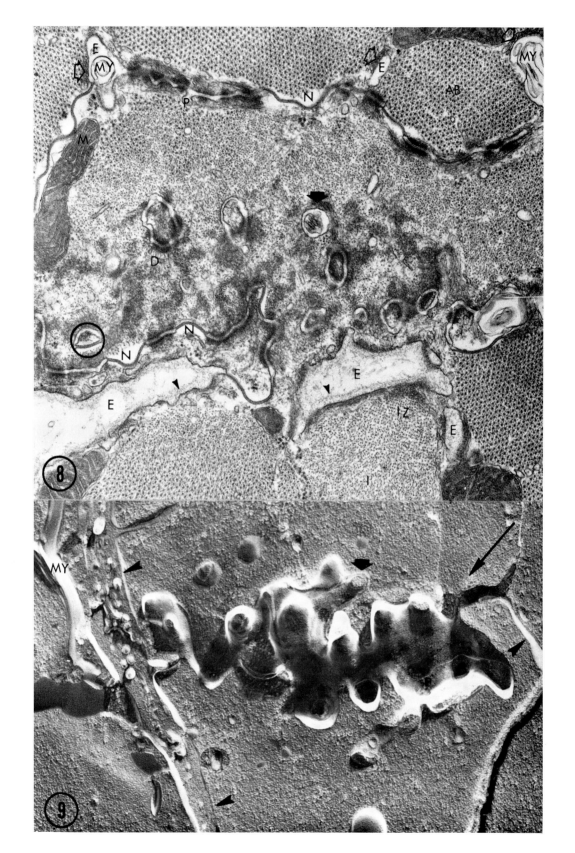

PLATE 10. High magnification of two junctional complexes in dog cardiac muscle. The more delicately structured one (below) is a desmosome, the other one (above) is a ff. adherens. Note fibrillar profiles inserting into these junctional complexes as opposed to the nexus which is usually surrounded by clear cytoplasm (cf. Plate 6). X 150,000.

PLATE 11. This picture illustrates a higher magnification of a transverse section through a finger-like prong of an intercalated disc (cf. Plate 9) between two cells (1,2) in a dog heart at a level quite similar to the encircled area of Plate 8. As in Plate 8, part of the profile is a nexus (between large arrow heads). However, the tissue of Plate 11 has not been stained and, as a result, the cell membranes of either cell forming the nexus are just barely discernible (small arrow heads). Instead the gap (not visible in Plate 8) which is seen in the nexus of Plate 12, and which is accessible from the extracellular space, has been filled with an electron-dense tracer (appearing black in the illustration), in this case lanthanum. The open arrow indicates an area where extracellular space filled with lanthanum but away from the nexus (cf. Plate 8), has been cut obliquely. X 160,000.

PLATE 12. High magnification of part of a nexus between two cells (1,2) comparable to a region marked by brackets in Plate 6. The two plasma membranes approach each other leaving a roughly 2 mµ gap (arrow) that can be filled with an electron-dense material (Plate 11,13). X 400,000.

PLATE 13. High magnification of a tangential (grazing) section through the region of the gap of the nexus revealing a hexagonal lattice (arrow) with a center-to-center spacing of about 10 mµ. Many of the hexagons contain an electron-dense dot in the center. The lanthanum appears to trace the hexagons seen in the freeze-etch preparations (Plate 14, insets 2 & 3). The central electron-dense dot in the hexagons defies explanation at the moment but may point to a more complex geometry of cell apposition in the region of the gap, perhaps not unlike that seen in honey combs. X 180,000.

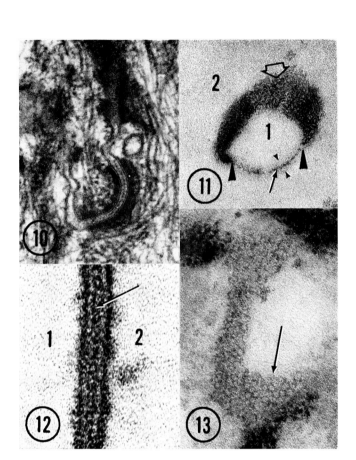

PLATE 14. This figure represents two complimentary replicas of freeze-etched fracture surfaces of the nexus of dog cardiac muscle. The encircled region on the left shows particles (A face, 38, high relief granules, 66) that on the right shows pits (10 mµ center-to-center, B face, 38, reverse low relief, 67). The long arrow heads indicate areas where the fracture planes are reversed. The fracture surfaces show "defects" of an otherwise rather regular hexagonal pattern (less regular on the left, cf. optical diffraction pattern) shown in different magnifications of the same area (insets 1,2,5, 6, arrows). The particles of the A face seem to have been pulled out of the pits and in the case of the "defects" may have been carried along into the B face. A particles and B pits match. To facilitate matching, a series of particles were marked with black dots (inset 5, cf. inset 1) that match with dots put over the complimentary pits in inset 6 (cf. inset 2). The open arrow in inset 1 shows a region where the pits of the B face are split (cf. inset 8, open arrow); the adjacent row of pits just above can be matched with a complimentary row of particles in inset 2 (small arrows). Inset 8 is a drawing of one possible topographic relationship between different fracture faces. The particles of the A face fit the pits of the B face. C is the cytoplasmic face of the inner leaflet of one of the unit membranes forming the nexus and is not shown in the freeze-etch preparations. Black arrows point to the presumed position of the outer leaflet of each unit membrane; two apposed outer leaflets form the gap (small arrow, cf. inset 7) seen in high resolution pictures of stained material (inset 7, long arrows, cf. Plate 12). Freeze-etch images equivalent to face D (small arrow heads) are controversial (low relief, ref. 67). The small arrow in inset 8 indicates the space presumably occupied by lanthanum (·48). Continuity of the A particles across the gap (X) cannot be ruled out, nor is it clear whether the A particles of cell I and II oppose each other center-to-center (cf. insets 7 & 8). However, a moire composed of two interfering hexagonal lattices (resulting in a 5 mµ lattice spacing instead of the usually reported 10 mµ spacing) has only rarely, if ever, been seen in flat sheets of negatively stained isolated nexuses (3). MR_1 and MR_2 are the complimentary faces of the split unit membranes just outside the nexus. X 88,000. Insets 1 & 2: 140,000.

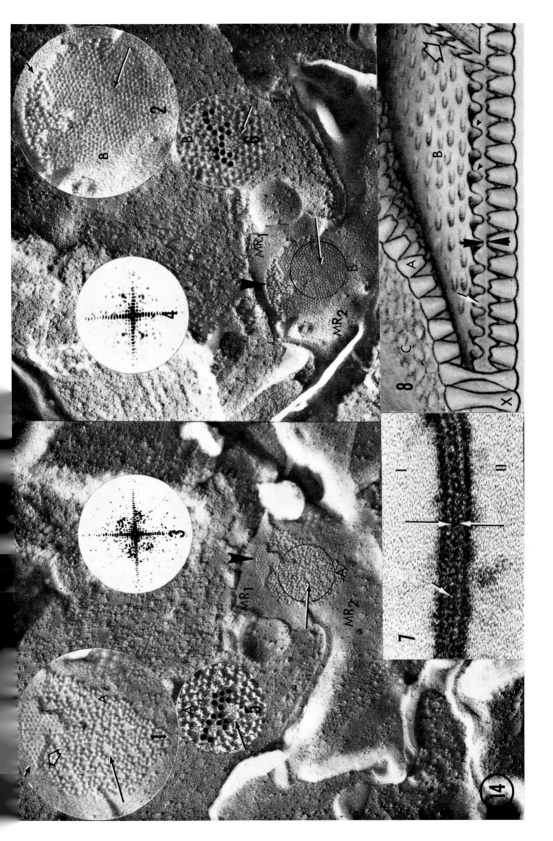

PLATE 15. Longitudinal section through dog papillary muscle. The SR (arrow head) is a network of tubules continuous across the limits of the sarcomere. The SR is commonly associated with coated vesicles (long arrow). X 37,000. Inset: 40,000.

PLATE 16. Longitudinal freeze-etch preparation of dog cardiac muscle. The fracture plane exposes an inside surface, studded with particles (HR granules 66), of the outer leaflet of the SR unit membrane. The complimentary surfaces (small arrows) are generally rather smooth except for the encircled region (cf. higher magnification, circular inset) in which, presumable due to a lower shadow angle, pits (small arrow head) can be identified as opposed to the elevated particles nearby (double arrow head). These relationships have yet to be checked out with complimentary replicas, since the topographic relation between particles and pits (A and B faces) was proved only for the nexus. The SR membranes show lead phosphate reaction product, presumably released by an ATPase during cytochemical experiments, which produces a cribriform pattern (arrow head of rectangular inset) suggesting deposition of electron-dense material, as a quasi-intracellular negative stain, around particles of similar size as those seen in the freeze-etch preparation. M = Mitochondrial image. T = Transverse tubules. Brackets = Interior coupling. X 35,000. Large arrow = Shadow direction. Rectangular inset: X 45,000.

PLATE 17. Longitudinal section of dog cardiac muscle demonstrating interior couplings (brackets) on transverse tubules (T). M = Mitochondria. Short arrows indicate what presumably are Z tubules (see text). Z = Z line. X 34,000. Inset: Peripheral coupling (bracket) on sarcolemma surrounding the whole fiber. E = Extracellular space. Junctional SR (long arrow) contains junctional granules (small arrow) and has junctional processes toward the sarcolemma (small arrow heads). SR = Large arrow head. X 50,000.

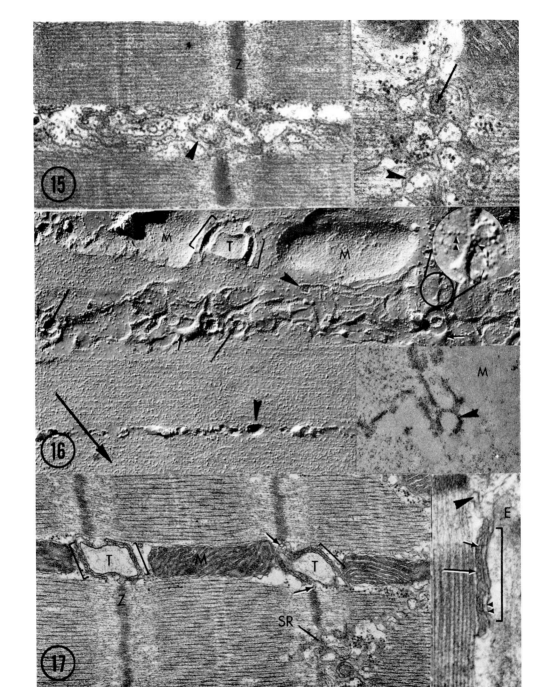

PLATE 18. Longitudinal section through 2 fibers (1,2) of papillary muscle of the hummingbird. Specialized portions of the SR (extended junctional SR, arrow heads) are seen at almost all Z lines and, morphologically, are identical to, as well as continuous with, junctional SR of peripheral couplings (short arrows, cf. Plate 17 and Fig. 1). Bird hearts have no transverse tubules and, thus, no interior couplings. However, junctional SR (as extended junctional SR) extends deep into the muscle fibers. M = Mitochondrion which, in this case, is almost three sarcomeres long. L = Lipofuscin granules. E = Extracellular space. BV = Blood vessel. X 7,000.

PLATE 19. Transverse section through hummingbird papillary muscle. Extended junctional SR (large arrow) surrounds Z lines (Z) and I bands (I) as well as mitochondria (M). The SR (small arrow) around A bands (A) does not contain junctional granules nor does it have obvious junctional processes. X 60,500.

PLATE 20. Grazing section through extended junctional SR (arrows) showing a dense matrix which reveals a faint cribriform pattern. Some of the images are attributable to material inside the extended junctional SR, others to projections of superimposed structures, including the junctional processes. X 80,000.

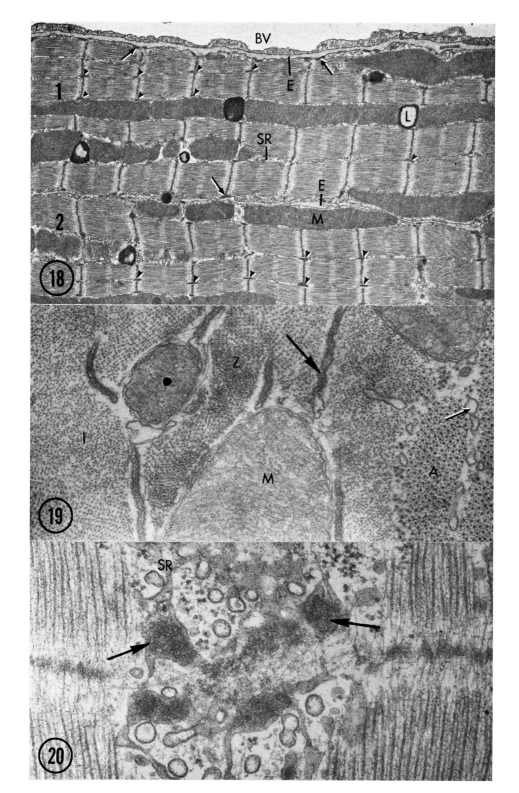

REFERENCES

1. ARNTZEN, C. J., R. A. DILLEY, and F. L. CRANE. A comparison of chloroplast membrane surfaces visualized by freeze-etch and negative staining techniques; and ultrastructural characterization of membrane fractions obtained from digitonin-treated spinach chloroplasts. J. Cell Biol. 43: 16-31, 1969.

2. BASKIN, R. J. Ultrastructure and calcium transport in crustacean muscle microsomes. J. Cell Biol. 48: 49-60, 1971.

3. BENEDETTI, E. L. and P. EMMELOT. Hexagonal array of subunits in tight junctions separated from isolated rat liver plasma membranes. J. Cell Biol. 38: 15-24, 1968.

4. BREEMEN, V. L. v. Intercalated disc in heart muscle studied with the electron microscope. Anat. Rec. 117: 49-63, 1953.

5. BULLIVANT, S. Particles seen in freeze-fracture preparations. In: Proceedings of the 27th annual Meeting of the Electron Microscopy Society of America, edited by C. J. Arceneaux. Baton Rouge: Claitor's Publ. Div., 1969, p. 206.

6. CHALCROFT, J. P. and S. BULLIVANT. An interpretation of liver cell membrane and junction structure based on observation of freeze-fracture replicas of both sides of the fracture. J. Cell Biol. 47: 49-60, 1970.

7. DEAMER, D. W. and R. J. BASKIN. Ultrastructure of sarcoplasmic reticulum preparations. J. Cell Biol. 42: 296-307, 1969.

8. DEWEY, M. M. and L. BARR. A study of the structure and distribution of the nexus. J. Cell Biol. 23: 553-585, 1964.

9. EBASHI, S. and M. ENDO. Calcium ion and muscle contraction. In: Progress in Biophysics and Molecular Biology 18, edited by J. A. V. Butler and D. Noble. London: Pergamon Press, 1968, p. 123-183.

10. EBASHI, S. and F. LIPMANN. Adenosine triphosphate-linked concentration of calcium ions in a particulate fraction of rabbit muscle. J. Cell Biol. 14: 389-400, 1962.

11. FARQUHAR, M. G. and G. E. PALADE. Junctional complexes in various epithelia. J. Cell Biol. 17: 375-412, 1963.

12. FAWCETT, D. W. and N. S. MCNUTT. The ultrastructure of the cat myocardium. I. Ventricular papillary muscle. J. Cell Biol. 42: 1-45, 1969.

13. FORSSMANN, W. G. and L. GIRARDIER. Untersuchungen zur Ultrastruktur des Rattenherzmuskels mit besonderer Berücksichtigung des sarkoplasmatischen Retikulums. Z. Zellforsch. 72: 249-275, 1966.

14. FORSSMANN, W. G. and L. GIRARDIER. A study of the T system in rat heart. J. Cell Biol. 44: 1-19, 1970.

15. FRANZINI-ARMSTRONG, C. and K. R. PORTER. The Z disc of skeletal muscle fibrils. Z. Zellforsch. 61: 661-665, 1964.

16. FRANZINI-ARMSTRONG, C. Studies of the triad. I. Structure of the junction in frog twitch fibers. J. Cell Biol. 47: 488-499, 1970.

17. GAY, W. A. and E. A. JOHNSON. An anatomical evaluation of the myocardial length-tension diagram. Circulation Res. 21: 33-43, 1967.

18. GORDON, A. M., A. F. HUXLEY, and F. J. JULIAN. The variation in isometric tension with sarcomere length in vertebrate muscle fibers. J. Physiol. 184: 170-192, 1966.

19. HACKENBROCK. C. R. Ultrastructural bases for metabolically linked mechanical activity in mitochondria. II. Electron transport-linked ultrastructural transformations in mitochondria. J. Cell Biol. 37: 345-369, 1968.

20. HASSELBACH, W. and M. MAKINOSE. Die Calciumpumpe der "Erschlaffungsgrana" des Muskels und ihre Abhängigkeit von der ATP-Spaltung. Biochem. Z. 333: 518-528, 1960.

21. HILL, A. V. On the time required for diffusion and its relation to processes in muscle. Proc. Roy. Soc., London, Ser. B. 135: 446-453, 1948.

22. HIRAKOW, R. Fine structure of Purkinje fibers in the chick heart. Arch. Histol. 27: 485-499, 1966.

23. HIRAKOW, R. Ultrastructural characteristics of the mammalian and sauropsidan heart. Amer. J. Cardiol. 25: 195-203, 1970.

24. HOWSE, H. D., V. J. FERRANS and R. C. HIBBS. A light and electron microscopic study of the heart of a crayfish, Procambarus clarkii (Giraud). I. Histology and histochemistry. J. Morphol. 131: 237-252, 1970.

25. HUXLEY, A. F. and R. NIEDERGERKE. Structural changes in muscle during contraction. Interference microscopy of living muscle fibers. Nature. 173: 971-973, 1954.

26. HUXLEY, A. F. and R. E. TAYLOR. Local activities of striated muscle fibers. J. Physiol. 144: 426-441, 1958.

27. HUXLEY, H. E. and J. HANSON. Changes in the cross-striations of muscle during contraction and stretch and their structural interpretation. Nature. 173: 973-976, 1954.

28. JAMIESON, J. D. and G. E. PALADE. Specific granules in atrial muscle cells. J. Cell Biol. 23: 151-172, 1964.

29. JEWETT, PAUL and J. R. SOMMER. The left bundle branch, Purkinje network, and working muscle of the newborn dog: ultrastructure. Forty-third Scientific Session of the American Heart Association, Atlantic City, N. J., 1970.

30. JEWETT, PAUL, J. R. SOMMER and E. A. JOHNSON. Cardiac muscle. Its ultrastructure in the finch and hummingbird with special reference to the sarcoplasmic reticulum. J. Cell Biol. 49: 50-65, 1971.

31. JOHNSON, E. A. and J. R. SOMMER. A strand of cardiac muscle: its ultrastructure and the electrophysiologic implications of its geometry. J. Cell Biol. 33: 103-129, 1967.

32. KAWAMURA, K. Electron microscope studies on the cardiac conduction system of the dog. Japan Circulation J. 25: 594-616, 1961.

33. KELLY, D. E. The fine structure of skeletal muscle triad junctions. J. Ultrast. Res. 29: 37-49, 1969.

34. KNAPPEIS, G. G. and F. CARLSEN. The ultrastructure of the Z disc in skeletal muscle. J. Cell Biol. 13: 323-335, 1962.

35. KREUTZIGER, G. O. Specimen surface contamination and loss of structural detail in freeze-fracture and freeze-etch preparations. In: Proceedings of the 27th Annual Meeting of the Electron Microscopy Society of America, edited by C. J. Arceneaux. Baton Rouge: Claitor's Publ. Div., 1969, p. 138.

36. LEAK, L. V. The ultrastructure of myofibers in a reptilian heart: the boa constrictor. Amer. J. Anat. 120: 553-582, 1967.

37. MARTINEZ-PALOMO, A., J. Alanis and D. Benitez. Transitional cardiac cells of the conductive system of the dog heart. Distinguishing morphological and electrophysiological features. J. Cell Biol. 47: 1-17, 1970.

38. MCNUTT, N. S. Ultrastructure of intercellular junctions in adult and developing cardiac muscle. Amer. J. Cardiol. 25: 169-183, 1970.

39. MCNUTT, N. S. and R. S. WEINSTEIN. Interlocking subunit arrays forming nexus membranes. In: Proceedings of the 27th Annual Meeting of the Electron Microscopy Society of America, edited by C. J. Arceneaux. Baton Rouge, Claitor's Publ. Div., 1969, p. 330.

40. MCNUTT, N. S. and R. S. WEINSTEIN. The ultrastructure of the nexus. A correlated thin-section and freeze-cleave study. J. Cell Biol. 47: 666-688, 1970.

41. MCNUTT, N. S. and D. W. FAWCETT. The ultrastructure of the cat myocardium. II. Atrial muscle. J. Cell Biol. 42: 46-67, 1969.

42. NAYLER, W. G. and N. C. R. MERRILLEES. Some observations on the fine structure and metabolic activity of normal and glycerinated ventricular muscle of toad. J. Cell Biol. 22: 533-550, 1964.

43. NELSON, D. A. and E. S. BENSON. On the structural continuities of the transverse tubular system of rabbit and human myocardial cells. J. Cell Biol. 16: 297-313, 1963.

44. OKAMOTO, HITOJI. An electron microscopic study of the specific granules in the atrial muscle cell upon the administration of agents affecting autonomic nerves. Arch. Histol. 30: 467-478, 1969.

45. PAGE, E. The occurrence of inclusions within membrane-limited structures that run longitudinally in the cells of mammalian heart muscle. J. Ultrast. Res. 17: 63-71, 1967.

46. PORTER, K. R. and G. E. PALADE. Studies on the endoplasmic reticulum. III. Its form and distribution in striated muscle cells. J. Biophys. and Biochem. Cytol. 3: 269-300, 1957.

47. PURKINJE, J. E. Mikroskopisch-Neurologische Beobachtungen. Arch. Anat. Physiol. 2: 281-295, 1845.

48. REVEL, J. P. and M. J. KARNOVSKY. Hexagonal array of subunits in intercellular junctions of the mouse heart and liver. J. Cell Biol. 33: C7-C12, 1967.

49. ROBERTSON, J. D. The occurrence of a subunit pattern in the unit membranes of club endings in Mauthner cell synapses in gold fish brains. J. Cell Biol. 19: 201-221, 1963.

50. SCHIEBLER, T. H. and H. H. WOLFF. Electronenmikroskopische Untersuchungen am Herzmuskel der Ratte Während der Entwicklung. Z. Zellforsch. 69: 22-40, 1966.

51. SCHIPP, R. and A. B.-v. WEHREN. Zur funktionellen Bedeutung der osmiophilen Granula in Herzorganen niederer Vertebraten. Z. Zellforsch. 108: 243-267, 1970.

52. SCHULZE, W. and A. WOLLENBERGER. Charakterisierung Zytochemische Lokalisation und von phosphatabspaltended Fermenten im sarkotubularen System quergestreifter Muskeln. Histochemie. 10: 140-153, 1967.

53. SIMPSON, F. O. The transverse tubular system in mammalian myocardial cells. Amer. J. Anat. 117: 1-18, 1965.

54. SIMPSON, F. O. and D. G. RAYNS. The relationship between the transverse tubular system and other tubules at the Z disc levels of myocardial cells in the ferret. Amer. J. Anat. 122: 193-208, 1968.

55. SJOSTRAND, F. S., E. ANDERSSON-DEDERGREN and M. M. DEWEY. The ultrastructure of the intercalated disc of frog, mouse and guinea pig cardiac muscle. J. Ultrast. Res. 1: 271-287, 1958.

56. SLAUTTERBACK, D. B. Mitochondria in cardiac muscle cells of the canary and some other birds. J. Cell Biol. 24: 1-21, 1965.

57. SOMMER, J. R. Purkinje fibers in the atrium (Abstract). Federation Proc. 27: 357, 1968.

58. SOMMER, J. R. Chicken cardiac muscle: a transitional stage between amphibian and mammalian cardiac muscle (Abstract). J. Cell Biol. 39: 127A, 1968.

59. SOMMER, J. R. and P. R. JEWETT. Cardiac muscle. In: Cardiac Hypertrophy, edited by N. Alpert. New York: Academic Press, 1971. In press.

60. SOMMER, J. R. and E. A. JOHNSON. Cardiac muscle. A comparative study of Purkinje fibers and ventricular fibers. J. Cell Biol. 36: 497-526, 1968.

61. SOMMER, J. R. and E. A. JOHNSON. Cardiac muscle. A comparative ultrastructural study with special reference to frog and chicken hearts. Z. Zellforsch. 98: 437-468, 1969.

62. SOMMER, J. R. and E. A. JOHNSON. Comparative ultrastructure of cardiac cell membrane specializations. A review. Amer. J. Cardiol. 25: 184-194, 1970.

63. SOMMER, J. R. and M. S. SPACH. Electronmicroscopic localisation of ATPase in myofibril and sarcoplasmic reticulum of normal and abnormal dog hearts (Abstract). Federation Proc. 22: 195, 1963.

64. SOMMER, J. R. and M. S. SPACH. Electronmicroscopic demonstration of adenosinetriphosphatase in myofibrils and sarcoplasmic membranes of cardiac muscle of normal and abnormal dogs. Amer. J. Path. 44: 491-505, 1964.

65. SOMMER, J. R. and R. L. STEERE. The nexus freeze-etched (Abstract). J. Cell Biol. 43: 136, 1969.

66. SOMMER, J. R. and R. L. STEERE. Cardiac sarcoplasmic reticulum and nexus: Freeze-etching and ATPase localization (Abstract). Federation Proc. 29: 390, 1970.

67. SOMMER, J. R. and R. L. STEERE. The nexus studies with complimentary freeze-etch replicas: a revisit (Abstract). J. Cell Biol. 47: 198A, 1970.

68. SPIRO, D. The fine structure and contractile mechanism of heart muscle. In: The Myocardial Cell. Structure, Function, and Modification by Cardiac Drugs, edited by S. A. Briller and H. L. Conn. Philadelphia: Univ. of Penn. Press, 1966.

69. STALEY, N. A. and E. S. BENSON. The ultrastructure of frog ventricular cardiac muscle and its relationship to mechanisms of excitation-contraction coupling. J. Cell Biol. 38: 99-114, 1968.

70. STEERE, R. L. Electron microscopy of structural detail in frozen biological specimens. J. Biophys. Biochem. Cytol. 3: 45-60, 1957.

71. STEERE, R. L. and M. MOSELEY. New dimensions in freeze-etching. In: Proceedings of the 27th Annual Meeting of the Electron Microscopy Society of America, edited by C. J. Arceneaux. Baton Rouge: Claitor's Publ. Div., 1969, p. 202.

72. STENGER, R. J. and D. SPIRO. The ultrastructure of mammalian cardiac muscle. J. Biophys. Biochem. Cytol. 9: 325-351, 1961.

73. TRUEX, R. C. and M. Q. SMYTHE. Comparative morphology of the cardiac conduction tissue in animals. Ann. N. Y. Acad. Sci. 127: 19-33, 1965.

74. WALKER, S. M. and G. R. SCHRODT. Connections between the T system and sarcoplasmic reticulum. Anat. Rec. 155: 1-10, 1966.

75. WEBER, A., R. HERZ and I. REISS. The nature of the cardiac relaxing factor. Biochem. Biophys. Acta 131: 188-194, 1967.

76. YAMAMOTO, T. Observations on the fine structure of the cardiac muscle cells in goldfish (Carassius auratus). In: Electrophysiology and Ultrastructure of the Heart, edited by T. Sano, V. Mizuhira and K. Matsuda. Tokyo: Bunkodo Co., 1967.

CALCIUM, AND THE CONTROL OF MUSCLE ACTIVITY

D. J. HARTSHORNE AND L. J. BOUCHER

Chemistry Department
Carnegie-Mellon University
Pittsburgh, Pennsylvania 15213

Calcium appears to be involved in the regulation of contraction and relaxation in all types of muscle (9,10,34,45). Most of the investigations into the mechanism of regulation have been done using skeletal muscle since this is the most accessible tissue. It is hoped however that in all muscles there is a common basic theme, and that the knowledge gained with skeletal muscle may be applied tentatively towards an understanding of the biochemistry of the other muscle types. The work described below has been done predominately with rabbit skeletal muscle.

Our present knowledge of the regulatory process stems from the observations of many investigators (see reviews by Ebashi and Endo [9] and Weber [53]) but in particular from the discoveries that (a) actomyosin-type ATPase activity requires minute amounts of calcium (52), and (b) that a protein component other than actin and myosin is involved (7). This "protein component" Ebashi called "native tropomyosin." The usual preparation of tropomyosin (1), however, lacked biological activity and it was subsequently found (11,12) that an additional factor named troponin was required. Both of these components are essential for the calcium-dependent response of the actomyosin system (13,22). About equal proportions of troponin and tropomyosin are required for maximum effect, and it has been shown using the analytical ultracentrifuge that the two components form a physically distinct complex. At a weight ratio of 1.0:1.37, tropomyosin to troponin, a single sedimenting species was observed which sedimented faster than either troponin or monomeric tropomyosin (22,24).

It is now widely accepted that any mechanism of muscle contraction must be compatible with the sliding filament model (28,29,33). In this scheme, contraction occurs when the two sets of filaments interdigitate, and there is no marked reduction in the lengths of either the thick or thin filaments. Clearly, in order that tension could be developed there must be some kind of interaction between the two sets of filaments. There have been several suggestions as to how this is done but the most popular version is that developed by H. E. Huxley and collaborators (for review see 31). In this hypothesis, direct physical contact is made between actin and the myosin cross-bridges. In support of the idea, it is known that the actin-binding ability and ATPase activity is found in the "head" of the myosin molecule (37) and that this forms the terminal part of the cross-bridge. Furthermore, X-ray evidence indicates that the cross-bridges are positioned differently in the contracted and relaxed states (30,32). Relaxed muscle is freely extensible and thus any interaction between the thick and thin filaments would not be expected. It would appear then that the troponin and tropomyosin function to regulate the actin-myosin interaction, i.e., to regulate the cross-bridge attachment to actin so that in the absence of calcium no connections are made, but in the presence of calcium the cross-bridges are allowed to interact with actin. This idea is consistent with the observations (15,41) that troponin and tropomyosin in the absence of calcium inhibit the ATPase activity of the heavy meromyosin subfragment I-actin complex by causing an increase in the apparent dissociation constant rather than an alteration of the maximum ATPase velocity. One way in which this could be done is to allow a calcium-dependent conformational change of troponin and/or tropomyosin to alter the actin-myosin interaction. Evidence that calcium elicits a conformational change in troponin which may be transmitted through tropomyosin to actin has been presented by Tonomora, et al., (50). Whatever the mechanism proves to be, however, it must require strategic positioning of the tropomyosin and troponin molecules. Thus, it is of considerable interest to determine the location of these proteins in the myofibril.

First, it is helpful to consider the interaction of the pertinent proteins "in vitro". These may be summarized as follows:

    a) Tropomyosin binds to actin (6,36,38)

    b) Troponin does not bind strongly to actin (12,16)

    c) Tropomyosin binds to troponin (11,22)

    d) Neither tropomyosin nor troponin bind to myosin (35).

Thus, the arrangement of these proteins might be expected to be in the order; troponin binding to tropomyosin which binds to actin, and the location could initially be limited to the thin filaments and Z-line.

Recently, selective extraction and recombination studies (48,49) have eliminated the Z-line from consideration and thus the thin filaments are the most likely location. This has been suspected for many years, but it has been difficult to estimate the stoichiometry of actin:tropomyosin:troponin because of the wide variation in the estimates of the myofibrillar content of these proteins, tropomyosin in particular (cf. 25). Using a combination of different analytical procedures we have recently estimated (25) that the percentage of tropomyosin in rabbit psoas muscle is 4.2%. Taking the content of actin as 20% (20,55) and the molecular weights of G-actin and tropomyosin as 46,000 (44) and 70,000 (27,58), respectively, this would result in a molar ratio of actin to tropomyosin of 7.2:1. If it is assumed that the actin molecules are arranged as a "strand," and all the evidence suggests this, then the distance occupied by 7.2 actin molecules is approximately 390 Å (7.2 X 54.6 Å [32]) which is close to the length of the tropomyosin molecule. Actin in the thin filament is actually double stranded (20) and thus every 390 Å one would find approximately 14 molecules of actin and 2 molecules of tropomyosin. Thus, the tropomyosin might be expected to be arranged as two end-to-end polymerized filaments, possible located on either side of the double actin helix, as suggested earlier by Hanson and Lowy (20) and more recently by Moore, et al., (39). Troponin is not as well defined as tropomyosin and thus its estimation is subject to possibly greater error. However if we use the figures which are available, namely of approximately 5.4% contribution to the myofibrillar protein (25) and a molecular weight of 80,000 (9, Dreizen and Hartshorne, to be published) one would expect one molecule of troponin for each tropomyosin molecule. The precise orientation cannot be evaluated as troponin consists of more than one protein and the interactions of these sub-components has not been completely established. But, since troponin binds to tropomyosin, we may be reasonably sure that its distribution along the thin filament is the same as that for tropomyosin, i.e., 390 Å. This result was obtained several years ago by Ebashi and his co-workers (16,40) as was a model of the thin filament similar to that presented above (9,14).

Since the myofibrillar ATPase activity is regulated by calcium, we might expect that one or more of the myofibrillar proteins binds calcium. This expectation was fulfilled, and it was discovered by Fuchs and Briggs (17) and Ebashi, et al., (8,14) that troponin was the calcium receptive protein. The amount of calcium bound by troponin varies somewhat in different reports but is generally about 2-3 moles calcium/10^5 g protein. Also, it has been shown that the binding data for troponin is probably best explained by assuming two classes of binding sites (14) and the figure at 2-3 moles calcium/10^5 g protein represents binding only to the high affinity class of sites. Extending the binding data to the thin filament model presented above, it may be calculated that about 5 molecules of calcium are bound every 390 Å at a free calcium concentration of 10^{-6}M. This is similar to the value obtained by Fuchs and Briggs (17) and also to the amount of calcium found by Weber and Herz (54) to interact with myofibrils during super-precipitation.

As indicated earlier, troponin is heterogeneous and it is of interest to determine which of the components binds calcium. Before this question is answered, however, it would be helpful to describe briefly the function of the separated components. Troponin was originally separated into two functional parts (23) these were called troponin A and troponin B (Schaub and Perry [46] using a different method also separated troponin into two functionally distinct components.) Troponin A, by itself or with tropomyosin, did not influence the ATPase activity of actomyosin. On the other hand, troponin B inhibited the ATPase activity, but unlike the inhibition caused by the complex of troponin and tropomyosin, this was not sensitive to changes in the calcium concentration. In other words, even in the presence of up to 10^{-4}M calcium (21) the troponin B still inhibited the actomyosin. When tropomyosin was mixed with troponin B, the inhibition became greater although the sensitivity towards calcium was not restored. It was shown that the inhibition was maximum at an equal weight ratio of troponin B and tropomyosin and that at this ratio a single distinct sedimenting species was observed in the ultracentrifuge (26). The complex of troponin B and tropomyosin was therefore not regulated and formed a calcium insensitive inhibitor. The regulation of this complex is in fact that the function of troponin A, so that as troponin A is added to troponin B plus tropomyosin, the inhibition becomes sensitive to calcium.

TABLE 1. Properties of tropomyosin, troponin, troponin A and troponin B.

Property	Tropomyosin	Troponin (A + B)	Troponin A	Troponin B
Molecular Weight	70,000 (27,58)	80,000 (9)	18,500 (25), 22,000 (18)	≈60,000[B]
Dimensions	Rod, 490 Å X 20 Å (27)		Sphere, radius ≈ 17 Å	
Conformation	> 90% α-helix (4)	42% α-helix (26,47) 15% β-structure (26)	30% α-helix (26)[C]	40% α-helix (26)
"Subunits"	2 Equal (58)		][B,D]	2-3 (19,25)
Solubility	Water Soluble (1)	Water Soluble (11)	Water Soluble (23)	Water Insoluble Soluble at High Salt. (23)
Intrinsic Viscosity	0.5 dl/g (27,51)	0.12 dl/g (9) 0.15 dl/g (26)	0.05 dl/g (25)	
$S^\circ_{20,w}$	2.6 (2,27)	3.8 (26), 4.0 (9)	1.8 (26)[C]	3.8 (26)
$E^{1\%}_1$ cm.	2.9 (24)	≈ 4.5 (24)[E] 2.	2.35 (25)	≈ 5.0
$E_{278\ nm}/E_{260\ nm}$	2.6 (24)	1.4[E]	1.05 (25), 0.9 (18)	1.6 (26)
%Myofibrillar Protein	4.2% (25)	5.6% (25)	≈1%	≈4%
Distribution[F]	Bound to actin		Bound to troponin B	Bound to tropomyosin
Calcium binding[G]	Very Little (17)	2-3 moles/10^5 g (14,17,25)	8 moles/10^5 g (25)	?

Both troponin A and troponin B are heterogeneous, and of the two, troponin A was found to be easier to resolve. As suspected because of its function, troponin A bound most of the calcium associated with troponin, and the calcium-binding protein was purified and characterized (see Table 1). Its molecular weight is approximately 18,500 (25).

Troponin B (the inhibitor) has been more difficult to purify, largely because of its low solubility at low ionic strength. Preliminary evidence from our laboratory indicates that it is comprised of at least two components. This supports the contention of Greaser and Gergely (19) who have separated troponin into four components which they have called troponins 1 to 4. The molecular weights were found to be 15,000, 25,000, 35,000, and 22,000 for troponins 1 to 4, respectively. By recombining the separated components they found that a functional troponin was made with a mixture of troponin

---

TABLE 1. Footnotes.

- A. This table gives data accepted by the authors and as such is biased to their opinion. It is not intended as a complete bibliographic record.
- B. Dreizen and Hartshorne, to be published.
- C. Done with impure preparations.
- D. Purified troponin A.
- E. This property will vary depending on the preparative method and the figure given is an average value.
- F. There are alternative ways to allocate the distribution, but this represents the authors preference.
- G. These figures are for the binding of calcium to the high affinity sites only.

2, troponin 3 and troponin 4. Troponin 4 is presumably equivalent to our purified troponin A, and troponin B would therefore be expected to contain at least troponins 2 and 3. The necessity for this complicated mixture of components is not clear and more work on this system is required.

Troponin has a high Ca^{2+} binding constant ($\simeq 10^6$ M^{-1}) and information about the nature of this high affinity binding site would be of value in delineating the mechanism of the calcium trigger action. Initial experiments have centered around determining whether other cations show calcium-like activity. We chose to use biological activity as a measure of replacement of the calcium function. The system contained actomyosin, troponin and tropomyosin, and was made calcium-free by washing with EGTA and subsequent washing with solvents treated with Chelex-100. Under calcium-free conditions, the ATPase activity of the actomyosin was inhibited. Various concentrations of different metal ions were then added and the system assayed for removal of inhibition. A plot of the activity of the mixture versus the concentration of the added metal ion is given in Figure 1. The order of effectiveness being; Ca^{2+}, Sr^{2+}, Cd^{2+} >> Mn^{2+}, Zn^{2+}, Ce^{3+} > Sn^{2+}, Mg^{2+}, Ba^{2+}, Yb^{3+}, Eu^{3+}, La^{3+}. Sc^{3+}, Na^+, K^+. As expected Ca^{2+} and Sr^{2+} showed activity. Surprisingly, Cd^{2+} was also quite effective. The very slight activity for Mn^{2+}, Zn^{2+} and Ce^{3+} may be related to Ca^{2+} contamination of these salts. The metal ions on the extreme right of the order show negligible activity. In a parallel study, Fuchs (personal communication) has also found that Sr^{2+}, Cd^{2+} and, to a lesser extent, Mn^{2+} and Pb^{2+} displaced Ca^{2+} bound to troponin.

The principle of isomorphous replacement in biological systems (57) suggests that a functional metal ion may be replaced by another ion of similar size with retention of some activity. This proves to be so here since the divalent cations, Cd^{2+} ($r = 1.03$ Å) (the ionic radii given [56] are those for six coordinated metal ions.) and Sr^{2+} ($r = 1.20$ Å) show a Ca^{2+} ($r = 1.08$ Å) like sensitivity. Other divalent ions which were not effective are either too large or too small, e.g., Mg^{2+} ($r = 0.80$ Å) and Ba^{2+} ($r = 1.44$ Å). Monovalent ions like Na^+ ($r = 1.07$ Å), even though of

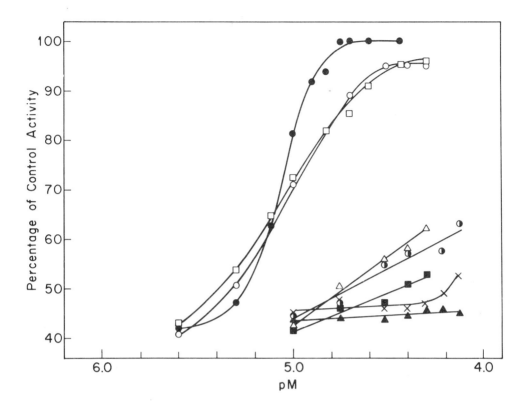

FIG. 1. Effect of various cations on the ATPase activity of actomyosin. Assay conditions; 2.5 mM ATP, 2.5 mM $MgCl_2$, 25 mM tris-HCl (pH 7.6). Desensitized actomyosin 0.7 mg., troponin 0.1 mg., tropomyosin 0.1 mg. Final volume of assay 2 ml. Temperature 25°C. Addition of cation varied. Calcium (●), strontium (□), cadmium (○), manganous (■), zinc (△), cerous (◐), sodium (▲), stannous (X). The chlorides were used. Precautions were taken to avoid contamination by calcium, these included the use of acid-washed glassware and the treatment of all solvents and ATP with Chelex-100. The proteins were dialyzed against 1 mM EGTA, 10 mM tris-HCl (pH 7.6), 0.5 mM dithiothreitol, and subsequently against several changes of calcium-free 10 mM tris-HCl (pH 7.6).

the proper size, were also ineffective. This is related to the weaker binding of the lower charged cation. A surprising result is that the trivalent rare earth ions like La^{3+} (r = 1.18 Å), Eu^{3+} (r = 1.03 Å) and Yb^{3+} (r = 0.95 Å) did not show any activity (5). Because of their enhanced charge and proper size, these ions would be expected to be quite effective.

Calcium is classified as a hard Lewis acid (42) and binds to hard base oxygen donors such as phenolate and carboxylate anions while only weakly interacting with the soft base nitrogen and sulfur donors. Thus, it is reasonable to assume that calcium binds to troponin via carboxylate functions. In general, there is an inverse relationship between binding and size for hard acids, i.e., small cations bind monodentate ligands more strongly than large cations (43). However, multidentate ligands can impose a metal ion-size requirement for favorable binding (43). The high binding constant of calcium to troponin may be related to the existence of a multidentate binding site on the protein which readily accepts ions of approximately the size of calcium. A feature of calcium is the existence of a variety of coordination numbers 6, 7, 8, etc., and stereochemistries that it can adopt (43). In fact, the metal is likely to accept the stereochemistry imposed on it by the donor atoms at the binding site of the protein. By and large, ligand binding to calcium is a result of favorable entropy since the corresponding enthalpy is generally small (3). The large positive entropy arises from the loss of coordinated water from the metal when chelation occurs. The binding of calcium to troponin should then be temperature insensitive providing of course that the protein remains native. Since calcium salts are easily dehydrated (43), calcium should also be at least partially dehydrated when it interacts with troponin.

It is not surprising that strontium can behave like calcium in our systems since the coordinating tendencies of the ions are similar. However, cadmium's coordination characteristics are significantly different from those of calcium. Cadmium is classified as a soft Lewis acid and as such binds preferentially to nitrogen and sulfur ligands (42). Nonetheless, cadmium binds quite strongly to oxygen atom donors. In fact, it forms more stable oxygen donor complexes than does calcium (3). Thus, it is not unreasonable that cadmium could bind to the troponin at the same site as calcium. The isomorphous replacement shown here suggests that the calcium binding site is size and charge specific.

ACKNOWLEDGMENTS

This work was supported by grants (to D. J. Hartshorne) from the National Institutes of Health (HE-09544 and GM-46407) and the National Science Foundation (GB-8388).

REFERENCES

1. BAILEY, K. Tropomyosin: a new asymmetric protein component of the muscle fibril. Biochem. J. 43: 271-279, 1948.

2. BAILEY, K., H. GUTFREUND, and A. G. OGSTON. Molecular weight of tropomyosin from rabbit muscle. Biochem. J. 43: 279-281, 1948.

3. CHRISTENSEN, J. J., and R. M. IZATT. Handbook of Metal Ligand Heats and Related Thermodynamics Quantities. New York: Marcel Dekker, 1970.

4. COHEN, C., and A. G. SZENT-GYORGYI. Optical rotation and helical polypeptide chain configuration in α-proteins. J. Amer. Chem. Soc. 79: 248, 1957.

5. DARNALL, D. W., and E. R. BIRNBAUM. Rare earth metal ions as probes of calcium ion binding sites in proteins. J. Biol. Chem. 245: 6484-6486, 1970.

6. DRABIKOWSKI, W., and J. GERGELY. The effect of the temperature of extraction on the tropomyosin content in actin. J. Biol. Chem. 237: 3412-3417, 1962.

7. EBASHI, S. Third component participating in the superprecipitation of "natural actomyosin". Nature 200: 1010, 1963.

8. EBASHI, S., F. EBASHI, and A. KODAMA. Troponin as the Ca^{++}-receptive protein in the contractile system. J. Biochem. 62: 137-138, 1967.

9. EBASHI, S., and M. ENDO. Calcium ion and muscle contraction. In: Progress in Biophysics and Molecular Biology 18, edited by J.A.V. Butler and D. Noble. Oxford: Pergamon Press, 1968, p. 123-183.

10. EBASHI, S., H. IWAKURA, H. NAKAJIMA, R. NAKAMURA, and Y. OOI. New structural proteins from dog heart and chicken gizzard. Biochem. Z. 345: 201-211, 1966.

11. EBASHI, S., and A. KODAMA. A new protein factor promoting aggregation of tropomyosin. J. Biochem. 58: 107-108, 1965.

12. EBASHI, S., and A. KODAMA. Interaction of troponin with F-actin in the presence of tropomyosin. J. Biochem. 59: 425-426, 1966.

13. EBASHI, S., and A. KODAMA. Native tropomyosin-like action of troponin on trypsin treated myosin B. J. Biochem. 60: 733-734, 1966.

14. EBASHI, S., A. KODAMA, and F. EBASHI. Troponin. I. Preparation and physiological function. J. Biochem. 64: 465-477, 1968.

15. EISENBERG, E., and W. W. KIELLEY. Native tropomyosin: effect on the interaction of actin with heavy meromyosin and subfragment-I. Biochem. Biophys. Res. Commun. 40: 50-56, 1970.

16. ENDO, M., Y. NONOMURA, T. MASAKI, I. OHTSUKI, and S. EBASHI. Localization of native tropomyosin in relation to striation patterns. J. Biochem. 60: 605-608, 1966.

17. FUCHS, F., and F. N. BRIGGS. The site of calcium binding in relation to the activation of myofibrillar contraction. J. Gen. Physiol. 51: 655-676, 1968.

18. GREASER, M. L., and J. GERGELY. Calcium binding component of troponin. Federation Proc. 29: 463, 1970.

19. GREASER, M. L., and J. GERGELY. Isolation of three functionally required proteins from troponin. Abstracts Soc. Gen. Physiol., Woods Hole, Mass., 1970.

20. HANSON, J., and J. LOWY. The structure of F-actin and of actin filaments isolated from muscle. J. Mol. Biol. 6: 46-60, 1963.

21. HARTSHORNE, D. J. Interactions of desensitized actomyosin with tropomyosin, troponin A, troponin B, and polyanions. J. Gen. Physiol. 55: 585-601, 1970.

22. HARTSHORNE, D. J., and H. MUELLER. Separation and recombination of the ethylene glycol bis (β-aminoethyl ether)-N,N'-tetraacetic acid-sensitizing factor obtained from a low ionic strength extract of natural actomyosin. J. Biol. Chem. 242: 3089-3092, 1967.

23. HARTSHORNE, D. J., and H. MUELLER. Fractionation of troponin into two distinct proteins. Biochem. Biophys. Res. Commun. 31: 647-653, 1968.

24. HARTSHORNE, D. J., and H. MUELLER. The preparation of tropomyosin and troponin from natural actomyosin. Biochim. Biophys. Acta 175: 301-319, 1969.

25. HARTSHORNE, D. J., and H. Y. PYUN. Calcium binding by the troponin complex, and the purification and properties of troponin A. Biochim. Biophys. Acta. In press.

26. HARTSHORNE, D. J., M. THEINER, and H. MUELLER. Studies on troponin. Biochim. Biophys. Acta 175: 320-330, 1969.

27. HOLTZER, A., R. CLARK, and S. LOWEY. The conformation of native and denatured tropomyosin B. Biochemistry 4: 2401-2411, 1965.

28. HUXLEY, A. F., and R. NIEDERGERKE. Structural changes in muscle during contraction. Interference microscopy of living muscle fibres. Nature 173: 971-973, 1954.

29. HUXLEY, H. E. Electron microscope studies of the organisation of the filaments in striated muscle. Biochim. Biophys. Acta 12: 387-394, 1953.

30. HUXLEY, H. E. Structural difference between resting and rigor muscle; evidence from intensity changes in the low-angle equatorial X-ray diagram. J. Mol. Biol. 37: 507-520, 1968.

31. HUXLEY, H. E. The mechanism of muscular contraction. Science 164: 1356-1366, 1969.

32. HUXLEY, H. E., and W. BROWN. The low-angle X-ray diagram of vertebrate striated muscle and its behavior during contraction and rigor. J. Mol. Biol. 30: 383-434, 1967.

33. HUXLEY, H., and J. HANSON. Changes in the cross-striations of muscle during contraction and stretch and their structural interpretation. Nature 173: 973-976, 1954.

34. KATZ, A. M. Contractile proteins of the heart. Physiol. Rev. 50: 63-158, 1970.

35. KOMINZ, D. R., and K. MARUYAMA. Does native tropomyosin bind to myosin? J. Biochem. 61: 269-271, 1967.

36. LAKI, K., K. MARUYAMA, and D. R. KOMINZ. Evidence for the interaction between tropomyosin and actin. Arch. Biochem. Biophys. 98: 323-330, 1962.

37. LOWEY, S., H. S. SLAYTER, A. G. WEEDS, and H. BAKER. Substructure of the myosin molecule. I. Subfragments of myosin by enzymic degradation. J. Mol. Biol. 42: 1-29, 1969.

38. MARTONOSI, A. Studies on actin. VII. Ultracentrifugal analysis of partially polymerized actin solutions. J. Biol. Chem. 237: 2795-2803, 1962.

39. MOORE, P. B., H. E. HUXLEY, and D. J. DEROSIER. Three-dimensional reconstruction of F-actin, thin filaments and decorated thin filaments. J. Mol. Biol. 50: 279-295, 1970.

40. OHTSUKI, I., T. MASAKI, Y. NONOMURA, and S. EBASHI. Periodic distribution of troponin along the thin filament. J. Biochem. 61: 817-819, 1967.

41. PARKER, L., H. Y. PYUN, and D. J. HARTSHORNE. The inhibition of the ATPase activity of the subfragment I-actin complex by troponin plus tropomyosin, troponin B plus tropomyosin and troponin B. Biochim. Biophys. Acta 223: 453-456, 1970.

42. PEARSON, R. G. Hard and soft acids and bases, HSAB, Part II. Underlying Theories. J. Chem. Ed. 45: 643-648, 1968.

43. PHILLIPS, C. S. G., and R. J. P. WILLIAMS. Inorganic Chemistry II. New York: Oxford University Press, 1965, p. 80-85.

44. REES, M. K., and M. YOUNG. Studies on the isolation and molecular properties of homogeneous globular actin. Evidence for a single polypeptide chain structure. J. Biol. Chem. 242: 4449-4458, 1967.

45. RÜEGG, J. C. Smooth muscle tone. Physiol. Rev. 51: 201-248, 1971.

46. SCHAUB, M. C., and S. V. PERRY. The relaxing protein system of striated muscle. Resolution of the troponin complex into inhibitory and calcium ion-sensitizing factors and their relationship to tropomyosin. Biochem. J. 115: 993-1004, 1969.

47. STAPRANS, I., and S. WATANABE. Optical properties of troponin, tropomyosin and relaxing protein of rabbit skeletal muscle. J. Biol. Chem. 245: 5962-5966, 1970.

48. STROMER, M. H., D. J. HARTSHORNE, H. MUELLER and R. V. RICE. The effect of various protein fractions on Z- and M-line reconstitution. J. Cell Biol. 40: 167-178, 1969.

49. STROMER, M. H., D. J. HARTSHORNE, and R. V. RICE. Removal and reconstitution of Z-line material in a striated muscle. J. Cell Biol. 35: C23-C28, 1967.

50. TONOMURA, Y., S. WATANABE, and M. MORALES. Conformational changes in the molecular control of muscle contraction. Biochemistry 8: 2171-2176, 1969.

51. TSAO, T.-C., K. BAILEY, and G. S. ADAIR. The size, shape and aggregation of tropomyosin particles. Biochem. J. 49: 27-36, 1951.

52. WEBER, A. On the role of calcium in the activity of adenosine 5'-triphosphate hydrolysis by actomyosin. J. Biol. Chem. 234: 2764-2769, 1959.

53. WEBER, A. Energized calcium transport and relaxing factors. In: Current Topics in Bioenergetics 1, edited by D.R. Sanadi. New York: Academic Press, 1966, p. 203-254.

54. WEBER, A., and R. HERZ. The binding of calcium to actomyosin systems in relation to their biological activity. J. Biol. Chem. 238: 599-605, 1963.

55. WEBER, A., R. HERZ, and I. REISS. The role of magnesium in the relaxation of myofibrils. Biochemistry 8: 2266-2271, 1969.

56. WHITTAKER, E. J. W., and R. MUNTUS. Ionic radii for use in geochemistry. Geochim. Cosmochim. Acta 34: 945-956, 1970.

57. WILLIAMS, R. J. P. The biochemistry of sodium, potassium, magnesium, and calcium. Quart. Rev. 24: 331-365, 1970.

58. WOODS, E. F. Molecular weight and subunit structure of tropomyosin B. J. Biol. Chem. 242: 2859-2871, 1967.

CHEMISTRY OF MUSCULAR CONTRACTION: MYOSIN

PAUL DREIZEN

Department of Medicine and Program in Biophysics
State University of New York Downstate Medical Center
Brooklyn, New York 11203

INTRODUCTION

The myofibril of cardiac muscle and skeletal muscle contains an interlocking array of thick filaments, comprised of myosin and thin filaments, comprised of actin, tropomyosin and troponin (13,25,26,31). Cross-bridges extend outwards from the thick-filament core (26,31), and the contractile event involves a coupling of cyclic interactions between cross-bridges and thin-filament sites with the enzymatic hydrolysis of adenosine triphosphate (ATP) by myosin. This paper will review recent studies on the structure and function of myosin, and describe some aspects of the temperature dependence of myosin ATPase which may have significance for cardiac function in hibernating mammals.

STRUCTURE OF MYOSIN

Electron microscopic studies. The myosin molecule has a rodlike tail, approximately 1300 Å in length and 20 Å in diameter, which terminates in a globular head, roughly 200 Å in length and 50 Å in diameter (32,56), that appears to be bilobular (47,59).

Myosin is cleaved by trypsin into two fragments, light meromyosin and heavy meromyosin (57). Light meromyosin exhibits the association properties of myosin (51) and forms the thick filament core (32), whereas heavy meromyosin exhibits the ATPase and actin-binding properties of myosin (51) and forms the cross-bridges (32). On electron microscopy, light meromyosin comprises two-thirds of the rodlike tail of myosin, and heavy meromyosin comprises the rest of the rodlike portion and the globular head of myosin (32,56). Heavy meromyosin can be digested further with trypsin (54) or papain (44) into a smaller fragment, termed subfragment-1 (54), that retains ATPase activity and actin-binding properties. Subfragment-1 appears as a globular fragment on electron microscopy, and each particle of subfragment-1 has been identified with one lobe of the globular head of myosin (47,49).

Subunit composition. Analysis of the subunit composition of myosin depends on precise determination of its molecular weight, and this matter has been controversial for nearly twenty years, the essential problem being that apparently myosin undergoes irreversible aggregation (62). For example, in sedimentation equilibrium experiments on rabbit skeletal myosin (23), the weight-average molecular weight increases from 500,000 to 800,000 during prolonged ultracentrifugation over 180 hours, reflecting the occurrence of continuous aggregation throughout the period of centrifugation. Dimers and other low n-mers of myosin may be selectively sedimented from monomer in high speed sedimentation equilibrium experiments, permitting characterization of the monomer distribution at low radial distances and low protein concentrations. Experiments so performed indicate that preparations of myosin contain a predominant monomer component of 468,000 molecular weight and additional aggregated protein, the proportion of which varies with the method of preparation and increases with the duration of storage (11, 19).

Rabbit skeletal myosin liberates light and heavy components during prolonged treatment in 6.7M urea (65) and at alkaline pH (39); but it was believed at first that any low molecular weight material, if present, represented contaminant proteins or fragments of myosin. This interpretation rested largely on physical chemical studies (37) that were originally taken to imply dissociation of myosin by 5 M guanidine into 3 equivalent subunits.

However, detailed analysis (12) of ultracentrifugal data on myosin in 5 M guanidine demonstrated, quite unambiguously, that purified preparations of myosin are in fact heterogeneous in 5 M guanidine. High-speed sedimentation equilibrium experiments indicate dissociation of myosin by 5 M guanidine into a heavy component of about 210,000 to 220,000 mol weight and a light component (20,000 mol weight) that comprises 10 to 15 percent of the total protein (11,20). Following removal of the light component, sedimentation equilibrium experiments on purified heavy component in 5 M guanidine indicate a molecular weight of 212,000 ($\pm$ 5,000) for the heavy chains and confirm their weight-equivalence (26).

These results indicate that myosin contains 2 heavy chains (212,000 mol weight) and several light chains of average molecular weight about 20,000. But the ultracentrifugal experiments on myosin in 5 M guanidine do not yield precise information on the stoichiometry of light and heavy components, and this technical problem prompted experiments on the effect of alkali on myosin (20). At pH 11, myosin is dissociated into a predominant heavy component, with a sedimentation coefficient close to that of native myosin, and a trailing component that comprises 12 percent of the total protein. Sedimentation equilibrium experiments also indicate the presence of 12 percent light component and yield molecular weight values of 420,000 for the heavy component and 20,200 for the light component (Fig. 1). Assuming 470,000 molecular weight for myosin, these data indicate dissociation of approximately three light chains from the intact heavy chain core (20,23). This result is supported (less exactly) by the difference in molecular weight between intact myosin and heavy alkali component and other evidence that approximately 3.5 C-terminal isoleucine residues of native myosin belong to the light chains (20,23).

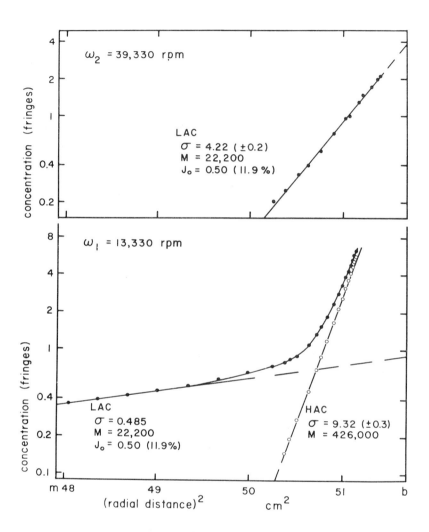

FIG. 1. Data from a representative sedimentation equilibrium experiment on myosin in 0.1 M Na_2CO_3, 0.4 M KCl, pH 11.0. Top, equilibrium at 39,330 rpm: bottom, equilibrium at 13,300 rpm. (●) observed fringe displacement from meniscus position, light alkali component (LAC); and (○) for heavy alkali component (HAC). (M) molecular weight, (m) meniscus, (b) bottom and (J_o) average concentration in fringes. $\sigma = 2d(\log J)/d(r^2)$, where J is concentration in fringes, and r is radial distance (23).

Comparable stoichiometry and molecular weight for the light chains of myosin have been reported from experiments based on dissociation of myosin by acetylation (42). On chromatography, the proportion of light component is not significantly changed on single passage of myosin through DEAE-Cellulose (23,27,42), DEAE-Sephadex (57,67), or cellulose phosphate (23). However, there is extensive denaturation of myosin during three chromatographic cycles on cellulose phosphate, and the residual soluble myosin (about 10% of the unchromatographed myosin) contains only 2 light chains (23).

Other studies show that the light chains are dissociated from the intact heavy chain core of myosin on heat treatment (22,23), in 2 M guanidine (23), and in concentrated salt solutions at neutral pH (16,18,22). The salt effect is of special interest, for anions and cations may be listed in a Hofmeister series with respect to their dissociating effect on myosin. Thus, light chains are fully dissociated from myosin by 1 M KSCN and 4 M LiCl, but are not dissociated by KCl or NaCl solutions at salt concentrations to saturation. In some instances, as on treatment of myosin in 4.7 M NH_4CL at 4°C (16,18) and in 0.4 M KCl at temperatures above 4°C (22,23), there is incomplete, time-dependent dissociation of light component from myosin. Following subunit dissociation of myosin, light and heavy components undergo reassociation on dialysis against 0.4 M KCl at pH 7 and 4°C, provided that the period of denaturation was brief and thiol groups were protected (18,23). Light and heavy components both aggregate during prolonged dissociation and on storage in 0.4 M KCl at pH 7 (18,23).

Proteolytic Fragments. Lowey and Cohen (45) suggested that light meromyosin has a double-stranded α-helical conformation, based on comparison of amino acid content and hydrodynamic estimates of length. Ultracentrifugal studies (20) indicate that light meromyosin remains intact at pH 11, without release of any light chains. Heavy meromyosin has a more complex structure and is dissociated at pH 11 into a remnant of the heavy chain core ($\sim$ 250,000 mol weight) and most, if not all, of the light component of intact myosin (20). The light component liberated on acetylation (42) is also found in heavy meromyosin, as are the C-terminal isoleucine residues of the light chains (20).

Subfragment-1 comprises 70-80 percent of heavy meromyosin and has molecular weight about 110,000 (68); this stoichiometry would suggest that heavy meromyosin yields two particles of subfragment- (62). Ultracentrifugal experiments (64) indicate that papain-prepared subfragment-1 is dissociated by 5 M guanidine and at pH 11 into remnants of one heavy chain, 85,000 in weight, and one light chain, 18,000 in weight. Trypsin-prepared subfragment-1 is a more degraded particle, comprising a remnant of one heavy chain (65,000 molecular weight), one light chain (18,000 molecular weight), and considerable peptide material.

Figure 2 summarizes the present evidence on the structure of myosin and the subunit interactions on which the proposed structure is based. Native myosin comprises an axial core of 2 heavy polypeptide chains, each of 210,000 molecular weight, that terminate in a globular region also containing three light polypeptide chains, having average weight 20,200. Each subfragment-1 particle contains remnants of 1 light chain and 1 heavy chain; an additional light chain appears to be less tightly bound to myosin as indicated by selective dissociation in 4.7 M NH_4Cl and during repeated chromatography on cellulose phosphate.

Light chain heterogeneity. The light chains are electrophoretically heterogeneous, with 3 or 4 major bands for the light component dissociated from carboxymethylated myosin at pH 11 (20) and from acetylated myosin (41,42). Comparable electrophoretic patterns are obtained after dissociation of light chains from chemically unmodified myosin at pH 11, in 2 M guanidine and in concentrated salt solutions, indicating that electrophoretic heterogeneity derives at least in part from true structural differences among the light chains (17,18). Nevertheless, prolonged alkaline and salt treatment does result in modification of the electrophoretic pattern (18,19) and aggregation of light chains (8,18,23), with extensive heterogeneity of the light component.

Locker and Hagyard (43,44) have described different electrophoretic patterns for light component dissociated by acetylation of myosin from red, white, and cardiac muscle; and comparable electrophoretic variation is obtained for the light alkali component of red and white skeletal myosin that was not chemically modified (19, 21 and our unpublished observations).

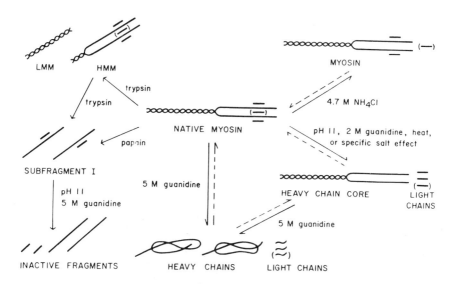

FIG. 2. Schematic diagram of subunit structure and interactions of myosin. Chain lengths are drawn proportional to atomic mass, and differ from the actual dimensions of rodlike globular portions of myosin as determined from electron microscopy (8).

Figure 3 shows characteristic patterns for the light chains of red, intermediate, and white rabbit skeletal myosin; the specific electrophoretic pattern varies somewhat in different animals. Ultracentrifugal experiments indicate that the proportion of light component is comparable in myosin of red and white muscle, implying that the various muscles contain myosin molecules having different distributions of light chains, but the same proportion of light chains per myosin molecule (21, and unpublished observations).

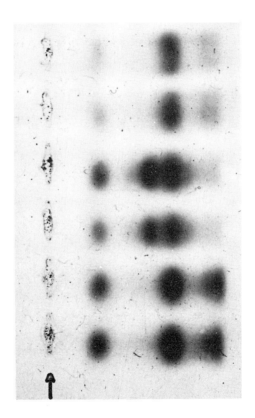

FIG. 3. Cellulose acetate electrophoresis of light alkali component from rabbit skeletal myosin. 0.75 M sodium barbital, pH 8.6. Light chains from white muscle (top), intermediate muscle (middle), and red muscle (bottom). Arrow indicates origin; anode to right.

A fraction of the total light component which is dissociated from myosin in 4.7 M NH_4Cl contains predominantly protein migrating as a fast electrophoretic band on cellulose acetate (17,18). The NH_4Cl dissociated protein comprises approximately one mole per mole of myosin, is found in preparations of red and white skeletal myosin as well as cardiac myosin, and is not essential for the ATPase activity of myosin (see below). It has been suggested that this fraction represents a distinct protein, termed myonin, that is closely bound to myosin proper and possible may be involved in the interaction of myosin with sites on the thin filament.

MYOSIN-NUCLEOTIDE INTERACTION

In general, those denaturing conditions which result in dissociation of light chains from myosin also lead to irreversible inactivation of ATPase. Thus, on storage of myosin at temperatures above 25°C, there is loss of ATPase and light chain dissociation, with comparable values of activation enthalpy (55 kcal/mole) and activation entropy (100 e.u.) for both processes (22,23). Similarly, myosin ATPase is irreversibly denatured on alkaline treatment above pH 10 (52), and loss of enzymatic activity occurs over the same range in pH at which light chains are dissociated from myosin (11,20,23). The loss of ATPase precedes light chain dissociation and is presumably due to alkaline hydrolysis of critical residues at or near the ATPase site (23).

A close relationship between dissociation of light chains from myosin and rapid irreversible inactivation of ATPase was also found in a series of experiments on myosin in concentrated salt solutions (6,9,22). Thus, treatment of myosin in those salt solutions which fully dissociate light chains from the heavy chain core is accompanied by rapid irreversible loss of ATPase at or slightly below the transition concentration for light chain dissociation. However, some of the light component does not appear essential for myosin

ATPase, as shown by nearly full recovery of Ca-ATPase following the dissociation by 4.7 M NH_4Cl of approximately one mole light chain per mole of myosin (9,16-18).

More direct evidence of a dependence of myosin ATPase on subunit interactions has been obtained in recombination experiments on myosin. After dissociation of myosin in 2 M KCl - 0.01 M ATP, pH 11, about 10 to 15 percent of the original ATPase activity is recovered in reconstituted myosin, but significantly less ATPase activity is recovered in protein reconstituted with less than 12 percent light component. In experiments involving dissociation of myosin in 4 M LiCl and subunit fractionation by citrate salting-out (6,7,9), fully reconstituted myosin retains about 70 percent of control ATPase, provided that thiol groups are protected; but protein with less than 8 percent light component shows a sharp fall in ATPase, and light chains alone exhibit no activity. Following chromatographic fractionation of myosin in 4 M LiCl (61), about 30 percent ATPase is recovered in reconstituted myosin, and again no activity is recovered in purified light and heavy components. Finally, recombination experiments involving LiCl-citrate fractionation of NH_4Cl-treated myosin (2 heavy chains and 2 light chains) confirm that recovery of Ca-ATPase requires a molar ratio of essential light to heavy chains (7,9).

The recombination experiments implicate an interaction between essential light and heavy chains in the ATPase activity of myosin. One might suppose that residues from essential light and heavy chains participate at or near the hydrolytic site. Alternatively one subunit, say the heavy chain, may contain the hydrolytic site, and the adjacent subunit (essential light chain, according to this model) may stabilize the unique conformation required for enzymatic activity.

The ATPase activity of white muscle myosin exceeds that of red muscle myosin (2,44,60), and there are significant differences between the light and heavy chains of the two types of myosin. Hybridization experiments on red and white skeletal myosin indicate that the level of ATPase depends on light chain and probably also heavy chain composition (21, and unpublished observations). Fetal myosin exhibits less ATPase activity than adult myosin (63), and this difference, too, is presumable related to specific subunit composition (unpublished observations of J. Dow and A. Stracher). The overall evidence would support the hypothesis that myosin comprises an isoenzyme system, in which specific ATPase activity varies with subunit composition.

Myosin and heavy meromyosin contain about 1.6 binding sites for ADP (38,46). Following LiCl-citrate fractionation of myosin, recombination experiments indicate comparable binding of ADP to reconstituted myosin, with or without the NH_4Cl dissociated light chain present (7-9). The number of ADP binding sites is diminished in protein containing fewer light chains, and heavy chain core and light chains alone do not exhibit significant binding. Thus, nucleotide binding involves some kind of interaction between light and heavy chains. The simplest explanation would be that nucleotide provides a direct bridge between essential light and heavy chains within each protomer, but this interpretation does not exclude the possibility of weak interaction of nucleotide with light or heavy chains alone.

The subunit structure of myosin is stabilized by ATP to alkaline (6,23) and salt (6,7) dissociation, and this effect appears to depend on the interaction of myosin with the terminal phosphate groups of ATP (8). Electrophoretic experiments show that chain phosphates stabilize the interaction of the two essential light chains with the heavy chains (10). These findings, taken in conjunction with the recombination experiments on ATPase and ADP binding, would suggest that the triphosphate part of ATP bridges essential light and heavy chains in the region of the hydrolytic site (Fig. 4). Alternatively, one might hypothesize an allosteric relationship between nucleotide binding and stabilization of subunit interactions, but this explanation would not be consistent with the lack of catalytic activity in light or heavy chains alone.

HYPOTHESIS I: triphosphate bridge

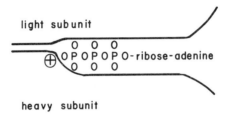

HYPOTHESIS II: conformational change

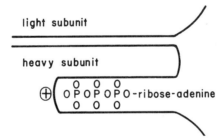

FIG. 4. Alternative models for nucleotide binding to myosin protomer. Representation is schematic and fails to show the non-linearity of the phosphate chain due to the tetrahedral orientation of σ orbitals of phosphorous, or the possibility of free rotation about the diester linkages of ribose (19).

MYOFIBRILLAR ORGANIZATION OF MYOSIN

The original electron microscopic findings on muscle (26,31) have been substantiated by low-angle X-ray diffraction studies (25,34,35,38,39) which provide details of the structure of the thick filament during relaxation and contraction. The thick filament contains myosin molecules with their elongate rodlike portions in staggered array along the thick filament and their globular heads protruding as cross-bridges at 143 Å intervals along the filament axis. In the resting state, the cross-bridges are helically arranged with 120° angular rotation per cross-bridge pair, so that six cross-bridges are found per 429 Å helical repeat (25,35). Analysis of stoichiometric and molecular weight data on the major myofibrillar proteins would suggest that each cross-bridge contains one myosin molecule (8).

During isometric contraction and shortening on the order of 5 to 10 percent, there is no change in the 143 Å cross-bridge repeat, but there is a large decrease in the intensity of off-meridional layer lines. These latter have been interpreted by Huxley and Brown (35) to imply disruption of the helical arrangement of cross-bridges and some longitudinal disorder of the individual cross-bridges, with more substantial changes in their azimuthal and possible radial positions. During rigor, a condition in which ATP is depleted and cross-bridges interact directly with thin filament sites, the overall helical arrangement is disordered, although the 143 Å interval between cross-bridges in maintained (25,35). Changes in the equatorial reflections indicate that cross-bridges move from a position alongside the thick filament core during the living state to the vicinity of the thin filaments during rigor (34). The thin filament periodicities do not change significantly during contraction or rigor (25,35), and the attachment of subfragment-1 to actin involves a unique orientation of both molecules (53).

Huxley and Brown (35) and Hanson (25) have interpreted the X-ray diffraction findings to provide essential confirmation of the sliding-filament concept since the constancy of cross-bridge and actin periodicities in relaxed and contracted states implies that filament length also remains constant during contraction. It seems to have been generally assumed, although never directly proven, that the sliding together of thick and thin filaments is a linear movement in which cross-bridges of the same azimuthal angle move along a single thin filament. But there is no compelling reason to suppose that cross-bridge advance during contraction need involve linear movement only. Indeed, Huxley and Brown (35) have noted that the pattern of cross-bridge attachments might be changing continuously, so that a rotational disorder is imposed on the cross-bridges in a given thick filament. There are several ways in which a longitudinal advance of cross-bridges might involve concomitant rotational movement (8).

a) <u>Linear model</u>: unsynchronized circumferential and possible radial movement of the cross-bridges, in conjunction with repetitive longitudinal interactions of the cross-bridges with actin sites. This explanation has been suggested by Huxley and Brown (35); however, it is difficult to visualize a geometrical arrangement of cross-bridges and thin-filament sites which permits cross-bridges to undergo clockwise and counter-clockwise rotation at random. Also, the selective loss in intensity of the 429 Å reflection with respect to the 143 Å reflection is not simply attributable to a purely random process.

b) <u>Screw model</u>: a turning and advance of the rigid thick-filament core generated by an oriented rotational turning of cross-bridges; that is, a screw motion of a thick filament within its surrounding envelope of stationary thin filaments. However, a screw motion is difficult to reconcile with the marked decrease in intensity of the 429 Å reflection during active contraction. Nor does a screw motion appear likely on grounds of symmetry, since concomitant advance of cross-bridges at opposite ends of a thick filament would involve clockwise rotation of cross-bridges at one end and counterclockwise rotation of cross-bridges at the other end. This would necessitate an antisymmetrical arrangement of myosin molecules within each half-filament, an arrangement that does not appear in accord with evidence on the structural organization of the mid-filament region (55).

Torsional model: a rotational (isometric) or spiral (isotonic) thrust of a cross-bridge as it interacts with a thin-filament site, with resultant distortion of the thick filament core. Successive cross-bridges would be activated as they are brought into precise registry with thin-filament sites, and the contractile movement would have the characteristics of a torsional elastic wave proceeding symmetrically from opposite ends of a thick filament. According to this model, contraction would virtually involve winding of the thick-filament core into an elastic coil. During isometric contraction and also slight shortening, the 143 Å reflection might be expected to remain nearly constant, whereas the 429 Å helical repeat would be lost insofar as this pattern represents a time-and-space average over different filaments.

As previously noted (8), there are several findings which are indicative of a drastic structural rearrangement of myosin molecules within the thick-filament core during contraction and these findings are somewhat suggestive of a torsional model for contraction.

a) Striated muscle is birefringent, with anisotropic (A) bands and isotropic (I) bands. The birefringence of A bands may be identified with anisotropy of the thick filaments. Muscle exhibits both intrinsic and form birefringence (66); intrinsic birefringence derives from anisotrophy in polarization of the individual myosin molecules, presumably their rodlike helical portions within the thick-filament core, and form birefringence derives from structural organization of the myosin molecules within the thick-filament core. In glycerol-extracted muscle fibers, both form birefringence and intrinsic birefringence are diminished on shortening (66), suggesting that contraction is accompanied by change in the helical conformation of myosin and rearrangement of myosin molecules within the thick-filament core.

b) The shortening of glycerinated muscle is accompanied by loss of helical conformation on wide-angle X-ray diffraction (49) and by thermodynamic properties suggestive of a cooperative structural transition during contraction (48,50). These findings have been interpreted in terms of a melting of the ordered arrangement of myofibrillar proteins (48-50). Although this interpretation has been disputed vigorously (cf. discussions by H. E. Huxley and R. E. Davies in reference 48), the essential evidence remains unexplained according to the conventional sliding filament theory. It would seem though that the phenomena described by Mandelkern and associates (48-50) are attributable, at least in part, to the disruption of the native α-helical conformation of light meromyosin during contraction. This effect might possibly result from torsional distortion of the thick filament during contraction or, alternatively, a longitudinal elastic wave generated by cross-bridge interactions, among other explanations; however, the available evidence does not allow a unique molecular interpretation.

c) The heat of shortening represents a well-known physiological phenomenon (28,30) whose molecular basis has been elusive. The magnitude of shortening heat is proportional to the extent of shortening at constant load (30); shortening heat is not directly related to the enzymatic hydrolysis of ATP (4), and there is no net shortening heat over a complete contraction-relaxation cycle (3). From thermodynamic considerations, this would imply that shortening heat may be identified with a loss of entropy during contraction, and a commensurate gain in entropy during relaxation. This kind of cycle does not have any ready explanation in terms of a simple sliding-filament concept, nor of a phase transition involving the myofibrillar proteins; but the evidence would be consistent with elestic coiling of the thick-filament core during contraction, and uncoiling during relaxation. In similar manner, a negative heat of lengthening (29) might be attributed to unwinding of the partially coiled thick filament from its resting state. A statistical mechanical model for entropy exchange during contraction of striated muscle is described elsewhere (5).

d) Finally, there is some recent evidence that would seem to favor a torsional model for contraction. Although the average length of thick filaments does not change during shortening of sarcomeres to 1.6 µ (26,31) (that is, shortening of muscle to about two-thirds of resting length), it is well-known that muscle may shorten to less than two-thirds of resting length and this would entail some shortening of thick filaments during contraction. In this respect, Gordon, A. F. Huxley, and Julian (24) have reported that vertebrate skeletal muscle may shorten to 30 to 40 percent of rest length before tension is abolished, and shortening to this extent would result in decrease of the average length of thick filaments from 1.6 µ to approximately 1.0 µ. Shortening of thick filaments appears to be accompanied by the formation of contraction bands and these have been attributed by H. E. Huxley (33) to forcing up crumpled thick filaments against the Z-line; however, detailed studies have not been presented on sarcomeres shortened below 1.6 µ. In an unpublished reinvestigation of this problem, L. Herman and I have confirmed that thick filaments may shorten appreciably during marked contraction but there seems to be a proportionate decrease in the periodicity between cross-bridges in the shortened thick filaments. This finding is not consistent with a sliding-filament concept based on cyclic movement of cross-bridges about a rigid thick-filament core, but the evidence would be consistent with elastic coiling of the thick filament during contraction.

ASPECTS OF MYOSIN ATPase RELATING TO HIBERNATION

The rate of ATP hydrolysis by myosin appears to play a unique role in the contractile process. For example, the myofibrils of fast muscle split ATP at a rate two to three times the rate at which the myofibrils of slow muscle split ATP (15), and this difference has been related to the specific ATPase activity of myosin isolated from fast and slow muscles (2,15,60). Even more striking evidence for the crucial physiological role of myosin is obtained from the work of Bárány (1) that myosin ATPase activity varies more than two orders of magnitude in different vertebrate and invertebrate muscles, and that specific myosin ATPase varies quantitatively with the velocity of shortening in the different muscles. These differences are attributable to variations in the primary structure of myosin, implicating phylogenetic differences in myosin in the regulation of the velocity of shortening.

Under customary assay conditions for myosin ATPase, the rate of ATP hydrolysis by purified myosin is less than the rate of ATP hydrolysis by intact myofibrils. Thus, the level of Ca^{++}-activated myosin ATPase is about one-tenth, that of EDTA-activated myosin ATPase about one-fourth, and that of Mg^{++}-activated actomyosin ATPase about one-tenth of the rate of splitting by the myofibril itself (4). Eisenberg and Moos (14) have shown from kinetic analyses that fully complexed actomyosin (obtained by increasing the ratio of actin to myosin and extrapolating the results to infinite actin ratio) splits ATP at nearly the same rate as do myofibrils. Although the level of myofibrillar ATPase depends on the interaction of thin filament proteins with myosin, there is no evidence as yet for any specificity of actin, tropomyosin, or troponin in regulating myosin ATPase activity; however, this possibility has not really been explored in depth.

Myosin ATPase is activated by increasing temperature with values for Q_{10} on the order of 2 and values for $\Delta H^{\ddagger}$ about 10 kcal/mole. These values are fairly typical of other mammalian enzyme systems. Myosin ATPase does not exhibit any extensive variation in temperature dependence that would be comparable with the variation in absolute levels of myosin ATPase from different muscles. For example, similar values for Q_{10} are found for myosin ATPase in poikilothermic animals, such as frog and tortoise, and homeothermic animals, such as rabbit (1). This is not too surprising, since activation energy represents a fundamental property of the enzymic mechanism for ATP hydrolysis.

Nevertheless, there is some indication that the temperature dependence of myosin ATPase may be subject to adaptive change. H. D. Kim and I had the opportunity to study the contractile proteins of a benthic fish, Corypheroides sp., during the recent expedition of the Alpha-Helix to the Galapagos. These fish were recovered from a depth of 1,200 fathoms, where the pressure is about 3,000 psi and the temperature is close to 0°C. Studies on the isolated proteins are still in progress, but some preliminary data on the temperature dependence of myosin ATPase may be mentioned. The Arrhenius plot for Ca^{++}-activated myosin ATPase is not dissimilar to that for Ca^{++}-activated myosin ATPase of rabbit muscle, with a Q_{10} value of 1.8. However, EDTA-activated myosin ATPase exhibits a somewhat different dependence, with a Q_{10} value of 1.4 at 0°C to 15°C, and no change in ATPase activity on increase from 25°C to 36°C. Although Ca^{++}-activated ATPase and EDTA-activated ATPase do not represent true biological activities of myosin, these findings do demonstrate the existence of a myosin molecule which, at least under certain conditions, has an altered temperature sensitivity.

If one may attach any adaptive significance to the temperature dependence of myosin ATPase, then the activation of myosin ATPase on increase from 0°C to 37°C provides obvious advantage for homeothermic animals. The coin is reversed in the case of hibernating mammals where a decrease in body temperature from 37°C to close to 0°C is accompanied by a marked decrease in heart rate, with a concomitant decrease in cardiac output but no significant change in stroke volume (36). This would suggest that the cardiac contractile proteins of hibernating mammals may be nearly as efficient at a temperature close to 0°C as at 37°C, despite an expected eight-fold decrease in myosin ATPase over the same range of temperature, assuming Q_{10} of 2. If so, the cardiac contractile proteins of hibernating mammals may contain some adaptive mechanism for myosin ATPase to function equally as well at 0°C as at 37°C.

A priori, thermal adaptation might be achieved by any of several possible mechanisms:

a) The cardiac myosin of hibernating mammals might have a low activation energy for ATPase, with a Q_{10} value not greatly in excess of one. According to this hypothesis, myosin might split ATP at a rate relatively independent of body temperature. This kind of phenomenon would represent an extreme form of the variation in temperature dependence for myosin ATPase that was found in the benthic fish Corypheroides.

b) Cardiac myosin might exist in two different forms: a high temperature form comparable with the myosin of non-hibernating mammals, and a low temperature form capable of functioning close to 0°C. The change from one form to another might be achieved by synthesis of new myofibrils or alternatively - turnover of light chains on pre-existing thick filaments.

The light and heavy chains of myosin are held together by non-covalent interactions, and light and heavy chains appear to be synthesized independently (58). If the light chain population were to be largely replaced within several days of onset of hibernation, then there might be sufficient change in the level of myosin ATPase to result in thermal adaptation.

c) The troponin-tropomyosin-actin system may possibly compensate for the temperature dependence of myosin ATPase. However, this alternative does not seem too likely in that the essential conformational changes involved in the activation of myosin ATPase would have to be drastically different in hibernating and non-hibernating ammals.

d) A decrease in temperature from 37°C to close to 0°C would increase the extent of hydrogen bonding in water. This would tend to increase the dielectric constant of the medium between thick and thin filaments, with decrease in electrostatic force between filaments and perhaps increased contractile force per cross-bridge interaction. As a rough estimate of this effect, the dielectric constant of free water increases from 74 at 37°C to 88 at 0°C. Although the dielectric constant of the interfilamentous space in muscle in unknown, it seems improbable that the temperature dependence of the dielectric constant could approach that required to compensate for an eight-fold change in myosin ATPase. Moreover, a simple change in dielectric constant would not explain the difference between hibernating and non-hibernating mammals.

One or more of these mechanisms may possibly be implicated in the adaptation of hibernating mammals to thermal change. The matter is open to experimental inquiry, and perhaps some hard evidence rather than speculation may be forthcoming at the next Symposium on Mammalian Hibernation.

ACKNOWLEDGMENTS

The research in this paper was supported by research grants from the Health Research Council of New York City (U-1365), the National Institutes of Health (AM-06165), and the New York Heart Association.

The author is a Career Scientist of the Health Research Council of New York City.

REFERENCES

1. BÁRÁNY, M. ATPase activity of myosin correlated with speed of muscle shortening. In: The Contractile Process, Proceedings of a Symposium, sponsored by the New York Heart Association. Boston: Little, Brown and Company, 1967, p. 197-218.

2. BÁRÁNY, M., K. BÁRÁNY, T. RECHARD, and A. VOLPE. Myosin of fast and slow muscles of the rabbit. Arch. Biochem. Biophys. 109: 185-191, 1965.

3. CARLSON, F. D., D. J. HARDY, and D. R. WILLKIE. Total energy production and phosphocreatine hydrolysis in the isotonic twitch. J. Gen. Physiol. 46: 851-882, 1963.

4. DAVIES, R. E., M. J. KUSHMERICK, and R. E. LARSON. ATP, activation, and the heat of shortening of muscle. Nature 214: 148-151, 1967.

5. DREIZEN, P. Molecular basis of muscular contraction. II. A model for the contraction of striated muscle. Trans. N. Y. Acad. Sci. In press.

6. DREIZEN, P., and L. C. GERSHMAN. Structure and function of myosin. II. Dependence of ATPase on subunit composition. Biophys. J. 9: A236, 1969.

7. DREIZEN, P., and L. C. GERSHMAN. Structure and function of myosin. II. Myosin-nucleotide interaction (Abstract). Abstracts, Third Intern. Biophys. Congress, Cambridge, England, 183, 1969.

8. DREIZEN, P., and L. C. GERSHMAN. Molecular basis of muscular contraction Myosin. Trans. N. Y. Acad. Sci. 32: 170-203, 1970.

9. DREIZEN, P., and L. C. GERSHMAN. Relationship of structure to function in myosin II. Salt denaturation and recombination experiments. Biochemistry 9: 1688-1693, 1970.

10. DREIZEN, P., A. L. CAPULONG, and L. C. GERSHMAN. Structure and function of myosin II. Subunit interactions in resting and activated states. Biophys. J. 10: 80a, 1970.

11. DREIZEN, P., L. C. GERSHMAN, P. P. TROTTA, and A. STRACHER. Myosin. Subunits and their interactions. In: The Contractile Process, Proceedings of a Symposium, sponsored by the New York Heart Association, Boston: Little, Brown and Company, 1967, p. 85-118.

12. DREIZEN, P., D. J. HARSHORNE, and A. STRACHER. The subunit structure of myosin. I. Polydispersity in 5 M guanidine. J. Biol. Chem. 241: 443-448, 1966.

13. EBASHI, S., and M. ENDO. Calcium ion and muscle contraction. In: Progress in Biophysics and Molecular Biology 18, edited by J. A. V. Butler and D. Noble. London: Pergamon Press, 1968, p. 123-183.

14. EISENBERG, E., and C. MOOS. The adenosine triphosphatase activity of acto-heavy meromyosin. A kinetic analysis of actin activation. Biochemistry 7: 1486-1489, 1968.

15. GERGELY, J., D. PRAGAY, A. F. SCHOLZ, J. C. SEIDEL, F. A. SRETER, and M. M. THOMPSON. Comparative studies on white and red muscle. In: Molecular Biology of Muscular Contraction, edited by S. Ebashi, F. Oosawa, T. Sikine, and Y. Tonomura.

16. GERSHMAN, L. C., and P. DREIZEN. Structure and function of myosin. I. Subunit structure. Biophys. J. 9: A235, 1969.

17. GERSHMAN, L. C., and P. DREIZEN. Structure and function of myosin. I. Light chain heterogeneity (Abstract). Abstracts, Third Intern. Biophys. Congress, Cambridge, England, 183, 1969.

18. GERSHMAN, L. C., and P. DREIZEN. Relationship of structure to function in Myosin. I. Subunit dissociation in concentrated salt solutions. Biochemistry 9: 1677-1687, 1970.

19. GERSHMAN, L. C., and P. DREIZEN. Structure and function of myosin. In: Cardiac Hypertrophy, edited by N. Alpert. New York, Academic Press. In press.

20. GERSHMAN, L. C., P. DREIZEN, and A. STRACHER. Subunit structure of myosin. II. Heavy and light alkali components. Proc. Natl. Acad. Sci. U.S. 56: 966-973, 1966.

21. GERSHMAN, L. C., D. H. RICHARDS, and P. DREIZEN. Structure and function of myosin. I. Evidence for an isozyme concept. Biophys. J. 10: 80a, 1970.

22. GERSHMAN, L. C., A. STRACHER, and P. DREIZEN. Subunit interactions of myosin. In: Symposium on Fibrous Proteins, edited by W. G. Crewther. New York: Plenum Press, 1968, p. 150-153.

23. GERSHMAN, L. C., A. STRACHER, and P. DREIZEN. Subunit structure of myosin. III. A proposed model for rabbit skeletal myosin. J. Biol. Chem. 244: 2726-2736, 1969.

24. GORDON, A. M., A. F. HUXLEY, and F. J. JULIAN. The variation in isometric tension with sarcomere length in vertebrate muscle fibers. J. Physiol. 184: 170-192, 1966.

25. HANSON, J. Recent X-ray diffraction studies of muscle. Quart. Rev. Biophys. 1: 177-216, 1968.

26. HANSON, J., and H. E. HUXLEY. The structural basis of contraction in striated muscle. In: S. E. B. Symposium IX. Fibrous Proteins, edited by R. Brown and J. F. Danielli. New York: Academic Press, 1955, p. 228-264.

27. HARTSHORNE, D. J., and A. STRACHER. Effects of detergents on muscle proteins. Biochem. Z. 345: 70-79, 1966.

28. HILL, A. V. The heat of shortening and the dynamic constants of muscle. Proc. Royal Soc., London Ser. B. 126: 136-195, 1938.

29. HILL, A. V. Production and absorption of work by muscle. Science 131: 897-903, 1960.

30. HILL, A. V. The effect of load on the heat of shortening of muscle. Proc. Royal Soc., London, Ser. B. 159: 297-318, 1964.

31. HUXLEY, H. E. Muscle Cells. In; The Cell, Biochemistry, Physiology, Morphology. Specialized Cells: Part I, edited by J. Brachet and A. R. Mirsky. New York: Academic Press, 1960, vol. I, p. 365-481.

32. HUXLEY, H. E. Electron microscopic studies on the structure of natural and synthetic protein filaments from striated muscle. J. Mol. Biol. 7: 281-308, 1963.

33. HUXLEY, H. E. Structural evidence concerning the mechanism of contraction in striated muscle. In: Muscle, edited by W. M. Paul, E. E. Daniel, C. M. Kay, and G. Monkton. Oxford: Pergamon Press, 1965, p. 3-28.

34. HUXLEY, H. E. Structural difference between resting and rigor muscle; evidence from intensity changes in the low-angle equatorial X-ray diagram. J. Mol. Biol. 37: 507-520, 1968.

35. HUXLEY, H. E., and W. BROWN. The low-angle X-ray diagram of vertebrate striated muscle and its behavior during contraction and rigor. J. Molec. Biol. 30: 383-434, 1967.

36. JOHANSSON, B. W. Heart and circulation in hibernators. In: Mammalian Hibernation III, edited by K. C. Fisher, A. R. Dawe, C. P. Lyman, E. Schönbaum, and F. E. South. Edinburgh: Oliver and Boyd, 1967, p. 200-218.

37. KIELLEY, W. W., and W. F. HARRINGTON. A model for the myosin molecule. Biochim. Biophys. Acta 41: 401-421, 1960.

38. KIELY, B., and A. MARTONOSI. The binding of ADP to myosin. Biochim. Biophys. Acta 172: 158-170, 1969.

39. KOMINZ, D. R., W. R. CARROLL, E. N. SMITH, and E. R. MITCHELL. A subunit of myosin. Arch. Biochem. Biophys. 79: 191-199, 1959.

40. KOMINZ, D. R., E. R. MITCHELL, T. NIHEI, and C. M. KEY. The papain digestion of skeletal myosin A. Biochemistry 4: 2373-2381, 1965.

41. LOCKER, R. H., and C. J. HAGYARD. A correlation of various small subunits of myosin. Arch. Biochem. Biophys. 120: 241-244, 1967.

42. LOCKER, R. H., and C. J. HAGYARD. Small subunits in myosin. Arch. Biochem. Biophys. 120: 454-461, 1967.

43. LOCKER, R. H., and C. J. HAGYARD. Variations in the small subunits of different myosins. Arch. Biochem. Biophys. 122: 521-522, 1967.

44. LOCKER, R. H., and C. J. HAGYARD. The myosin of rabbit red muscles. Arch. Biochem. Biophys. 127: 370-375, 1968.

45. LOWEY, S., and C. COHEN. Studies on the structure of myosin. J. Molec. Biol. 4: 293-308, 1962.

46. LOWEY, S., and S. M. LUCK. Equilibrium binding of adenosine diphosphate to myosin. Biochemistry 8: 3195-3199, 1969.

47. LOWEY, S., H. S. SLAYTER, A. G. WEEDS, and H. BAKER. Substructure of the myosin molecule. I. Subfragments of myosin by enzymic degradation. J. Mol. Biol. 42: 1-29, 1969.

48. MANDELKERN, L. Some fundamental mechanisms of contractility in fibrous macromolecules. In: The Contractile Process, Proceedings of a Symposium, sponsored by the New York Heart Association. Boston: Little, Brown and Company, 1967, p. 29-60.

49. MANDELKERN, L., A. S. POSNER, A. F. DIORIO, and K. LAKI. Mechanism of contraction in the muscle fiber-ATP system. Proc. Natl. Acad. Sci. U.S. 45: 814-819, 1959.

50. MANDELKERN, L., and E. A. VILLARICO. The effect of salts and of A-5-TP on the shortening of glycerinated muscle fibers. Macromolecules 2: 394-401, 1969.

51. MIHÁLYI, E., and A. G. SZENT-GYÖRGYI. Trypsin digestion of muscle proteins. III. Adenosinetriphosphatase activity and actin-binding capacity of the digested myosin. J. Biol. Chem. 201: 211-219, 1953.

52. MOMMAERTS, W. F. H. M., and I. GREEN. Adenosinetriphosphatase systems of myosin. III. A survey of the adenosinetriphosphatase activity of myosin. J. Biol. Chem. 208: 833-843, 1954.

53. MOORE, P. B., H. E. HUXLEY, and D. J. DEROSIER. Three-dimensional reconstruction of F-actin, thin filaments and decorated thin filaments. J. Molec. Biol. 50: 279-295, 1970.

54. MUELLER, H., and S. V. PERRY. The degradation of heavy meromyosin by trypsin. Biochem. J. 85: 431-439, 1962.

55. PEPE, F. A. The myosin filament. I. Structural organization from antibody staining observed in electron microscopy. J. Molec. Biol. 27: 203-236, 1967.

56. RICE, R. V. Electron microscopy of macromolecules from myosin solutions. In: Biochemistry of Muscular Contraction, edited by J. Gergely. Boston: Little, Brown and Company, 1964, p. 41-58.

57. RICHARDS, E. G., C.-S. CHUNG, P. APPEL, and H. S. OLCOTT. Release of small components from purified myosin. Federation Proc. 26: 727, 1967.

58. SARKAR, S., and P. H. COOKE. In vitro synthesis of light and heavy polypeptide chains of myosin. Biochem. Biophys. Res. Commun. 41: 918-925, 1970.

59. SLAYTER, H. S., and S. LOWEY. Substructure of the myosin molecule as visualized by electron microscopy. Proc. Natl. Acad. Sci. U.S. 58: 1611-1618, 1967.

60. SRETER, F. A., J. C. SEIDEL, and J. GERGELY. Studies on myosin from red and white skeletal muscle of the rabbit. I. Adenosine triphosphatase activity. J. Biol. Chem. 241: 5772-5776, 1966.

61. STRACHER, A. Evidence for the involvement of light chains in the biological functioning of myosin. Biochem. Biophys. Res. Commun. 35: 519-525, 1969.

62. STRACHER, A., and P. DREIZEN. Structure and function of the contractile protein myosin. Current Topics in Bioenergetics 1, edited by D. R. Sanadi. New York: Academic Press, 1966, p. 153-202.

63. TRAYER, I. P., and S. V. PERRY. The myosin of developing skeletal muscle. Biochem. Z. 345: 87-100, 1966.

64. TROTTA, P. P., P. DREIZEN, and A. STRACHER. Studies on subfragment-1, a biologically active fragment of myosin. Proc. Natl. Acad. Sci. U.S. 61: 659-666, 1968.

65. TSAO, T. C. Fragmentation of the myosin molecule. Biochim. Biophys. Acta 11: 368-382, 1953.

66. WEBER, H. H., and H. PORTZEHL. Muscle contraction and fibrous muscle proteins. Advances in Protein Chemistry 7: 161-252, 1952.

67. WEEDS, A. G. Small sub-units of myosin. Biochem. J. 105: 25C-27C, 1967.

68. YOUNG, D. M., S. HIMMELFARB, and W. F. HARRINGTON. On the structural assembly of the polypeptide chains of heavy meromyosin. J. Biol. Chem. 240: 2428-2436, 1965.

CREDITS

The editors and authors are grateful to the following publishers, organizations and individuals for permission to use materials previously published:

Figure 1: The American Society of Biological Chemists Inc.; L. C. Gershman, A. Stracher and P. Dreizen, The Journal of Biological Chemistry 244: 2726-2736, 1969. Copyright The American Society of Biological Chemists Inc., 1969.

Figure 2: The New York Academy of Sciences; P. Dreizen and L. C. Gershman, <u>Transactions of the New York Academy of Sciences</u> 32: 183, 1970. Copyright The New York Academy of Sciences, 1970.

Figure 4: Academic Press Inc.; L. C. Gershman and P. Dreizen, "Structure and function of myosin," in <u>Cardiac Hypertrophy</u>, edited by N. Alpert. New York: Academic Press, 1971. In press. Copyright Academic Press Inc., 1971.

MEMBRANE PROCESSES AND ACTIVATION
IN MAMMALIAN CARDIAC MUSCLE

WILLIAM W. SLEATOR

Department of Physiology and Biophysics
University of Illinois, Urbana, Illinois 61801

INTRODUCTION

It was my assignment for this occasion to deal with "Cardiac Membrane Physiology" at the cellular level with special attention to the action potential, flux measurements, and excitation-contraction coupling. The amount of work currently being done and the information available in these fields is too large even to survey in the time and space available. Therefore I have selected, from among those of which I have some direct knowledge through personal involvement, three more limited topics which may be of some general interest, and in part even slightly relevant to the subject of this symposium.

The first of the topics concerns the nature of the cardiac action potential, with special consideration of the mechanism of the plateau which is the principal feature distinguishing it from action potentials of nerve and skeletal muscle. Then I will take up a question in which there is currently a great deal of interest, namely, that of excitation-contraction coupling in cardiac muscle. It is now becoming apparent that it is this process which is mainly responsible for the differences in contractile behavior between cardiac and skeletal muscle. Finally, I will describe recent experiments with cardiac muscle from two species not normally studied by zoologists, the chimpanzee and Homo sapiens. Although neither of these species is a hibernator, it

may be that the results to be presented will be of more interest to this
audience than those on the lower mammals. For one thing, they provide a
directly visible and striking example of several important features of the
cardiac action potential and of the excitation-contraction coupling process.
Moreover, and more relevantly, they make obvious a fundamental qualitative
change in behavior at the cellular level produced by decreasing the temperature a few degrees.

THE CARDIAC MUSCLE ACTION POTENTIAL

Since the early nineteen-fifties, when Silvio Weidmann carried out his
pioneering work on Purkinje fibers with intracellular microelectrodes (14,15),
it has been recognized that the depolarization phase of the cardiac action
potential consists of a rapid but brief influx of sodium ions. As established
a few years before by Hodgkin, Katz and Huxley (6,5), with the help of giant
axons from the squid, this sodium current, driven by the large concentration
gradient for sodium across the cell membrane as well as by the electrical
potential difference, is made possible by a sudden great increase in the
conductance of the cell membrane for sodium (hereafter referred to as gNa).
This conductance change is triggered by depolarization to a threshold, and
the sodium current is inactivated (cut off) by an almost equally rapid fall
in gNa which occurs after approximately 1 msec, independently of membrane
voltage. (This is one of the main results of Hodgkin and Huxley [5] which
would have been almost impossible to establish without the use of a voltage
clamp). In squid nerve as well as in those other nerve fibers which have
been so studied (myelinated frog nerve, for example) the initial depolarization is followed, after some delay, by a slower and smaller increase in
potassium conductance, gK. This increases the rate at which K^+ ions can
flow out of the cell (driven by the outward potassium concentration gradient)

and accelerates the repolarization process. Thus, in squid nerve for example, repolarization to resting potential takes only slightly longer than depolarization.

It is in this respect that cardiac muscle action potentials differ most markedly from those of nerve and skeletal muscle. Generally speaking, mammalian cardiac action potentials range from 100 to 300 msec in duration. For much of this time, the membrane is in the vicinity of zero potential and drifting slowly downward toward resting potential. This period is referred to as the plateau phase, and it is followed by one of rapid repolarization which, though subject to great variability, is normally one or two orders of magnitude slower than depolarization. The nature of the ionic permeability or conductance changes responsible for the plateau and rapid repolarization phases have not yet been established with complete security. Sometime ago Weidman (16) demonstrated that in Purkinje fibers total membrane resistance increased during the plateau. Nevertheless it was desirable, because of the great variations of cellular structure and function between different parts of the same heart (e.g., Purkinje fibers and atrial muscle) and between species, to experimentally determine these conductance changes in normal contractile muscle of one of the laboratory animals commonly used for physiological and pharmacological study.

Recently Fozzard and Sleator (1) carried out membrane resistance measurements with two microelectrodes in guinea pig atria. One microelectrode was used to record the intracellular potential and the other for injection of constant current pulses during diastole and at various times during the action potential. The contribution of chloride to membrane current was estimated by measurement of resistance after replacement of chloride by acetylglycine (to which the membrane is highly impermeable), and it was concluded that chloride conductance changes did not play a significant role in the action potential. Making use of the resulting data on changes in total membrane resistance during the action potential plateau, and of certain reasonable assumptions, it was possible to calculate relative values of the sodium and potassium conductances separately at various times during the plateau and repolarization phases. [The assumptions used concerned (a) the nature of intercellular connections in the two-dimensional sheet of atrial muscle, (b) the constancy of phase from one point to another in the small atrium, and (c) that the time rate of change of voltage was small enough, after depolarization, for capacitance currents to be neglected.]

The results of these calculations provide the following analysis of the action potential of guinea pig atrial muscle and corresponding tissues of other mammals. Since absolute values of conductance could not be obtained because of geometrical complexity, all conductances are expressed in terms of gK for the resting membrane gKr.

(A) gNa (sodium conductance) at rest: was found to average about 2% of gKr.

(B) gNa during the action potential: the earliest reliable measurements, obtained about 20 msec after depolarization near the peak of the action potential, show gNa still somewhat larger than gKr. gNa then fell rapidly to values between 40% and 10% of gK during the course of the plateau, and reached its normal resting value near the time the cell membrane was fully repolarized.

(C) gK during the action potential: the earliest measurements (20 msec after depolarization) showed a fall in gK to about 50%; 50 msec later near the middle of the plateau gK reached a minimum of 20-25%; it remained near this until the end of the plateau when it increased steeply during rapid repolarization, reaching 100% of resting value at about the time of complete repolarization (200 msec).

These results can be summarized with the statement that the plateau of the cardiac action potential (responsible for its long duration relative to that of nerve and skeletal muscle) is due to the more gradual fall in sodium conductance (which "tails off" for 100 msec or more) and to an extended period of greatly reduced gK, which decreases the rate at which potassium can flow outward through the membrane; this holds the membrane potential relatively constant and delays rapid repolarization until gK and gNa have begun to approach their resting values.

Additional evidence that the plateau is due to a period of low gK is provided for atrial muscle by experiments with acetylcholine (ACh). It has been known for many years (4,13) that ACh increases the permeability of atrial cells to potassium. Such an increase, if present both at rest and during the action potential, would preserve the ability of the membrane to carry the potassium outward current and thus speed up repolarization. This is indeed found to be the case for guinea pig, and for all other mammalian atria we have tested including human atrium. ACh at physiological concentration (10^{-6} M or less) will generally shorten the AP markedly, and can reduce it by 90% (to about 20 msec in guinea pig).

RELATIONS BETWEEN ACTION POTENTIAL AND CONTRACTION STRENGTH

From the point of view of this paper, the above analysis of the cardiac action potentials is important because of its bearing on our understanding of the excitation-contraction coupling process. It is well known that individual contractions of most of the conventional cardiac muscle preparations, from fine trabeculae to whole hearts, may vary greatly in strength as a result of irregular rhythms, e.g., rest intervals, frequency changes, extra systoles, etc. A systematic study of these phenomena some years ago

(10) led us to conclude that the changes in contraction strength could not be directly attributed to changes in action potential duration or shape. However, it also became clear that reducing the duration of the action potential for a period extending over a number of beats did result in greatly weakened contractions.

Because the action potentials in this multicellular preparation were recorded from a single cell and the contractions from the whole atrium, there were complications (such as that of diffusion time for the ACh) and these experiments did not settle the question as to whether the reduced contraction strength was a direct result of the shortened action potential. Therefore, the following definitive experiment was carried out: after a steady state had been established at a constant frequency of one beat per second, a rest interval of 40-80 secs was interposed, which would allow ample time for penetration of the ACh throughout the tissue. In the control run (without ACh) the first beat after such a rest interval had a contraction typically 60-100% stronger than standard, elicited by an action potential differing only slightly from normal (the difference is a slightly shorter plateau). After this large post-rest beat, contractions were smaller than standard and built up with a typical staircase effect over about 20 beats.

Application of an effective dose of ACh just after the start of such a rest interval led to the following result: the strength of the first contraction after the interval was nearly the same as that of the control run (i.e., 60% greater than standard). The second beat was somewhat smaller than that in the control, and thereafter contractions declined rapidly toward zero. However the action potential of the first post-rest beat had a duration characteristic of a full ACh effect. It was shortened to about 20% of normal duration, as were all subsequent action potentials. This indicated that ACh diffuses throughout the tissue and exerts its effect on the cell membranes. However, the fact that the post-rest contraction was nearly normal in magnitude shows that such an action potential is capable of eliciting a full-blown contraction, and that the cause of the small steady state contractions in the presence of ACh is not due to the inability of short action potentials to trigger the contractile mechanism.

The necessity of explaining these results and those of a number of other experiments had led us to develop the following picture of the excitation-contraction (E-C) coupling process in mammalian atrial muscle:

The action potential has two (at least) different and distinct functions:

(A) It is the signal which turns on or <u>triggers</u> the contraction. For this function any action potential will do which completely depolarizes the membrane (e.g., the action potential can be shortened to 10% of normal duration, and still work as a trigger).

(B) The action potential causes some change within the cell which determines the strength of <u>subsequent</u> contractions. For this function its shape and, especially, its duration are very important.

    (a) Long duration action potentials: subsequent contractions large.

    (b) Short action potentials: subsequent contractions small.

At present there is little doubt that the specific thing accomplished by long duration action potentials is the admission of more calcium into the cell. This result has been substantiated by several workers (2,8) by means of direct measurements of calcium influx under various conditions. Experiments like that described above indicate, however, that the calcium which enters during a particular action potential, whatever its duration, does not contribute to the strength of the contraction triggered by that action potential. Its effect appears in the second subsequent beat and persists through the next 5-20 activations.

It might be appropriate, before finishing the description of E-C coupling in heart muscle, to summarize the current views of this process in skeletal muscle. E-C coupling has been studied longer and more intensively in skeletal muscle, and in the last few years most of the major steps have been established quite firmly in some detail (see Symposium in Federation Proceedings, Vol. 24, No. 5, September 1965). The essential E-C coupling events in skeletal muscle may be outlined as follows:

(A) The action potential depolarizes the cell membrane. The influence of membrane depolarization is transmitted, probably electronically, to the interior of the cell along the well defined T-tubules, which are numerous, regularly spaced and open to the outside medium.

(B) When the depolarization rises above some threshold value (between -40 and 0 mV), free Ca is released from a store in some component of the sarcoplasmic reticular (S-R) system, and begins to diffuse into the space around the contractile elements. (The Ca-release process is still far from being understood). When the membrane repolarizes after about 2 msec, this flow of Ca out of the reservoirs stops.

(C) The free Ca causes the contractile elements to contract with an intensity roughly proportional to the amount of Ca released during the depolarization.

(D) After repolarization, the sarcoplasmic reticular elements recollect the free calcium from the myoplasm, and sequester it over the next 60-100 msec; it is the time course of this process that determines relaxation time of the twitch.

This picture provides a very satisfactory explanation for many of the observations on skeletal muscle: for example, when a second action potential occurs immediately, a second "unit" of Ca is immediately released, and the subsequent contraction is approximately doubled (summation); or, when the duration of a single action potential is increased as with zinc (9), the resulting twitch is greater because of the greater time during which Ca can flow out of the S-R. One might summarize by saying that in skeletal muscle the release of Ca, and thus the strength of the contraction, is directly controlled by the action potential (turned on by depolarization, and off by repolarization). Contraction does not appear to be limited by the amount of Ca available in the S-R reservoir, when triggered by normal AP's at least at frequencies up to that of complete tetanus. (However, contractures of skeletal muscle fibers evidently do act by depleting the stores of calcium, (7).

As already indicated, the behavior of mammalian heart muscle is qualitatively different from this in several important respects. For example, when a second action potential occurs after a minimum interval (and here the refractory period is more than 150 msec instead of 5-10 msec), it produces only a negligible contraction compared to standard. Considerable time (depending on previous frequency) must elapse before a full strength contraction can be elicited. If more than a standard interval elapses, the contraction is larger than standard (post-rest beat).

These observations and a number of others can be summed up and explained as follows: They support the hypothesis that in mammalian cardiac muscle the contraction strength is limited by the <u>amount</u> of calcium available for release from the S-R reservoir at the time of the action potential, rather than by the duration or number of depolarizations. This "amount available" varies with time depending on the history of the preparation, thus accounting for the strong dependence of contraction on frequency and regularity of interval between beats. We must also take special notice of the fact that at short intervals, a second nearly normal action potential produces a very small contraction even after complete relaxation of the first contraction; this is what eliminates the possibility of significant summation in heart. It appears that in this tissue the site <u>into</u> which calcium is collected to produce relaxation is not the same site as that <u>from</u> which calcium is released by the AP. One of the goals of our future research is to further test these hypotheses and others that are effective in explaining the difference in behavior between heart and skeletal muscle, and whenever possible to attach more specific meaning to the concepts involved (e.g., the nature and location of the intracellular sites out of which the calcium is released, and into which it is sequestered).

HUMAN AND CHIMPANZEE ATRIAL MUSCLE

The remainder of this paper will be concerned with the physiological properties of human and chimpanzee atrial muscle, as observed and recorded under laboratory conditions similar to those used for the experiments already described (11,12). This may be of more interest than some of the other things I could discuss because it speaks for the value of comparative studies, and brings into the picture results from a species heretofore largely ignored by comparative physiologists. The results provide a striking argument against the previously widespread assumption of both physiologists and physicians, that human heart muscle is essentially the same as that of dogs.

Results of work with this tissue also provide a graphic illustration of the analysis of the cardiac action potential described above, since here we find that the changes in membrane conductances g_{Na} and g_K responsible for the shape of the action potential are clearly separated in the action potential as recorded. We will also find an example of important qualitative changes in behavior at the cellular level which can develop as a result of temperature changes of 5 or 10 degrees.

For these experiments atrial muscle was obtained from cardiac patients undergoing open-heart surgery with the support of the heart-lung machine (11). Small strips of this muscle were set up in the conventional muscle bath containing oxygenated Krebs bicarbonate solution. When this tissue was stimulated at regular intervals it did not usually approach a steady state of constant contraction, but instead alternated periodically between two phases with markedly different contraction strengths. These regular variations occurred in most of the specimens studied and will be referred to as "cycling".

The nature of the large contractions became clear when records with intracellular electrodes were obtained from this tissue. An example of such a record is shown in Fig. 1. Here, as in all such records, when the tissue goes into high phase contractions the action potential acquires a second spike which arises spontaneously from the low plateau. Study of many such records shows that the second spike of an action potential does not contribute anything to the peak tension of the contraction triggered by that action potential. The effect of the second spike does appear in the first component of the following contraction: Its strength is increased by the presence of the previous second spike. This effect is cumulative through 5 to 10 action potentials so that the full contraction strength is not attained until at least five beats after high phase starts. Similarly, the effect persists to the same extent at the end of high phase, and contraction strength does not reach the steady low phase level until more than five action potentials without second spikes have occurred. Thus, we must infer that an intracellular process is going on which can accumulate some effect of the second spike over this period whether or not the tissue is active, and can put it to use in subsequent contractions. In this respect the process is similar to that in paired pulse stimulation where comparable delays occur between the start of the double stimuli and the peak contraction strength attained. Recent experiments in which the concentration of calcium in the bathing medium has been changed from normal in both directions, or strontium substituted for calcium, have provided evidence that the second spike produces its effect by controlling the entrance of calcium into the cell (12).

Another unusual feature of this tissue is the manner in which its properties change with temperature. The effects described and illustrated so far can usually be seen only at temperatures below 33° C. At 37° C the action potentials have a more conventional shape; i.e., the plateaus are shorter in duration, and at a higher (less negative) voltage. These relations can be seen in Fig. 2. As the temperature falls below 37° C, the plateau not only becomes longer in duration, but also shifts to a lower (more negative) voltage; at 24° C (top row) it is nearer to resting potential than to zero. This behavior is quite different from that of other kinds of cardiac muscle we have studied. In general, when temperature is lowered, the plateau simply becomes longer in duration, but its mean voltage stays the same. It is precisely the property of human atrial cell membranes which produces this fall in plateau voltage with temperature that is responsible for the unique behavior of human and chimpanzee tissue. With a long duration plateau at a low enough voltage, reactivation of sodium conductance can occur, and if the potential is near threshold at this time, another spike will be triggered. One can view this process as similar in effect to holding the membrane near threshold voltage with a polarizing current; in many excitable tissues this induces multiple firing. Second spikes rarely occur above 33° C presumably because the plateau voltage is not low enough nor its duration long enough for sodium conductance to recover.

As indicated above for guinea pig atria, information about the state of the membrane during the plateau can be obtained from experiments with

MEMBRANE PROCESSES IN CARDIAC MUSCLE

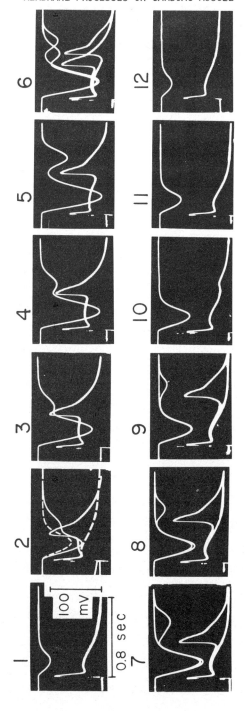

FIG. 1. Isometric contractions and transmembrane action potentials throughout a cycle of which the high phase started after a 10 sec rest interval. Total duration of high phase (frames 2 through 11) about 30 sec. Electrode in same cell from frames 2 through 12 (right atrium of 12 yr old with aortic stenosis).

Frame 1, control, low phase steady state before interval. Frame 2, the second response (broken lines) and the third response (solid lines) after the interval. (The first response, not shown, had an action potential similar to that of the third, but a contraction smaller than the second). Frames 3 to 5, records every 2 or 3 sec; every response double; the second spike of the action potential gradually moves out (away from the first) along the plateau, while the two components of the contraction become more separate, and the first component increases in strength. Frame 6, superposition of two consecutive beats; second spike alternates between two positions near end of plateau; strength of contraction (first component) remains near maximum. Frames 7 to 9, superposition of consecutive beats; alternation between single and double responses with second spike becoming later; decrease in strenght of both components of contraction. Frame 10 (2 sec after frame 9), second spike replaced by a "hump" with barely perceptible contractile response. First component of contraction same as frame 9. Frame 11 (4 sec after frame 10), second spike completely gone throughout preparation; contraction still above low phase level. Frame 12 (10 sec after frame 11), action potential and contraction complete return to low phase steady state. (Compare with frame 1).

acetylcholine. It has been established for many kinds of heart muscle that
ACh increases the potassium conductance of the cell membrane. The plateau
represents a period when gK has a value lower than that for the "resting"
membrane (1). Thus the ability of ACh to shorten or abolish the plateau in
these species can be explained on the basis of its augmentation of gK.
Human atrial tissue is quite sensitive to ACh, and it is evident from these
experiments that moderate doses of ACh can completely eliminate the effects
of changes in gK from the AP, so that what we are left with is due entirely
to the changes in gNa (a "pure" sodium spike). Similar experiments have
been done with adenosine (3) which also increases gK in some kinds of cardiac
muscle. Like ACh, adenosine can completely abolish the plateau in human
atrial muscle, and the action potential "spike" which remains is indistinguish-
able from that seen in the presence of ACh. Taken together, these results
provide persuasive evidence that the plateau in human atrial muscle is due
to a period in which the membrane has a low potassium conductance.

Another finding of great interest is that atrial muscle from chimpanzees
manifests the same complex pattern of behavior as that of human beings.
Although no intensive systematic study of heart muscle from chimpanzees has
yet been carried out, we have worked with atria from a number of different
chimpanzees and have confirmed in detail that this muscle tissue is almost
identical in physiological properties to human tissue. This applies both to
the ability of the cell membranes to generate second spikes, and also to the
tendency to alternate regularly (to cycle) between "high phase" periods
when action potentials have double spikes, and "low phase" periods when
action potentials are single and contractions relatively small. (A number of
attempts to induce double spiked action potentials and/or cycling in a variety
of the more ordinary laboratory species, including several kinds of monkeys,
have been uniformly unsuccessful).

MEMBRANE PROCESSES IN CARDIAC MUSCLE

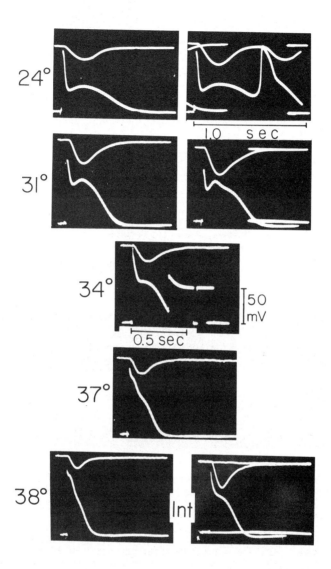

FIG. 2. Effects of temperature changes on action potentials and isometric contractions. Right atrial muscle from 5 yr old with IVSD. Temperature increased from 24° to 38° C. 24° C. Typical low and high phase action potentials and contractions. 31° C. Rest interval, preceding frame 2. 38° C. Rest interval, 20 sec, between frames. Changes form of action potential.

A series of records taken from one of the chimpanzee experiments is shown in Fig. 3. (Experimental conditions were the same as those normally used with human atrium: temperature 30° C). Though all the chimpanzees showed both double spiked action potentials and some capacity to cycle, there appeared to be considerably more variation from one individual to another than in human

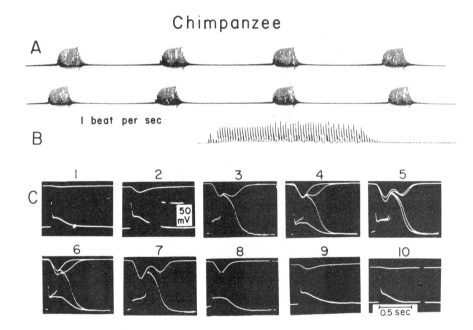

FIG. 3. Behavior of right atrial muscle from chimpanzee. (Stimulation as 1 per sec in oxygenated Krebs' solution at 30° C). Row A, cycling pattern similar to that seen in human atrium. Row B, one cycle from a different preparation showing alternation between single and "double" contractions at end of high phase. Row C, isometric contractions (upper trace) and transmembrane action potentials (from the same preparation as row B). Frame 1, low phase. Frame 2, single action potential at start of high phase. Frame 3, early in high phase. Frames 4 and 6, alternation of single and double spikes in last half of high phase. Frames 8 to 10, return to low phase; note decreases in duration and height of plateau.

beings. In row A, isotonic contractions were recorded during a period of very regular cycling having a pattern which has been seen only occasionally in human tissue where the two phases are usually of about equal length. The transitions to and from high phase both take five or six beats which is quite typical. In row B, another preparation from the same animal was recorded at higher paper speed. From the middle of the high phase, here lasting about 60 sec, there is irregular alternation between single and double action potentials. As expected, the single contractions are generally larger since they are <u>preceded</u> by action potentials with double spikes. Again the transitions last about five beats, indicating that the effect of a double action potential persists about this long in a regularly beating muscle.

Row C shows isometric contractions (upper trace, tension increase downward), and transmembrane action potentials recorded during transitions into and out of high phase. The variations in time between first and second spikes are clearly shown (frames 4 and 5, or 6 and 7) as well as the frequent alternation between single and double spiked action potentials (frames 4, 5 and 6 are such superpositions of two successive sweeps).

The over-all picture presented by these results could easily be mistaken for that of human tissue. However, one feature of this record is unusual among human specimens, i.e., the duration of the second spike, here about 200 msec or more than eight times the duration of the first spike. Such long second spikes have only showed up in human records once or twice among 200 different specimens and then are only probable when the subject is young. (The chimpanzee of Fig. 3 was about 10 years old, which is fully mature). Since we don't yet know the ionic mechanism of the second spike the possible significance of this difference is not clear. However, it seems probably that the second spike represents an extended period of high gNa and is not dependent on a concomitant fall in gK. Thus when the second spike is longer, more Na^+ comes in, and the number of beats during which a cell can remain in high phase before saturation with sodium (or exhaustion) is more limited. This effect could account for the shorter duration of high phase characteristic of chimpanzee muscle.

It may well be that the applicability, or even relevance, of the results presented above to the central questions faced by those interested in hibernation and hypothermia has been more obscure than apparent. However, before closing I would like to express the optimistic view that as a deeper and more exact understanding emerges of these questions, their connections with other problems such as the physiology of hibernation will also become clearer and their study more rewarding.

REFERENCES

1. FOZZARD, H. A., and W. SLEATOR. Membrane ionic conductances during rest and activity in guinea pig atrial muscle. Amer. J. Physiol. 212: 945-952, 1967.

2. GROSSMAN, A., and B. F. FURCHGOTT. The effects of frequency and calcium concentration on ^{45}Ca exchange and contractility on the isolated guinea pig auricle. J. Pharmacol. Exptl. Ther. 143: 120-130, 1964.

3. GUBAREFF, T.de, and W. SLEATOR. Effects of caffeine on mammalian atrial muscle and its interaction with adenosine and calcium. J. Pharmacol. Exptl. Ther. 148: 202-214, 1965.

4. HARRIS, E. J., and O. F. HUTTER. The action of acetylcholine on the movements of potassium ions in the sinus venosus of the heart. J. Physiol. 133: 58P-59P, 1956.

5. HODGKIN, A. L., and A. F. HUXLEY. A quantitative description of membrane current and its application to conduction and excitation in nerve. J. Physiol. 117: 500-544, 1952.

6. HODGKIN, A. L., and B. KATZ. The effect of sodium ions on the electrical activity of the giant axon of the squid. J. Physiol. 108: 37-77, 1949.

7. HODGKIN, A. L., and P. HOROWICZ. Potassium contractures in single muscle fibers. J. Physiol. 153: 386-403, 1960.

8. LITTLE, G. R., and W. W. SLEATOR. Calcium exchange and contraction strength of guinea pig atrium in normal and hypertonic media. J. Gen. Physiol. 54: 494-511, 1969.

9. SANDOW, A., S. R. TAYLOR, and H. PREISER. Role of the action potential in excitation-contraction coupling. Federation Proc. 24: 1116-1123, 1965.

10. SLEATOR, W., R. F. FURCHGOTT, T. de GUBAREFF, and V. KRESPI. Action potentials of guinea pig atria under conditions which alter contraction. Amer. J. Physiol. 206: 270-282, 1964.

11. SLEATOR, W., and T. DE GUBAREFF. Transmembrane action potentials and contractions of human atrial muscle. Amer. J. Physiol. 206: 1000-1014, 1964.

12. SLEATOR, W., and T. DE GUBAREFF. Spontaneous paired pacing in human atrial muscle. In: Paired Pulse Stimulation of the Heart, edited by P.F. Cranefield and B.F. Hoffman. New York: The Rockefeller University Press, 1968, p. 107-119.

13. TRAUTWEIN, W., and J. DUDEL. Zum Mechanismus der Membranwirkung des Acetylcholin an der Herzmuskelfaser. Arch. Ges. Physiol. 266: 324-334, 1958.

14. WEIDMANN, S. The electrical constants of Purkinje fibres. J. Physiol. 118: 348-360, 1952.

15. WEIDMANN, S. _Electrophysiologie der Herzmuskelfaser_. Bern: Huber, 1956.

16. WEIDMANN, S. Effect of current flow on the membrane potential of cardiac muscle. _J. Physiol._ 115: 227-236, 1951.

THE MECHANICAL PROPERTIES OF THE ISOLATED PAPILLARY MUSCLE FROM THE THIRTEEN LINE GROUND SQUIRREL

NORMAN R. ALPERT, BURT B. HAMRELL AND WILLIAM HALPERN

Department of Physiology
University of Vermont
Given Medical Building
Burlington, Vermont

The geometry of the heart is complicated and therefore it is difficult to describe the in vivo mechanical behavior of the contractile and non-contractile elements precisely. Two biological models which allow this complex architecture to be simplified are the isolated papillary muscle (16) and trabecular carneae (70) preparations. When these are quickly and carefully dissected from the animal and oxygenated in an appropriate solution they provide a long cylindrical piece of heart muscle that is biologically stable for substantial periods of time. Mechanical studies on the in situ papillary muscle in the canine left ventricle give credence to the use of isolated muscle preparations for the evaluation of myocardial performance (17). Isolated cardiac muscle strips thus may be studied in much the same manner as isolated skeletal muscles and can afford the basis for a complete description of the mechanical properties of the myocardium. They allow measurements of the length-tension, force-velocity, series and parallel elasticity, and active state parameters.

The framework for carrying out experiments on isolated heart muscle has been in terms of the three element model either in the Maxwell or Voigt configuration (Fig.1). This three-element model was used extensively in the analysis of skeletal muscle activity by A. V. Hill and others after 1938. Since it plays a major role in the interpretation of the data for isolated cardiac muscle preparations, it is important to understand the basis for the use of the three-element model in skeletal muscle physiology.

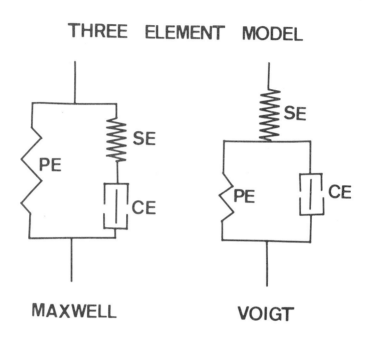

FIG. 1. The three element model of muscle.

Until the three-element model was generally adopted the interpretation of mechanical and thermodynamic data was based on the visco-elastic model (Fig. 2). The model evolved to explain some of the following experiments. Hartree and Hill (32) found that stretching passive muscle caused the production of extra heat. The amount of heat liberated increased with the velocity of stretch and was thought to arise from the degradation of mechanical energy due to viscous forces resisting the length changes. Subsequently a description of the work performed by the human forearm as a function of velocity easily fit the model of a spring in a viscous medium (34). Gasser and Hill (25) performed a series of experiments based on the model in which they showed that the spring constant increased elevenfold and the damping coefficient sixteen-fold when the muscle was excited. In the model the contractile elements resided in the spring which on excitation became a new elastic body.

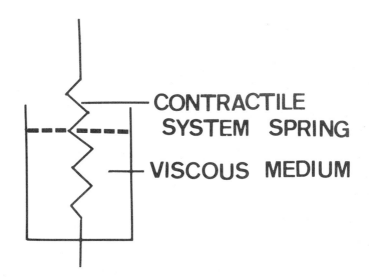

FIG. 2. The viscoelastic model of muscle.

Although the viscoelastic model was not generally abandoned until 1938 many experimental observations did not fit it. Fenn (22,23) found the total energy liberated (heat + work) by excited muscle depended on the work performed by the muscle. With the visco-elastic model one would expect heat + work to be a constant. Levin and Wyman (43) found nonlinear work velocity curves contrary to that expected from the model Bouchaert et al. (7) found an undamped elastic component in muscle. Finally Fenn and Marsh (24) showed that velocity and force of contraction were not linearly related but could best be described by the exponential relationship:

$$P = P_o e^{-aV} - kV \qquad (1)$$

where P is force, P_o is maximum tension in an isometric tetanus, V is the velocity of shortening for load P, a and k are constants. These experiments suggested that the nonlinearity in the force velocity curve was related to the intrinsic characteristics of the force generator rather than viscosity.

Hill (35) confirmed the nonlinear nature of the force velocity relationship and showed the data was well approximated by a hyperbolic curve

$$(P + a)(V + b) = b(P_o + a) \qquad (2)$$

where P is the force, V is the velocity of shortening, P_o is the maximum isometric force. P_o, b and a are constants in this equation. Equation b (above) for the mechanical events is identical with the thermal relationship showing that the rate of extra energy liberation during shortening (P + a)V, was linearly related to the load.

$$(P + a)V = b(P_o - P) \qquad (3)$$

Equations b and c established the identify of the mechanical and thermal constants. These experiments lead to the acceptance of the three-element model and it has proved a useful concept for evaluating muscle mechanics (Fig. 1).

Thus studies describing the mechanical behavior of muscle have to uncover the properties of the contractile, series elastic and parallel elastic elements. The contractile element (CE) has been characterized in terms of (a) active state, (b) force-velocity and (c) length-tension. The techniques employ a variety of maneuvers which analytically or mechanically eliminate the elastic elements.

For skeletal muscle the following is held to be true. When the "active state" is defined as the force which would be developed by the CE remaining at a constant length, there is (a) an abrupt transition from rest to a high plateau level of activity equivalent to tetanic force, (b) a brief plateau and (c) a rapid exponential decline to rest levels (Fig. 3) (18,36,37,46,57). The active state in frog sartoris at 0° C is brought into full activity within 10 msec, remaining at a plateau value for 50 msec and then decays exponentially over the following 200 to 300 msec. These values are halved for each 10° C rise in temperature (41,42,46). The active state is first greater than the twitch, falls through the peak of the twitch and is then less than the twitch force. When the "active state" is defined in terms of the time course of the maximum unloaded contractile element velocity there is (a) an abrupt increase in activity and (b) a slow fall in activity which parallels the decline in twitch force in terms of intensity and rate of fall (41,42).

During isometric contraction the contractile element (CE) shortens and extends the passive series elastic element (SE) (Fig. 1). The isometric myogram (Fig. 3) is determined by the force velocity characteristic of the CE and the stress-strain curves of the SE. The SE of frog sartorius is stretched 2% of rest length at maximum isometric tension and the shape of the total stress-strain curve is invariant with time (41). The force velocity curve conforms to a rectangular hyperbole and is readily fit by Hill's classic

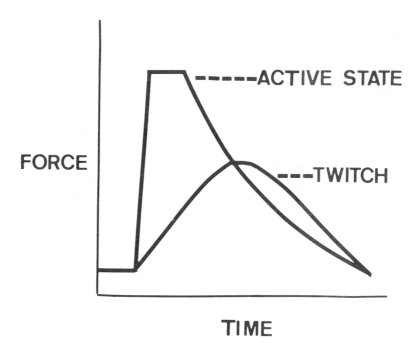

FIG. 3. The active state curve and isometric myogram (twitch) curves.

formulation (Equation 2) (35) or by an exponential relationship (Equation 1) (24). The length tension curve has been related to the thick and thin filament overlap. As the muscle length is increased the active tension rises to a peak and then shows a linear decay (Fig. 4) (28). Parallel elasticity (PE) does not come into play until the muscle is stretched to a point where it can develop maximum isometric tension. These findings in skeletal muscle based on the three-element model place all of the intrinsic viscosity in the contractile element. As will be developed subsequently it is important to note that the basis for abandoning the viscoelastic model does not imply that there is no intrinsic viscosity. In the application of this model to heart

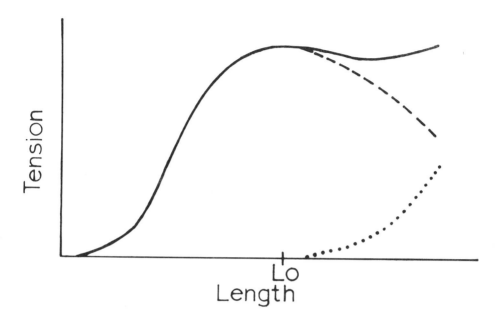

FIG. 4. The length-tension curve. The upper solid line reflects the total tension, the dotted line the passive tension and the dashed line the active tension. L_o is the length at which maximum active tension occurs.

muscle it is extremely important to note the following differences between skeletal and heart muscle:

  a) Heart muscle cannot be tetanized.
  b) It behaves as a functional syncytium.
  c) The active state has extremely slow onset and slow decay.
  d) The parallel elastic elements come into play in heart muscle throughout the entire length-tension relationship and seldom can be neglected.

e) The series compliant elements are stiffer in skeletal muscle than heart muscle.

f) Skeletal muscle contractile elements develop more force.

g) Fibers in skeletal muscle are parallel whereas in heart muscle the geometry is more complex.

These differences must be kept in mind when considering mechanical data developed in myocardial preparations.

In an attempt to define "active state" in cardiac muscle using techniques similar to those employed in skeletal muscle (18,36,37,42,46,57,58) Brady (8-12) has applied quick stretches through the time course of the twitch of rabbit right ventricular papillary muscles. His findings indicate that the maximum instantaneous isometric force sustained by the muscle is similar to the peak isometric force that would have developed at the stretched length. The degree to which this finding reflects contractile element function is dependent on an assumption of a three-element model (Fig.1.) in which the parallel elastic element is relatively extensible compared to the activated contractile element. This can be achieved in cardiac muscle only by working at muscle lengths considerably below L_{max}. A viscous element would become manifest as a force dependent on the speed of the quick stretch perturbation. This added complexity is difficult to evaluate in Brady's approach. Although slow stretches resulted in less stress for a fixed amount of strain more time was available for contractile element interaction during the length perturbation; both slow and fast stretches terminated at 50 msec after the stimulus. Since cardiac muscle cannot be tetanized and hence must be studied under non-steady state conditions the pre-history of the contractile unit prior to the amount of instantaneous intervention must be quantitated.

Brady (9) found that quick stretches applied during the first half of the rising phase of an isometric twitch elicited the responses described above. Quick stretches carried out later than 200 msec after the stimulus in a rabbit right ventricular papillary muscle preparation at 22°C resulted in instantaneous tension levels lower than predicted by the length-tension relationship and the subsequent isometric tension development at the new stretched length was lower than that following earlier quick stretches.

Using a quick release technique and observing tension redevelopment, Brady (9) noted that the peak rate of tension redevelopment did not occur until approximately 150 msec after the stimulus. This fact coupled with the similar appearance of maximal isotonic quick release shortening velocity at approximately the same time led Brady to suggest that "active state" in heart muscle is slow in onset and that its intensity is of the same order of magnitude as peak twitch tension for a contractile unit length corresponding to the final attained external muscle length.

The problems inherent in applying a three-element model to cardiac muscle have been reviewed by Brady (10). The problems in modeling revolve around the lack of a steady-state phenomenon such as tetany; the relatively high resting tension, and the probable slow progression of the intensity of contractile element interaction through the time course of a twitch. Thus a length and/or force perturbation must be uniformly rapid relative to the maximum rate of interfilament interaction in order to "freeze" myocardial activity at a particular contractile element length.

In this regard the techniques mentioned by Brady's discussion, small rapid stretches and releases and quick releases, are rather gross perturbations when compared to the estimated 100 Å range of motion of a crossbridge.

Brady concludes that no one three-element model explains all the data obtained from myocardial preparations. This conclusion is not surprising in view of some of the anatomical features of the myocardium including probable sarcomere buckling at short muscle lengths (26,27) and a possible spiral orientation of myocardial fibers angled at about 30°C around the longitudinal axis of the papillary muscle (59).

Both Edman et al. (18-21) and Sonnenblick (65,66) have attempted quantification of the intensity of contractile element interaction by obtaining load-velocity data at multiple points during the time course of a twitch either using simple afterloaded contractions (65) or a quick release technique (18-21,66). Both groups have redefined "active state" in terms of velocity of contractile element shortening in a lumped simple three-element model and have obtained results similar to Brady indicating a slow onset of contractile element interaction. It is difficult to compare these authors' data quantitatively because of differences in experimental conditions including: temperature, stimulus parameters, perfusate composition, animal and lever system. In several instances these factors are not described in sufficient detail to assess their influence on the data.

Brady has also explored the approach of stretching the right ventricular rabbit papillary muscle during the time course of a twitch. The stress-strain relationship, previously determined, is used to program an electromagnetic ergometer loaded in series with the muscle (10) which controls the degree and time course of the stretch. The validity of this attempt to study activated cardiac muscle at "constant" contractile element length is dependent on the assumption of the three-element model. If in fact the contractile element is arranged in series or parallel with a viscous element (11) the interpretation of Brady's data would be much more difficult.

Assumptions regarding the interrelationships of the mechanical components of heart muscle have been critically evaluated by Pollack (54,55). He stresses the importance of considering the instantaneous relationship of the parallel and series elastic components as ultrastructural reorientation takes place during contraction. Several authors have proposed either the maximal theoretical rate of external muscle shortening when loading has become infinitely small, V_{max}, as a unique index of the state of the contractile element in cardiac muscle (62) and/or the velocity at a very low load (18,19,21,65). As Pollack illustrates, the loading characteristics of the contractile element and the manner in which the velocity of muscle shortening is reflected at the sarcomere depend on the model chosen. Part of Pollack's criticism (54,55) centers around earlier work carried out utilizing simple "afterloaded" isotonic contractions; the muscle interacts with a simple lever system with appropriate restraints such that one passive muscle length is established by a fixed load, additional loads ("afterload") may be added which the muscle encounters only after activation. Although activation by supramaximal square wave stimuli applied by massive parallel plate electrodes located near the muscle is almost instantaneous, because of the constraints of the afterloaded technique, heavier loads are lifted later in the time course of the twitch. With this experimental technique, the initial velocity of shortening is measured at varying times during the twitch.

In an attempt to obtain "instantaneous" data Jewell and Wilkie (45) introduced the quick release technique. At a given time after the onset of isometric tension development the muscle is released to a previously determined afterload. This technique was briefly mentioned in the discussion of active state and has been used extensively in an attempt to define the force-velocity relationship in cardiac muscle (9,12,18-21,33,50-52,66). Data gathered by these authors differ considerably both qualitatively and quantitatively. The fundamental problem seems to involve the degree to which the force-velocity relationship can be approximated by a rectangular hyperbole as originally described for skeletal muscle (35). Obviously the quantitative assessment of the intercept on the ordinate, i.e. velocity at "zero" load, is dependent on the analysis of the experimentally determined points. Edman consistently has published hyperbolic configurations (18-21) whereas Sonnenblick's group has published curves of varying configuration depending on the type of experiment (52,63,64) or the conditions (71).

In contrast to Sonnenblick's and Edman's findings (18-21,63-66) Hefner's data (33,50) is not accounted for by a rectangular hyperbolic function. Furthermore Hefner's estimate of V_{max} varied with length. Sonnenblick has proposed in the above quoted work that V_{max} is independent of external muscle length when other factors influencing the contractile state of myocardium are constant. The latter suggests that V_{max} is a unique property of the intensity of myofilament interaction independent of the degree of overlap or the history of the activated sarcomere up to the moment of study. This group has recently analyzed their data in terms of the force-velocity-length-time relationships (13). The latter approach attempts to take into consideration the instantaneous relationship of contractile element shortening and extension of the series elastic component. The success of this approach is inextricably related to the assumed muscle model which is not clearly stated but is presumably a lumped three element type. At this juncture in time and with the type of approach described in a later section of this review, it is appropriate to begin to try to discover and quantitate the cardiac muscle model components and their interfaces and then test them in a suitable experimental setting.

Cardiac muscle manifests significant passive or resting tension at all lengths where developed tension is manifest. Developed or active tension is a function of external muscle length and, in a manner analogous to Huxley's work in skeletal muscle (39), Spiro and Sonnenblick have attempted to demonstrate the relationship of resting sarcomere length to subsequent peak tension development upon activation (67,69). Cardiac muscle presents a challenge in this regard because of the previously mentioned data regarding fibril buckling (26,27) and spiral orientation of the fibers (59) which branch extensively. The lack of a suitable single fiber preparation has hampered this type of work.

Series compliance of myocardium represents approximately 5 to 10% of muscle length at peak tension (33,50,52,72). As discussed previously, the appropriate localization of the latter in relation to the passive length-tension relationship needs further clarification and must await techniques that take into consideration any viscous components that may be present and that incorporate both the viscous and elastic components into a quantitative relationship with the contractile element.

The complexity of heart and skeletal muscle make it necessary to extend the standard approaches for evaluating the mechanical properties. Physical properties such as compliance and viscosity observed at the macroscopic level reflect primary and secondary bondings occurring at the molecular and atomic levels (60). Thus, a spring with internal damping possesses a certain stiffness and coefficient of viscosity. It could also be described by attributes such as stress-strain and force-velocity characteristics but these are indirect measures which do not necessarily yield to structural interpretations. The contractile apparatus of muscle is a complex system of interacting distributed functional components and as such has made the task of uncovering its fine structure a difficult one. The results of most experimental interrogations of muscle mechanics are expressed in terms of the rate of movement and force generation from which, as we have seen, lumped-parameter models are constructed to explain the observations and to serve as clues to its functioning at the ultrastructure level. The contractile element (CE) remains expressed primarily by force-velocity relationship; and viscous elements are often invoked to fit the data to a particular model, but they remain empirical and unquantified.

A method which uncovers a model, at the same time quantifying the parameters, has been developed and used by this laboratory (29,30). This stochastic signal, or pseudo-random white noise (PRWN), perturbation technique also has the advantage over the conventional quick-release afterloaded isotonic and other test methods in that the testing signal perturbs the length by only about 0.05% of L_{max}. In other words, disturbance to the cross-bridge formation in minimal. For example, in skeletal muscle, releases greater than about 100 Å at the sarcomere level, or about 0.4% L_{max}, are likely to destroy cross-bridge formation and give rise to erroneous mechanical measurements (53). Although, the distance a single cross-bridge in cardiac muscle can move and remain attached in not known, obviously, the smaller the movement the less destructive is the test.

The deterministic nature of the muscle to respond with a unique change in tension to an imposed displacement change, or vice versa, underlies all approaches to muscle mechanics studies. And it is through analysis of these input-output relationships that the muscle preparation is ultimately described. Several of these methods employ step-function quick or controlled releases with subsequent analysis of the results in the time domain (i.e. inputs and outputs are time-dependent variables) (14,38,40); others, for investigating skeletal and insect muscle use a series of sinusoidal perturbations with analysis of the results proceeding in the frequency domain (6,34,71). Formal mathematical relationships demonstrating the equivalence of the transient (time) and harmonic (frequency) methods for analyzing viscoelastic bodies, and in particular as applied to muscle, have been published (2,65). The pseudo-random vibrational signal employed in our technique is essentially a more general signal than either the stepfunction or a group of sinusoids in that it includes any and all possible signals. Thus, the application of this stochastic signal vibration to a muscle is tantamount to subjecting the specimen simultaneously to a large number of harmonic signals. In this way, every possible natural mode of the physical system is excited and the intrinsic properties of its elements and their arrangement can be extracted from the total response. This type of system identification has been applied to many physical systems and the mathematics and limitations of the method are fully developed (3-5,48,49,56).

The pseudo-random white noise (PRWN) method in regard to muscle requires that during each pseudo-random test period the system behaves linearly and contains no time-varying parameters. The validity of this criterion is subject to test by a statistical function in the course of the analysis. Thus, for each frequency component of the PRWN input displacement signal, the output force response bears a certain causal amplitude and phase relationship to the input. The total assembly of these output-input relationships define the dynamic stiffness modulus transfer function for the muscle. From this function, an equivalent viscoelastic model with quantifiable parameters is derivable without *a priori* assumption, thereby providing another significant advantage to the method.

One pseudo-random period is made to follow another with great rapidity (e.g., 34 msec) so that changes in the muscle system transfer function from the diastolic or resting state throughout the course of contraction can be determined. The derived model was invariant with time for the papillary muscle of the ground squirrel so that the measurements revealed the changing nature of the parameters of the model brought about by stimulation, contraction and relaxation. These results are compared with the more classical isometric and isotonic determinations of CE, SE and PE characteristics.

The methods for studying the discrete mechanical properties of cardiac muscle described above have not been applied to myocardium from mammalian hibernators. Published material includes studies of atrial preparations (15,40,47), strips of right ventricular free wall (61), trabecular muscles (68), and isolated heart preparations (45). These preparations are not optimum for the quantification of mechanical parameters primarily because of the wide variation in fiber orientation. Papillary muscle fibers approximate a parallel orientation with reference to the long axis of the muscle to a degree discussed earlier.

Hibernating mammals survive core body temperature levels of 5° C during hibernation and when not in the hibernating state (44). The isolated hearts of mammalian hibernators continue to function under appropriate conditions at temperatures of 10° C and less (44). Hearts from non-hibernators irreversibly deteriorate at these low temperatures. The primary question is in what way do the fundamental mechanisms of the mechanical response to temperature change of hibernator myocardium differ from non-hibernator cardiac tissue.

METHODS

Pseudo-random white noise perturbations: A mature thirteen-line ground squirrel captured in Florida (<u>Citellus tridecemlineatus</u>) was anesthetized with ether and the heart rapidly removed and dissected in Krebs-Ringer's solution (Na^+ 152, K^+ 3.6, Cl^- 135, HCO_3^- 25, Mg^{++} 1.2, $H_2PO_4^-$ 1.3, SO_4^{--} 1.2, Ca^{++} 5.0 meq/l and glucose 5.6 mM) oxygenated with a gas mixture of 95% O_2, 5% CO_2 resulting in a pH of 7.4. A pair of left ventricular papillary muscles were rapidly dissected free, their ends clamped in aluminum tubular clips, and each was vertically suspended in separate température controlled chambers filled with the solution described above and linked through the chamber floor to a capacitance force transducer (29,30).

The upper end of one of the muscles was linked via a thin section of straight stainless steel tubing to a specially constructed vibrator with a stiffer suspension than previously employed, 3 Kg/mm (a modified Ling Electronics Model 203 Shaker). Electronic feedback of the displacement and velocity signals resulted in effective transmission of vibrational energy to about 1200 Hz. The muscle was equilibrated at 29° C while being stimulated with a square wave pulse of 1 msec duration 10-20% above threshold values and applied every five seconds. Pre-load was increased by lengthening the muscle until peak active tension was reached. Then, in order to minimize the effects of parallel elastic elements, the length was reduced until peak isometric tension was about 20% of maximum (1.6 grams); this length was maintained for the remainder of the experiment in which only temperature was altered.

At each temperature several isometric contractions were recorded on FM tape with and without the PRWN perturbations. The complex nature of the PRWN displacement and twitch tension signals is presented in Figure 5. In these experiments displacements were $\pm$ 8 μ peak-to-peak (or less), and the PRWN period was 34.1 msec. The twitch without the stochastic perturbations always superimposed itself through the mean of the response with the noise showing that the PRWN signal did not alter formal function. Also recorded were: the rate of change of tension; a signal coincident with the stimulus; and a clock signal derived from the pseudo-random noise generator (Hewlett-Packard Model 3722A).

Isometric measurements and isotonic measurements. The other left ventricular papillary muscle was placed in a similar bath but its upper end was linked to the long arm of a 25:1 mechanical advantage light magnesium level optically linked to a photocell displacement transducer system linear

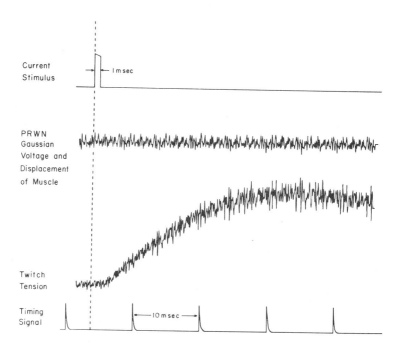

FIG. 5. The (PRWN) displacement record is seen in the middle record. The lower force trace was recorded during a twitch with one end of the muscle subjected to PRWN displacement perturbations while the other end of the muscle is attached to the isometric force transducer. The top trace is a recording of the applied stimulus.

over a range > 2.0 mm. The force and displacement signals were electrically differentiated with respect to time via an R-C circuit with a time constant of 1.5 ms. Recordings were obtained from a Tektronix 564 series oscilliscope for a Beckman Dynograph. The muscle was allowed to equilibrate for one hour under isotonic conditions with a load of 0.1 gm/mm^2 and stimulus parameters as described above. Length in this muscle was changed by altering

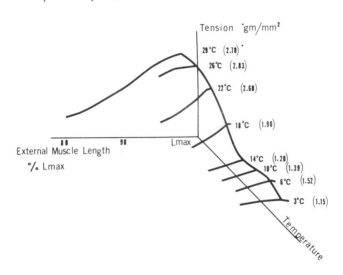

FIG. 6. Length-tension plane is parallel to the plane of the page whereas the tension-temperature plane is normal to the page. The solid line in the latter plane joins the peak tension plots at L_{max} at each temperature. Weight 6.22 mg. Cross sectional area at L_{max} 0.75 mm^2 stimulus: 1/5 seconds, 0.8 ms square wave, ∼ 50 ma.

the passive load; isometric force at each length was determined by fixing the lever position by means of suction applied to a small contact area on the upper surface of the long arm of the lever. The force velocity relationship was obtained using the quick-release technique described by Jewell and Wilkie (41). The suction restraint on the lever was released at the interval of peak rate of isometric tension development by programmed activation of

TABLE 1. The isometric twitch.

		26°C	22°C	18°C	14°C	10°C	6°C	3°C
Force (gm/mm^2)	Passive	0.47	0.47	0.47	0.47	0.47	0.46	0.44
	Active	2.76	2.02	1.37	1.03	1.20	1.44	1.05
Peak $\frac{dp}{dt}$ $(\frac{gm/mm^2}{sec})$		26.30	14.10	7.51	3.95	2.74	2.08	1.41
Latency (ms)		10	20	30	40	60	90	120
Time to Peak Tension (ms)		175	235	330	480	720	1160	1280
Time to Half Relaxation (ms)		330	450	620	885	1330	2120	2580

Same muscle as in Fig. 6 but at a passive tension of 0.44 to 0.47 gm/mm^2. Weight 6.22 mg. Length at this passive tension was 8.1 to 8.2 mm. Stimulus: 1/5 seconds, 0.8 ms, $\sim$ 50 ma.

a three-way solenoid valve that was arranged to: (a) vent the system to atmosphere and (b) close the sub-atmospheric pressure channel (connected to a vacuum pump).

RESULTS

Papillary muscle from the thirteen-line ground squirrel continues to develop force as the temperature is lowered from 29° to 3° C. There is a fall in peak active tension at L_{max} from 29° to 14° C. Below 14° tension tends to plateau as the temperature is lowered. The length-tension-temperature relationship is illustrated in Figure 6. Data from the isometric measurements are summarized in Table 1. The peak active force (P) is halved

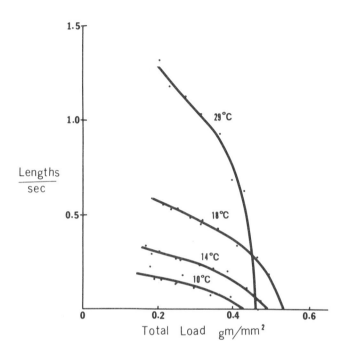

FIG. 7. The force velocity relationship as a function of temperature. Stimulus: 1/5 sec; 0.8 msec square wave; ~ 50 ma. Muscle weight 3.35 mg, cross sectioned area 1.0 mm^2, preload 0.16 gm/mm^2. The abscissa reflects both the preload and afterload.

as the temperature is decreased from 26° to 13° C while the maximum rate of force development dP/dt exhibits a six-fold decrease. From 14° to 3° C both P and dP/dt max plateau. Concomittantly time to peak tension (TPT) increases eight-fold as the temperature is lowered from 26° to 3° C. As the temperature is decreased from 29° to 10° C the maximum velocity of contraction falls rapidly. Maximum force increases from 29° to 19° C and then decreases as the temperature is lowered to 10° C (Fig.7). The stress-

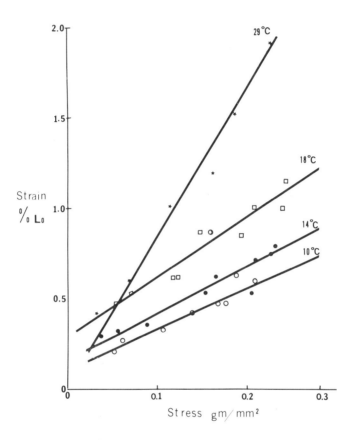

FIG. 8. The stress-strain curve of papillary muscle evaluated by quick release to predetermined afterload. The release was triggered at dP/dtmax and thus the stress-strain relationship is measured substantially below peak isometric tension.

strain relationship evaluated by the quick release technique to predetermined afterload at the time of dP/dt max is illustrated in Figure 8.

Three periods were selected for computer analysis (University of Vermont Computation Center, IBM System 360/44) using the PRWN method: (a) rest prior to stimulation; (b) the interval bracketing the maximum rate of change of tension; and (c) the interval around the peak of the twitch. The log stiffness vs log frequency curve could be approximated by a low frequency and a high frequency horizontal line connected in the intermediate frequency

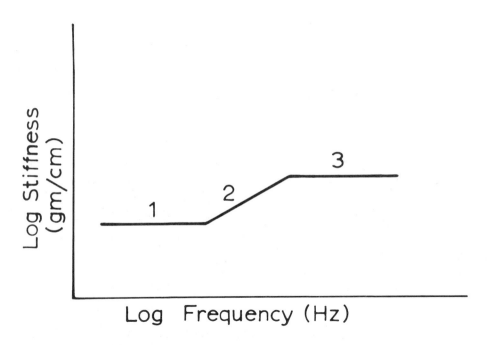

FIG. 9. The log stiffness versus the log frequency curves are obtained from analysis of displacement and force records during the application of PRWN perturbations to the muscle (cf. 29,30). Plateau 1 reflects stiffness of elements A and D, the slope 2 reflects viscous element C and plateau 3 reflects the stiffness of element A (see in Fig. 10).

range by a sloping straight line (Fig.9). This implies that the muscle looks like one spring (or elastic element) at low frequencies, another spring at high frequencies, and a combination of two springs and a viscous element elsewhere. This behavior is consistent with the model shown in Figure 10. Assuming that stiffness B is small (parallel elastic element), values for the elements, A, C, and D were evaluated from the scaled results. The summary of this data appears in Table 2.

The coefficient of damping, C, of the viscous element and the stiffness of the elastic element, D, both increased as a function of twitch tension at all temperatures. For this reason their properties are believed to derive from the tension generating elements of the contractile protein, i.e., the cross-bridges. On the other hand, the elastic element, A, did not change during contraction. The stiffness of A and D and the coefficient of damping, C, increased as a function of decreasing temperature (Table 2).

DISCUSSION

It is well known that lowering the body temperature in non-hibernating mammals may lead to death. In contrast hibernators are capable of surviving core body temperatures as low as 5° C. Isolated heart preparations from this latter group will function at temperatures well below those where mechanical responses cannot be elicited from non-hibernating mammalian heart preparations. The ability of this heart tissue to function at low temperatures may result from differences in membrane properties, excitation-contraction coupling interfaces or contractile properties. Cat and rabbit papillary muscles cease functioning when the temperature falls below 18° C whereas the thirteen-line ground squirrel papillary muscle develops force at temperatures as low as 3° C.

The peak force of the isometric twitch decreases as a function of temperature from 29° to 14° C. From 14° to 5° C there is a plateau and then

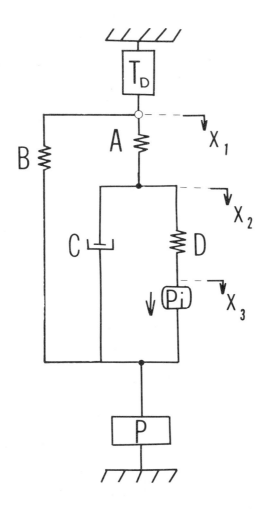

FIG. 10. A model of the thirteen-line ground squirrel papillary muscle derived from the PRWN method. P_i is a mechanical force source which simply constrains the force at the point of its inclusion to be some prescribed function of time. Its mechanical impedance is a short-circuit; that is, a link of zero compliance between the lower parts of elements D and C. Thus, P_o is not the conventional representation of the CE and any compliance or viscosity associated with this force generator appears elsewhere in the elements of the model. A is the classical series compliance, D the compliance and C the viscosity associated with the contractile element. B is the parallel elastic element. T_D is a time delay used in the analyses. P is the external force.

a secondary fall as the temperature is lowered to 3° C. The peak force
depends on (a) the intensity duration rate of decay of the active state;
(b) the force velocity curves reflected by the active state; and (c) the
compliance of the series elastic elements. From the force velocity data,
Figure 7, one can see that the intensity of the active state has decreased
as the temperature is lowered. This is compensated for by an increase in
the duration of active state as reflected in increases in time to peak
tension as well as time to half relaxation. In addition the compliance
decreases as a function of temperature, which for any given active state
configuration would tend to increase peak tension. The peak force of
2.8 gm/mm^2 is somewhat lower than that found for rabbit or cat papillary
muscle (1,12,31,62).

Series compliance measurements, carried out by the quick release
technique, were made at low forces to coincide with the time of maximum
dP/dt. At these low forces there is a linear relationship between stress
and strain with an increase in stiffness as the temperature is decreased.
The results are qualitatively and quantitatively comparable to those
previously reported for rabbit and cat papillary muscle (1,31,52). It is
important to note that quick release measurements at low temperatures may
be better indices of different compliant elements than quick release measurements at higher temperatures. At the higher temperatures and the normal quick
release velocities, the coefficient of damping of element C is .005 gm/cm/
sec and does not interfere with series arrangement of elements A and D
(Fig. 10). In this situation the compliance would be greater than that of
the most compliant element D (Table 2). As the temperature is lowered the
coefficient of damping of element C increases dramatically so that the
compliance of element D would be masked by the viscous element, C, in the

TABLE 2. Isometric twitch (PRWN analysis).

TEMP (C)	A gm/cm	REST		dP/dtmax		Pmax	
		C gm/cm/sec	D gm/cm	C gm/cm/sec	D gm/cm	C gm/cm/sec	D gm/cm
29.0	171	0.025	28	0.055	74	0.072	113
21.9	158	0.040	48	0.065	123	0.074	81
18.1	920	0.160	222	0.290	374		
14.0	1080					0.340	460

A, C and D are elements of the muscle derived from the PRWN method seen in Figure 10. The values obtained for dP/dtmax and Pmax were derived from analysis of the PRWN period which included the time of maximum rate of change of force dP/dtmax and the maximum force, Pmax, respectively.

quick release experiments. Under these circumstances, element A is the main contributor to SE and since it is stiffer than D the SE stiffness would increase. In addition the stiffness of both elements A and D increase as the temperature is decreased. From the PRWN method quantitative data on A, D, and C as a function of temperature was obtained. All increase as the temperature is lowered. During activiation of all temperatures A remained constant whereas D and C increased in direct proportion to the force developed. From these results we feel that element A's properties reflect a noncontractile part of the muscle, such as the Z lines, intercalated discs or stray tendons. The coefficient of damping, C, of the viscous elements and the stiffness of the elastic element, D, both increased as a function of twitch tension at all temperatures. For this reason their properties are believed to derive from the tension generating elements of the contractile proteins and could readily explain the increase in stiffness and viscosity reported by Gasser and Hill (25). In the PRWN analysis element B has been neglected. The compliance of this element was found to be invariant with temperature. Since the experiments were carried out substantially below L_{max} the contribution of B was considered to be minimal and thus could be safely neglected. It is necessary to recognize that the muscle system described is nonlinear because of changes of parameters C, and D, with force. It is important to note that even at rest there appears to be some crossbridge activity imparting a resting viscosity and stiffness to elements C and D. Table 3 compares the modulus of elasticity measured by the PRWN and quick release methods at several temperatures.

The force velocity curves were measured at constant length by means of quick release to varying afterloads. The initial length chosen for the measurements was 20% below L_{max} in order to minimize the contribution of PE to the measurements. The curves obtained showed a characteristic decrease of velocity as the force was increased although none of these curves could be readily approximated by a rectangular hyperbola. Although no attempt was made to extrapolate the force velocity curves to V_{max} (0 load) it is apparent that at low loads velocity decreases as the temperature is lowered. The shape and magnitude of these curves are similar to those found for cat muscle although the temperature range studied here is necessarily greater.

Several factors may be involved in the decrease in the force velocity relationship. These are alterations in the force generating system and increases in intrinsic viscosity. From the derived muscle model (Fig.10) one can analyze and quantitate the factors which contribute to changes in myocardial performance as the temperature decreases. The force velocity curves (Fig.7) indicate that the velocity of contraction decreases at any given load as the temperature is lowered. Is this change a function of alterations in A, C and D as a function of temperature or, Pi, the force generating system? The relationship of internal velocity of shortening to the force generator, Pi, and the damping element, C, is seen in Equation 4[1]. A decrease in the force generator strength, Pi, or an increase in the value of the coefficient of damping, C, or both could explain the velocity change. The force generator Pi is related to changes in the rate of change of force, dP/dt, and the ratio, C/A, Equation 5[1]. Both C/A and dP/dt decrease as the temperature falls. Thus Pi must decrease as the temperature is decreased. This occurs despite the increase in C as the temperature is lowered. From the relationship seen in Equation 5[1] it is obvious that following activation the force generator, Pi, first is greater than P, equals P at the peak of the twitch and falls below P during relaxation. From the model and the data, the PRWN method allows for a detailed evaluation of Pi, A, C, and D during activity of the ground squirrel papillary muscle. These data and

---

[1] Figure 9 includes the internal force generator Pi in an assumed position among the viscous element C and the elastic elements D and A. A detailed discussion of the construction of this model is found in reference (31). When force equations for this model are solved for isometric contraction (where motion of $x_1 = 0$) and P is the external force one finds:

$$x_2 = V = (Pi - P)/C \qquad (4)$$

where P is the measured total tension and that

$$\tilde{Pi} = P + \frac{C}{A} \dot{P} \qquad (5)$$

where $\dot{P} = dP/dt$.

TABLE 3. Change in series compliance with temperature.

Temperature C°	Modulus of Elasticity gm/mm^2 $\Delta L/L$	
	QR	PRWN
29	11	29
22		38
18	19	147
14	26	
10	30	

The moduli were obtained from the relatively linear slopes of the stress-strain curves derived from quick release measurements (QR) and analysis of the PRWN data where elements A, C and D were evaluated.

the data derived from standard techniques help explain the alteration in contraction of the isolated ground squirrel papillary muscle which occur as the temperature is lowered.

REFERENCES

1. ABBOTT, B. C., and W. F. H. M. MOMMAERTS. A study of inotropic mechanisms in the papillary muscle preparation. J. Gen. Physiol. 42: 533-551, 1959.

2. ALFREY, T., and P. DOTY. The methods of specifying the properties of visco-elastic material. J. Appl. Phys. 16: 700-713, 1945.

3. APTER, J. T., and E. MARQUEZ. Correlation of visco-elastic properties of large arteries with microscopic structure. V. Effects of sinusoidal forcings at low and resonance frequencies. Circulation Res. 22: 393-404, 1968.

4. BENDAT, J. S. Interpretation and application of statistical analysis for random physical phenomena. IEEE Trans. Bio-Med. Eng. 9: 31-43, 1962.

5. BENDAT, J. S., and A. G. PIERSOL. In: Measurement and Analysis of Random Data. New York: John Wiley and Sons, 1966.

6. BOETTIGER, E. G. In: Recent Advances in Invertebrate Physiology, edited by B. T. Scheer. Eugene, Oregon: Oregon University Publications, 1957, p. 117.

7. BOUCKAERT, J. P., L. CAPELLEN, and J. DEBLENDE. The viscoelastic properties of frog's muscle. J. Physiol. 69: 473-492, 1930.

8. BRADY, A. J. Time and displacement dependence of cardiac contractility: problems in defining the active state and force-velocity relations. Federation Proc. 24: 1410, 1965.

9. BRADY, A. J. Onset of contractility in cardiac muscle. J. Physiol. 184: 560, 1966.

10. BRADY, A. J. The three element model of muscle mechanics: its applicability to cardiac muscle. Physiologist 10: 75-86, 1967.

11. BRADY, A. J. A measurement of the active state in papillary muscle (Abstract). Physiologist 10: 130, 1967.

12. BRADY, A. J. Active state in cardiac muscle. Physiol. Rev. 48: 570-600, 1968.

13. BRUTSAERT, D. L., and E. H. SONNENBLICK. Force-velocity-length-time relations of the contractile elements in heart muscle of the cat. Circulation Res. 24: 137-148, 1969.

14. BUCHTHAL, F., and E. KAISER. Factors determining tension development in skeletal muscle. Acta Physiol. Scand. 8: 38-74, 1944.

15. BURLINGTON, R. F., J. T. MAHER, and E. T. ANGELAKOS. Effect of temperature on contractility of isolated atria from a hibernator and a non-hibernator. Life Sciences 7, Part 1: 449-452, 1968.

16. CATTELL, M., and H. GOLD. The relation of rhythm to the force of contraction of mammalian cardiac muscle. Amer. J. Physiol. 133: 236-237, 1941.

17. CRONIN, R., J. A. ARMOUR, and W. C. RANDALL. Function of the in-situ papillary muscle in the canine left ventricle. Circulation Res. 26: 67-75, 1969.

18. EDMAN, K. A. P., D. W. GRIEVE, and E. NILSSON. Studies of the excitation-contraction mechanism in the skeletal muscle and the myocardium. Archiv. Ges. Physiol. 290: 320-334, 1966.

19. EDMAN, K. A. P., and E. NILSSON. The dynamics of the inotropic changes produced by ouabain and increased contraction rate. Acta Physiol. Scand. 63: 507-508, 1965.

20. EDMAN, K. A. P., and E. NILSSON. The mechanical parameters of myocardial contraction studied at a constant length of the contractile element. Acta Physiol. Scand. 72: 205, 1968.

21. EDMAN, K. A. P., and E. NILSSON. The dynamics of the inotropic change produced by altered pacing of rabbit papillary muscle. Acta Physiol. Scand. 76: 236-247, 1969.

22. FENN, W. O. A quantitative comparison between the energy liberated and the work performed by isolated sartorius muscle of the frog. J. Physiol. 58: 175-203, 1923.

23. FENN, W. O. The relation between work performed and energy liberated in muscular contraction. J. Physiol. 58: 373-395, 1924.

24. FENN, W. O., and B. S. MARSH. Muscular force at different speeds of shortening. J. Physiol. 85: 277-297, 1935.

25. GASSER, H. S., and A. V. HILL. The dynamics of muscular contraction. Proc. Roy. Soc., London, Sec. B. 96: 398-437, 1924.

26. GAY, W. A. Cinemicrophotographs presented at the 1969 Scientific Sessions of the American Heart Association.

27. GAY, W. A., and E. A. JOHNSON. An anatomical evaluation of the myocardial length-tension diagram. Circulation Res. 21: 23, 1967.

28. GORDON, A. M., A. F. HUXLEY, and F. J. JULIAN. The variation in isometric tension with sarcomere length in vertebrate muscle fibres. J. Physiol. 184: 170-192, 1966.

29. HALPERN, W. An evaluation of the use of pseudo-random white noise signals in the investigation of the mechanical properties of skeletal muscle. (Ph.D. Thesis.) University of Vermont, Burlington, 1969.

30. HALPERN, W., and N. R. ALPERT. Measurement of nonlinear viscous and elastic elements in isometrically contracting frog sartorius muscle using stochastic length perturbations (Abstract). Federation Proc. 29: 655, 1970.

31. HAMRELL, B. B., and N. R. ALPERT. Myocardial quick-release shortening: stress and contractile element velocity interaction. (Abstract). Federation Proc., 1971. In press.

32. HARTREE, W., and A. V. HILL. The four phases of heat production of muscle. J. Physiol. 54: 84-128, 1920.

33. HAFNER, L. L., and T. E. BOWEN. Elastic components of cat papillary muscle. Amer. J. Physiol. 212: 1221-1227, 1967.

34. HILL, A. V. The maximum work and mechanical efficiency of human muscles and their most economical speed. J. Physiol. 56: 19-41, 1922.

35. HILL, A. V. The heat of shortening and the dynamic constants of muscle. Proc Roy. Soc., London, Ser. B 126: 136-195, 1938.

36. HILL, A. V. The abrupt transition from rest to activity in muscle. Proc. Roy. Soc. London, Ser. B 136: 399-420, 1949.

37. HILL, A. V. The 'plateau' of full activity during a muscle twitch. Proc. Roy. Soc. London, Ser. B 141: 498-503, 1953.

38. HUXLEY, A. F. Muscle structure and theories of contraction. Progress in Biophysics and Biophysical Chem. 7: 255-318, 1957.

39. HUXLEY, H., and J. HANSON. Changes in the cross-striations of muscle during contraction and stretch and their structural interpretation. Nature 173: 973-976, 1954.

40. ILLANES, A., and J. M. MARSHALL. The effects of ouabain on isolated atria of the ground squirrel; comparison with rat and rabbit atria. Arch. Exptl. Path. Pharmak. 248: 15-26, 1964.

41. JEWELL, B. R., and D. R. WILKIE. An analysis of the mechanical components in frog's striated muscle. J. Physiol. 143: 515-540, 1958.

42. JEWELL, B. R., and D. R. WILKIE. The mechanical properties of relaxing muscle. J. Physiol. 152: 30-47, 1960.

43. LEVIN, A., and J. WYMAN. The viscous elastic properties of muscle. Proc. Roy. Soc., London, Ser. B 101: 218-243, 1927.

44. LYMAN, C. P. Circulation in mammalian hibernation. Handbook of Physiology. Circulation. Washington, D. C.: Amer. Physiol. Soc., 1965, sect. 2, vol. III, p. 1967-1989.

45. LYMAN, C. P., and D. C. BLINKS. The effect of temperature on the isolated hearts of closely related hibernators and non-hibernators. Cellular Comp. Physiol. 54: 53-62, 1959.

46. MACPHERSON, L., and D. R. WILKIE. The duration of the active state in a muscle twitch. J. Physiol. 124: 292-299, 1954.

47. MARSHALL, J. M., and J. S. WILLIS. The effects of temperature on the membrane potentials in isolated atria of the ground squirrel, Citellus tridecemlineatus. J. Physiol. 164: 64-76, 1962.

48. MILSUM, J. H. Biological Control Systems Analysis. New York: McGraw-Hill Book Co., 1966.

49. MISHKIN, E., and L. BRAUN (eds.). Adaptive Control Systems. New York: McGraw-Hill Book Co., 1961.

50. NOBLE, M. I. M., T. E. BROWN, and L. L. HEFNER. Force-velocity relationship of cat cardiac muscle, studied by isotonic and quick-release techniques. Circulation Res. 24: 821-833, 1969.

51. PARMLEY, W. W., D. L. BRUTSAERT, and E. H. SONNENBLICK. Effects of altered loading on contractile events in isolated cat papillary muscle. Circulation Res. 24: 521-532, 1969.

52. PARMLEY, W. W., and E. H. SONNENBLICK. Series elasticity in heart muscle: its relation to contractile element velocity and proposed muscle models. Circulation Res. 20: 112, 1967.

53. PODOLSKY, R. J., A. C. NOLAN, and S. A. ZAVELAR. Cross-bridge properties derived from muscle isotonic velocity transients. Proc. Natl. Acad. Sci. U.S. 64: 504-511, 1969.

54. POLLACK, G. H. Analysis of the cardiovascular system: pulmonary arterial hemodynamics and cardiac muscle mechanics. (Ph.D. Thesis.) Pittsburgh, University of Pennsylvania, 1968.

55. POLLACK, G. H. Maximum velocity as an index of contractility in cardiac muscle: a critical evaluation. Circulation Res. 26: 111-127, 1970.

56. RFX, R. L., and G. T. ROBERTS. Correlation, signal averaging and probability analysis. Hewlett-Packard Journal 21, No. 3: 2, 1969.

57. RITCHIE, J. M. The duration of the plateau of full activity in frog muscle. J. Physiol. 124: 605-612, 1954.

58. RITCHIE, J. M. The effect of nitrate on the active state of muscle. J. Physiol. 126: 605-612, 1954.

59. RODBARD, S. (quoted in a personal communication by Bernard C. Abbott).

60. ROSENTHAL, D. Introduction to Properties of Materials. Princeton, New Jersey: D. Van Nostrand Co., Inc., 1964.

61. SMITH, D. E., and B. KATZUNG. Mechanical performance of myocardium from hibernating and non-hibernating mammals. Amer. Heart J. 71: 515-521, 1966.

62. SONNENBLICK, E. H. Force-velocity relations in mammalian heart muscle. Amer. J. Physiol. 202: 931-939, 1962.

63. SONNENBLICK, E. H. Implications of muscle mechanics in the heart. Federation Proc. 21: 975-990, 1962.

64. SONNENBLICK, E. H. Determinants of active state in heart muscle: force, velocity, instantaneous muscle length, time. Federation Proc. 24: 1396, 1965.

65. SONNENBLICK, E. H. Instantaneous force-velocity-length determinants in the contraction of heart muscle. Circulation Res. 16: 441-451, 1965.

66. SONNENBLICK, E. H. Active state in heart muscle: its delayed onset and modification by inotropic agents. J. Gen. Physiol. 50: 661-676, 1967.

67. SONNENBLICK, E. H. Correlation of myocardial ultrastructure and function. Circulation 38: 29-44, 1968.

68. SOUTH, F. E. Rate and magnitude of tension production of ventricular muscle from hibernating and non-hibernating muscles. In: Proceedings of the First Annual Rocky Mountain Bioengineering Symposium, edited by G. J. D. Schock. Colorado: USAF Academy, 1964, p. 251-257.

69. SPIRO, D., and E. SONNENBLICK. Comparison of the ultrastructural basis of the contractile process in heart and skeletal muscle. Circulation Res. Supplement II: 14-37, 1964.

70. ULLRICK, W. C., and W. V. WHITEHORN. A muscle column preparation from the rat's left ventricle. Circulation Res. 4: 499-501, 1956.

71. WHITE, D. C. S. Rigor contraction and the effect of various phosphate compounds on glycerinated insect flight and vertebrate muscle. J. Physiol. 208: 583-605, 1970.

72. YEATMAN, L. A., W. W. PARMLEY, and E. H. SONNENBLICK. Effects of temperature on series elasticity and contractile element motion in heart muscle. Amer. J. Physiol. 217: 1030-1034, 1969.

CURRENT APPROACHES TO THE STUDY OF THE DYNAMIC GEOMETRY OF THE LEFT VENTRICLE IN CONSCIOUS ANIMALS

I. Left Ventricular Dimensional Measurements in Conscious Instrumented Dogs.
II. Instantaneous Changes in External Diameter and Left Ventricular Volume in Thirteen-Lined Ground Squirrels.

EDWARD WM. HAWTHORNE and FRANCES KRAFT-HUNTER

Department of Physiology and Biophysics,
College of Medicine, Howard University, Washington, D.C. 20001

INTRODUCTION

Our experiences and problems have been extensive in the development and use of transducing devices for the continuous recording of ventricular dimensions from intact animals (5,13,14,15,20,40). Consequently the first purpose of this communication is to describe and characterize the types of instantaneous dimensional changes that can be continuously monitored in conscious instrumented dogs during single cardiac cycles and for periods of time extending over many months. In addition we will explore the possibilities in the use of these measurements for estimating left ventricular volume. Secondly we wish to present preliminary observations from our efforts to adapt these techniques for use in animals with very small left ventricles, such as the thirteen-lined ground squirrel. The ultimate objective is to make a beat to beat analysis of the dynamic geometrical and mechanical changes occurring in the left ventricle of these animals as they enter and arouse from deep hibernation.

As Lyman (21) has pointed out "the student of hibernation is limited in the techniques at his disposal by the pecularities of hibernation itself.

Because the onset of hibernation is unpredictable and, in our present state of knowledge, cannot be induced, chronic methods of recording are a necessity when studying the entrance into hibernation or the hibernating state itself."

1. LEFT VENTRICULAR DIMENSIONAL MEASUREMENTS IN CONSCIOUS INSTRUMENTED DOGS

There are three major factors which affect the performance of the ventricles at any given ejection period. These are (a) the initial end-diastolic stretch or length of the ventricular myocardial fibers, (b) the load resisting shortening of the ventricle during ejection, and (c) the state of myocardial contractility at the time. It is necessary to have a knowledge of the instantaneous dimensional changes of the ventricles occurring during the cardiac cycle in order to properly characterize any one of these three factors. The estimation of the degree of initial end-diastolic stretch or the length of the ventricular myocardial fibers at end-diastole requires measurement of at least end-diastolic volume and pressure. The dimensional measurements involved in estimating end-diastolic volume of the left ventricle are the magnitudes of the equatorial internal and external radii, wall thickness and the length of the major semiaxis. In normal states the volume dimension is related to end-diastolic pressure and may be extrapolated from a knowledge of the end-diastolic pressure-volume relationship (28). Yet this relationship varies with wall thickness and internal cross-sectional area of the ventricle. Furthermore, some fixed reference point is required to deal with quantifying the amount of initial end-diastolic stretch that is present in the myocardial fibers from time to time in the same animal and between different animals. For this a knowledge of the unstretched length of the ventricle (end-diastolic volume or cross-sectional area) at zero load or zero diastolic pressure would be of great value.

Estimation of end-systolic volumes also requires dimensional measurements. This volume is of considerable importance, for the magnitude of this dimension and consequently end-systolic myocardial fiber length is determined by the load resisting shortening and the state of myocardial contractility (9,17,39).

The load resisting shortening during ejection has often been defined as being either the aortic pressure of the aortic impedance to outflow against which left ventricular ejection takes place. More correctly, however, it can be defined as the midwall tensile stress which must be overcome for ejection to occur and proceed to end-systole. One measure of this afterload could be the average hoop stress in the wall of the left ventricle at each instant during contraction. Average hoop stress may be defined as the force per unit area in the circumferential direction at the equator of the left ventricle. It is calculated for a thick walled ellipsoidal structure such as the left ventricle by the formula below:

$$\sigma_h = \frac{PR_i}{H} \left[\frac{1}{1 + \frac{H}{2R_i}\left(\frac{R_i}{B}\right)} - 2 \right] \left[1 - 1/2\left(\frac{R_i}{B}\right)^2 \right] \quad (1)$$

where H is wall thickness, R_i is internal radius and P is left ventricular pressure all in appropriate units.

Contractility is most difficult to define. However, it is generally believed that the contractile properties of cardiac muscle can be characterized and the contractile state of cardiac muscle defined in terms of force, velocity, length and time (10,26,28,36). Currently, estimations are made of velocity of contractile element shortening and force in terms of hoop stress. And plots of the hyperbolic force-velocity-length relations using instantaneous dimensional and pressure data and the rates of change of these obtained during a contraction are used to evaluate the contractile state.

Following the pioneer studies of Rushmer and his associates (30-34) a considerable number of studies by many investigators have been made that were designed to define more clearly the dynamic geometrical changes occurring in the left ventricle during the cardiac cycle. Various techniques and approaches have been used. Investigators have developed techniques for the analog recording of the instantaneous changes that occur during the cardiac cycle in left ventricular wall thickness (5,9), left ventricula internal diameter (1,19,26), external diameter (15,33,41,42), base to apex length (6,15,31), external cross-sectional area (13,25), and external circumference (7,10,13,14,15,31,36,38). Additionally the architecture of the left ventricle has been analyzed in a variety of other ways (24,28). Most of these measurements have been used in attempts at measuring instantaneously the changes in ventricular volume during the cardiac cycle.

The use of indicator dilution techniques (17,37,38) and angiocardiographic methods (4,6,8,11,12,18,22,23,34,35,36) are the two chief means of estimating left ventricular volumes aside from techniques involving the use of dimensional measurements. The use of instantaneous dimensional measurements for ventricular volume estimation have largely been restricted to studies in experimental animals whereas the other two methods have been used extensively in man and animals.

The use of the indicator dilution method for estimating ventricular volumes, that is end-diastolic and end-systolic Volumes as well as stroke volume, was first developed by Holt and his associates (16). The technique involves securing "step-function dilution curves" at the aortic root, as the indicator is "washed-out" of the left ventricle after it has been injected into the ventricle during the previous diastole. The angiocardiographic method involves taking a series of X-ray pictures of the heart at high-speed, usually 60 frames/second, and during this process injecting rapidly a bolus of radioopaque dye into the ventricle. The films are taken over the "wash out" period of the injected dye. Ideally biplane cineradiography is used. Neither of these two methods are capable of permitting continuous monitoring of ventricular volume changes that occur beat by beat over any extended period of time.

Certainly for making continuous volume measurements over extended periods of time, in laboratory animals, the use of appropriate electronic transducers for monitoring of various ventricular dimensions is the only available approach.

Preparation of Instrumented Dogs. The dogs used vary in age, sex and breed. Their body weight varies between 20 and 30 kilograms. All dogs are given doses of sodium ampicillin (500 mgms) twice daily for the three consecutive days just prior to surgery.

Immediately prior to anesthetizing each animal, they are given intramuscular injections of atropine sulfate (1 mgm/kg) and xylocaine (200 mgms). These are given primarily to offer some protection against the development of ventricular premature contractions and ventricular fibrillation during the course of the surgery, since the heart is manipulated during this time. All animals are anesthetized with pentabarbital sodium using a dose of 30 mgm/kg body weight. The anesthetic is injected intraperitoneally. Once the dog is asleep the left chest is shaved and the animal is placed on the operating table so that he is lying on his right side. Sterile surgical procedure is used from this point onward.

A standard left thoracotomy is performed using an electrocautery, and the left chest is opened widely. Respiration is maintained artificially by using a respirator to supplement intermittent positive pressure respiration. The lungs are protected with warm moist towels. The left lungs are then gently retracted ventrally and a 2 cm section of the thoracic aorta adjacent to the aortic arch is dissected free of its mediastinal reflection. A purse string, 0.5 cm diameter, is sewn in this section and blood flow through the region of the purse string is interrupted by pinch clamping with a "J" shaped Potts clamp. The aorta is opened within the purse string and a 6.5 mm diameter implantable pressure cell (Konigsberg Instruments, Pasadena, California) inserted by stretching the opening to accommodate the larger cell. The purse string is drawn tightly around the lead wire of the cell and is used to anchor the transducer adjacent to the internal wall of the vessel. The entire 2 cm section of isolated aorta is then reinforced with a snuggly fitted Debakey seamless dacron arterial graft.

The pericardium is then widely opened and 1% butyn sulfate solution is sprayed over the exposed myocardial surface. A two cm diameter purse string suture is sewn around the apex of the left ventricle. A six inch long straight needle, threaded with 12 inches of 00 silk suture is passed across the diameter of the left ventricle with the surgeon guiding the path of the needle so that it passes across the diameter of the left ventricle near its equator, and in the anterior to posterior direction. The suture is then hooked with a special valvulotome which is inserted thru the apex of the left ventricle for this purpose. The hooked suture is then brought out through the apex of the left ventricle as the assistant tightens the purse string. The suture is cut and the lead wires of each coil (for internal diameter measurements) tied, one coil to each of the cut suture ends, and pulled back into position against the left ventricular wall. A 00 silk suture is used to close puncture holes in the left ventricle and anchor the coils against the endocardial wall. A 6.5 mm implantable pressure transducer is inserted three cm into the left ventricle via the opening in the apex and anchored in that position with OO silk suture. An inflatable cuff of appropriate diameter is then placed around the root of the previously isolated ascending aorta. At this point a pair of coils are coaxially attached to the epicardium of the left ventricle at its equator for later measurement of external diameter. Lead beads of different sizes are then inserted subendocardially as diagrammed in Figure 1.

The pericardium is loosely closed with interrupted stitches. A thoracic drainage tube is inserted through the sixth intercostal space. Ribs four and five are apositioned with five or six No. 5 sternal sutures. The overlying muscles are closed wither with No. 2 catgut or with No. 2 silk using interrupted mattress stitches. The pneumothorax is removed by applying suction to the thoracic drainage tube using 12 to 15 cm of water negative pressure. This is continued for several hours. Five hundred to one thousand cc of 10% canine plasma, and dextran-saline solution, is usually given the dog at this point. The lead wires from all the pressure cells and the coils are then passed subcutaneously from the dorsal end of the chest incision to the exterior of the back of the dog's neck.

Postoperative care consists of daily injections of 500 mg sodium ampicillin, 1 gm sodium methicillin and 2 gm of sulfadimethoxine for the first postoperative week. This was followed by 4 to 6 weeks of oral therapy using 500 to 1000 gm of sodium ampicillin per day. The thoracic drainage tube is left in place for approximately ten days and aspirated daily with gentle suction. For the first 48 to 72 hours the dog is regularly monitored and maintained on regular dosages of xylocaine.

<u>Measurements of Wall Thickness Changes</u>. Left ventricular wall thickness is measured by use of pair of coils of fine coated copper wire with attached lead wires. The coils are coaxially placed and attached one on each side of

appropriate sections of the left ventricular wall near its equator and in an area devoid of major epicardial coronary vessels. One of the pair of coils is used as the exciting coil and is energized by an oscillating current of a chosen constant frequency, voltage and amperage. The oscillating magnetic field thus produced induces an EMF in the opposite coil. The magnitude of this induced current is solely a function of the separation of the two coils at any instant and therefore provides the measure of the wall thickness at any instant (13,25,41). Since the amount of induced current in the secondary coil is nonlinearily as well as inversely proportional to the distance between the two coils, the induced current is passed through a log amplifier (General Radio) and the output is used as the measure of wall thickness. After amplification, rectification and demodulation this provides a d.c. voltage which is linearly and inversely proportional to the distance between the two coils, over the range of distance that they are used.

Measurements of External Circumference or Changes. External left ventricular circumference was measured by means of a mercury-in-silastic strain gauge, positioned about the equator of the left ventricle (the technique for preparing and using these gauges has been described (14).

Measurement of Aorta-Apex Length. The electromagnetic induction system used for this measurement has been previously described (15). Essentially in these experiments a primary coil, consisting of 300 turns of fine coated (#38) wire, is sewn to the apex of the left ventricle. A secondary coil, consisting of 6 turns of no. 27 polyvinyl-coated wire, is wound at the root of the aorta, just above the coronary arteries. When the primary coil is excited at 20 kHz, using an appropriate current strength a magnetic field is generated and it induces in the secondary coil an electromagnetic force with voltage inversely proportional to the distance from the source of the magnetic field. The EMF induced in the secondary coil is received into a null' detector, filtered at 20 kHz, demondulated, amplified, and the log output is recorded.

Measurement of Left Ventricular Internal and External Diameter. Left ventricular internal and external diameters have been chosen as the dimensions to be monitored for long term studies in dogs and other animals because they involve the minimal dimensional measurements for estimating ventricular volume and average hoop stress. The transducing devices used for these measurements are pairs of coils of fine coated copper wire. The diameter of each coil is 6 mm. The coils are rugged in construction and rarely break down after implantation. The technique of measuring the changes in internal and external diameter are the same as described above for measuring wall thickness. The primary difference is the coaxial placement of the pairs of coils. Figure 1 diagramatically shows the endocardial placement of the pair of coils (B and B^1) for measuring internal diameter and the epicardial placement of the pair of coils (A and A^1) for measurements of external diameter.

Also shown in Figure 1 is the location of lead beads (black dots) which are used to calibrate the diameter and internal volume measurements using biplane cineradiography (23) when this is desired.

Measurements of Pressure. Left atrial, left ventricular, and aortic pressure are measured in chronic animals through use of implantable pressure transducers (Konigsberg Instruments, Pasadena, California) that are 6.5 cm in diameter and which have an adequate frequency response. The site of location of the aortic and ventricular implantable pressure transducers is shown in Figure 1.

Estimation of Ventricular Volume using Left Ventricular Diameter. Previous studies in our laboratories have involved measurements simultaneously of the dimensions of base to apex length, external circumference, left ventricular wall thickness, and these measurements have also been made along with estimation of left ventricular length using biplane cinefluorography.

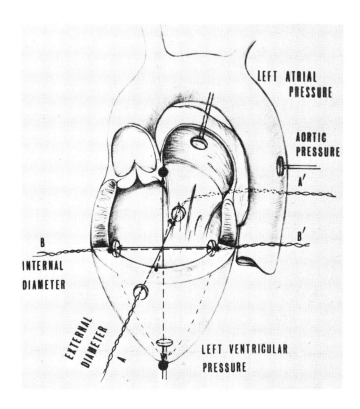

FIG. 1. Diagram showing placement of transducing devices for measuring changes in left ventricular internal diameter (B,B') and external diameter, aortic, left atrial and left ventricular pressures.

From these types of studies we have observed that the axis ratio of the left ventricle remains nearly constant during the period of ejection. As a result we are able to substitute a K value for axis ratio in equation 2 below. This permits estimation of the instantaneous changes in left ventricular volume during the ejection phase of each beat where:

$$V_i = 4/3 \pi R_i^3 \text{ (B)} \qquad (2)$$

and V_i = internal volume, R_i = internal radius and B = axis ratio.

In the present series the axis ratio at end-diastole and end-systole is determined for a particular animal and the mean of these two values is then used for the value (B) in equation 2.

<u>Typical Recordings from Conscious Instrumented Dogs</u>. Figure 2 shows the simultaneous changes in left ventricular wall thickness and external circumference that are observed during normal cardiac cycles in conscious dogs.

In this case wall thickness changes were monitored using induction coils coaxially placed across the ventricular wall. Wall thickness measurements can also be derived from simultaneously monitoring changes in internal and external diameter. The transducing devices may either be two pairs of sonar micrometers (1) or two pairs of induction coils.

It is seen that wall thickness begins to increase with the onset of ventricular systole and peak thickness is achieved only at end-ejection. Approximately 9% of the total wall thickening occurs during isovolumic contraction, while 35% occurs during the ejection phase.

Figure 3 shows the typical change occurring in left ventricular external circumference and aorta to apex length in conscious dogs during a cardiac cycle. Analysis of dimensional changes of this type reveals that only slight changes occur in the axis ratio during the period of ejection. Since ejection begins at the end of isovolumic contraction the major changes in shape have already occurred and the ventricle apparently decreases in size concentrically during the ejection phase.

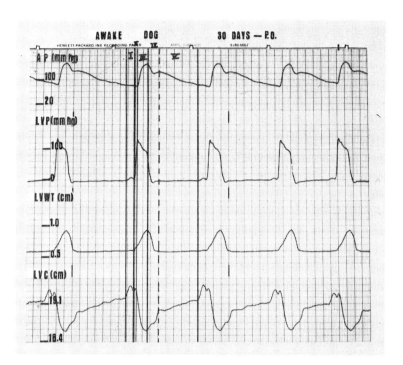

FIG. 2. Simultaneous changes in left ventricular circumference (LVC), left ventricular wall thickness (LVWT), left ventricular pressure (LVP) and aortic pressure (AP). I = atrial contraction; II = isovolumic contraction; III = ejection; IV = rapid filling; V = diastasis.

In both Figures 2 and 3 the isovolumic expansion of left ventricular circumference is clearly shown as has been previously described by Rushmer (31), Hawthorne (5,13,14) and others (38,41).

It is still technically difficult to measure simultaneously, using analog transducing devices, the changes occurring in left ventricular length, left ventricular wall thickness, left ventricular circumference, left ventricular external diameter, and left ventricular internal diameter. However, by

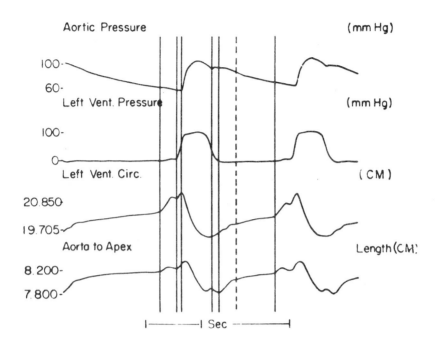

FIG. 3. Changes in aorta to apex length during the cardiac cycle.

combining the use of biplane cinefluorographic techniques for identifying the changes in linear dimensions between metal markers with simultaneous monitoring of changes in internal diameter, or external circumference and wall thickness, it is possible to follow for short periods the simultaneous changes occurring in all these dimensions during the cardiac cycle. Initial studies of this nature have been reported (15).

Figure 4 shows the type of recordings obtained for an awake instrumented dog two months after surgery. In this case internal diameter of the left ventricle (LVID) was monitored simultaneously with left atrial pressure

(LAP), left ventricular pressure (LVP), aortic pressure (AP), dP/dt, heart rate and the electrocardiogram (ECG). The typical sinus arrythmia seen in conscious dogs in good clinical condition is present. Dogs instrumented in this way may be followed for many months.

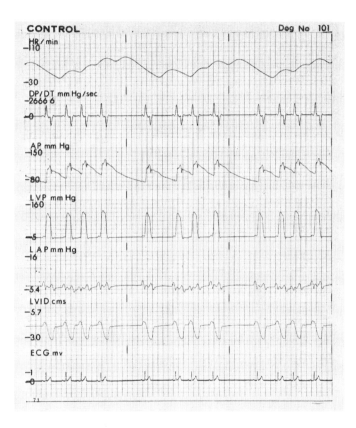

FIG. 4. LVID = left ventricular internal diameter; LAP = left atrial pressure; LVP = left ventricular pressure; AP = aortic pressure. (Cf. text for description).

Figure 5 is a recording from another conscious instrumented animal. This dog was instrumented as diagrammed in Figure 1. Both external and internal diameters are recorded as well as other variables. This animal has a moderate degree of mitral insufficiency. The characteristic pattern

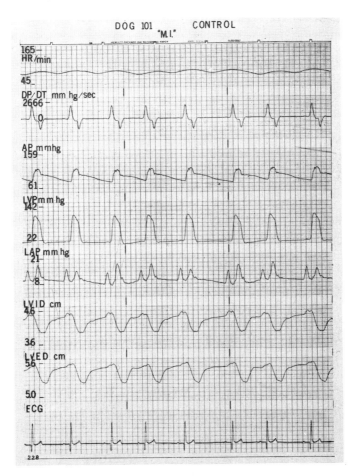

FIG. 5. "M.I." = mitral insufficiency; LVED = left ventricular external diameter. The other abbreviations are as in fig. 5. (See text for description).

of change in each primary variable is seen. From recordings of this type a large number of derived variables can be calculated for each cardiac cycle. These include estimations of end-diastolic, end-systolic and stroke volumes, ejection fraction, total peripheral resistance, stroke work, stroke power, wall stresses and the velocity of contractile element shortening.

II. INSTANTANEOUS CHANGES IN EXTERNAL DIAMETER AND LEFT VENTRICULAR VOLUME IN THIRTEEN-LINED GROUND SQUIRRELS

Only within the past ten to fifteen years have measurements of the cardiac output of small animals been made. Kent and Popovic, in 1963, developed a technique of cannulating the aorta and the right ventricle to obtain cardiac output determinations in unanesthetized rats. This same technique Popovic (27) has applied to hibernating ground squirrels. In 1958, Bullard (2,3) obtained cardiac output determinations by using dye dilution curves.

As yet techniques have not been developed for continuous monitoring of ventricular dimensional changes or for the purpose of providing beat by beat estimations of end-diastolic volume, end-systolic volume, stroke volume and ejected fraction. Our studies were designed to determine the feasibility of continuously monitoring changes in external left ventricular diameter in anesthetized open-chest ground squirrels. This was recognized as a beginning approach to the development of miniature transducing devices in order to apply the techniques developed for continuous monitoring of ventricular dimensions and pressures in dogs and larger animals to the study of ground squirrels and other animals with very small hearts.

Methods. Thirteen-lined ground squirrels (Citellus tridecemlineatus) weighing 230-310 gm were used. They were anesthetized with nembutal (approximately 0.05 mgs/gm of body weight intraperitoneally). The chest was opened in the midline and the heart removed from the pericardium. Positive-pressure ventilation was maintained by a modified Harvard respirator. A pair of induction coils approximately 4 mm in diameter were placed coaxially across the minor semiaxis of the left ventricle. They were used to measure the instantaneous changes in the external diameter of the left ventricle, as previously explained earlier in this paper. Lead II of the electrocardiogram was recorded.

At the end of each experiment selected cardiac cycles were analyzed and the calibrated external diameters were used to estimate external end-diastolic and end-systolic ventricular volume by using the following formulas:

$$V_{E_{eivc}} = 4/3 \pi R_{E_{EIVC}}^3 \frac{L/2}{R_{E_{EIVC}}} \tag{3}$$

$$V_{E_{es}} = 4/3 \pi R_{E_{es}}^3 \frac{L/2}{R_{E_{es}}} \tag{4}$$

External end-diastolic volume was taken as the calculated volume at the end of isovolumic contraction. This volume was calculated as $4/3 \pi$ times the external radius cubed times the left ventricular axis ratio. The length of the ventricle was measured from the beating heart with calipers. Since the ventricle was most often in diastole than in any other phase, this measurement was taken to be the end-diastolic or maximum length of the ventricle. This could have been checked if desired by sewing metal markers on the base and apex of the ventricle and taking a long X-ray film of the chest.

In the above formula $V_{E_{EIVC}}$ is the external volume at the beginning of ejection, and $R_{E_{EIVC}}$ is the external radius at the beginning of ejection. L is the length of the ventricle from base to apex.

Internal volume was obtained by subtracting left ventricular muscle volume from the calculated external volume. Ellipsoidal geometry was assumed.

The assumption was made that the axis ratio of the left ventricle did not change significantly during ejection.

Results. Figure 6 shows a recording of the instantaneous changes in left ventricular external diameter taken from an anesthetized open-chest ground squirrel. This record is qualitatively similar to the type recorded from anesthetized open-chest dogs. Characteristically in the dog, the heart is small and underfilled in anesthetized open-chest preparations. This situation is usually indicated by the large increases in diameter or circumference that occur during the period of isovolumic contraction. In this regard, records obtained from anesthetized open-chest animals are quite different from those seen in conscious instrumented animals.

In this figure all the phases of the cardiac cycle are recognizable. The heart rate in this anesthetized squirrel was initially between 450 and 500 per minute. When the chest was first opened, rates within this range were observed in 6 other ground squirrels as well. The heart rate decreased with time after opening the chest and this apparently was due to the cooling effect of having the body cavity open plus the trauma of the gauge placement.

The estimated values for stroke volume and cardiac output do not differ significantly from those reported in the literature. It is of interest that the calculated ejection fractions are similar to those found in anesthetized dogs.

Axis ratios were determined by measuring the base to apex length of the beating left ventricle using calipers. It was assumed that this measured length was a good estimate of the end-diastolic length of the ventricle. In one ground squirrel with a heart rate of 30/min, EDV was 1.77 cc, ESV was

1.18 cc and stroke volume was .61 cc/beat. The animal had a cardiac output of 18.3 cc and an ejection fraction of 34%.

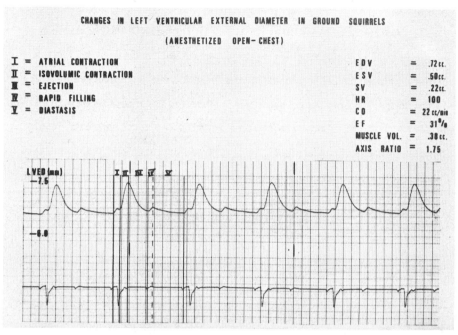

FIG. 6.

Discussion. The preliminary observations clearly show that continuous external left ventricular diameter measurements can be made on ground squirrel hearts. Techniques are already available for measuring aortic and ventricular pressures in these animals (2,27). It seems possible therefore to prepare these animals using sterile surgery, for study in the conscious state.

The use of ventricular muscle mass and the calculated external left ventricular volume changes for estimating internal volume changes seems feasible and would have validity in short term experiments where the ventricle muscle would not have hypertrophied.

The use of metal markers as shown in Figure 1 would permit more accurate determination of the axis ratios during ejection by use of biplane cineradiography.

It appears therefore, that in the near future it will be possible to make continuous estimations of ventricular volume, pressure, wall tensile stress, equatorial wall thickness, and internal and external radius changes in conscious, intact ground squirrels. Thus, it will be possible to evaluate in more detail the changes in cardiac dynamics and force-velocity-length changes in these hibernators as left ventricular function is altered with lowering of body temperature. Studies are underway in our laboratory to help in achieving this objective.

REFERENCES

1. BISHOP, V. S., L. D. HOROWITZ, H. L. STONE, H. F. STEGALL, and E. J. ENGELKEN. Left ventricular internal diameter and cardiac function in conscious dogs. J. Appl. Physiol. 27: 619-623, 1969.

2. BULLARD, R. W. Cardiac output of the hypothermic rat. Amer. J. Physiol. 196: 415-419, 1959.

3. BULLARD, R. W., and G. E. FUNKHOUSER. Estimated regional blood flow by rubidium 86 distribution during arousal from hibernation. Amer. J. Physiol. 203: 266-270, 1962.

4. CHAPMAN, C. B., O. BAKER, and J. H. MITCHELL. Left ventricular function at rest and during exercise. J. Clin. Invest. 38: 1202-1213, 1959.

5. COTHRAN, L. N., W. C. BOWIE, J. E. HINDS, and E. W. HAWTHORNE. Left ventricular wall thickness changes in unanesthetized horses. In: Factors Influencing Myocardial Contractility, edited by R.D. Tanz, F. Kavaler, and J. Roberts. New York: Academic Press, 1967, p. 163-176.

6. DAVILA, J. C., and M. E. SANMARCO. An analysis of the fit of mathematical models applicable to the measurement of left ventricular volume. Amer. J. Cardiol. 18: 31-42, 1966.

7. DAVILA, J. C., M. E. SANMARCO, and C. M. PHILLIPS. Continuous measurement of left ventricular volume in the dog. I. Description and validation of a method employing direct external dimensions. Amer. J. Cardiol. 18: 574-583, 1966.

8. DODGE, H. T., H. SANDLER, D. W. BALLEW, and J. D. LORD. The use of biplane angiocardiography for measurement of left ventricular volume in man. Amer. Heart J. 60: 762-776, 1960.

9. DOWNING, S. E., and E. H. SONNENBLICK. Cardiac muscle mechanics and ventricular performance: force and time parameters. Amer. J. Physiol. 207: 705-715, 1964.

10. FRANKLIN, D. L., R. L. VAN CITTERS, and R. F. RUSHMER. Left ventricular function described in physical terms. Circulation Res. 11: 702-711, 1962.

11. GREENE, D. G., R. CARLISLE, C. GRANT, and I. L. BUNNELL. Estimation of left ventricular volume by one-plane cineangiography. Circulation 35: 61-69, 1967.

12. GRIBBE, P., L. HIRVONEN, J. LIND, and C. WIGELIUS. Cineangiocardiographic recordings of the cyclic change in volume of the left ventricle. Cardiologia 34: 348-366, 1959.

13. HAWTHORNE, E. W. Instantaneous dimensional changes of the left ventricle in dogs. Circulation Res. 9: 110-119, 1961.

14. HAWTHORNE, E. W. Symposium on measurements of left ventricular volume. Part III. Dynamic geometry of the left ventricle. Amer. J. Cardiol. 18: 566-573, 1966.

15. HINDS, J. E., E. W. HAWTHORNE, C. B. MULLINS, and J. H. MITCHELL. Instantaneous changes in the left ventricular lengths occurring in dogs during the cardiac cycle. Federation Proc. 28: 1351-1357, 1969.

16. HOLT, J. P. Estimation of residual volume of the ventricle of the dog's heart by two indicators dilution technics. Circulation Res. 4: 187-195, 1956.

17. HOLT, J. P. Regulation of the degree of emptying of the left ventricle by the force of ventricular contraction. Circulation Res. 5: 281-287, 1957.

18. HOOD, W. P., W. J. THOMPSON, C. E. RACKLEY, and E. L. ROLETT. Comparison of calculations of left ventricular wall stress in man from thin-walled and thick-walled ellipsoidal models. Circulation Res. 24: 575-582, 1969.

19. HORWITZ, L. D., V. S. BISHOP, H. L. STONE, and H. F. STEGALL. Continuous measurement of internal left ventricular diameter. J. Appl. Physiol. 24: 738-740, 1968.

20. HUNTER, F. K., L. N. COTHRAN, and E. W. HAWTHORNE. The mechanical characteristics of hypertrophied heart muscle. In: Cardiac Hypertrophy edited by N. Alpert. New York: Academic Press, in press.

21. LYMAN, C. P. Circulation in mammalian hibernation. In: Handbook of Physiology, Circulation, edited by W.F. Hamilton and P. Dow. Washington, D.C.: Amer. Physiol. Soc., 1965, sect. 2, vol. III, chapt. 56, p. 1967-1989.

22. MCDONALD, I. G. The shape and movements of the human left ventricle during systole. A study in cineangiography and by cineradiography of epicardial markers. Amer. J. Cardiol. 26: 221-230, 1970.

23. MITCHELL, J. H., and C. B. MULLINS. Dimensional analysis of left ventricular function. In: Factors Influencing Myocardial Contractility, edited by R.D. Tanz, F. Kavaler, and J. Roberts. New York: Academic Press, 1967, p. 177-188.

24. NINOMIYA, I., and M. F. WILSON. Analysis of ventricular dimension in the unanesthetized dog. Circulation Res. 16: 249-260, 1965.

25. PERRY, S. L. C., and E. W. HAWTHORNE. Left ventricular volume measured by electromagnetic plethysmography. J. Nat. Med. Assn. 50: 117-119, 1958.

26. PIEPER, H. P. Catheter-tip instrument for measuring left ventricular diameter in closed-chest dogs. J. Appl. Physiol. 21: 1412-1416, 1966.

27. POPOVIC, V. Cardiac output in hibernating ground squirrels. Amer. J. Physiol. 207: 1345-1348, 1964.

28. ROSS, J., J. W. COVELL, E. H. SONNENBLICK, and E. BRAUNWALD. Contractile state of heart characterized by force-velocity relations in variably afterloaded and isovolumic beats. Circulation Res. 18: 149-163, 1966.

29. ROSS, J., E. H. SONNENBLICK, J. W. COVELL, G. A. KAISER, and D. SPIRO. The architecture of the heart in systole and diastole: technique of rapid fixation and analysis of left ventricular geometry. Circulation Res. 21: 409-421, 1967.

30. RUSHMER, R. F. Continuous measurements of left ventricular dimensions in intact unanesthetized dogs. Circulation Res. 2: 14-21, 1954.

31. RUSHMER, R. F. Length-circumference relations of the left ventricle. Circulation Res. 3: 639-644, 1955.

32. RUSHMER, R. F., and D. K. CRYSTAL. Changes in configuration of the ventricular chambers during the cardiac cycle. Circulation 4: 211-218, 1951.

33. RUSHMER, R. F., D. L. FRANKLIN, and R. M. ELLIS. Left ventricular dimensions recorded by sonocardiometry. Circulation Res. 4: 684-688, 1956.

34. RUSHMER, R. F., and N. THAL. The mechanics of ventricular contraction: a cinefluorographic study. Circulation 4: 219-228, 1951.

35. SANDLER, H., and H. T. DODGE. Left ventricular tension and stress in man. Circulation Res. 13: 91-104, 1963.

36. SANMARCO, M. E., and J. C. DAVILA. Continuous measurement of left ventricular volume in dogs: estimation of volume-dependent variables using the ellipsoidal model. In: Factors Influencing Myocardial Contractility, edited by R.D. Tanz, F. Kavaler, and J. Roberts. New York: Academic Press, 1967, p. 199-214.

37. SANMARCO, M. E., K. FRONEK, C. M. PHILLIPS, and J. C. DAVILA. Continuous measurement of left ventricular volume in the dog. II. Comparison of washout and radiographic techniques with the external dimension method. Amer. J. Cardiol. 18: 584-593, 1966.

38. TAYLOR, R. R., H. E. CINGOLANI, and R. H. MCDONALD. Relationships between left ventricular volume, ejected fraction, and wall stress. Amer. J. Physiol. 211: 674-680, 1966.

39. TAYLOR, R. R., J. W. COVELL, and J. ROSS. Volume tension diagrams of ejecting and isovolumic contractions in left ventricle. Amer. J. Physiol. 216: 1097-1102, 1969.

40. WALKER, M. L., E. W. HAWTHORNE, and H. D. SANDLER. Methods for assessing performance for the intact hypertrophied heart. Cardiac Hypertrophy edited by N. Alpert. New York: Academic Press, 1971. In press.

41. WILSON, M. F. Left ventricular diameter, posture and exercise. Circulation Res. 11: 90-95, 1962.

42. WILSON, M. F., I. NONOMIYA, and W. M. CALDWELL. Left ventricular dimension-flow relations in unanesthetized dogs. In: <u>Factors Influencing Myocardial Contractility</u>, edited by R.D. Tanz, F. Kavaler and J. Roberts. New York: Academic Press, 1967, p. 149-161.

EXCITATION AND CONTRACTION MECHANISMS IN HEART MUSCLE DURING HIBERNATION AND HYPOTHERMIA

Understanding how heart muscle in hibernating animals continues to function at body temperatures where hearts from non-hibernators cease to perform remains one of the intriguing goals in modern biology. Solving this problem requires answers at the ultrastructural, electrical, chemical, mechanical and thermodynamic levels. Most of our knowledge of muscle function comes from experiments carried out on skeletal muscle. What can be carried over to heart muscle in general and hibernating heart muscle specifically?

A knowledge of the structure of heart muscle from hibernators as well as non-hibernators during different functional states is essential. Ultrastructural analysis and correlation of this structure with function is an initial prerequisite to evaluating the similarities between the heart and skeletal muscle and more importantly, the differences. To begin with, there is no single fiber preparation on which mechanical and anatomical studies can be carried out. Thus it is difficult to eliminate the sampling errors. Where is the sample taken from and, once chosen, how do the various fixing procedures alter the structure? A second important problem in evaluating heart muscle is the general lack of conformity of sarcomere lengths, and the phenomenon of buckling which occurs during shortening. These problems make the all important consideration of mechanical performance as a function of sarcomere length guess work at best. Until direct observation of a single fiber sarcomere length can be made during excitation we must rely on lumped statistical approaches to relating length to force. The results are only as good as the extent and accuracy of the sampling. Anatomical quantitation in heart remains a difficult but important goal.

Thus anatomical uncertainty must be considered when evaluating mechanical and electrophysiological data. We always are dealing with bundles of fibers with a complex geometry when studying heart muscle. For the mechanical measurements, non-homogeneity must be taken into account. For the electrophysiological measurements, the possibility of many types of low resistance pathways between cells makes an intimate knowledge of these mandatory in evaluating any electrophysiological result on the muscle. Differences found in different parts of the heart as well as differences between species must be catalogued.

During the past two decades much has been learned regarding the biochemical basis of muscle contraction. Combined anatomical and biochemical studies have localized the major proteins constituting the contractile machinery and assigned specific roles to them. Since the sliding filament hypothesis is now generally accepted, and force generation is thought to occur by the cyclic interaction of the myosin cross-bridges with actin, it is essential to understand the strategic positioning of the tropomyosin-troponin complex in this system. We need to develop detailed knowledge of the physical changes which occur in the troponin, tropomyosin, actin and myosin complex during activation and inactivation. All estimates of the stoichiometry is based on studies of skeletal muscle. Parallel studies must be carried out on heart muscle from hibernators and non-hibernators. Data must be developed regarding the binding of Ca^{++} to the trigger protein troponin and the subsequent configurational changes which occur. The temperature effects on this process along with simultaneous studies on excitation contraction-coupling parameters might clarify some of the problems.

Myosin plays one of the major roles in the contractile process. Evidence supports the hypothesis that myosin contains an isoenzyme system in which specific ATPase activity varies with subunit composition. Since contractile velocity and performance is closely correlated with myosin ATPase activity, it is essential to study the thermal adaptation of this aspect of the myocardial myosin system. Thus the myosin from hibernators could have a peculiar temperature dependence or exist in a high and low temperature form. Since the transformation of one set of light chain to another can occur rapidly because of the rapid turnover rate of these subunits, this turnover and its effects on the myosin ATPase must be investigated under the variety of physical circumstances suggested using hibernator and non-hibernators.

A major advance in the understanding of membrane behavior in skeletal muscle resulted from the application of voltage clamp techniques to the muscle. Translating this approach to heart muscle immediately presented a number of difficulties. The fibers are small and a number of low resistance pathways exist between the fibers, while the clefts are irregularly placed and of varying depth. The voltage seen at the electrode thus may not be a good sample of all the fibers in the bundle. In addition, transients are poorly defined and substantial differences between ventricular and Purkinje fibers exist which must be explained. Nonetheless the application of these techniques to heart muscle and appropriate modeling to explain the results must be continued along with studies of the temperature effects on normal and hibernating hearts.

Mechanical studies on muscle strips from hearts employ techniques which were developed for skeletal muscle. Evaluating results obtained while using these techniques involves the assumption of specific muscle models and the knowledge of sarcomere lengths. It is important to determine the efficacy of a given model, as well as the sarcomere length independently, before the standard techniques are used to evaluate the mechanical behavior. Carrying out these experiments on the myocardium of hibernating animals should provide finer details of the mechanical behavior and allow correlation of this with the chemistry and anatomy. Finally, all of these data must be assembled into the intact heart, and in this regard analysis of the dynamic geometry of the left ventricle of small animals is important. Preliminary studies using all this type of analysis were reported at these meetings and future experiments should provide important data on the difference between the dynamic geometry of hearts from hibernators and non-hibernators. Hopefully at the next meeting of the hibernation group many of these questions will be answered.

           Symposium Committee on Cardiac Muscle

          Norman R. Alpert, Chairman
          Joachim R. Sommer
          D. J. Hartshorne
          Paul Dreizen
          William W. Sleator
          Edward W. Hawthorne

CNS AND THERMOREGULATION

CORTICAL AND SUBCORTICAL ELECTRICAL ACTIVITY IN HIBERNATION AND HYPOTHERMIA
A Comparative Analysis of the Two States

LJ. T. MIHAILOVIĆ

Institute of Pathological Physiology
Faculty of Medicine
University of Belgrade, Yugoslavia

INTRODUCTION

It is my task to review the work on spontaneous and evoked cortical and subcortical electrical activity of the brain during natural hibernation and artificially induced hypothermia, to critically examine the present status of our knowledge in this expanding field of research and attempt to make a comparative analysis of the two states.

SPONTANEOUS AND EVOKED CORTICAL AND SUBCORTICAL ELECTRICAL ACTIVITY IN HIBERNATION

For a physiologist and in particular for a neurophysiologist the problem of hibernation is not exhausted by a study of thermoregulation but acquires an increasing interest as a problem of "hibernal sleep". Thus it is considered as an active and highly coordinated state which presumes the persistence of functional organization of the various levels and systems of the brain. With this in view we shall approach the study of electrical phenomenology of hibernation.

<u>Spontaneous electrical activity</u>. Fifty years elapsed between the final investigations of Merzbacher (54,55), which brought to a close the neurophysiological contributions to hibernation studies during the last

century, and the renewed interest in such work by Chatfield, Lyman and their collaborators in the United States and simultaneously by Kayser's group in Europe. Chatfield et al. (14), using the golden hamster, and Rohmer et al., (69), using the European ground squirrel, were the first to record the spontaneous electrical activity of the neocortex during arousal from hibernation, thus opening the way for future studies of the brain in hibernation by means of modern neurophysiological methods.

The studies of Chatfield, Lyman and their associates (13,14,47,48) deserve particular attention not only because of their pioneer character but also because they succeeded, essentially from an analysis of spontaneous electrocorticograms (ECG) alone, in drawing important conclusions about the functional state of both non-specific activitating and specific relay systems (the diffuse thalamocortical and ascending reticular activating system). Subsequently, their conclusions were largely confirmed and elaborated upon by other authors who directly explored their functional state.

In a study of cortical electrical activity of the golden hamster during arousal from hibernation, Chatfield et al. (14) found that in the rostral area of the neocortex the first signs of electrical activity did not appear until the temperature of the cortex reached 16°C and in most cases 19-23°C. The initial slow, continuous activity increased during warming both in amplitude and frequency and became increasingly more complex due to the superimposition of spindle bursts. Finally at about 29°C the high frequency, low amplitude desynchronized activity took place. From this study they inferred that different levels of the CNS were not equally resistant to the influence of low temperatures. According to them the reticular activating system was most sensitive to cold because it was the last to appear during arousal from hibernation. Next in order of sensitivity was the thalamocortical system. The spino-bulbo-thalamo-cortical relay systems and the cortex, together with the descending pathways are less sensitive than the previous two systems, while the peripheral nerves are most resistant since they are capable of conducting impulses to 2°C (13).

Independently and almost at the same time, Kayser with his co-workers (37,69) discovered that in the European ground squirrel, unlike the golden hamster, the first signs of electrical activity can be recorded at only 5°C. The dispute as to the lowest temperature at which the electrical activity may appear during arousal from hibernation was reconciled by a new experiment of Lyman and Chatfield (47) who showed that in the marmot (Marmota monax) the first spontaneous electrical activity may be recorded at 11°C, i.e., at a temperature at which in the ground squirrel there is clear spontaneous activity, while in the cerebral cortex of the golden hamster there is absolute electrical silence. The existence of significant differences in the sensitivity of the CNS of different species of hibernators to the influence of cold, which explained the controversial results, has been repeatedly confirmed in numerous subsequent studies.

Again, the first data on subcortical electrical activity of the brain during arousal from hibernation were reported by Chatfield and Lyman (15). Comparing systematically the electrical activity picked up separately from a variety of subcortical and cortical structures in golden hamsters at different temperatures, the authors found that the electrical activity was largely confined to components of the limbic system and their presumed efferent pathways. Olfactory bulbs, corpus striatum, thalamus and anterior hypothalamus were conspicuously quiescent. On the basis of these findings and the obviously agitated state of the hamster as full arousal from hibernation was attained, they concluded that the process of arousal is initiated by afferent impulses from the periphery which activate the limbic system either directly or indirectly.

Further studies by Raths (68) on the European hamster (<u>Cricetus cricetus</u> <u>L</u>.), Mokhin et al. (62) on a Russian type of squirrel (suslik), Andersen et al. (1) on the bat (<u>Sicista betulina Pallas</u>) Shtark (71-74) and Putkonen et al. (67) on the European hedgehog (<u>Erinaceus europeus</u> <u>L</u>.) and our studies on the European ground squirrels (<u>Citellus citellus</u> <u>L</u>.) (56), although performed under different conditions and in different species, largely confirmed the general observations and concepts of Chatfield and co-workers as regards the development of electrical activity in a given structure of the brain, the order of activation of different parts of the brain and, particularly, the significance and function of the limbic system in the process of arousal from hibernation.

Investigations of Strumwasser (81-84), Shtark (71-74) and South et al. (79) were carried out under more appropriate ecophysiological conditions and are nearly the only studies in which cortical and subcortical electrical activity and brain temperatures were recorded simultaneously throughout all the stages of the hibernating cycle. In contrast to the results of others, Strumwasser, using <u>C</u>. <u>beecheyi</u>, was the first, and remains the only, investigator to report persistent electrical activity in the sensory-motor cortex throughout the entire period of deep (6°C) hibernation. Thus a long lasting controversy was initiated regarding the existence of continuous electrocortical activity during maintained hibernation. The issue threatened to deviate the efforts of investigators to sterile ground. Its only consequence was to delay the search for the fundamental central mechanisms governing the process of hibernation. Fortunately the controversy was terminated by Kayser himself

(34,36) who pointed out that neither continuous electrical activity nor silence lasting for the duration of a period of hibernation could be seen in the cortex of the European ground squirrel.

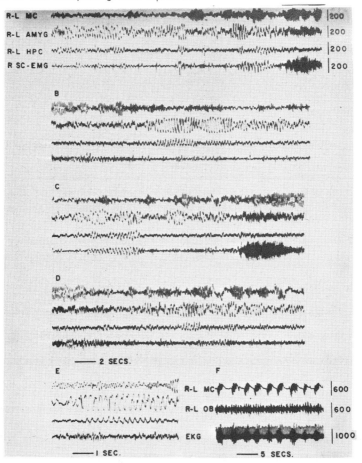

FIG. 1. Activity in the brain correlated with shivering during arousal from hibernation. A-D - continuous strips; E is a faster record; F is from different squirrel showing the difference in the activity of the motor cortex during arousal shivering. AMYG - amygdaloid nucleus; HPC - hippocampus; L and R - left and right side of the brain respectively; MC - motor cortex; OB - olfactory bulb; SC - sensory cortex; calibration in µV (82).

The entrance into hibernation of C. beecheyi was described by Strumwasser (82) as being characterized by a 10 hz motor cortex discharge which preceded and continued during shivering and by an amygdaloid-hippocampal 5 hz rhythm following shivering as the motor cortex discharge was starting to disintegrate (see Fig. 1, A - D). Thus, the hippocampo-amygdaloid complex appeared to be involved with the motor cortex in a mutually inhibitory mechanism correlated with the control of shivering, which in turn was supposed to act as a brake, slowing down the rate of fall of temperature as the animal entered hibernation. These correlates could not be seen during the arousal shivering indicating that a different controlling system was active here (see Fig. 1F). The maintained motor cortex discharge and electrical activity in the septum and hypothalamic areas during deep hibernation as opposed to the relative silence of the lateral geniculate and mesencephalic reticular formation (MRF), was considered to be related to basic homeostatic mechanisms at low temperatures. Periodic bursts of activity seen in generally inactive areas were thought to be needed to keep the particular area functioning. Touching the animal induced the appearance of growing electrical activity in the MRF which was associated with increased and continuous muscle activity. The persistence of electrical activity in the sensory-motor cortex and the ability to desynchronize at temperatures as low as 6.1°C is consistent with complex coordinated behavior required for auditory discrimination of focusing of attention which was described. It is unfortunate, however, that the transmission of signals along the specific and nonspecific pathways to the cortex was not investigated in these intriguing experiments.

In a more recent study (85,86) it was demonstrated that the rectified and integrated electrical activity recorded from the septum pellucidum of C. beecheyi during a 50-day period exhibited a circadian oscillation. In

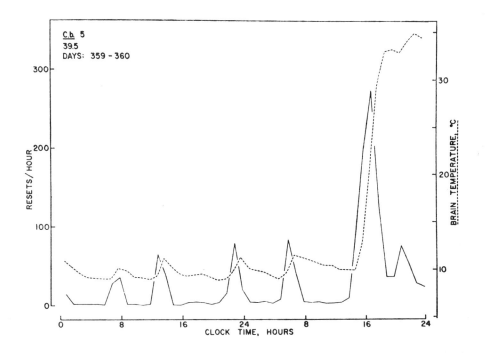

FIG. 2. Simultaneous recording of brain temperature and integral of amygdala EEG as a function of clock time. An arousal is evident on the second day, before hour 16. (86).

deep hibernation similar oscillations of electrical output from amygdala within a 4-9 hour period preceded the small oscillations in the temperature of the brain by about 20 minutes, and appeared to anticipate arousal from hibernation when their amplitude became critically large (Fig. 2).

Consistent in most major respects with the results of Strumwasser concerning electrical activity during hibernation are the findings of South et al. (79). In an attempt to test experimentally the notion implied by numerous authors (18,29,48,88,90) that hibernation is entered from sleep or may represent an extension of it, they made computer analyses (autopower spectra) of cortical and subcortical electrical activities of the yellow bellied marmot (Marmota flaviventris) during normothermic arousal, sleep, and throughout all stages of both natural hibernation and artificial hypothermia. During the first stage of entry into hibernation, as the brain temperature dropped to about 25°C the electrical activity was not changed remarkably. As in normothermic sleep, the distribution of time in which the animal appeared to be in slow sleep remained close to 80%; suggesting that at least the initial stage of entry into hibernation may well be a physiological extension of sleep. As the brain temperature approached 20°C, oscillator-like potentials arose dramatically from the preoptic area and disappeared with equal rapidity at about 15°C. The background activity of the MRF diminished slowly but on occasion it increased secondary to bursts of fusiform activity from the cortex. This second stage of entry into hibernation was further characterized by a progressive reduction in power, primarily in the higher frequencies. During deep hibernation, at brain temperatures of 7-8°C, the most prominent features were occasional spindles originating from the frontal cortex and spikes or trains of spikes from the occipital cortex which were associated with increased background activity in the hypothalamus and MRF. Termination of hibernation and the beginning of arousal was characterized by asynchronous cortical activity accompanied by increased activity in the hypothalamus and MRF which were observed before the first detectable rise in brain temperature. Cortical activation proceeded with the involvement of subcortical activity especially with respect to an increase in power at higher frequencies. As the brain approached 20°C, the oscillator-like potentials in the preoptic area reappeared at a frequency of 7-10 hz which also was pronounced in both the cortex and MRF and disappeared before 27°C was attained. At temperatures

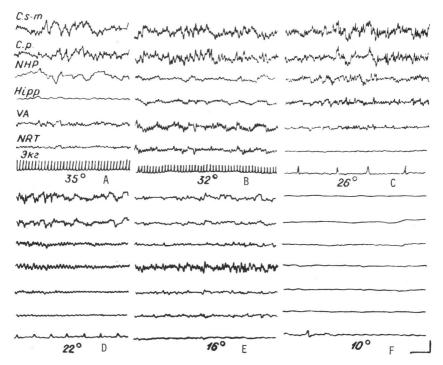

FIG. 3. Electrical activity of various regions of the brain during the initial (a,b), and terminal stage of entry into hibernation (c,d,e) and in deep hibernation (f). C.s-m - sensory motor cortex; C.p. - posterior commissure; NHP - posterior hypothalamus; VA-n. - ventralis anterior of the thalamus; NRT-n. - reticularis tegmenti (75).

above 35°C it was difficult to differentiate arousal from entry, except for the fact that during arousal there was a stronger tendency for low frequency synchrony between 2 and 5 cps in the cortex, hypothalamus and MRF.

The extensive investigations by Shtark (71-74), which are summarized in his recent monograph (75), showed that the brain of the European hedgehog also functioned as a well organized system during hibernation. The entry into hibernation was characterized by a gradual decrease in the amplitude of spontaneous electrical activity which was most prominent in the sensory motor cortex (Fig. 3). As the brain temperature fell to 23°C low voltage alpha-like activity appeared in all investigated brain structures. Between 22° and 19°C there was a reduction in the spindle frequency involving the cortex, MRF and some other subcortical formations. Below 16°C the successive extinction of all electrical activity followed an orderly sequence: first in the neocortex, then the MRF, medial thalamus, lateral thalamus, olfactory bulb and finally the hippocampus and related limbic structures. In deep hibernation, maintained at 8-10°C, there was complete electrical silence in all investigated parts of the brain. Arousal from hibernation proceeded with a faster reappearance of electrical activity following an order the reverse of that seen during entry into hibernation. It should be pointed out that during so called "test arousals" the electrical activity invariably appeared first in the hippocampus and disappeared last in this structure during subsequent test drops or final entry into hibernation. Consistent with this observation is the most recent finding of the same author that unit activity from hippocampal neurons could be picked up at temperatures as low as 6°-8°C (75).

Evoked electrical activity. It is perhaps superfluous to remark that spontaneous electrical activity represents an overall expression of autorhythmicity of neuronal aggregates which is maintained largely by dynamogenic influences of the afferent inputs from peripheral receptors along with impulses which are received by a region of the CNS from other brain structures or systems (57). From the functional point of view, however, spontaneous electrical activity is an index which cannot give but a general impression on the state of various cerebral structures and their functional organization during the process under consideration. A study of transmission of signals produced by appropriate physiological stimulation of the receptors and, to a lesser degree, of responses of the CNS to electrical stimuli offers far greater possibilities for dynamic analysis and precise evaluation of the functional state of specific pathways between the receptor and the area of projection as well as the zones of convergence which can control or modify afferent signals (28).

Except for the gross behavioral and EEG responses to sound and touch observed by Strumwasser (82) in C. beecheyi there have been no systematic studies of evoked potentials in hibernation, most likely because external stimuli may interrupt entry into the state of dormancy or initiate arousal.

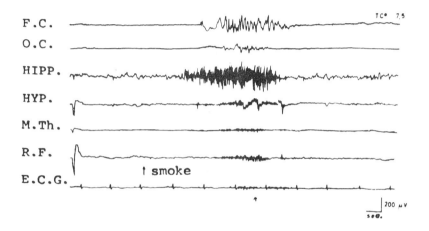

FIG. 4. Activity evoked by a puff of smoke directed at the European ground squirrel arousing from hibernation. F.C. - frontal cortex; O.C. - occipital cortex; HIPP. - hippocampus; M.Th. - midline thalamus; R.F. - mesencephalic reticular formation. Explanation in the text (56).

In our laboratory, potentials evoked by adequate physiological stimuli and the transmission of signals along specific and nonspecific pathways were investigated in European ground squirrels during successive stages of arousal from hibernation (56). A puff of smoke directed at an animal whose temperature was 7.5°C invariably induced fusiform bursts of high voltage fast activity in the hippocampus. This preceded by a few seconds the appearance of a burst in the frontal cortex which in turn was occasionally followed by visible muscular movement. The response evoked by an auditory stimulus was recorded in and near the cochlear nucleus at 6.1°C. The response was in the form of a sharp spike which retained its shape and amplitude throughout the arousal. At the level of MRF it could not be picked up until the temperature had risen to about 17°C. In the area of parietal cortex it was first seen at a temperature of 8-9°C as a barely visible positive deflection and, somewhat later, as a rapid biphasic spike which signalled the arrival of the afferent volley and depolarization of the soma and basilary dendrites of the cortical pyramidal cells (Fig. 5).

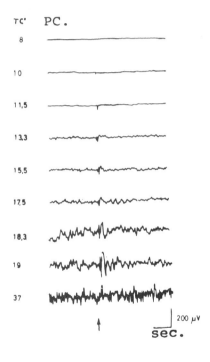

FIG. 5. Evolution of the response evoked by an auditory stimulus in the parietal cortex of the brain of European ground squirrel arousing from hibernation. PC-parietal cortex. For explanation see the text. (56).

The critical frequency of stimulation at which it could be elicited was very low at first and increased with the rise in temperature. Further evolution of the response was characterized by: (a) a progressive decrease in latency and an increase in amplitude; (b) variations in the form - beginning at about 12°C a secondary slow wave of gradually increasing amplitude becomes apparent; evidencing the depolarization of apical dendrites of efferent cortical cells and involvement of the diffuse thalamocortical system; (c) the occurrence above 17°C of a slow consecutive afterdischarge as an expression of the reverberation of impulses along thalamocortical synapses (12) and a marked tendency for the diffusion of signals to the frontal and occipital areas of the cerebral cortex; (d) the disappearance of sensory after-discharges on further rewarming along with a progressive attenuation of the evoked responses (especially of the secondary component) and restriction of its appearance to the area of parietal cortex, signalling the arousal of the ascending activating system. It finally disappears in the intricate and fully desynchronized electrical activity of the awake animal.

Potentials evoked by visual stimuli were first recorded at considerably higher temperature (above 18°C) (56). Their evolution was hardly distinguishable from that of the response elicited by an auditory stimulus, suggesting the universality of the changes in response, regardless of the modality of sensory stimulus.

Modifications of evoked potentials described above are in full accordance with the results of studies in which the responses of the sensory-motor cortex to electrical stimulation of the contralateral sciatic nerve were investigated in anaesthetized hibernators subjected to artificial cooling and subsequent rewarming (14,76,77). The sequence of evoked-potential changes seen by Shtark and Daniliuk in European hedgehog (76) during induction of hypothermia appears essentially as the inverse image of what had been seen during arousal of the ground squirrel. Of considerable interest was the demonstration (77) that under the same experimental conditions (i.e., cooling a hedgehog under chloralose anaesthesia) local postsynaptic potentials of apical dendrites of cortical pyramidal cells (evoked by direct electrical stimulation of the cortical surface) remained well preserved down to 11°C - a temperature at which the soma of pyramidal neurons had already been found to be electrically inexcitable (Fig. 6).

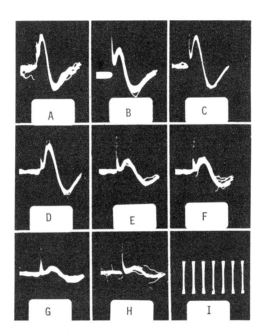

FIG. 6. Changes of dendritic potentials in the cerebral cortex of the European hedgehog during induced hypothermia. Records at cortical temperatures of A-32°; B-29°; C-27°; D-23°; E-20°; F-18°; G-14°; and H-12°C. Calibration 250 μV. Time marker 20 msec. (77).

The studies of evoked potentials proved nonspecific-polysynaptic pathways to be more sensitive to low temperature than were the specific-oligosynaptic ones. This would be the inferred conclusion from the changes seen in the electrocortical activity, i.e., the appearance of desynchronization in the first instance and of spindle bursts in the second (14). Changes of electrical activity evoked by local application of strychnine (14,56,75) also corroborate these conclusions.

The value of these inferences was further confirmed in a more recent study of the interaction of the diffuse thalamocortical system and the brain stem reticular activating system during arousal from hibernation (57). The experiments were based on the observations of Moruzzi and Magoun in 1949 (63) that the recruiting response evoked by electrical stimulation of intralaminar thalamic nuclei could be reduced or completely blocked by intercurrent electrical stimulation of the MRF. Considering this phenomenon as an index which would enable more precise evaluation of the functional states of the diffuse thalamocortical system and reticular activating system respectively, we subjected it to systematic analysis during arousal from hibernation. It was found that single electric shocks applied to the intralaminar thalamic nuclei at temperatures below 8°C failed to induce any response in the frontal cortex. At a temperature of 8.5°C, however, only the first in a series of 0.5/sec. stimuli was effective in eliciting the secondary component of the evoked response. With the rise in temperature the critical frequency increased so that at 11°C each stimulus in a sequence of 1/sec. was found to be capable of evoking the cortical response. This phenomenon, first described by Forbes et al. (23) in the cat under deep barbiturate anaesthesia and observed subsequently by Chatfield et al. (14) in the anaesthetized and artificially cooled golden hamster as well, was identified as the phenomenon of Wedensky (91). Accordingly, it was interpreted as being due to the refractoriness of neuronal components responsible for the secondary component of the evoked response. Above 15°C the longer trains of repetitive stimuli produced a series of waxing and

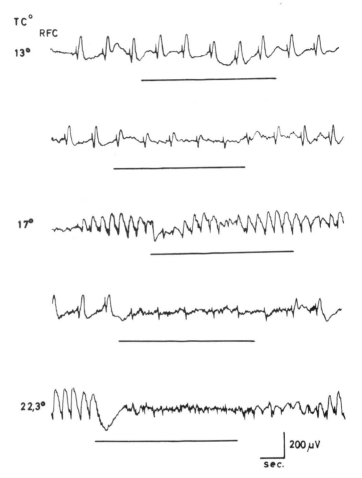

FIG. 7. Effects of intercurrent electrical stimulation of MRF (full line below the records) on recruited responses evoked by single shock stimuli applied to intralaminar thalamic nuclei in the European ground squirrel arousing from hibernation. RFC-right frontal cortex. Further explanation in the text. (57).

waning sequences characteristic of the activity of the diffuse thalamo-cortical system. At 17°C the intercurrent stimulation of the MRF, hitherto ineffective, induced a partial, and and above 20°C a complete abolition of the secondary wave of the response to repetitive stimulation of intralaminar thalamic nuclei, rendering the primary component unaffected. At still higher temperatures the effect persisted even after the reticular stimulation had been discontinued (Fig. 7).

It is evident that in spite of obvious differences in the temperature ranges at which the characteristic electrical events were recorded, the studies reviewed have demonstrated the existence of many common and important features in the electrical activity and the functional organization of the brain during hibernation. Although they revealed no specific electroencephalographic sign by which the entry into hibernation could be predicted they all tend to emphasize, to different extents, the important role of the hippocampus and the related structures of the limbic system with respect to the processes of entry into hibernation and especially in the initiation of arousal. The studies indicated that at any given stage of hibernation there is a dynamic highly coordinated and patterned interaction between the various brain levels and the main specific and non-specific functional systems. The rigorous control of the progressive inhibition and facilitation of entry and arousal lends credence to the notion that a given interaction pattern may be critically significant with respect to the regulation of the homeostatic mechanisms at any particular stage of hibernation.

The experimental evidence seems to permit a visualization of the sequence of electrical events during the different stages of hibernation. The first stage of entry into hibernation is characterized by a gradual reduction of the afferent input via the multisynaptic network of the non-specific reticular activating system. This results in a release of the influence of the diffuse thalamocortical system which now, through recruitment at synaptic junctions, facilitates synchronization of the postsynaptic electrogenesis which is maintained by the autorhythmicity of cortical neuronal aggregates and the arrival of afferent impulses via the cold-resistant specific oligosynaptic pathways. At this stage synchronized electrical activity is generalized throughout the brain. As temperature drops further there is a more profound suppression of impulse transmission through polysnaptic pathways until there is a functional exclusion of the diffuse thalamocortical system. This is associated with and followed by a progressive diminution of spontaneous electrical activity in the neocortex, MRF and a majority of subcortical structures. Final entry into hibernation is characterized by: (a) reduction of the primary component of responses evoked by external stimuli - their disappearance from the neocortex suggesting its deafferentation; (b) a decreased frequency of bursts in the frontal cortex; and (c) restriction of the electrical activity of the hippocampus and related parts of the limbic system to a minimum. Throughout this and later stages of deep hibernation these limbic structures apparently assume a critical importance in the maintenance of the basic homeostatic mechanisms at a minimal energy expenditure level. Long term records during deep hibernation have revealed intermittent bursts of a typical pattern originating most prominently from the frontal cortex but present in other structures as well. Although it has not been possible to demonstrate any particular electrical phenomenon as being the signal of arousal from hibernation, there does seem to be an agreement that the limbic system along with its hippocampal formation might be involved specifically.

Several experimentally demonstrated properties of the hippocampus suggest it could possibly function as a neural trigger for the thermogenesis of arousal as well as act as a controller of the other activating mechanisms involved during the various stages of arousal. Some of these properties are: (a) the high degree of resistance of hippocampal neurons to cold (75); (b) the presence in the closely related amygdala of circadian oscillations in activity which, reportedly, anticipate arousal when their amplitude becomes critically large (85); (c) prompt reactivity to adequate external stimuli (56); (d) appearance first in the hippocampus of continuous and well modulated activity at the earliest stages of both partial and definite arousals (56,72,75); (e) extremely rapid increase in its excitability as arousal is underway (56); and (f) the preferential orientation of its discharges in the descending direction - toward the hypothalamus and MRF (31,43,56). Once initiated through either endogenous or exogenous activation of the limbic system the process of arousal proceeds apparently following essentially the sequence of events, in reverse, of that occurring during entry into hibernation.

SPONTANEOUS AND EVOKED CORTICAL AND SUBCORTICAL ELECTRICAL ACTIVITY IN ARTIFICIALLY INDUCED HYPOTHERMIA

In this section the studies will be reviewed in which the effects of progressive hypothermia on spontaneous and evoked electrical activity were investigated in non-hibernating homeothermic animals.

<u>Spontaneous electrical activity</u>. Reports on functional alterations of the nervous system associated with artificially induced hypothermia have been traced back to the middle of the last century (45). It was not until 1949, however, that the electrical activity of the brain was investigated under the conditions of reduced body temperature by Ten Cate et al. (89). In curarized rats they found the electroencephalogram unaltered at temperatures between 39°-32° C. When the temperature declined below about 30°C the amplitude of the EEG gradually decreased. The electrical activity slowed progressively and virtually disappeared at body temperatures of 18-20° C. Numerous subsequent studies (10,35,42) on rats refrigerated under similar conditions gave essentially comparable results.

Recording spontaneous EEG activity from isolated cat heads artificially perfused with cooled blood, Gaenshirt et al. (25) noted that within the range of 38°-20° C the voltage increased and the frequency decreased, both exponentially, as the temperature declined (Fig. 8). Below 32° C both

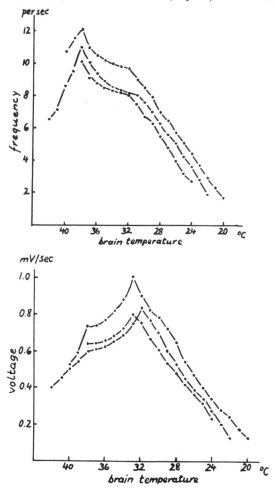

FIG. 8. The relationship between mean frequencies and mean voltages of the EEG and the brain temperature at progressive stages of induction of hypothermia. (25).

frequencies and voltages decreased following a nearly linear regression. In both parameters of the EEG the critical break occurred at about 33°C and was attributed to metabolic changes (enzymatic blockade) in the brain. The changes produced by cooling were reversible, suggesting the maintenance of the structural metabolic units even when the cortex was electrically inert. Suda et al. (87) also observed that in cats subjected to progressive uniform cooling of the brain there was diminution of spontaneous electrical activity which was preceded by a stage of increased amplitude without a change in frequency. This coincided with a hyperreactivity of the brain to incoming afferent volleys. The general depressant action of cold on the amplitude and frequency of brain waves was confirmed in unanaesthetized curarized dogs (11,16,4) as well as in canine decapitate preparations (4). In monkeys cooled by ice water immersion to a rectal temperature of 20°C, Scott et al. (70) observed that as the temperature fell there was a progressive diminution of amplitude with no change in frequency. At 30°C low voltage alpha waves were present but the record was dominated by delta waves arising in frontal areas. At 20° C only a slight wavering at 2 - 3 cps could be seen. During rewarming the rhythms reappeared at temperatures 1 - 2 degrees above that at which they were found to disappear. Frontal cortical activity recovered last and subsequently the EEG was normal (9). The electrical activity recorded in man through the nasopharyngeal leads was also found to cease around 20°C and reappear at about the same temperature on rewarming (19).

Of special interest is the observation of Ferrari and Amantea (22) that in curarized dogs subjected to progressive hypothermia, spontaneous epileptiform seizures occurred at temperatures between 34 - 32°C. At first of generalized grand mal type, somewhat later the convulsive activity assumed the pattern of "recruited" response with a typical waxing and waning which was suggestive of a "centroencephalic" genesis of the seizures, and of a maintained activity by the diffuse thalamocortical system. The lowest temperature at which the epileptiform seizures were observed was 22°C. The existence of the phenomenon has been confirmed in curarized rats (26), cats (30,44) and dogs (46). It might be argued that the hyperirritability could have been caused by the depolarizing action of curare (12,21). The facts that the phenomenon was also seen in decapitate dogs (4) and in rabbits cooled by ice water immersion (64) would militate against such an objection.

Surprisingly little information is available on simultaneously recorded spontaneous electrical activity of cortical and subcortical structures during induced hypothermia. Except for the two papers by Massopust and co-workers (50,51) no such studies have come to our attention. These authors compared the effects of cooling on the cortex and anterior MRF in barbiturized cats and guinea pigs respectively. While cortical activity fell, gradually reaching the isoelectric point at 25°C, the MRF retained good amplitude to about 23° C buccal temperature. In another paper the effects of hypothermia on spontaneous electrical activity of cortical (frontal, parietal and occipital) and subcortical (anterior MRF, hypothalamus and dorsomedial thalamic nucleus) structures were compared under two barbiturates (pentobarbital Na, Thyamylal Na) and one relaxant (gallamine triethiodide). This study reconfirmed a well known fact that spontaneous activity of the central nervous system during induced hypothermia is differentially modified by anaesthetics and pharamacodynamic agents (4,8,34,49). The most striking observation was the well preserved activity in the anterior MRF at intracerebral temperatures of 26° C in animals given gallamine. Such was not seen when animals were given barbiturates. On rewarming, changes in electrical activity mirrored in reverse those which occurred during cooling. In the cortex potentials reappeared in the sequence of frontal, parietal and, finally, occipital areas. Thus, except for the rather dubious conclusion that the MRF is most resistant to cooling, these studies contributed little to our understanding of functional organization and cortico-subcortical relationships in hypothermia. This however was amply done in studies of evoked electrical activity.

Evoked electrical activity. The work of Koella and Ballin (38) deserves attention similar to that paid to the hibernation study of Chatfield et al. (14) because it was the first to provide information about the functional state of specific and non-specific systems of the brain during artificial hypothermia. In cats anaesthetized with Dial, they recorded evoked electrical activity of the cortex at body temperature levels between 38° C and 25° C. It was shown that generalized cortical "arousal", as produced by dipping of the tail into hot water, was reduced and eventually abolished as the temperature was gradually lowered and reappeared if control conditions were reestablished. Acoustically evoked potentials in the auditory area persisted throughout the temperature range. These evoked potentials were pronounced and at low temperatures could be recorded from larger cortical areas than at the control temperature. Thus if temperature was gradually lowered, nociceptive and acoustically evoked potentials revealed a pattern similar to that observed by French et al. (24) with respect to increasing the depth of barbiturate anaesthesia. It was concluded that the multisynaptic "activating" system which is involved in the production of the generalized cortical response to nociception is very sensitive to low temperature and anaesthesia whereas these factors hardly influence the oligosynaptic pathways involved in the transmission of the acoustic potentials. The differential reactivity of oligosynaptic and multisynaptic pathways to temperature have been investigated further by studying cortical

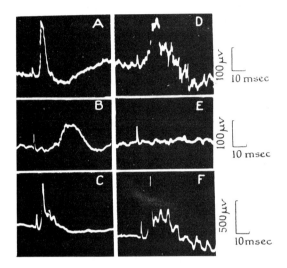

FIG. 9. A, B, and C - Control evoked potentials at 37°C: A - postero-ventral lateral nucleus of the thalamus; B - periaqueductal midbrain reticular formation; C - cervical dorsal column. D, E and F show the responses at 37°C. Note the disappearance of the evoked potential in the midbrain reticular formation (80).

and subcortical evoked potentials in unanesthetized dogs (4) and cats (4,30). Studying the transmission of auditory signals, Benoit (4) observed a considerable facilitation of cortical evoked potentials at 33-30° C. There was a marked accentuation of the secondary component and along with it, a diffusion of the signal far beyond the primary auditory area. At about 30° C the evoked potential was found to be followed by a slow after-discharge which was most conspicuous within the primary auditory area but visible outside it as well. Below 23°C the progressive depression of cortical evoked response ensued. No parallel modifications of potentials picked up at the cochlear nucleus were seen. Extinction of the auditory signal was found to occur in the MRF at much higher temperatures than in the specific thalamic relay nucleus. Differential sensitivity to cold of the two structures or systems was fully confirmed in a subsequent study by Stevenson et al. (80) (Fig. 9). Decreases in amplitudes and increases in latencies of the evoked responses of the retina, central visual pathway and the optic cortex were observed in cats subjected to progressive cooling and were thought to be due to the direct effect of cold on the reactivity and "conduction velocity of the neurochemical systems." The phenomena were reversible within the range of temperatures employed (52,53).

The long known phase of hyperresponsiveness of the central nervous system (3,27) and hyperreflexia (20) occurring in hypothermia was the subject of an extensive analysis by Brooks and his associates (7,40,87). These studies have been thoroughly reviewed elsewhere (39) and will not be detailed here. Suffice it to say that a transient stage of hyperresponsiveness occurring between the temperatures of 35-25° C was demonstrated at

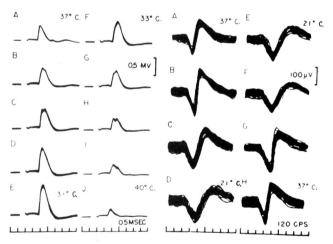

FIG. 10. The effects of cooling on the monosynaptic reflex elicited by gastrocnemius nerve stimulation (left) and on cerebral cortical potentials evoked by stimulation of contralateral superficial radial nerve (right). Note the simultaneous augmentation in amplitude and duration of the spinal reflex and cortical evoked potentials that occur as a result of cooling (39).

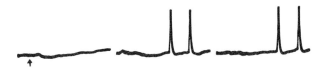

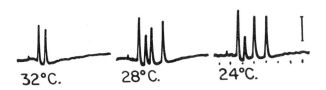

FIG. 11. Individual interneuron activity recorded by stimulation of gastrocnemius (top) and tibialis posterior (bottom). Time = 5 msec. Amplification 2 mV. Note the change of monosynaptic to polysynaptic reflex as hypothermia develops. (39).

practically all levels of the central nervous system. It was thought not to be due to an increased excitability but to: (a) repetitive discharge probably due to failure of accommodation and prolongation of the excitatory potentials; (b) the greater participation of interneurons and (c) increased recruitment at synaptic junctions because of these. Spread of activation, overflow of excitation into systems not normally invaded and epileptiform activity may be explained by such a mechanism. This also could account for a general failure of confinement of activity to normal pathways, changes in the reaction patterns of the nervous system and the loss of specificity (38,39) during hypothermia.

Analysis of works reviewed above reveals the existence of some dissimilarities with respect to the temperature levels at which modifications occur and at which all spontaneous and/or evoked electrical activity disappear.

It seems however that the differences are related more to the diversity of cooling techniques, anaesthetics and pharmacodynamic agents employed than to animal species. There are, however, many points on which there is a general agreement permitting some generalizations concerning the effects of artificially induced hypothermia on electrical activity of the central nervous system.

    a. In general, cooling exerts a depressant action on both spontaneous and evoked electrical activity of the brain. There is, however, a stage of hyperresponsiveness demonstrated practically at all levels of the neural axis that precedes a stage of general depression and may result in convulsive activity of the brain.

    b. There is a marked differential sensitivity to hypothermia of the various cortical regions and subcortical structures of the brain. Transmission of signals along the non-specific polysynaptic pathways is blocked at considerably higher temperatures than along the specific, oligosynaptic pathways.

    c. There is always a critical temperature below which the amplitude and frequency of brain waves diminish following a nearly linear regression until all electrical activity ceases. The critical level does not appear to be species dependent.

    d. Changes in electrical activity induced by cooling are fully reversible unless hypothermia is maintained too long. On rewarming, the activity of various subcortical and cortical structures reappears following, in order, the reverse of that characterizing its disappearance during progressive hypothermia.

    e. The effects of cooling on spontaneous and evoked electrical activity of cortical and subcortical structures of the brain and its functional systems are closely comparable to those of progressive barbiturate anaesthesia. This seems to justify the term "cold narcosis."

A COMPARATIVE ANALYSIS OF THE TWO STATES

There have been several studies on the relative effects of induced hypothermia on identical cortical and subcortical areas of hibernators and non-hibernators (14,33,35,50,53). Except for the general ability of the nervous system of hibernators to function and survive at considerably lower temperatures (13,32,53,65,78), no fundamental qualitative differences between the two have been uncovered. The most striking finding remains the excellently preserved activity of the anterior MRF of hibernators at low temperatures (50).

It should be emphasized a survey of the experimental material reveals several important differences. The existence of a hyperresponsive stage has been shown to occur in both hibernation and hypothermia. However, the epileptiform seizures, frequently observed during induction of hypothermia, have not been reported to occur during entry into hibernation. The absence of epileptiform activity during the hyperresponsive phase suggests the presence of an active restraining mechanism which would confine neural activity to normal pathways subserving entrance into hibernation.

Of greatest value and relevance for the main task of this paper, i.e. a comparison of the electrical activity in hibernation and hypothermia, are the studies in which the effects of artificially induced hypothermia on brain waves were investigated in hibernators of the same species, the electrical activity of which had already been recorded during one or more stages of natural hibernation. Unfortunately only few such studies are available.

In the initial works of Chatfield et al. (14) and Kayser et al. (37) only electrocortical activity was studied. No differences were found between the golden hamster and the ground squirrel arousing from hibernation and rewarming from induced hypothermia.

South et al. (79), in their previously mentioned study explored the electrical activity and autopower spectra in yellow-bellied marmots during all stages of natural hibernation and artificial hypothermia. Cooling induced by means of the closed vessel technique (2) resulted in a general non-specific depression of the central nervous system relative to what had been seen during the entry into hibernation in the same animals (Fig. 12). The occurance between 20° - 15° C of high voltage 6 hz oscillator potentials originating in the preoptic area during entrance into hibernation occurred during hypothermia but was not as well organized. The subsequent progressive decrement of cortical and MRF activity was more pronounced than during the entry into hibernation. During maintained hibernation the central nervous system was far from quiescent and retained a highly organized character, but during 1.5 - 2 hours of hypothermia (6 - 8° C) very little activity was recognizable except for infrequently occuring bursts. At no time was the organized activity such as that seen during hibernation apparent in any of the investigated brain structures. Rewarming from hypothermia preceded in a fashion not strongly differentiated from arousal from hibernation except for a slight overall depression in power spectra.

Comparing the dynamics of changes in electrical activity of cortical and subcortical structures of the brain of European hamsters during arousal from the state of natural hibernation and subsequent rewarming from induced hypothermia, Raths (68) observed several characteristic differences between the two processes (Fig. 13). Arousal from hibernation initially involved an intense activation of the rostral midbrain (8° C), then the

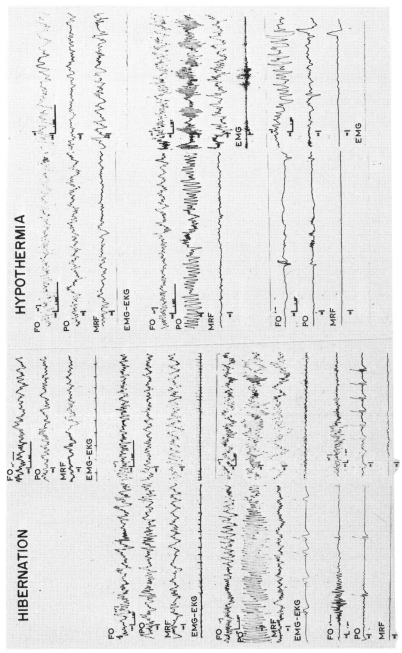

FIG. 12. Spontaneous electrical activity of the brain of the yellow-bellied marmot during natural hibernation and artificially induced hypothermia. FO: fronto-occipital cortex; PO: preoptic area; MRF: mesencephalic reticular formation; EMG-EKG: activity from two needle electrodes in dorsum of neck. EKG artifacts at lower temperatures due to unipolar recording. Left hand columns: entry or induction. Right hand columns: arousal. Levels and directions of change in temperature as indicated. Note changes in calibration (79).

CORTICAL AND SUBCORTICAL ACTIVITY 517

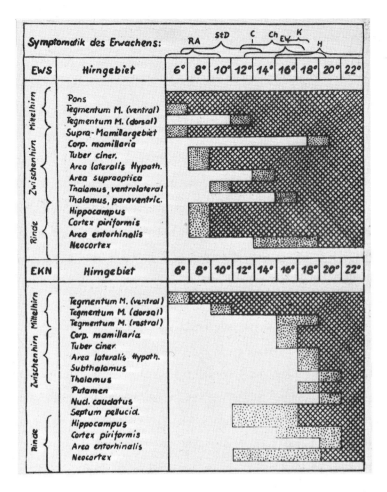

FIG. 13. The diagram summarizing findings on electrical activity of the brain during arousal from natural hibernation (top) and rewarming from artificially induced hypothermia (bottom). Dotted areas - activity in bursts. Crosshatched areas - the duration of the activity at increasing body temperatures (68).

rhinencephalon (9° - 10° C), followed by the tuber cinereum and the lateral hypothalamus (10° - 12° C). The next in the order of activation was the anterior hypothalamus (16° - 17° C) preceding the appearance of electrical activity in the mamillary bodies (21° C). Rewarming from hypothermia, on the other hand, being far more dependent on external temperature, appeared as a clearly passive process up to 18° C. The activity of rhinencephalon and hypothalamus was at first negligible and remained so for a long time. The activity in the anterior hypothalamus did not follow the activity in the tuberal and lateral part as was the case during arousal from hibernation but appeared somewhat before or at the same time as activity in the other hypothalamic areas and the mamillary bodies. In contrast to arousal from hibernation the electrical activity during rewarming from induced hypothermia appeared in all recorded structures within the same range of temperature (17° - 22° C). The changes in neocortical activity were found to be comparable.

We (56) also compared spontaneous and evoked cortical and subcortical potentials of European ground squirrels during arousal from hibernation and during rewarming from hypothermia induced by the closed vessel technique. Three animals, in which the process of arousal from hibernation had already been recorded, were kept awake for some time at room temperature and then artificially cooled to 5° C. During spontaneous rewarming the electrical activity was recorded under identical experimental conditions as in arousal from hibernation until intensive shivering made further recording virtually impossible. A comparative analysis of the two states revealed numerous similarities but there were some differences in the basic characteristics of electrical activity which were seen in all the animals.

The arousal from hibernation proceded through the clearly distinguishable and orderly sequence of electrical events (Fig. 14A). Simultaneously with the rapid development of continuous electrical activity in the hippocampus, there arose autogenous bursts in the posterior hypothalamus and MRF in an integrated patterned interaction. Prior to this, activity in these structures had been quite random.

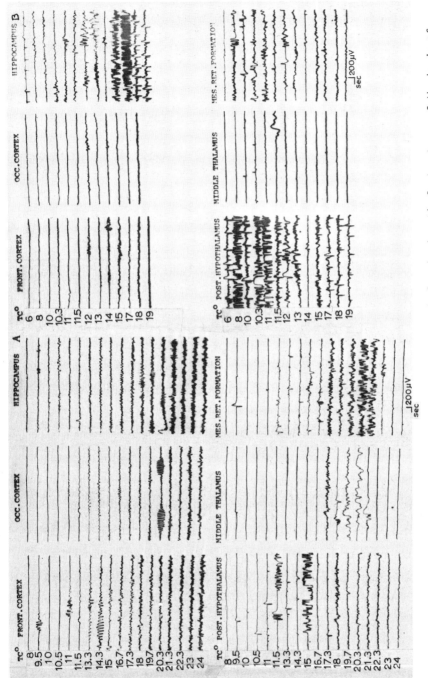

FIG. 14. Spontaneous electrical activity of various cortical and subcortical structures of the brain of a single European ground squirrel arousing from natural hibernation (A) and rewarming from the artificially induced hypothermia (B). Detailed explanation in the text (56).

Initial arousal was also characterized by an increasing frequency of typical bursts in the frontal cortex which were occasionally followed by muscular activation. The appearance soon afterwards of the primary component of the evoked response in the cortex indicated the recovery of transmission along specific oligosynaptic pathways and increasing reafferentation of the cortex. As the temperature rose further there was an overall increase in the electrical activity of the subcortical formations, especially the posterior hypothalamus, and the first appearance in the neocortex of continuous activity. Coincident with the appearance of the secondary component of the evoked response, signaling reestablishment of impulse transmission along polysynaptic pathways, there was a bilaterally synchronous appearance of the recruiting type of spindle bursts indicating the entry of the diffuse thalamocortical system into the process of electrogenesis. With the establishment of thalamocortical interaction, the functional organization of the brain during arousal from hibernation acquires a new and very important dimension. Conditions are thus created which favor a rapid rise in the general level of facilitation along with an increase in the ability of the higher levels of organization to impose integrated control on the lower. Intensified postsynaptic electrogenesis, through enhanced recruitment at synaptic junctions, was evidenced by: (a) a remarkable increase in the amplitude of the secondary component of evoked response; (b) the appearance of a slow after-discharge; (c) the diffusion of signals to the widespread areas of the cerebral cortex and (d) the predominance of synchronized forms of spontaneous electrical activity. Coincident with this stage of apparent hyperresponsiveness and spread of excitability was the rather sudden appearance in the anterior hypothalamus and midline thalamic nuclei of long lasting sequences of high voltage activity. These were occasionally synchronous with the activity in the MRF. The terminal stage of arousal from hibernation was characterized by a gradual quieting down and eventual disappearance of burst activity in subcortical structures. This was replaced by continuous activity on the one hand, and the entry into the process of electrogenesis of the reticular activating system on the other. This was reflected by: (a) initially partial, and later total abolition of recruited responses in the cortex to stimulation of the MRF; (b) restriction of the field from which an evoked response could be picked up to the primary receiving area and (c) a gradual preponderance of low voltage fast components in the electrocorticogram which culminated in the complete desynchronization and full behavioral arousal of the animal.

In contrast to arousal from hibernation, rewarming from deep hypothermia was characterized by: (a) a considerable delay in the appearance of the characteristic initial cortical bursts and an overall delay of continuous spontaneous activity in the investigated structures; (b) a very marked and, in contrast to hibernation, uncorrelated paroxysmal activity in the posterior hypothalamus and brain stem which was retained in the former structure to end of the experiment; (c) an early appearance (beginning at 6° C) of isolated spikes and, subsequently, trains of spikes in the hippocampus which, between 13° and 20° C, developed into a long lasting high voltage seizure activity, which was never observed during arousal from hibernation (Figs. 14B and 15) and (d) the absence from both the anterior hypothalamus and midline thalamic nuclei of the transient stage of high voltage activity regularly seen within the temperature range of 15° - 20° C during arousal from hibernation (Fig. 14B and 16).

Thus the non-specific delay, an overall depression of electrical activity and the lack of succession in reappearance of electrical activity of various brain structures during rewarming from hypothermia as opposed to the arousal from hibernation, seems to be a common finding in all three

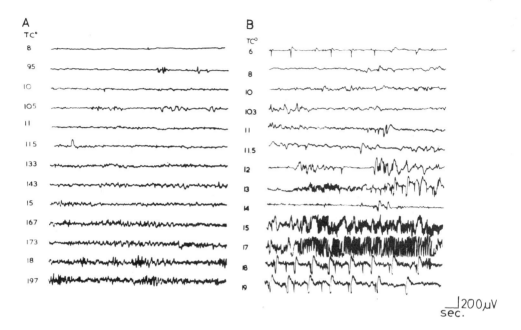

FIG. 15. Spontaneous electrical activity of the hippocampus of a single European ground squirrel arousing from natural hibernation (A) and rewarming from artificially induced hypothermia (B). Note that in contrast to arousal from hibernation, intense seizure activity occurred during rewarming from hypothermia (56).

of the aforementioned studies. It should be pointed out that although not seen by South et al. (79), seizures in the hippocampus during rewarming from hypothermia were clearly present in the record published by Raths (68) but were not mentioned or commented on by the author (Fig. 17). The sudden occurrence of transient activity in the preoptic area and anterior hypothalamus during arousal from hibernation has also been commonly seen. During rewarming from induced hypothermia it either appeared after a considerable delay and was markedly less organized and distinguishable (68,79) or did not appear at all (56). The functional significance of this activity is far from clear. South et al. (79) considered it to have been probably a function

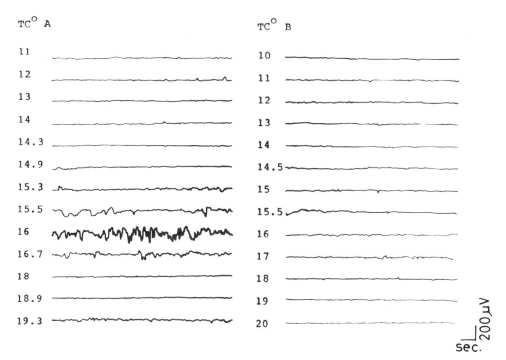

FIG. 16. Spontaneous electrical activity in the anterior hypothalamus of a single European ground squirrel arousing from natural hibernation (A) and rewarming from artificially induced hypothermia (B). Note that in contrast to arousal from hibernation, the transient activity occurring at temperatures between 15 - 18° C was absent during rewarming from hypothermia (56).

of the brain temperature rather than the manifestation of the thermoregulatory controller. It is not inconceivable that the preoptic activity might have been the expression of the activation of some inhibitory mechanism related to the control of thermoregulation or of excitability levels during arousal. Its sudden appearance during the stage of apparent hyperresponsiveness during arousal from hibernation and its disorganization, or total absence, during rewarming from artificial hypothermia within a temperature range at which the hippocampus gets involved in the most intense seizure activity, seems compatible with the above tentative interpretation (cf. Fig. 14B and 16). This also does not seem inconsistent with the most recent

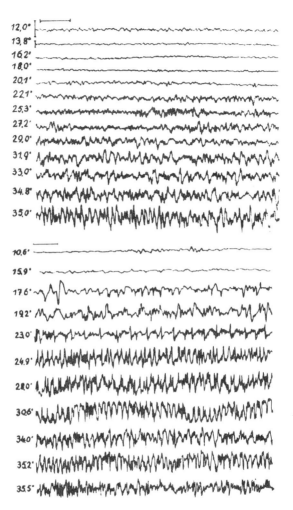

FIG. 17. Bipolar recordings from posterior hippocampus of a single European hamster during arousal from hibernation (top) and rewarming from artificially induced hypothermia (bottom). Note that in contrast to arousal from hibernation, seizure activity occurred during rewarming from hypothermia. (68).

finding that thermal stimulation (heating) of the preoptic area and anterior hypothalamus in Citellus tridecemlineatus resulted in a decrease of metabolism and body temperature (92).

A comparative analysis of the experimental evidence shows the existence of remarkable differences in the phenomenology of electrical activity of cortical and subcortical structures during hibernation and hypothermia. Since they were seen in the same animals of the same species during the two states they cannot be attributed to the action of cold alone. The fact that essentially the same differences were described in the experiments of Raths (68) in which hypothermia was induced by ice-water immersion, as in those of South et al. (79) and of ours (56) would militate against the argument that they could be due to residual effects of hypoxia and hypercapnia caused by the use of the closed vessel technique (56,79). Nevertheless, such a possibility must not be excluded.

Natural hibernation is undoubtedly an active process regulated and characterized by highly dynamic and coordinated interactions of various brain levels along with the main functional specific and non-specific systems of the brain. These are rigorously timed with respect to successive inhibition and facilitation during both entry and arousal. At any given stage of hibernation the activity of a particular organizational level of the brain or of the patterns of interaction may be critically significant for the maintenance and control of the homeostatic mechanisms. Such a functional organization of the brain accounts for the retention of the high specificity of the reaction patterns of the nervous system during hibernation.

Artificially induced hypothermia, on the other hand, largely appears as a passive state characterized by a profound overall depression of the central nervous system and a poorly coordinated succession of inhibition and facilitation of the various levels and functional systems of the brain during induction and rewarming. This could account for the general failure of confinement of activity to normal pathways and a partial loss of the specificity of the reaction patterns of the brain.

There is no evidence that hypothermia reduces the resistance to cold of a given neural structure. That is why we are inclined to believe that reasons for the differences in electrical activity of the brain during hibernation and hypothermia should be sought in variabilities of the dynamic properties of the functional organizations of the brain during the two states.

In full accord with this view is the conclusion of Shtark (75) that the "programmed" successive activation and extinction of function in the four main levels of the brain (brain stem, thalamus, archipalium and neocortex) represent the neurophysiological basis upon which the homeostatic optimums at the various stages of natural hibernation depend.

CONCLUSION

Despite obvious similarities, the experimental evidence demonstrates the existence of essential differences in the phenomenology of cortical and subcortical electrical activity of the brain in hibernation and artificially induced hypothermia. These indicate that the two states are not equivalent. The dichotomy seems to arise from differences between them in the dynamic properties of the functional organization of the various levels of the brain and its main specific and non-specific systems. Further investigation of these dynamic properties, especially during the stage of hyperresponsiveness of the brain, might throw more light on the basic nature of the differences. Rather than invoking qualitatively different mechanisms by way of explaining differences between hibernation and hypothermia, it would appear more germane to focus on differential adaptation of already available mechanisms within the central nervous system.

REFERENCES

1. ANDERSEN, P., K. JOHANSEN, and J. KROG. Electroencephalogram during arousal from hibernation in the birchmouse. Amer. J. Physiol. 199: 535-538, 1960.

2. ANDJUS, R. K. L'application de l'anesthesie hypoxique en hypophysectomie. Arch. Biol. Sci., Belgrade 2: 19-31, 1959.

3. BARRON, D. H., and B. H. C. MATTHEWS. Dorsal root potentials. J. Physiol. 94: 27P-29P, 1938.

4. BENOIT, O. Contribution à l'Étude des Effets de l'Hypothermie sur quelques Structures Cérébrales. Lyon: Imprimerie Emmanuel Vitto, 1956.

5. BREMER, F. Considerations sur l'origin et la nature «ondes» cérébrales. Electroencephalog. Clin. Neurophysiol. 1: 177-193, 1949.

6. BREMER, F. Some Problems in Neurophysiology. London: Athlone Press, 1953.

7. BROOKS, C. M., K. KOIZUMI, and J. L. MALCOLM. Effects of changes in temperature on reactions of the spinal cord. J. Neurophysiol. 18: 205-216, 1955.

8. CAHN, J., and R. PIERRE. Contrôle electroencéphalographique de l'hibernation artificielle. Anesth. Analg. 11: 568-582, 1954.

9. CALLAGHAN, J. C., D. A. McOWENS, J. W. SCOTT, and W. G. BIGELOW. Cerebral effects of experimental hypothermia. Arch. Surg. 68: 208-215, 1954.

10. CAZZULO, C. L., and A. GUARESCHI. EEG patterns (conduction and integration limits) during deep artificial hypothermia in rats. Electroencephalog. Clin. Neurophysiol. 6: 532, 1954.

11. CAZZULO, C. L., A. GUARESCHI, and A. BEDUSCHI. Electrical activity of the brain in dogs during artificial hypothermia. Electroencephalog. Clin. Neurophysiol. 6: 532, 1954.

12. CHANG, H.T. Similarity in action between curare and strychnine on cortical neurons. J. Neurophysiol. 16: 221-233, 1953.

13. CHATFIELD, P. O., A. F. BATTISTA, C. P. LYMAN, and J. P. GARCIA. Effects of cooling on nerve conduction in a hibernator (golden hamster) and non-hibernator (albino rat). Amer. J. Physiol. 155: 179-185, 1948.

14. CHATFIELD, P. O., C. P. LYMAN, and D. P. PURPURA. The effects of temperature on the spontaneous and induced electrical activity in the cerebral cortex of the golden hamster. Electroencephalog. Clin. Neurophysiol. 3: 225-230, 1951.

15. CHATFIELD, P. O., and C. P. LYMAN. Subcortical electrical activity in the golden hamster during arousal from hibernation. Electroencephalog. Clin. Neurophysiol. 6: 403-408, 1954.

16. CIER, J. F., O. BENOIT, and M. TANCHE. De quelques facteurs influençant l'activité electrique du cortex cerebrale dans l'hypothermie. J. Physiologique 48: 458-461, 1956.

17. ČUPIĆ, D., L. KRAŽLIĆ, B. BELESLIN, AND LJ. MIHAILOVIĆ. Seasonal variations in the content of free amino acids in the brain of hibernator (Citellus citellus). Acta Med. Iugoslav. 19: 19-24, 1966.

18. DEMENT, W. C. Sleep. In: Mammalian Hibernation III, edited by K.C. Fisher, A.R. Dawe, C.P. Lyman, E. Schoenbaum and F.E. South. Edinburgh: Oliver and Boyd, 1967, p. 175-199.

19. DREW, C. E. Profound hypothermia in cardiac surgery. Brit. Med. Bull. 17: 37-42, 1961.

20. FAY, T., and L. W. SMITH. Observations on reflex responses during prolonged periods of human refrigeration. Arch. Neurol. Psychiat., 45: 215-222, 1941.

21. FELDBERG, W., and K. FLEISCHHAUSER. The site of origin of the seizure discharge produced by tubocurarine acting from the cerebral ventricles. J. Physiol., 160: 258-283, 1962.

22. FERRARI, E., and L. AMANTEA. Convulsive electrocortical discharges in hypothermic dog. Electroencephalog. Clin. Neurophysiol. 7: 441, 1955.

23. FORBES, A., A. F. BATTISTA, P. O. CHATFIELD, and J. P. GARCIA. Refractory phase in cerebral mechanisms. Electroencephalog. Clin. Neurophysiol. 1: 141-175, 1949.

24. FRENCH, J. D., M. VERZEANO, and H. W. MAGOUN. A neural basis of the anesthetic state. Arch. Neurol. Psychiat., 69: 519-529, 1953.

25. GAENSHIRT, H., W. KRENKEL, and W. ZYLKA. The electrocorticogram of the cat's brain at temperatures between 40°C and 20°C. Electroencephalog. Clin. Neurophysiol. 6: 409-413, 1954.

26. GIRETTI, M. L., F. S. RUCCI, and M. LA ROCCA. Effects of lowered body temperature on hyperoxic seizures. Electroencephalog. Clin. Neurophysiol. 27: 581-586, 1969.

27. GRUNDFEST, H. The augmentation of the motor root reflex discharge in the cooled spinal cord of the cat. Amer. J. Physiol. 133: 307, 1941.

28. HAGBARTH, K. E., and D. I. B. KERR. Central influences on spinal afferent conduction. J. Neurophysiol. 17: 295-307, 1954.

29. HAMMEL, H. T. Temperature regulation and hibernation. In: Mammalian Hibernation III, edited by K. C. Fisher, A. R. Dawe, C. P. Lyman, E. Schoenbaum and F. E. South. Edinburgh: Oliver and Boyd, 1967, p. 86-96.

30. JOUVET, M., O. BENOIT, J. COURJON, and M. TANCHE. The EEG during hypothermia in paralyzed cats (cortical and subcortical changes, electrical responses to auditory stimuli and convulsive discharges). Electroencephalog. Clin. Neurophysiol. 8: 708, 1956.

31. KAADA, B. R. Somato-motor, autonomic and electrocorticographic responses to electrical stimulation of 'rhinencephalic' and other structures in primates, cat and dog. Acta Physiol. Scand. 24: 1-285, 1951.

32. KAHANA, L., W. A. ROSENBLITH, and R. GALAMBOS. Effect of temperature change on round-window response in the hamster. Amer. J. Physiol. 163: 213-223, 1950.

33. KAYSER, C. Le sommeil hivernal, probleme de thermoregulation. Rev. Can. Biol. 16: 303-389, 1957.

34. KAYSER, C. The Physiology of Natural Hibernation. New York: Pergamon Press, 1961.

35. KAYSER, C., and G. HIEBEL. L'hibernation naturelle et artificielle des hibernants et l'hypothermie généralisée expérimentale du rat et de quelques hibernants. Presse Medicale 60: 1699-1702, 1952.

36. KAYSER, C., and A. MALAN. Central nervous system and hibernation. Experientia 19: 441-451, 1963.

37. KAYSER, C., F. ROHMER, and G. HIEBEL. L'EEG de l'hibernant. Léthargie et réveil spontané du spermophile. Essai de reproduction de l'EEG chez le spermophile réveillé et le rat blanc. Rev. Neurol. 84: 570-578, 1951.

38. KOELLA, W. P., and H. M. BALLIN. The influence of temperature changes on the electrocortical responses to acoustic and nociceptive stimuli in the cat. Electroencephalog. Clin. Neurophysiol. 6: 629-634, 1954.

39. KOIZUMI, K., C. M. BROOKS, and J. USHIYAMA. Hypothermia and reaction patterns of the nervous system. Ann. N.Y. Acad. Sci. 80: 449-456, 1959.

40. KOIZUMI, K., J. L. MALCOLM, and C. M. BROOKS. Effect of temperature on facilitation and inhibition of reflex activity. Amer. J. Physiol. 179: 507-512, 1954.

41. KRŽALIĆ, L., D. ČUPIC, and LJ. MIHAILOVIC. Changes of aspartic acid in various cerebral structure of hibernating and aroused ground squirrels. Arch. Intern. Physiol. Biochim. 73: 817-825, 1965.

42. LEMAITRE, M. Der Einfluss der Temperatur auf das Electrocorticogramm der Ratte. Z. Biol. 106: 426-435, 1954.

43. LIBERSON, W. T., and K. AKERT. Hippocampal seizure states in guinea pig. Electroencephalog. Clin. Neurophysiol. 7: 211-222, 1955.

44. LIPP, J. A. Effects of deep hypothermia on the electrical activity of the brain. Electroencephalog. Clin. Neurophysiol. 17: 46-51, 1964.

45. LOUGHEED, W. M. Central nervous system in hypothermia. Brit. Med. Bull. 17: 61-65, 1961.

46. LOURIE, H., T. G. HOLMES, W. WEINSTEIN, H. G. SCHWARTZ, and J. L. O'LEARY. Observations on selective brain cooling in dogs. Arch. Neurol. 3: 163-176, 1960.

47. LYMAN, C. P., and P. O. CHATFIELD. Hibernation and cortical electrical activity in the woodchuck (Marmota monax). Science 117: 533-534, 1953.

48. LYMAN, C. P., and P. O. CHATFIELD. Physiology of hibernation in mammals. Physiol. Rev. 35: 403-425, 1955.

49. McMURREY, J. D., W. F. BERNHARD, J. A. TAREN., and E. A. BERING. Studies on hypothermia in monkeys. I. The effects of hypothermia on the prolongation of permissible time of total occlusion of the afferent circulation of the brain. Surg. Gynec. Obstet. 102: 75-86, 1956.

50. MASSOPUST, L. C., L. R. WOLIN, and J. MEDER. Spontaneous electrical activity of the brain in hibernators and non-hibernators during hypothermia. J. Exptl. Neurol. 12: 25-32, 1965.

51. MASSOPUST, L. C., L. R. WOLIN, M. S. ALBIN, and J. MEDER. Evoked responses from the eye and visual pathways in the hypothermic cat. Exptl. Neurol. 10: 383-392, 1964.

52. MASSOPUST, L. C., M. S. ALBIN, H. W. BARNES, R. MEDER, and H. W. KRETCHMER. Cortical and subcortical responses to hypothermia. Exptl. Neurol. 9, 249-261, 1964.

53. MASSOPUST, L. C., and L. R. WOLIN. Evoked potentials from the visual system in hypothermic hibernators and non-hibernators. Exptl. Neurol. 14: 134-143, 1966.

54. MERZBACHER, L. Untersuchungen an Winterschlafenden Fledermäusen. II. Mittheilung. Die Nervendegeneration während des Winterschlafes. Die Beziehungen zwischen Temperatur und Winterschlaf. Arch. Ges. Physiol. 100: 568-585, 1903.

55. MERZBACHER, L. Allgemeine Physiologie des Winterschlafes. Ergeb. Physiol. 3: 214-258, 1904.

56. MIHAILOVIC, LJ. On driving, prolongation and synchronisation of convulsive activity of the brain. Glas. Serbian Acad. Sci., Belgrade, 14: 91-135, 1958.

57. MIHAILOVIC, LJ., B. D. BELESLIN, D. ČUPIC, and D. POPESKOVIC. Interaction between the diffuse thalamo-cortical system and the brain stem activating system during arousal from hibernation. Iugoslav. Physiol. Pharmacol. Acta 5: 289-294, 1969.

58. MIHAILOVIC, LJ., B. BELESLIN, D. POPESKOVIC, and D. ČUPIC. O spontanoj iizvanaj električkoj aktimosti različstih partikalnih i supkortikalnih struktura mozga tekunice (Citellus citellus L.) u procesu budenja iz zimskog sna. Acta Med. Iugosl. 22: 165-228, 1968.

59. MIHAILOVIC, LJ. T., L. KRŽALIC, and D. ČUPIC. Changes of glutamine, glutanic acid, and GABA in cortical and subcortical brain structures of hibernating and fully aroused ground squirrels (Citellus citellus). Experentia 21: 709-713, 1965.

60. MIHAILOVIC, L. T., L. KRŽALIC, D. ČUPIC and B. BELESLIN. Changes of cerebral glutamine, glutamic acid and GABA during arousal from hibernation. Experentia 21: 100-102, 1965.

61. MIHAILOVIC, L., B. PETROVIC-MINIC, S. PROTIC and I. DIVAC. Effects of hibernation on learning and retention. Nature 218: 191-192, 1968.

62. MOKHIN, K. M., B. A. SAAKOV, L. M. KISELEVA, and E. V. LOGINOVA. K izucheniiû sezonnoi aktivnosti fiziologicheskikh funktsii malogo suslika. Rostov-on-the-Don, Gosudarstvennyi naacho-issledonatelskii protivochumnugo institut, Trudy, 121-129, 1963.

63. MORUZZI, G., and H. W. MAGOUN. Brain stem reticular formation and activation of the EEG. Electroencephalog. Clin. Neurophysiol. 1: 455-473, 1949.

64. NOELL, W. K., S. A. BRILLER, and W. B. BRENDEL. Effects of cold exposure on brain activity. Federation Proc. 11: 114, 1952.

65. POPOVIC V. Lethargic hypothermia in hibernators and nonhibernators Ann. N.Y. Acad. Sci. 80: 320-331, 1959.

66. PROTIC, S., B. MINIC-PETROVIC, and L. MIHAILOVIC. The R.N.A. and protein content in various brain structures of ground squirrels during arousal from hibernation. Iuogslav. Physiol. Pharmacol. Acta 6: 521-527, 1970.

67. PUTKONEN, P., H. S. SARAJAS, and P. SUOMALAINEN. Electrical activity of the olfactory bulb and cerebral cortex in hedgehogs arousing from hibernation. Ann. Acad. Sci. Fenn. Ser. A.V. 106: 1-11, 1964.

68. RATHS, P. Die bioelektrische Hirntätigkeit des Hamsters im Verlaufe des Erwachens aus Winterschlaf und Kältenarkose. Z. Biol. 110: 62-80, 1958.

69. ROHMER, F., G. HIEBEL and C. KAYSER. Recherches sur le fonctionnement du système nerveux des hibernants; les ondes cerebrales pendant le sommeil hibernal et le réveil; étude sur le spermophile. Compt. Rend. Soc. Biol. 145: 747-752, 1951.

70. SCOTT, J. W., D. MACQUEEN, and J. C. CALLAGHAN. The effects of lowered body temperature on the EEG. Electroencephalog. Clin. Neurophysiol. 5: 465, 1953.

71. SHTARK, M. B. Elektrofiziologicheskoe issledovanie zimnei spiachki. Fiziol. Zh. 47: 942, 1961.

72. SHTARK, M. B. Elektricheskaia aktivnost razlichnykh otdelov golovnogo mozga zimnespiashchikh. Fiziol. Zh. 49: 943-951, 1963.

73. SHTARK, M. B. Nekotorye osobennosti mozgovoi deiatel'nosti v usloviiakh nizkikh temperatur. In: Golovnoi Mozg i Reguliatsiia Funktsii Kiev: Academii Nauk, 1963, p. 124-126.

74. SHTARK, M. B. Interrelationship of electric and metabolic activity in the olfactory bulb of hibernating animals. Federation Proc. Transl. Suppl. 23: 238-240, 1964.

75. SHTARK, M. B. Mozg zimnespiashchikh. Novosobirsk: Nauka, 1970.

76. SHTARK, M. B., and V. P. DANILYUK. O vyzannykh potentsialakh v kore bol'shikh polusharii zimnespyashchikh. Dokl. Acad. Nauk SSSR 151: 740-743, 1963.

77. SHTARK, M. B., and V.P.O. DANILYUK. Dendritic potentials in the cerebral cortex of mammals hibernating in winter. Bull. Exp. Biol. Med. (Eng. Trans.) 59: 111-113, 1965.

78. SOUTH, F. E. Phrenic nerve-diaphragm preparations in relation to temperature and hibernation. Amer. J. Physiol. 200: 565-571, 1961.

79. SOUTH, F. E., and J. E. BREAZILE. Sleep, hibernation and hypothermia in yellow-bellied marmot (M. Flaviventris). In: Depressed Metabolism, edited by X.J. Musacchia and J.F. Saunders. New York: Elsevier Publ. Co., 1969.

80. STEVENSON, G. C., W. F. COLLINS, C. T. RANDT, and T. D. SAURWEIN. Effects of induced hypothermia on subcortical evoked potentials in the cat. Amer. J. Physiol. 194: 423-426, 1958.

81. STRUMWASSER, F. Factors in the pattern timing and predictability of hibernation in the squirrel, Citellus beecheyi. Amer. J. Physiol. 196: 8-14, 1959.

82. STRUMWASSER, F. Thermoregulatory, brain and behavioural mechanisms during entrance into hibernation in the squirrel, Citellus beecheyi. Amer. J. Physiol. 196: 15-22, 1959.

83. STRUMWASSER, F. Regulatory mechanisms, brain activity and behaviour during deep hibernation in the squirrel, Citellus beecheyi. Amer. J. Physiol. 196: 23-30, 1959.

84. STRUMWASSER, F. Some physiological principles governing hibernation in Citellus beecheyi. In: Mammalian Hibernation, Proceedings of the First International Symposium on Natural Mammalian Hibernation, edited by C.P. Lyman and A.R. Dawe. Cambridge, Mass.: Museum Comp. Zool., Harvard Coll., 1960, P. 285-318.

85. STRUMWASSER, F., J. SMITH, J. GILLIAM, and F. R. SCHLECHTE. Identification of active brain regions involved in the process of hibernation. XVI Intern. Congr. Zool. 2: 53-56, 1963.

86. STRUMWASSER, F., F. R. SCHLECHTE, and J. STREETER. The internal rhythms of hibernators. In Mammalian Hibernation III, edited by K.C. Fisher, A.R. Dawe, C.P. Lyman, E. Schoenbaum and F.E. South. Edinburgh: Oliver and Boyd, 1967, P. 110-139.

87. SUDA, I. K., K. KOIZUMI, and C. M. BROOKS. Analysis of effects of hypothermia on central nervous system responses. Amer. J. Physiol. 189: 373-380, 1957.

88. SUOMALAINEN, P. Hibernation and sleep. In: The Nature of Sleep, edited by G. E. W. Wolstenholme and M. O'Connor. Boston: Little, Brown and Co., 1961, P. 307-321.

89. TEN CATE, J., G. P. M. HORSTEN, and L. J. KOOPMAN. The influence of body temperature on the EEG of the rat. Electroencephalog. Clin. Neurophysiol. 1: 231-235.

90. TWENTE, J. W. and J. A. TWENTE. Regulation of hibernating periods by temperature. Proc. Nat. Acad. Sci., U.S. 54: 1058-1061, 1965.

91. WEDENSKY, N. E. Die Errengung, Hemmung und Narkose. Arch. Ges. Physiol. 100: 1-144, 1903.

92. WILLIAMS, B. A., and HEATH, J. E. Responses to preoptic heating and cooling in a hibernator Citellus tridecemlineatus. Amer. J. Physiol. 218: 1654-1660, 1970.

CREDITS

The editors and authors are grateful to the following publishers, organizations and individuals for permission to use materials previously published:

Figure 1: The American Physiological Society; F. Strumwasser, American Journal of Physiology 196: 15-22, 1959.

Figure 2: Oliver and Boyd; F. Strumwasser, F. R. Schlechte and J. Streeter, "The internal rythms of hibernators," in Mammalian Hibernation III, edited by K. C. Fisher, A. R. Dawe, C. P. Lyman, E. Schoenbaum, and F. E. South. Edinburgh: Oliver and Boyd, 1967, p. 110-139.

Figure 4: L. J. Mihailović, Srpska Akademeja Nauka I Umetnosti, Belgrade. Glas. Odeljenje Medicinskih Nauka. Belgrade. 14: 91-135, 1958.

Figure 5: L. J. Mihailović, Srpska Akademeja Nauka I Umetnosti, Belgrade. Glas. Odeljenje Medicinskih Nauka. Belgrade. 14: 91-135, 1958.

Figure 7: L. Mihailović, B. D. Beleslin, D. Čupić and D. Popesković, Juogoslavica Physiologica et Pharmacologica Acta Belgrade 5: 289-294, 1969.

Figure 8: Elsevier Publishing Company, Amsterdam, The Netherlands; H. Gaenshirt, W. Krenkel and W. Zylka, Electroencephalography and Clinical Neurophysiology 6: 409-413, 1954.

Figure 9: The American Physiological Society; G. C. Stevenson, W. F. Collins, C. T. Randt and T. D. Saurwein, American Journal of Physiology 194: 423-426, 1958.

Figure 10: The New York Academy of Sciences; K. Koizumi, C. M. Brooks and J. Ushiyama, <u>Annals of the New York Academy of Science</u> 80: 449-456, 1959. Copyright The New York Academy of Sciences, 1959.

Figure 11: The New York Academy of Sciences; K. Koizumi, C. M. Brooks and J. Ushiyama, <u>Annals of the New York Academy of Science</u> 80: 449-456, 1959. Copyright The New York Academy of Sciences, 1959.

Figure 12: American Elsevier Publishing Company, Inc., F. E. South and J. E. Breazile, "Sleep, hibernation and hypothermia in yellow-bellied marmot (M. flaviventris)," in <u>Depressed Metabolism</u>, edited by X. J. Musacchia and J. F. Saunders. New York: American Elsevier Publishing Company, Inc., p. 277-312, 1969.

Figure 13: Urban & Schwarzenberg, München, Berlin, Wein; P. Raths, <u>Zeitschrift Biologie</u> 10: 62-80, 1958.

Figure 14: L. J. Mihailović, <u>Srpska Akademeja Nauka I Umetnosti, Belgrade. Glas. Odeljenje Medicinskih Nauka.</u> Belgrade. 14: 91-135, 1958.

Figure 15: L. J. Mihailović, <u>Srpska Akademeja Nauka I Umetnosti, Belgrade. Glas. Odeljenje Medicinskih Nauka.</u> Belgrade. 14: 91-135, 1958.

Figure 16: L. J. Mihailović, <u>Srpska Akademeja Nauka I Umetnosti, Belgrade. Glas. Odeljenje Medicinskih Nauka</u>, Belgrade. 14: 91-135, 1958.

Figure 17: Urban & Schwarzenberg, München, Berlin, Wein; P. Raths, <u>Zeitschrift Biologie</u> 10: 62-80, 1958.

ANALYSIS OF CORRELATES BETWEEN LEVELS OF CONSCIOUSNESS AND ACTIVITY OF THE CENTRAL NERVOUS SYSTEM.

L. M. N. BACH

Division of Biomedical Sciences
School of Medical Sciences
University of Nevada, Reno, Nevada 89507

Anyone with the temerity to attempt a paper with the word consciousness in the title must himself, <u>a priori</u>, exhibit a profound measure of unconsciousness for there are few other terms so beset with traps both semantic and logical. One may view consciousness as a hypothesis, an epiphenomenon or a measureable function depending upon one's persuasion. Frequently described as the responsiveness of a normal, waking subject - a view which is neither sufficient nor necessary - it can only be studied effectively by the physiologist in the form of a particular type of behavior, often described in animals as "behavioral vigilance."

One must first ask, then, what behavior manifests consciousness? This question is immediately followed by another, what kind of neural process produces this behavior? In humans, at least, consciousness involves the clear indication of experience of awareness by the subject to the observer which may be verbalized or indicated by some non-verbal behavior. In ascending the phylogenetic scale, one is impressed by the degree to which behavior becomes less stimulus-bound, suggesting that brain activity is less completely controlled by afferent inputs and consonant with the observation that behavior becomes less predictable. In this respect, consciousness appears to involve a longer latency between input and output as it becomes more apparent, the inference being that some sort of "hold" is placed upon the input before activity occurs. In experiments with human patients, Libet (13) has demonstrated that electrical stimulation of the brain will produce a conscious perception, remarked upon by the patient, only after an interval of 0.5-1.0 seconds of continuous stimulation: lesser intervals are believed to produce unconscious reactions.

Most writers about consciousness suggest that the effect involves ideas about the future integrated with memories of the past coupled with the act of adjustment to complexities of the present. One view, expressed by Hebb (6), suggests that most of the activity of the "silent" areas of the cortex concerns the complex interlinking of these various processes.

Some of the observable processes associated with consciousness include:

A) The conclusion that it is a neural process:
B) Its occurrence in animal forms other than humans including
   a) the act and shifting of attention,
   b) manipulation of abstract ideas (animal communication, e.g.),
   c) the capacity for expectancy (tool use by animals),
   d) self-awareness and awareness of others (play), and
   e) ethical values (rescue by dolphins of distressed companions);
C) Its variable quality (sleep-wakefulness);
D) Dependence upon the level and shifting character of sensory input and perception; and
E) A dependence upon memory.

These characteristics, collated by Mountcastle (19), do not of course identify consciousness but merely reflect some of the observable consequences and permit us to both measure and manipulate consciousness to some limited extent.

Among the most qualified and experienced observers of consciousness in man are the clinical neurologists. Plum and Posner (22) have provided a very useful description of variations in depth of consciousness. <u>Alert wakefulness</u> indicates an immediate, full and appropriate response to sensory stimulation. <u>Lethargy</u> indicates a state of drowsiness, inaction and indifference in which responses to stimulation may be delayed or incomplete and which may require stronger than usual stimulation. The onomatopoeitic word <u>obtundation</u>, is descriptive of a state of even duller indifference which maintains wakefulness but little more. <u>Stupor</u> or <u>torpor</u>, refers to a state from which the subject can be aroused only by vigorous and continuous external stimulation. <u>Coma</u> designates states in which neither the psychological or motor response to stimulation is observable and, in its most profound form, represents the loss of all reflexes except respiration, perhaps swallowing, and the failure of thermoregulation.

SLEEP AND WAKEFULNESS

Certainly the most common alteration in consciousness receiving neurophysiological attention in recent years is the periodic variation known as sleep and wakefulness. The spectrum between sleep and wakefulness consists of several stages and types of behavior and activity but it is difficult to secure agreement as to what combination of these constitute any one level of sleep and wakefulness. A number of functional alterations have been described for various levels of sleep associated with high voltage slow waves (HVSW) of the EEG. These include: (a) a marked reduction or absence of the ability to respond, remember, or learn in connection with sensory stimuli; (b) EEG tracings with an obvious content of HVSW; (c) behaviorally, the subject is quiet and exhibits no "voluntary" movement; (d) most reflexes show a high threshold and are either depressed or absent, an effect which extends to muscle tone; (e) body temperature is reduced as much as 2° C below normal and the BMR is decreased by ten percent; (f) there is a reduction in sympathetic activity as indicated by a decreased blood pressure, a decreased heart rate and eyelid closure due to relaxation of Müller's muscle; (g) convergence of the eyes, pupilloconstriction - but continued reactivity to light - and upward deviation of the eyes; (h) decreased respiratory ventilation, a slight increase in plasma pCO_2 and an increased threshold of the inspiratory center to CO_2; (i) decreased urine production; and (j) a decrease in tear and gastrointestinal secretions and motility.

Sleep-wake cycles must be considered a special case of circadian rhythm which is uniquely brain-regulated in mammals; Richter (23) recently reported that circadian activity-rest rhythms in rats appear to be controlled at a hypothalamic level, since destruction of the ventromedian nucleus reduces these regular rhythms to randomized bursts of activity and rest of varying durations. In normal rats the circadian activity-rest or sleep-wake cycle is endogenous but can easily be entrained by 12 hour light - 12 hour dark cycles.

Kleitman (12) has described three types of sleep-wake rhythms in an ontogenetic series. The first type, the basic or metabolic rhythm is observable in the fetus (and in many larvae) exhibiting a 60 minute cycling at birth and persisting through an increasingly long cycle of 90 minutes in the adult. A new second type of cycle appears at birth which Kleitman relates to visceral activity and which has a period of 2-3 hours. The third and latest cycle to develope is the well-known 16 hour wake-8 hour sleep cycle which is largely a learned behavior.

Recently Sterman et al. (26) described a circadian distribution of various sleep levels. There are many criteria for measuring depth of sleep, such as alterations of amplitude and latency of sensory evoked potentials, increasing with increased levels of wakefulness, facility in developing or exhibiting operant conditioning, which increases with wakefulness and by alterations in the EEG pattern and various somatic and visceral signs described above. In their study, the sleep-wake periods in cats consisted of: (a) a 28% portion of wakefulness during which time the EEG showed a desynchronized "alert" pattern, and the animals were either engaged in "vital" activities, were active or quiet; (b) a 15% portion in which the EEG exhibited HVSW or a "drowsy" pattern and the animals were quiet; (c) a 42% portion featuring spindle bursts in the EEG and behavioral sleep; and (d) an activated EEG pattern of desynchronization for 15% of the period when REM sleep was evident.

REM (Rapid Eye Movement) sleep is an aspect which has received an enormous amount of attention because of its association with dreaming in humans. As the name indicates, among the features of this phase of sleep are: (a) rapid eye movements; (b) a desynchronized EEG reminiscent of the waking pattern; (c) decreased tone in the neck muscles; (d) increased somatic movement generally but decreased muscle tone as well; (e) an increase in brain temperature over the waking state; (f) an increased depth and decreased rate of respiration, approaching periodic breathing; (g) a decrease in heart rate, but many others claim tachycardia; and (h) a marked decrease in blood pressure sufficient to produce cerebral ischemia in preparations with carotid sinus deafferentation - although several workers claim that marked rises in blood pressure can occur with REM sleep. It is generally recognized that REM sleep is much deeper than HVSW sleep requiring a very intense stimulus to awaken the subject; if a lesser stimulus is employed REM sleep may convert to HVSW sleep. Appropriate stimulation of the brainstem reticular arousal system (ARAS) can convert HVSW sleep to

REM sleep, or if stronger, can produce awakening. It has been noted by Lisk and Sawyer (14) that when lights are turned off for rats exhibiting light-dark cycles of sleep-wake, REM sleep is the immediate consequence. Others have noted an oxygen dependency for REM sleep in which the amount and duration of REM is reduced during periods of low pO_2; increased pO_2 levels appear to increase the speed of onset of REM (9).

CENTRAL NERVOUS SYSTEM ACTIVITY

The natural supposition that wakefulness is associated with increased neuronal activity in the brain has been vitiated by Evarts et al. (3) and others who show, for the visual cortex, for example, that some neurons show increased frequency of discharge and others show decreased frequency during sleep. The main effect seems to be that during wakefulness there is a much greater degree of lability and variability in neuronal activity. An overall increase in activity of brain neurons is most prominent during the deep REM sleep, next most evident during HVSW sleep and least likely during wakefulness.

There have been several attempts to find the initial event or process which sets off a sleep phase. There are repeated searches for sleep "substances", the latest by Monnier (17), but most of these attempts have not proven convincing. Some alteration in peripheral sensory input is often invoked as the initiator of sleep. While it is true that slow frequency stimulation of Group II muscle afferents can produce sleep, as will similar stimulation of carotid sinus afferent nerves, the fact is that deafferentation or complete destruction of the major sensory pathways at the midbrain level completely fail to alter the normal sleep-wake cycles in any way.

Most attention has been directed to the central nervous system and the brainstem reticular formation, or ascending reticular arousal system (ARAS) as it is frequently termed. It is a well known story that lesions involving the midbrain ARAS cause coma and a sustained HVSW EEG. It is well established that lesions either in the midbrain ARAS or the immediately rostral posterior hypothalamus will produce coma; the integrity of these two areas is essential for normal wakefulness and, as shown originally by Hess (8) and more recently by Feldman (4), stimulation in these areas with a high frequency current will produce both behavioral and EEG arousal. The "wakefulness" area found in these regions appears to be interrupted by lesions and the coma which results may be a consequency of its destruction. The specific structure responsible for this has been identified as the nucleus reticularis pontis oralis which projects its high frequency output through the diffuse thalamic projection system. Just inferior to this area is found the nucleus reticularis pontis caudalis which Jouvet (11) identifies as the primary source of REM sleep, also characterized by a desynchronized EEG but "paradoxically" by a very deep sleep as well. Jouvet further relates this nucleus to the adjacent noradrenalin-sensitive locus cereolus which combination is essential for REM sleep. This system also projects its activity through the diffuse thalamic projection system to the cortex but, unlike HVSW sleep, REM sleep appears to be independent of any cortical feedback. The caudalis nucleus may discharge in bursts along with the motor cortex to account for the combined episodes of eye movement and general somatic movement during REM sleep; in the interburst intervals, the nucleus appears to generate, along with the cortex, presynaptic inhibition and direct motor neuron depression in cord ventral horn cells. A lesion just below this portion of the pons, sparing both pontile reticular nuclei, results in sustained insomnia, in contrast to the more rostral midbrain or hypothalamic lesions which interrupt the output of the pontile reticular nuclei and result in somnolence. The sustained wakefulness which can originate from the pons appears normally to be counterpoised by activity originating in the solitary nucleus of the medulla which Moruzzi (18) has shown to be important for the production of HVSW and the synchronous activity of the EEG. To some extent this may be due to inhibition of the pontile ARAS although there may also be some direct effect upon the diffuse thalamic projection system as well. Although high frequency stimulation of this structure can produce arousal, it seems that the nucleus solitarius and its projection through the ARAS and thalamic systems is more often normally related to HVSW sleep since it is closely related to the serotonin-containing raphe neurons associated with this type of sleep and one finds that the neurons of the nucleus itself begin discharging just before behavioral sleep begins and are otherwise quite inactive during wakeful states. Low frequency stimulation of this nucleus produces behavioral sleep as does carotid body afferent nerve stimulation which leads to the same nucleus.

All of these brainstem influences upon sleep and wakefulness feed into the thalamic diffuse projection which Hess (8) originally showed was capable of producing normal sleep behavior through low frequency stimulation (in the massa intermedia-centre medianum); this is accompanied by HVSW activity of the EEG.

Another important diencephalic area for production of sleep and wakefulness is in the lateral preoptic area. Nauta (21) originally showed lesions in this region resulted in insomnia, a finding more recently confirmed by Hernandez-Peon (7) who also found this area to be sensitive to noradrenalin which evokes wakefulness and to acetyl choline which produces sleep.

Observations by McGinty and Sterman (16) also confirm those of Nauta. Both of these workers as well as Hernandez-Peon (7) have observed that stimulation of the lateral preoptic region can produce sleep, an effect which is also found with stimulation of the basomedial nucleus of the amygdala which projects to the lateral preoptic region. In both these instances sleep is produced by slow frequency stimulation whereas wakefulness can be produced by high-frequency current. Jouvet (10) claims; although it has been disputed; that the amygdala and hippocampus, acting through Nauta's limbic-midbrain system, exercise a controlling influence upon REM sleep originating in the caudal pontile reticular nucleus. If true, this would be gratifying to those who view REM sleep as a function of the emotional and vegetative brain.

The neocortex also has an important role in the regulation of consciousness through its function as the interacting terminal for the diffuse thalamic projection. All metabolic and circulatory deficiencies which interfere with consciousness appear to have their effect at this level. In human patients this often takes the form of a progressive loss of consciousness starting with a slight reduction in alertness or awareness. Initially, the patient just appears preoccupied or uninterested and prefers to lie quietly, neither reading or talking; he concedes that thinking or reading is too much effort. Matters continue downhill as he progresses to increased episodes of drowsiness, reduced ability to respond, loss of memory, incorrect responses, etc.

Roldan et al. (24) observed alterations in electrical activity in various of these structures during sleep-wake cycles. In the natural sequence, a 10 minute interval (in cats) of HVSW sleep is associated with HVSW activity in neocortex, hippocampus and midbrain ARAS; at the last portion of this phase, ARAS threshold to sensory inputs begins to increase substantially. This phase is followed by a sudden two minute interval in which there is desynchronization of the neocortex, theta rhythm of the hippocampus and in the ARAS; a final one minute phase appears in which there is desynchronization at all levels and behavioral arousal.

It is quite apparent that 2, 4, 6, 10 ······n brain areas can be primarily identified as sources of wakefulness, sleep or both through stimulation or recording experiments. It is again no truism that the whole brain appears to be involved in varying degrees. One obviously must view a sequence of related structures interacting in some, as yet undescribed, fashion. For the moment, it is feasible to focus upon the thalamic diffuse projection system, which is regarded as a pacemaker of EEG and sleep-wake rhythms. The setting of this pacemaker appears to depend upon a balance of influences from the brainstem - the solitary nucleus, the pontile reticular nuclei, among others as well as inputs arriving from diencephalic and limbic levels.

HIBERNATION

There are several other states, such as habituation, "freezing" (fawns, opossum, various birds) in face of danger, hypnosis and hibernation which lend themselves well to comparison with the cited examples of levels of consciousness.

As with the wide variety of alterations in consciousness, there is a wide selection of hypothermic reactions and types of hibernation. Purely endogenous temperature-independent clocks of a circannian nature may produce hibernation, whereas in other forms some combination of low food supply and reduced body temperature may be sufficient (peromyscus, Nighthawks) or the absence of food with the absence of light (hummingbirds) may result in torpor even if the ambient temperature is normal. It is common to associate hibernation with cold, lack of food, seasonal changes, including alterations in light-dark ratios and to think about central nervous system activity in connection with these functions. A simplistic generalization of this nature is, of course, unwarranted.

It is quite common for hibernating forms to develop obesity prior to the onset of hibernation and it is well known that an obese animal will enter hibernation more readily and more completely than a thin form. Mrosovsky (20) has shown that a pre-obese hibernator behaves very much like a pre-obese form with hypothalamic ventromedian nucleus destruction. Neither animal will work for food but will gorge unremittingly if given food ad lib. Satinoff (25), who examined the effects upon hibernation, of ventromedian nucleus lesions, as well as lesions of the lateral hypothalamus (producing aphagia) and lesions of the preoptic region affecting temperature regulation found that these brain damaged animals went into the normal pattern of hibernation upon exposure to cold, with normal decreases in brain temperature but that it was not possible to arouse them afterwards by any means; each of these preparations died while hibernating. The conclusion was that none of these animals could recover normal temperature nor could they experience the normal and necessary periodic arousal which they (ground squirrels) apparently require to survive. This vital arousal is thought to be the consequence of some lethal metabolite which accumulates during hibernation. The notion is already familiar to us in connection with putative "sleep substances" accumulating because of prolonged wakefulness.

The classic work of Strumwasser (27) is well known to every student of hibernation although it has been recently disputed. He described "test-drops" of brain temperature which "tested" the level of metabolic activity during preparation for hibernation; these findings have recently been questioned, however. These oscillations of brain temperature were associated with levels of arousal, both of which showed increasingly excessive swings until the low level of wakefulness and low body temperature appropriate to hibernation were achieved. It would appear from this that a cyclic resetting of the temperature control in the preoptic region occurs in preparation for hibernation. The normal fluctuations in temperature, sleep-wake levels and food intake appear to be exaggerated as hibernation approaches. That this is an active process is suggested by the fact that heart rate and BMR both begin to decrease <u>prior</u> to the decrement in body temperature and thermogenesis which are associated more specifically with hibernation. The body temperature in the hibernator is always somewhat above ambient by an amount determined largely by the degree of insulation; if the ambient temperature drops too low, the animal arouses and its body temperature rises as well; the mechanism and brain areas involved in this critical response are unknown. Although temperature regulation persists in the hibernator, its sensitivity is less than normal since the response to administered pyrogens is defective. In this respect, there is a separation between the temperature-sensitive arousal and thermogenesis processes. Although the blood pressure remains fairly constant, heart rate decreases and respiration becomes erratic--almost periodic in character; the respiratory centers are still CO_2 sensitive to a normal degree.

Many observers have noted that the brain of the hibernator survives and functions normally at very low temperatures in contrast to the brains of non-hibernators. Homeotherms normally stop breathing at 19° C whereas the respiratory center of the hibernator continues to function even below 5° C; in the same fashion, nerve conduction in the non-hibernator stops at 9° C whereas the hibernator nerve continues to conduct below 3.5° C. Massopust et al. (15) recently shown that activity of the ARAS ceases at 23° C in the non-hibernator but persists to at least 16° C in the hibernator. Hammel (5) has accounted for effects of this sort by assuming temperature-independent neurons in the ARAS in contrast to temperature-dependent neurons in other brain areas; Chai and Wang (1) recently showed the existence of many temperature-dependent neurons in the ARAS however.

The electrical activity of the brain during hibernation shows a variety of patterns, including isoelectric activity, in various species and members of the same species. The classic work of Chatfield and Lyman (2) showed three types of activity upon arousal from hibernation. These were found only in the limbic system - not in the hypothalamus - (a) 17 /sec bursts of activity, (b) bursts of spikes, (c) continuous waves of activity. Others have noticed desynchronized activity, without behavioral arousal, with temperature increases above 6 degrees and 10 cps activity in the motor cortex, amygdala and hippocampus. Brain activity at 6 degrees seems similar to the arousal pattern but shows a smaller amplitude. Upon arousal, electrical activity returns to different brain areas at different temperatures.

SUMMARY

The wide variety of effects associated with consciousness in its different forms and levels, the large number of brain areas and kinds of electrical events which can be consistently measured have not provided any simple understanding of sleep or wakefulness, let alone consciousness.

The large number of types of hibernation and causes for them in different species and the more limited understanding of brain function in this activity leave much to be desired in our knowledge of the neural processes responsible for the onset of hibernation and the arousal from it.

Future work should aim at identifying the functional commonalities between hibernation and consciousness (coma, lethargy, sleep, [REM or HVSWO]). Recordings from structures commonly believed to be involved in sleep and wakefulness (thalamic diffuse projection, lateral preoptic region, pontile reticular formation nuclei, solitary nucleus, etc) should be made during different stages of hibernation and arousal. It would also be useful to employ various criteria of depth of sleep, operant conditioning, sensory evoked potentials, and EEG patterns, including incidence of REM and HVSW sleep during various stages and types of hibernation and arousal.

By the same token, the current controversy concerning the relationship between thermoregulation and food intake should be extended to studies of sleep and wakefulness. The interaction between these factors and the nature of arousal from hibernation or sleep remains unknown.

REFERENCES

1. CHAI, C. Y., and S. C. WANG. Cardiovascular and respiratory responses to cooling of the medulla oblongata of the cat. Proc. Soc. Exptl. Biol. Med. 134: 763-767, 1970.

2. CHATFIELD, P. O., and C. P. LYMAN. Subcortical electrical activity in the golden hamster during arousal from hibernation. Electroencephalog. Clin. Neurophysiol. 6: 403-408, 1954.

3. EVARTS, E. V., E. BENTAL, B. BIHARI, and P. R. HUTTENLOCHER. Spontaneous discharge of single neurons during sleep and waking. Science 135: 726-728, 1962.

4. FELDMAN, S. M., and H. J. WALLER. Dissociation of electrocortical activity and behavioral arousal. Nature 196: 1320, 1962.

5. HAMMEL, H. T. Regulation of internal body temperature. Ann. Rev. Physiol. 30: 641-710, 1968.

6. HEBB, D. O. Cerebral organization and consciousness. In: Sleep and Altered States of Consciousness, edited by S. S. Kety, E. V. Evarts, and H. L. Williams. Baltimore: Williams and Wilkins, 1967, p. 1-5.

7. HERNÁNDEZ-PÉON, R. Central neuro-humoral transmission in sleep and wakefulness. Prog. In Brain Res. 18: 96-117, 1965.

8. HESS, W. R. Functional Organization of the Diencephalon. New York: Grune and Stratton, 1954.

9. HUERTAS, J., and J. K. McMILLIN. Paradoxical sleep: effect of low partial pressures of atmospheric oxygen. Science 159: 745-746, 1968.

10. JOUVET, M. Neurophysiology of states of sleep. Physiol. Rev. 47: 117-177, 1967.

11. JOUVET, M. Biogenic amines and the states of sleep. Science 163: 32-41, 1969.

12. KLEITMAN, N. Sleep and Wakefulness. Chicago: Univ. of Chicago Press, 1963, p. 363-370.

13. LIBET, B. Cortical activation in conscious and unconscious experience. Perspect. Biol Med. 9: 77-86, 1965.

14. LISK, R. D., and C. H. SAWYER. Induction of paradoxical sleep by lights-off stimulation. Proc. Soc. Exptl. Biol. Med. 123: 664-667, 1966.

15. MASSOPUST, L. C., L. R. WOLIN, and J. MEDER. Spontaneous electrical activity of the brain in hibernators during hypothermia. Exptl. Neurol. 12: 25-32, 1965.

16. McGINTY, D. J., and M. B. STERMAN. Sleep suppression after basal forebrain lesions in the cat. Science 160: 1253-1255, 1968.

17. MONNIER, M., and L. HÖSLI. Humoral regulation of sleep and wakefulness by hypnogenic and activating dialysable factors. Prog. In Brain Res. 18: 118-123, 1965.

18. MORUZZI, G. Synchronizing influence of the brainstem and the inhibitory mechanisms underlying production of sleep by sensory stimulation. Electroencephalog. Clin. Neurophysiol. Supp. 13: 231-256, 1960.

19. MOUNTCASTLE, V. B. Sleep, wakefulness and the conscious state: intrinsic regulatory mechanisms of the brain. In: Medical Physiology, edited by V. B. Mountcastle. St. Louis: C. V. Mosby, 1968, vol. 2, p. 1315-1342.

20. MROSOVSKY, N. The adjustable brain of the hibernator. Sci. Amer. 218: 110-118, 1968.

21. NAUTA, W. J. H. Hypothalamic regulation of sleep in rats. J. Neurophysiol. 9: 285-316, 1946.

22. PLUM, F., and J. B. POSNER. Diagnosis of Stupor and Coma. Philadelphia: F. A. Davis, 1966, p. 1-41.

23. RICHTER, C. P. Sleep activities: their relation to the 24-hour clock. In: Sleep and Altered States of Consciousness, edited by S. S. Kety, E. V. Evarts, and H. L. Williams. Baltimore: Williams and Wilkins, 1967, p. 8-29.

24. ROLDAN, E., T. WEISS, and E. FIFKOVA. Excitability changes during the sleep cycle of the rat. Electroencephalog. Clin. Neurophysiol. 15: 775-785, 1963.

25. SATINOFF, E. Disruption of hibernation caused by hypothalamic lesions. Science 155: 1031-1033, 1967.

26. STERMAN, M. B., T. KNAUSS, D. LEHMANN, and C. D. CLEMENTE. Circadian sleep and waking patterns in the laboratory cat. Electroencephalog. Clin. Neurophysiol. 19: 509-517, 1965.

27. STRUMWASSER, F. Factors in the pattern, timing and predictability of hibernation in the squirrel. Amer. J. Physiol. 196: 8-30, 1959.

THE ROLE OF HYPOTHALAMIC MONOAMINES IN HIBERNATION AND HYPOTHERMIA

R. D. MYERS and T. L. YAKSH

Laboratory of Neuropsychology,
Purdue University
Lafayette, Indiana 47907

INTRODUCTION

While standing on Fifth Avenue in front of Macy's show-window, it is awfully easy to say, "Why didn't they set that alluring mannequin in the middle of the show-case, instead of to the side?" Whenever an outsider is looking in on any field, as a few of us here today are at hibernation, it is a simple matter to raise the question of why the central nervous system has generally been off in the periphery rather than focused in the center of one's experimental or theoretical thinking about hibernation. This, of course, is not meant to be a negative criticism, particularly to those endocrinologists who have contributed so much exciting and important work to the area. However, there is the distinct possibility that the myriad of humoral and other physiological changes related to the phenomenon of hibernation or deep hypothermia may in fact be a consequence of another mechanism. Do the chemical changes in the constituents of serum and other tissues parallel those that could be brought about by a neuronal mechanism which, on an evolutionary basis, operates selectively in the brain of a hibernator?

Rather than discuss the possible CNS mechanism for hibernation per se, we shall consider some interesting, if not thought-provoking, experimental data concerning the topic of hypothermia in the non-hibernating animal that may well reflect the possible cerebral mechanisms involved in the condition of hibernation. We believe, presently, that there are certain neurochemical systems of vital importance to the thermoregulatory mechanism of the ordinary mammal, including us primates, who are not, as yet, able to enter into the state of hibernation.

BACKGROUND OF THE NEUROCHEMICAL APPROACH TO THERMOREGULATION

Do certain biochemical systems in different regions of the brain control essential physiological and behavioral processes? If we are able to answer this question with some precision, the fundamental neural activity which regulates our vital functions can be understood. In fact, it may eventually be possible to modify body temperature and even induce hibernation by means of a centrally acting chemical agent. The applications of these concepts to medical and other sciences, as well as to the international programs for space exploration, are readily apparent (12).

Perhaps the most important discovery in this field, within the past decade, is the fact that certain endogenous substances, including biogenic amines and acetylcholine, are present in different concentrations within a number of structures in the brain-stem. If the resting levels of these substances are artificially elevated in a given region, rather remarkable physiological and behavioral changes occur. It appears that specific neural elements in the brain-stem are physiologically sensitive to the local application of endogenous factors such as monoamines or acetylcholine. As a result, these substances are considered now to be putative transmitters. A principal question which still persists, however, is how a chemical factor, if acting as a transmitter within a specific anatomical site, mediates or modulates a given physiological response.

Thermoregulation by a chemical thermostat. In 1963, it was discovered by Feldberg and Myers that several different endogenous substances within the brain-stem seemed to play an important role in the hypothalamic control of body temperature (4). When either serotonin (5-HT) or norepinephrine was injected into the cerebral ventricles or directly into the anterior hypothalamus of the conscious cat, hyper- or hypothermia, respectively, was produced (4,5). From these, and other experiments, a new theory unfolded in which it was proposed that 5-HT within the hypothalamus mediates the heat maintenance response, whereas norepinephrine, acting in functional opposition to 5-HT, activates the heat-loss pathway. Because of the apparent differences between species, in response to the monoamines, the usage of anesthetic agents, and even the ambient temperature of the experimental animal, much of our research has been carried out with the unanesthetized rhesus monkey under controlled environmental conditions (12).

From the results of a vast number of experiments, it would appear that a chemo-specific, anatomical "coding" of behavioral and physiological responses exists in the hypothalamus of the primate (21,25). To illustrate this concept, one amine, norepinephrine, seems to be utilized endogenously at different sites within this diencephalic structure for mediating more than one vital function (17). When the catecholamine is micro-injected into the anterior hypothalamus, the functional activity of the site is suppressed and hypothermia occurs (21). On the other hand, when the same amine is applied in an equivalent dose within the lateral hypothalamus, a fully satiated monkey eats voraciously and consumes an abnormal amount of food (14). When 5-HT is applied to either of these areas, the only response obtained regularly thus far is a rise in temperature when the indoleamine is injected directly into the anterior preoptic region (13).

To illustrate these changes in temperature in the conscious monkey, Figure 1 shows the hyperthermic effect of 6 µg 5-HT and the hypothermic

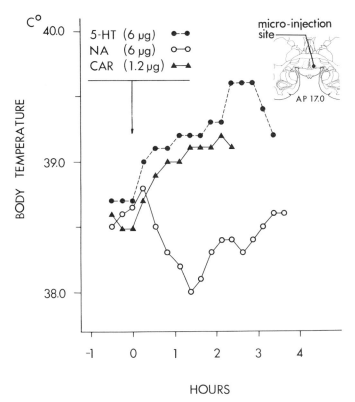

FIG. 1. Temperature responses of two monkeys following micro-injections at zero hour in the anterior hypothalamus (●) at AP 17.0 (inset) of 6 µg 5-HT (●----●); of 6 µg norepinephrine (o———o) and of 1.2 µg carbachol (▲———▲). 5-HT and norepinephrine were given in one animal and carbachol in the second (21).

effect of 6 µg norepinephrine given in the same site ventral to the anterior commissure at the coronal plane, AP 17.0. When acetylcholine was micro-injected at the same site, temperature rose in much the same way as with 5-HT. The hyperthermia following the application of 1.2 µg carbachol, again at the same site in the anterior, pre-optic region of the hypothalamus is also shown in Figure 1. Although these alterations in temperature were dose-dependent, a larger dose of 5-HT given at the same site often elicited a fall in temperature of short duration, which was then followed by a rise to fever level. A higher dose of norepinephrine, however, intensified not only the fall in temperature, but also the duration of hypothermia. If 5-HT and norepinephrine were given in the same dose range but at a slightly more caudal site in the normothermic monkey, the amines had virtually no effect on the animal's temperature.

An anatomical mapping of the sites in the monkey's hypothalamus at which a hypothermic response of at least 0.4°C followed within 1 hour after the application of norepinephrine is shown in Figure 2. In all sites, from coronal planes AP 13.0 to AP 16.0 inclusive, norepinephrine did not affect the monkey's temperature, but at AP 14.0 it produced hypothermia at three sites directly on the border of the ventromedial nucleus. At this level eating and drinking were elicited in twenty of thirty-seven experiments but it was difficult to determine whether these ingestive responses influenced the animal's temperature.

Acetylcholine (ACh), which is a principal neurotransmitter in the peripheral nervous system, may have a dual role in the central nervous system. ACh micro-injected into the lateral hypothalamus inhibits ingestive responses in the hungry monkey (14,25), but in the posterior hypothalamus, ACh evokes hyper- or hypothermia depending upon the site of the injection. An anatomical "mapping" of these temperature responses shows that the efferent cholinergic system of the primate seems to follow a trajectory which arises in the anterior hypothalamus and passes to the posterior area via the lateral and even medial regions.

Between the hypothalamus and the mesencephalon, the hypothermia induced by acetylcholine or carbachol is localized in sites adjacent to the

ADRENERGIC MICRO-INJECTION SITES

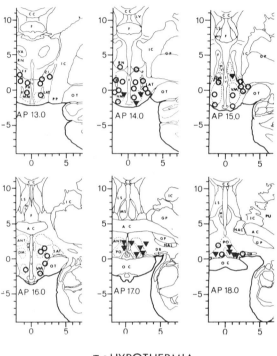

▼ = HYPOTHERMIA
o = NO RESPONSE

FIG 2. Anatomical mapping at six coronal (AP) levels of sites in the hypothalamus at which micro-injections of norepinephrine or epinephrine in doses of 1-12 μg produce hypothermia (▼). Sites at which these catecholamines cause no change in temperature are also indicated (o). AC anterior commissure; ANT anterior hypothalamic area; CC corpus callosum; DB diagonal band of Broca; DM dorsomedial nucleus; F fornix; FF fields of Forel; GP globus pallidus; IC internal capsule; LAT lateral hypothalamus; LV lateral ventricle; LS lateral septal nucleus; MM mammillary body; MS medial septal nucleus; NAC nucleus accumbens; OC optic chiasm; OT optic tract; PH posterior hypothalamic area; PO preoptic area; PP cerebral peduncle; PU putamen; PV paraventricular nucleus; RN reticular nucleus of the thalamus; VA anteroventral nucleus of the thalamus; ZI zona incerta; 3 V third ventricle. Horizontal and lateral scales are in mm. Vertical zero represents the stereotaxic zero point 10 mm above the inter-aural line (21).

mammillary body and ventral to the mammillary fasciculus princeps. Lateral to this region, however, cholinergic substances cause a rise in temperature when micro-injected in the ventral region of the zona incerta and in an area just dorsal to the cerebral peduncle. These opposite changes in temperature elicited at the junction of the hypothalamus and mesencephalon are illustrated in Figure 3 in which: 6 µg of a mixture of acetylcholine and eserine produced no response at site A; 6 µg of the same solution at site B produced a fall in temperature of over 1.0°C, which lasted 4 hours; 3 µg carbachol at site C evoked a hypothermia of over 1.5°C; and at site D, 6 µg carbachol produced a transient elevation in temperature. Eating, drinking, drowsiness, vomiting or alterations in respiratory rate occurred following only 15 per cent of the micro-injections at this anatomical level, and these responses could not be related to changes in temperature.

The anterior hypothalamic "thermostat" can also be affected by drugs and other agents. In fact, an anesthetic substance applied locally to this region will evoke an intense fall in temperature, which is dependent upon the dose of the anesthetic substance. Figure 4 illustrates the hypothermic responses produced by the anesthetic agent chloralose, which was micro-injected into the anterior hypothalamus of an anesthetized cat. When the temperature was rising, as the animal began to come out of anesthesia, chloralose in doses of 7, 14 or 32 micrograms produced a renewed hypothermia with the highest dose causing the temperature to fall to 32°C.

As a corollary to the studies involving the local elevation of an endogenous substance, we have also attempted to correlate the release of different transmitters during heat and cold stress. It is certain now that acetylcholine and other substances are spontaneously elaborated from different regions of the brain (1,2,6,10,11). Recently, we have found that at least two humoral factors of an unknown chemical nature are released

CHOLINERGIC MICRO-INJECTION SITES

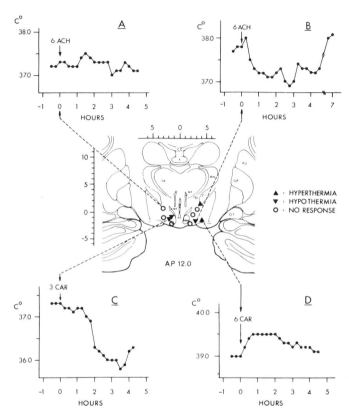

FIG. 3. Anatomical "mapping" at coronal level (AP) 12.0 of sites at the junction between the mesencephalon and posterior hypothalamus at which acetylcholine or acetylcholine-eserine mixture (ACh) in doses of 2-6 µg, or carbachol (CAR) in doses of 0.4-6 µg cause either hyperthermia (▲), hypothermia (▼) or no change in temperature (O). MF mammillary fasciculus princeps; MT mammillo-thalamic tract; other abbreviations and the stereotaxic scales are the same as in Fig. 2. In A, a micro-injection at site (O) of 6 µg ACh-eserine mixture at zero hour; in B, a micro-injection in site (▼) of 6 µg ACh-eserine mixture at zero hour, in C, a micro-injection in site (▼) of 3 µg carbachol at zero hour; and in D, a micro-injection at site (▲) of 6 µg carbachol at zero hour (21).

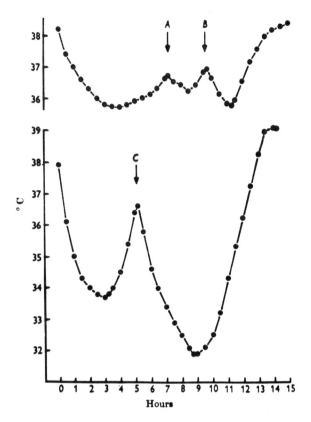

FIG. 4. Temperature responses of a cat anesthetized with intraperitoneal pentobarbitone sodium 33 mg/kg. At the arrows, micro-injections were made of chloralose: at A, 7 μg in 1 μl; at B, 14 μg in 2 μl; and at C, 32 μg in 4 μl. Infusion times were 46, 92 and 184 sec. respectively. The lower record was obtained a week after the upper record (Feldberg and Meyers, J. Physiol., 179: 509-517, 1965.)

from the hypothalamus and other regions of the monkey's brain during changes in the physiological state of the animal. When perfusate from the anterior hypothalamus of a cold-stressed, shivering donor monkey is transfused to the same site in a normothermic monkey, the recipient shivers and its temperature rises sharply. Conversely, if the same procedure is followed, with the exception that the donor is heated, the recipient's temperature usually falls (18). Since the temperature of the donor's perfusate is controlled, it would appear that direct evidence is provided which correlates the release of a specific neurotransmitter factor or factors in the hypothalamus with the central regulation of body temperature (17).

One of the active factors in the perfusate could be 5-HT. When a monkey is exposed to a cold air temperature, 5-HT release within its anterior hypothalamus increases from 4- to 24-fold. On the other hand, heating usually fails to alter the resting level of 5-HT and in some cases may suppress the normal output of this monamine by as much as 50 to 100 percent (23). In addition, during heat stress, the level of non-esterified fatty acids (NEFA) may double and even triple in the effluent collected from the anterior, pre-optic region. At the same time, blood levels of the fatty acids remain constant (23). Although the increase in NEFA in effluent collected from brain tissue has not been described previously in the literature, it may be that the elevation in the concentration of this substance may reflect an increase in norepinephrine metabolism within the rostral hypothalamus.

ALTERATION OF THE TEMPERATURE SET-POINT

If the theory is correct that the anterior hypothalamus contains the chemical "thermostat", then another area could possess the cells responsible for maintaining the "set-point." This notion seems to have some foundation in terms of a number of other related observations. For example, infants have no functional biogenic amine system and are unable to regulate against heat or cold stress; nevertheless, a temperature "set-point" of 37°C is apparently maintained by some other means.

Recently, a mechanism was discovered which may account for the phenomenon by which the body temperature of most mammals is intrinsically established at approximately 37°C (15). When sodium is perfused, in a calcium-free solution, through the cerebral ventricles of an unanesthetized cat, an intense hyperthermia is produced which is blocked by the addition of calcium to the perfusate (7). This finding led to the experiments of Myers and Veale (19) which resulted in the hypothesis that the mechanism for the set-point for body temperature depends on a constant balance between the concentration of sodium and calcium ions within the posterior

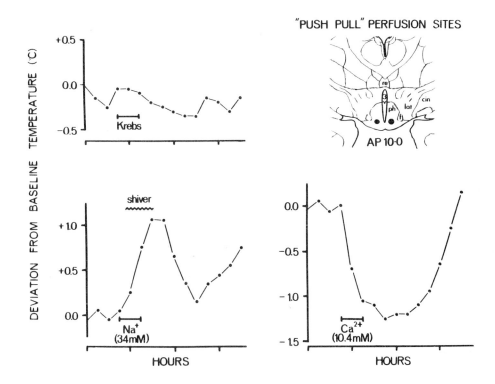

FIG. 5. Changes in the colonic temperature of an unanesthetized cat in response to the local perfusion of the posterior hypothalamus for 30 minutes of a Krebs solution alone (upper left); Krebs solution plus 34 mM excess sodium (lower left); Krebs solution plus 10.4 mM excess calcium (lower right). The site of each bilateral perfusion is designated by the dots in the inset. Shivering occurred as indicated (19).

hypothalamus (20). The evidence for this is as follows. When a solution containing sodium in excess of the normal physiological concentration is perfused within the posterior hypothalamus of a cat, by means of "push-pull" cannulae, a marked hyperthermia develops. On the other hand, an excess of calcium ions in a solution perfused at the same site causes a decline in temperature. As long as the ratio between the levels of

VENTRICULAR PERFUSIONS IN THE MONKEY

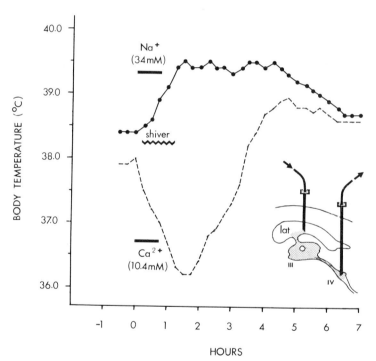

FIG. 6. Changes in the temperature of an unanesthetized rhesus monkey in response to the ventricular perfusion of artificial cerebrospinal fluid plus 34 mM excess sodium (<u>upper</u> <u>trace</u>) or 10.4 mM excess calcium (<u>lower</u> <u>trace</u>). The perfusions were at a rate of 0.1 ml per min and were separated by 48 hours. Inflow was in the lateral ventricle and effluent was collected from the fourth ventricle as shown by the inset (15).

sodium and calcium is maintained at the normal physiological concentrations, the temperature of the animal will remain stable. Figure 5 illustrates the deviations from baseline temperature following an excess in sodium and calcium ions.

A most important question is whether the temperature "set-point" can be raised or lowered in a species more closely related to man. This has been accomplished in recent experiments in which the sodium-calcium ratios were altered in a physiological solution which was perfused through the cerebral ventricles of the conscious rhesus monkey (15). As shown in Figure 6, excess sodium ions in the perfusate produced hyperthermia, whereas an excess in the normal amount of calcium evoked hypothermia. The onset, rate and magnitude of the calcium-induced fall in temperature depended on the rate and concentration of this essential cation (24). Figure 7 (top) illustrates the effect of calcium ions, in one monkey, perfused at 100 µl/min in three concentrations, at 48 hour intervals. Hypothermia was elicited when calcium was 10.1 mM or 22.7 mM in excess of that amount in the Krebs solution. However, in three experiments with a second monkey, 10.1 mM excess calcium was without effect, whereas 22.7 and 47.9 mM excess calcium produced a long-lasting hypothermic response which depended upon the concentration of calcium. Unlike the perfusion-dependent nature of the hyperthermic response to sodium, the temperature often continued to decline even after the perfusion with calcium had been terminated. As shown in Figure 7, excess calcium in a concentration of 47.9 mM caused a fall in temperature of 2.0°C during the interval of perfusion and an additional decline of 2.0°C during the following 45 minutes. In this case, the temperature of the animal did not return to the baseline level until over 4 hours had elapsed.

Within 3 to 5 minutes after the start of the perfusion of calcium, the ear vessels became dilated, the animal became flushed and the surfaces of the skin including the footpads and palms were warm to the touch. In several monkeys, the respiratory rate also increased, but this was not a

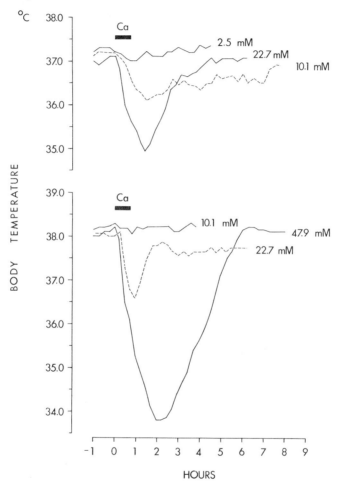

FIG. 7. Temperature records of two unanesthetized monkeys in response to the 40 minute perfusion (━━) from the lateral to the fourth ventricle at 100 µl/min of a Krebs solution containing 2.5, 10.1 or 22.7 mM excess calcium (top); and 10.1, 22.7 or 47.9 mM excess calcium (bottom) (24).

consistent response to excess calcium ions. When calcium in a low concentration was perfused, the monkey remained alert, but at the higher concentrations a general diminution of responses was produced that was characterized by the suppression of the withdrawal reflex to nociceptive stimuli, as well as the tendency for the orienting responses to auditory and visual stimuli to become sluggish. Within an hour after the perfusion had ended, vasoconstriction occurred and shortly thereafter, the animal began to shiver vigorously. An important finding in these experiments is that calcium did not ordinarily cause an "over-shoot" in body temperature, which is often observed following the hypothermia induced by an anesthetic (5). As the temperature of the animal approached the baseline, shivering decreased and abated altogether when the temperature was within 0.3 to 0.4°C of the pre-perfusion level (24).

If a new temperature "set-point" is, in fact, established as a result of altering the sodium-calcium balance in the brain-stem, then the primate should be able to thermoregulate around this new level of body temperature. To test this hypothesis, thermal stimuli were applied in the form of hot and cold water loads given intragastrically via a nasopharyngeal tube (22). Figure 8 illustrates the deep hypothermia induced by excess calcium ions in the artificial cerebrospinal fluid perfused from one lateral to the other lateral ventricle of a conscious monkey. The hypothermic temperature of approximately 31°C was sustained by repeated perfusions of calcium intermittently. After 6 hours, two water loads of 50 and 150 mls, each at a temperature of 58°C, were given intragastrically. In both cases, the animal became vasodilated for a short interval and its temperature rose. Then the temperature in both instances returned to the new "set-point" level following vasoconstriction.

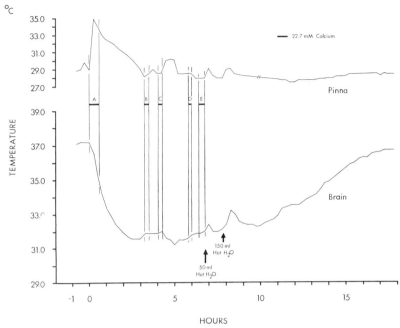

FIG. 8. Temperature records obtained from the pinna (top), and brain (bottom) of an unanesthetized monkey in response to the repeated perfusion (━━) from left to right lateral ventricle of a Krebs solution containing 22.7 mM excess calcium. The rate of perfusion was 200 µl/min (A) or 20 µl/min (B, C, D, E). At the arrows, 50 and 150 mls of 60°C water was loaded through a nasopharyngeal tube (22).

In a similar experiment portrayed in Figure 9, 50 and 150 mls of water chilled to 0°C, and given intragastrically during the calcium-induced hypothermia gave corresponding results. In this case, the monkey's temperature declined 0.5 and 1.0°C respectively and returned just to the new "set-point" level. Strong shivering appeared following the cold water load but abated suddenly as the monkey's body temperature approached the hypothermic "set-point." These two experiments, which have been repeated in different monkeys (22), clearly demonstrate that even though the temperature of the monkey is abnormally low, adequate thermoregulatory responses occur in response to cold and warm stimuli. This important finding lends further support to the hypothesis that a new "set-point" is actually established as a result of the shift in the ratio of sodium to calcium ions in the brain-stem of the monkey.

An observation of equal significance is that the functional integrity of the cells in the hypothalamus remains even during the period of hypothermia induced by a shift in the balance of ions in this structure. In

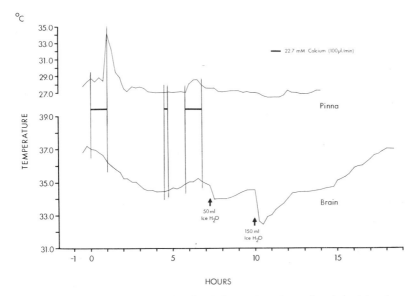

FIG. 9. Temperature records obtained from the pinna (top) and brain (bottom) of an unanesthetized monkey in response to the repeated perfusion (━━━) from the right to left lateral ventricle at 20 μl/min of a Krebs solution containing 22.7 mM excess calcium. At the arrows 50 and 150 ml of 0°C water was loaded through a nasopharyngeal tube (modified after 22).

an experiment in which excess calcium ions were perfused by means of push-pull cannulae inserted directly into the posterior hypothalamus of the monkey, the region implicated in the "set-point" function (15), normal physiological responses could still be elicited. When norepinephrine was micro-injected into the antero-lateral hypothalamus of the satiated monkey during the hypothermia, the animal began to consume food pellets immediately, and also drank water during the hour following the injection. This is shown in

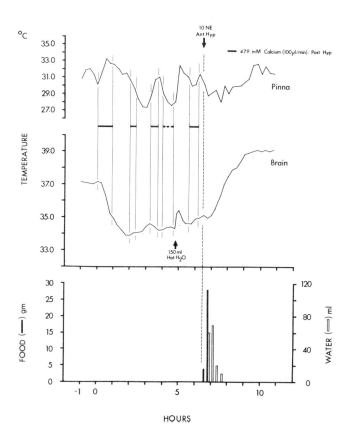

FIG. 10. Temperature records obtained from the pinna and brain along with food (left ordinate) and water (right ordinate) intakes in an unanesthetized monkey in response to repeated bilateral push-pull perfusions in the posterior hypothalamus of 47.9 mM excess calcium at the rate of 100 µl/min. At the lower arrow, 160 ml of 50°C water was administered with a nasopharyngeal tube. At the dotted line, 10 µg of norepinephrine was microinjected into the antero-lateral hypothalamus (modified after 22).

Figure 10. It would seem, therefore, that consummatory behavior can be elicited by direct activation of that region involved in the mediation of this response, and that there is no functional suppression associated with the re-setting of the "set-point", at least at this temperature, by distributing the ratio between sodium and calcium ions within the posterior hypothalamus.

CONTRIBUTION OF BASIC THERMOREGULATORY RESEARCH
TO THE PROBLEMS OF HIBERNATION

How can we better understand the fundamental mechanisms of hibernation such as those involved in the triggering, sustaining and arousal phenomena of this physiological state? Which aspects of research or normothermic functions are relevant to those processes in the hibernator?

First, it may be helpful to consider a synthesis of a large amount of research information based on experiments with the cat and monkey, so that conceptual relationships can be established between the two kinds of mammals, hibernator and non-hibernator. A schematic model of the hypothalamic control systems for temperature regulation in the non-hibernating primate (15) is presented in Figure 11. The temperature of the blood (8), a pyrogen or certain drugs such as anesthetics impinge upon 5-HT or norepinephrine (NE) containing cells of the anterior "thermostat." For body warming, 5-HT is released by these cells (16) in order to activate a cholinergic heat production pathway which passes through the posterior hypothalamus. In the posterior area, the "set-point" around which body temperature is regulated is maintained by an inborn balance in the ratio of sodium to calcium ions. The efferent cholinergic synapses for heat production and heat loss are also located in the posterior region, and may exist in the mesencephalon (16). Ascending monoaminergic nerve fibers follow a mesencephalic trajectory to the rostral hypothalamus. These fibers probably end on the cold and warm sensitive neurons which are those sensitive to 5-HT and norepinephrine. Differential thermal input from peripheral receptors (3) is most likely carried by these aminergic fiber systems. This model, which provides a framework for explaining many of the current experimental findings in this field, could serve as a basis

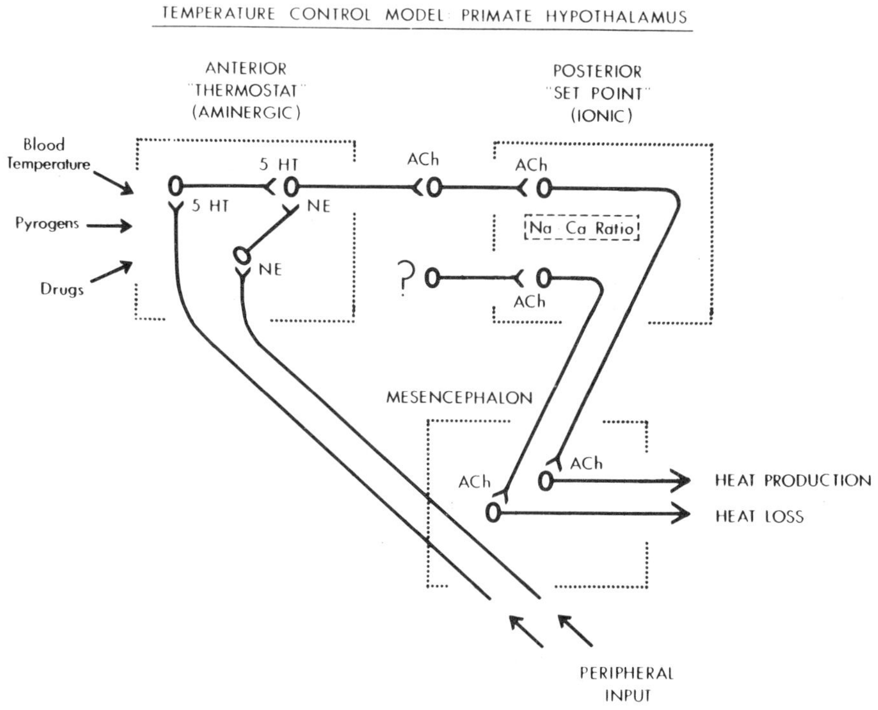

FIG. 11. Schematic diagram of a model to account for temperature regulation under normal conditions as well as during a pyrogen fever. Factors which affect the aminergic "thermostat" in the anterior hypothalamus are given, and the outflow to the posterior hypothalamic "set-point" is mediated by a cholinergic system which passes through the mesencephalon (15).

for research on the mechanisms in the brain responsible for the hibernation phenomenon. Several important points, however, must be stressed.

Localizing the onset mechanism. If the trigger or onset of hibernation is related to the downward shift in the temperature "set-point", then it is necessary to determine the relative concentrations of one essential cation to the other in the brain-stem. Based on research with the cat, it is entirely possible that the posterior area of the hypothalamus is delegated in some way to the initiation of hypothermia, with all of its attendant physiological changes, in the hibernating animal. Recently, Malan (9) demonstrated that small circumscribed lesions placed in the lateral aspect of the posterior hypothalamus abolished hibernation in the European hamster, C. cricetus. Not only did these lesions prevent the onset of hibernation, but the hamsters also gained weight and failed to build nests. During the months of November, December and January, the period of normal hibernation in this species, the lesioned hamsters also excreted norepinephrine in the urine in concentrations higher than the normal level.

Thus, we should like to make the following plea. To uncover the fundamental onset mechanism of hibernation, a precise anatomical localization of the neurochemical system is most assuredly required. Although the cerebral ventricles offer a route of investigation which not only is interesting but experimentally convenient, experiments in which the milieu of the ventricular fluid is altered constitute only a first approximation to the problem. In our present experiments, we are attempting to isolate the region or regions of the brain-stem which subserve the maintenance of the temperature "set-point."

Localization of the sustaining and arousal mechanism. It is well known that the hibernating animal possesses some capability for thermoregulation under the conditions of deep hypothermia. Because it is so clear that the action of a putative neurotransmitter in modulating the temperature of a homeotherm is restricted solely to the rostral half of the hypothalamus, we might expect that if a substance such as a monoamine also serves as a temperature transmitter in the hibernator it's locus of action would reside in an homologous diencephalic site. The sustaining mechanism as well as the arousal stimulus for the restoration of normothermia could both depend on the balance in the release of 5-HT or norepinephrine within the anterior hypothalamic area. It is important to recognize, however, that acetylcholine, amino acids, histamine or other substances should not be ruled out as central mediators or modulators in the hypothalamus of the hibernating animal. In fact, one could easily imagine that the ratio in the concentrations of a number of substances, one to another, in highly localized areas of brain tissue could well determine the specific characteristics of the hibernating state.

In summary then, we believe that the critical factors involved in the onset, maintenance and awakening from natural hibernation revolve about a modification of the ionic constituents as well as the activity of biogenic amines within the central nervous system. In both instances, these alterations are undoubtedly localized within specific regions of the hypothalamus, mesencephalon or other structures within the brain-stem. Our future progress and ultimate understanding of this remarkable phenomenon seems to depend on the elucidation of these fundamental neural systems.

REFERENCES

1. BELESLIN, D., E. A. CARMICHAEL and W. FELDBERG. The origin of acetylcholine appearing in the effluent of perfused cerebral ventricles of the cat. J. Physiol., 173: 368-376, 1964.

2. DELGADO, J. M. R., and L. RUBINSTEIN. Intracerebral release of neurohumors in unanesthetized monkeys. Arch. Int. Pharmacodyn. 150: 530-546, 1964.

3. IGGO, A. Temperature discrimination in the skin. Nature 204: 481-483, 1964.

4. FELDBERG, W., and R. D. MYERS. Temperature changes produced by amines injected into the cerebral ventricles during anesthesia. J. Physiol. 175: 464-478, 1964.

5. FELDBERG, W., and R. D. MYERS. Changes in temperature produced by micro-injections of amines into the anterior hypothalamus of cats. J. Physiol. 177: 239-245, 1965.

6. FELDBERG, W., and R. D. MYERS. Appearance of 5-hydroxytryptamine and an unidentified pharmacologically active lipid acid in effluent from perfused cerebral ventricles. J. Physiol. 184: 837-855, 1966.

7. FELDBERG, W., R. D. MYERS, and W. L. VEALE. Perfusion from cerebral ventricle to cisterna magna in the unanaesthetized cat. Effect of calcium on body temperature. J. Physiol. 207: 403-416, 1970.

8. HAYWARD, J. N., and M. A. BAKER. Diuretic and thermoregulatory responses to preoptic cooling in the monkey. Amer. J. Physiol. 214: 843-850, 1968.

9. MALAN, A. Controle hypothalamique de la thermorégulation et de l'hibernation chez le hamster d'Europe Cricetus cricetus. Arch. Sci. Physiol. 23: 47-87, 1969.

10. McLENNAN, H. The release of acetylcholine and 3-hydroxytyramine from the caudate nucleus. J. Physiol. 174: 152-161, 1964.

11. MITCHELL, J. F. The spontaneous and evoked release of acetylcholine from the cerebral cortex. J. Physiol. 165: 98-116, 1963.

12. MYERS, R. D. Is there a chemical thermostat in the brain? Naval Res. Rev. 20: 1-7, 1967.

13. MYERS, R. D. Discussion of serotonin, norepinephrine and fever. Adv. in Pharmacol. 6: 318-321, 1968.

14. MYERS, R. D. Chemical mechanisms in the hypothalamus mediating eating and drinking in the monkey. Annals N. Y. Acad. Sci. 157: 918-933, 1969.

15. MYERS, R. D. Hypothalamic mechanisms of pyrogen action in the cat and monkey. In: Ciba Foundation Symposium on Pyrogens and Fever. London: Ciba Foundation, 1971, p. 131-153.

16. MYERS, R. D., and D. B. BELESLIN. The spontaneous release of 5-hydroxytryptamine and acetylcholine within the diencephalon of the unanesthetized rhesus monkey. Exp. Brain Res., 1970. In press.

17. MYERS, R. D., and L. G. SHARPE. Chemical activation of ingestive and other hypothalamic regulatory mechanisms. Physiol. Behav. 3: 987-995, 1968.

18. MYERS, R. D., and L. G. SHARPE. Temperature in the monkey: Transmitter factors released from the brain during thermoregulation. Science 161: 572-573, 1968.

19. MYERS, R. D., and W. L. VEALE. Body temperature: possible ionic mechanism in the hypothalamus controlling the set-point. Science 170: 95-97, 1970.

20. MYERS, R. D., and W. L. VEALE. The role of sodium and calcium ions in the hypothalamus in the control of body temperature of the unanesthetized cat. J. Physiol. 1971. In press.

21. MYERS, R. D., and T. L. YAKSH. Control of body temperature in the unanaesthetized monkey by cholinergic and aminergic systems in the hypothalamus. J. Physiol. 202: 483-500, 1969.

22. MYERS, R. D., and T. L. YAKSH. Thermoregulation around a new set-point established in the monkey by altering the ratio of sodium to calcium ions within the hypothalamus. J. Physiol., 1971. In press.

23. MYERS, R. D., A. KAWA, and D. BELESLIN. Evoked release of 5-HT and NEFA from the hypothalamus of the conscious monkey during thermoregulation. Experientia 25: 705-706, 1969.

24. MYERS, R. D., W. L. VEALE, and T. L. YAKSH. Changes in body temperature of the unanaesthetized monkey produced by sodium and calcium ions perfused through the cerebral ventricles. J. Physiol., 1971. In press.

25. SHARPE, L. G., and R. D. MYERS. Feeding and drinking following stimulation of the diencephalon of the monkey with amines and other substances. Exp. Brain Res. 8: 295-310, 1969.

CREDITS

The editors and authors are grateful to the following publishers, organizations and individuals for permission to use materials previously published:

Figure 1: The Journal of Physiology; R. D. Myers and T. L. Yaksh, The Journal of Physiology 202: 483-500, 1969.

Figure 2: The Journal of Physiology; R. D. Myers and T. L. Yaksh, The Journal of Physiology 202: 483-500, 1969.

Figure 3: The Journal of Physiology; R. D. Myers and T. L. Yaksh, The Journal of Physiology 202: 483-500, 1969.

Figure 4: The Journal of Physiology; W. Feldberg and R. D. Myers, The Journal of Physiology 179: 509-517, 1965.

Figure 5: American Association for the Advancement of Science; R. D. Myers and W. L. Veale, Science 170: 95-97, 1970. Copyright American Association for the Advancement of Science, 1970.

Figure 6: The Ciba Foundation and Churchill Livingstone, Edinburgh and London; R. D. Myers, "Hypothalamic mechanisms of pyrogen action in the cat and monkey," in Ciba Foundation Symposium on Pyrogens and Fever. London: Churchill Livingstone, 1971, p. 131-153.

Figure 11: The Ciba Foundation and Churchill Livingstone, Edinburgh and London; R. D. Myers, "Hypothalamic mechanisms of pyrogen action in the cat and monkey," in Ciba Foundation Symposium on Pyrogens and Fever. London: Churchill Livingstone, 1971, p. 131-153.

A POSSIBLE MODEL FOR THERMOREGULATION DURING DEEP HIBERNATION

R. H. LUECKE and F. E. SOUTH

Space Sciences Research Center and
Departments of Chemical Engineering,
Veterinary Physiology, and of Physiology,
University of Missouri, Columbia, Missouri 65201

INTRODUCTION

The problem of thermoregulation at very low temperatures raises a difficulty that is a problem in logic almost as much as in physiology. The ability of any biomechanism to function at both high and low temperature implies that it has, to an important degree, been rendered independent of temperature. The existence of temperature regulation requires a sensor capable of being significantly affected by temperature change. The existence of temperature regulation during hibernation may be questioned on the basis that the temperature sensor is required to be both largely independent of changes in temperature and yet maintain a high sensitivity to them. Indeed, Hammel (6) estimated that the activity of those sensors which are most active during temperature increases would become inactive should the central temperature fall to below 30° to 32° C. The implication was that at low temperatures thermoregulation does not exist.

Furthermore, it has not been demonstrated that a temperature regulating system is required under hibernating conditions. Analysis of exothermic chemical reaction systems (1) shows that, for a wide range of physical parameters, there are three points for the heat generated by the chemical reaction to be in equilibrium with that transferred out of the system to the external environment (Fig. 1). These are points of intersection between the "S-shaped" reaction rate equation and the almost-linear transfer equation.

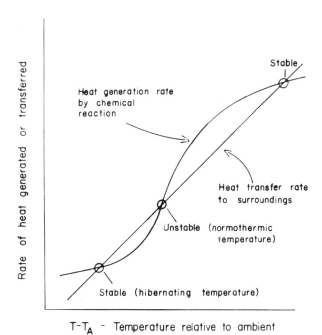

FIG. 1. Heat generation and removal diagram for an exothermic reaction system.

The intermediate equilibrium point represents the point of normothermia for most mammals. That this is an unstable equilibrium can be demonstrated by considering a small perturbation from the steady state. A small change in internal temperature would cause a greater increase or decrease in reaction rate (exponential) than could be compensated for by passive heat transfer; the net effect would be a further departure from equilibrium.

Stability in this type of system may be achieved by either controlling the effective concentration of the reactant or the rate of heat transfer with respect to the environment. Feedback thermoregulatory control systems of mammals is normally achieved by a combination of these methods.

The high and low temperature equilibrium states are inherently stable; the "flat" portions of the exponential curve have been reached. Consideration of small temperature perturbations from either of these temperature states shows that transferred heat changes more rapidly than generated heat. A perturbation to below either equilibrium point would result in a net heat accumulation which in turn would result in a subsequent return to the equilibrium point. The system is stable without feedback control. The high temperature equilibrium occurs because of exhaustion of fuel supply or because of secondary changes in the system. The low temperature point, analogous to the hibernal situation, represents an area of very low reaction rates. The temperature gradient needed to transfer the generated heat out of the system is also very low so the body temperature at this state is only slightly higher than that of the environment.

It is important to note that stability does not necessarily imply constancy of internal temperature. The abcissa in Figure 1 is a difference between internal and ambient temperatures; the internal temperature drifts passively with changes in the ambient. Such passive thermal "drifting" with environmental temperature is variously reported for hibernating mammals. For example, Lyman (10) cites that for a number of hibernators, including bats, dormice, and hamsters, body temperatures may passively follow that of the environment over the range of 3° to 15° C. Further, Hammel (6) hypothesized that for all species, "regulation of temperature during

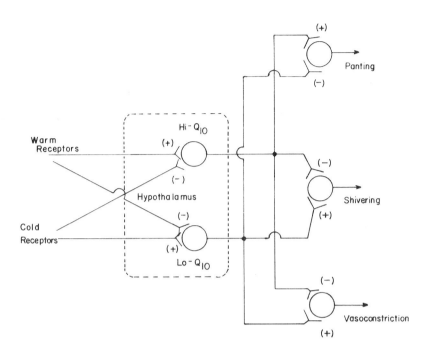

FIG. 2. Diagram of model for normothermic thermoregulating system for mammals.

hibernation...is no more than an equilibrium between heat loss and heat production which are in turn passive functions...of temperature".

Nonetheless, some evidence of low temperature regulation has been reported. Lyman (10) cites that a "residuum of temperature control" persists for bats, dormice and hamsters. At environmental temperatures below 3° C, increases in metabolic and heart rate may be observed which allow the animal to maintain the necessary temperature gradient between his body and the surroundings. In their classic study on marmots, Benedict and Lee (2) reported qualitatively graduated metabolic responses to decreased environmental temperatures. In the range of 2° to 18° C, the hibernating marmots maintain a differential of 1° to 4° C between rectal and environmental temperatures. The larger temperature differentials occurred at the lower environmental temperature. Heat production rates were higher at low temperature indicating the "retention of a vestige of chemical (thermo) regulation". Thermoregulation against freezing has obvious survival value.

A somewhat different point of view was reported by Strumwasser (12) in that he reported a surprisingly precise control of brain temperature (10.7 $\pm$ 0.05° C) while there were slow fluctuations in temperature between 7.8° and 8.4° C. He interpreted these data along with those obtained during entry into hibernation as indicating a rather precise control of body temperature during both stages of hibernation in Citellus beechyi.

THERMOREGULATION BY HIBERNANTS DURING NORMOTHERMIA

Hammel (5) proposed an operational thermoregulatory control law of the form: $R - R_0 = \alpha_R (T_{hypo} - T_{set,R})$

where $R - R_0$.....is the thermoregulatory response;

R_0 is the basal response level when $T_{hypo} = T_{set,R}$;

T_{hypo} is the actual temperature;

$T_{set,R}$ is the functional setpoint for the response R;

and α_R is the proportionality constant for the response $(R - R_0)$.

The equation states that a given thermoregulatory response is proportional to a signal which is a difference between the actual hypothalamic temperature and some threshold temperature for that particular response. The threshold temperature ($T_{set,R}$) adjusts as a function of other inputs to the system such as skin thermoreceptors, non-hypothalamic core receptors, state of arousal, exercise and cortical input. Without debating the ultimate correctness of the formulation, it must be admitted that it is at least a feasible candidate. The equation is capable of correlating much thermoregulatory data and hence, at least in its overall effects, must represent an approximation to the true relationship.

A physiological fine structure for this model was also proposed. The schematic diagram in Figure 2 shows hi-Q_{10} and lo-Q_{10} neurones located in the hypothalamus. These neurones synapse with motor neurons which may be either facilitated or inhibited in order to generate the appropriate thermoregulatory response. Axons of peripheral warm and cold receptors synapse with the sensory neurones and cause facilitation or inhibition. At equilibrium (thermoneutral zone) the outputs from the neurones are, in effect, balanced so that none of the thermoregulatory responses are activated. A change in temperature causes a relative increase in the firing rate of either the hi- or lo-Q_{10} receptors and an appropriate resultant thermolytic response. The important effect of skin temperature, etc. is, through facilitation or inhibition, to alter the temperature at which these regulatory responses occur.

Two questions must concern us here: (a) Do hibernators, under normothermic conditions possess a thermoregulatory control system similar to that of other mammals? and (b) To what extent is the normothermic regulatory system modified or replaced as the animal descends into hibernation?

In so far as the responsiveness of the preoptic area to thermal alteration is concerned, the answer to these first questions is probably in the affirmative. According to the results of Williams and Heath (13), metabolic rate and body temperature increased or decreased as the preoptic areas of C. tridecemlineatus was cooled or heated, respectively. Analogous responses were obtained recently in our laboratory (7) in which the electrographic responses of the marmot to preoptic temperature manipulation was studied under conditions of 20° and 5° C ambient temperature. The greatest deviation from what might be expected lay in the somewhat greater reliance on metabolic activity (chemical regulation) and with relatively low reliance on such mechanisms as panting and vasomotor activity. The control responses appeared to be less precise than that of most normothermic mammals - although it may be os some significance that the measurements were made during the autumn, prehibernal period.

THERMOREGULATION IN HIBERNATION

If one assumes the essential correctness of the normal thermoregulatory system then one is left with only a limited number of choices with respect to the possibility of any such system operating during hibernation. These might be:

a) Readjustment of the setpoint to a lower (hibernating) temperature which, once that temperature were obtained, would permit these systems to operate in a more or less "normal" manner. Continuous control during entry into the state might or might not be implied.

b) Abandonment of all control activity would be possible, assuming the ambient temperature was not too low, - the body temperature remaining at a fairly stable point as dictated by thermal exchange with the environment and minimal heat production.

c) Single-sided control might exist such that only defense against a lower limit would remain. Entry would be passive and control exerted only as lower limit was approached.

It is essential to the normothermic thermoregulatory model above that hi-Q_{10} neurones facilitate the motor neurones that increase heat loss, and simultaneously inhibit those that are responsible for heat loss and increased heat production. At the same time the lo-Q_{10} sensory neurones must inhibit and facilitate these activities in the opposite manner. If it is assumed that thermoregulation in hibernation is merely an extension of the normothermic model with an additional input for hibernating setpoint displacement, lowering setpoint would be accomplished by inhibition of the lo-Q_{10} and facilitation of the hi-Q_{10} neurones. As Hammel has observed, such a facilitation with continued activity down to 5° C may be regarded as rather unlikely. Accordingly there would be a lack of the "opposing force" necessary for proper operation of the low temperature regulatory system. Hammel (6) hypothesized that any low temperature regulation would be accomplished by "increased facilitation of lo-Q_{10} sensors by the reticular activating system possibly abetted by humoral agents." While quite adequate in so far as it goes, this description simply sidesteps the issue to a large extent.

Low temperature regulatory response may be shown to be graded to the degree of thermal stress and effective at maintaining approximate constancy of core temperature for a long term. Furthermore this control is improved with successive trials, i.e., progressive adaptation. To accommodate as many of the observed phenomena as possible, a new model for thermoregulation in hibernation is proposed here as shown in Figure 3. The normothermic model is altered only by the addition of an inhibitory channel for the lo-Q_{10} sensors that is activated in order to induce hibernation.

Since it has been shown that many tissues and organs have the peculiar ability to function at surprisingly low temperatures (cf. 11,14) it would not be unreasonable to assume that in the absence of active inhibition the lo-Q_{10} sensors of hibernators share this ability. If this is so, then the only way the hibernating temperatures could be reached and maintained would be by (a) increased facilitation of hi-Q_{10} sensors, (b) some kind of "disconnect" in the lo-Q_{10} portion of this system so that the firing of that neurone pool would be ignored, or (c) direct inhibition of the lo-Q_{10} sensors. Aside from many other difficulties, the first two of these alternatives are

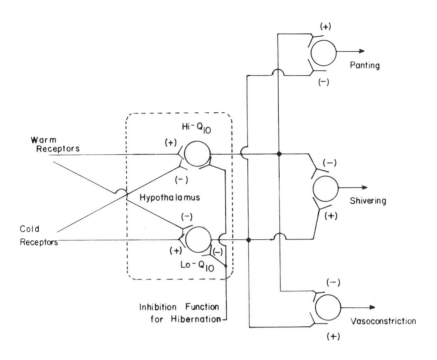

FIG. 3. Diagram of model proposes for thermoregulating system for mammalian hibernators.

physiologically wasteful. The last may be adapted to explain many of the observed low temperature thermoregulatory data. An assumption is added that the inhibitory function is somewhat more susceptible to failure at very low temperatures than are the lo-Q_{10} sensors themselves.

This assumption is a key point in the model and would explain the reactivation of heat producing responses as temperatures close to freezing are approached. Using this model, entry to hibernation occurs when the inhibitory process is activated. This would be the net equilavent to the belief that the entire control system is "switched off" during hibernation.

Note, however, that we propose an active inhibitory mechanism rather than the passive behavior of turning off a control process.

As extremely low temperatures are approached, the inhibitory function would begin to fail; the lo-Q_{10} sensory neurones would be, in effect, reactivated and an operational low temperature setpoint created. In contrast to the normthermic proportional control system, in this model the setpoint is created by gross displacement of the level of activity of the sensory neurones through intermittent action of the inhibitory function (Fig. 4).

Because of the multiplicity of factors involved, the point at which any given physiological function may fail is usually indeterminant - thus the operational low temperature setpoint may be viewed as being variable on both long and short term basis. The location of a line correlating sensory neurone firing rate with temperature may be viewed then, as analogous to the quantum mechanical explanation of electron orbits. At any given time the line may exist in one of two locations depending on a probability level that is a function of many factors such as ambient temperature or previous thermal history. Depending upon which of these probability clouds contains the correlation, the control output of heat production would tend, at low temperatures, to be an on-off situation at any given temperature. Whether the maximum or minimum activity occurred would be predictable only in the statistical sense.

Under conditions for which the inhibitory function is beginning to fail, the system would become more sensitive to other inputs. Thus a disturbance such as a high rate of decline of skin temperature might produce an output overriding the inhibition.

With the increased firing rate of the lo-Q_{10} sensory neurones resulting from the sustaining of inhibition, some degree of thermogenesis would be evoked. Vasoconstriction, if active, might not be detectable because of

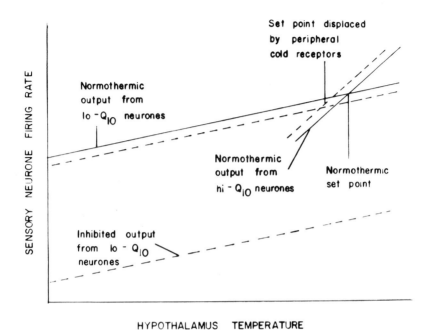

FIG. 4. Firing rate of hi - and lo - Q_{10} neurones in hypothalamus as functions of temperature.

low thermal gradient and minimal circulation rate. The increase in heat production would, of course, result in a sharp rise of brain temperature which in turn would permit renewed inhibitory activity of the lo-Q_{10} sensors. And again, the generation would be terminated allowing brain temperature to drop, repeating the cycle.

This type of action could produce a non-terminating series of cycles. Whether the cycles would be unstable and grow until arousal or whether a stable cycling system would result would depend on the degree of thermal stress. If environmental temperatures were low enough the cycles could grow in size until arousal from hibernation resulted. It's important to remember that the primary process active in this behavior is suspension of the fundamental mechanism that allows entry into, and endurance of, hibernation.

Should environmental temperatures be in an acceptable range, say 0.5-5°C, then the inhibitory function might be strengthened during a second cycle. The inhibited lo-Q_{10} sensor activity might be as great as before but the duration over which it is "switched on" would be reduced; thus this degree of overshoot in brain temperature would be reduced in the latest cycle. This control process would be frequency modulated rather than amplitude modulated. This type of action is termed "bang-bang" control, in that the controller is full on or full off.

A steady state of cycling could be reached when the inhibiting function is repeatedly overridden to only a small degree so that the subsequent brain temperature overshoot is small. This type of non-terminating but bounded cycling is called a "limit cycle". Given a small degree of adaptative ability of inhibitory function, this state could be reached for a rather broad range of low temperature conditions in an animal large enough to have significant thermal inertia. Brain temperature would not be a smooth, constant level however, but would tend to repeatedly cycle through a small temperature range.

LOW TEMPERATURE THERMOREGULATION EXPERIMENTS

In order to test these hypotheses, yellow-bellied marmots (M. flaviventris were fitted with chronic subcortical thermocouple leads, cortical and subcortical electrodes, and an EMG-EKG electrode on the dorsum of the neck along with a surface thermocouple. The animals were continously monitored as they entered and remained in hibernation at an ambient temperature of 5° C, unless altered for experimental purposes.

Several other workers have examined relationships between a forcing function of thermoregulatory system, such as ambient temperature, and one of the controller mechanism, such as heart rate or metabolic rate. Henshaw (8) for example, compared steady state metabolic rates of the little brown bat at various ambient temperatures and obtained a rough correlation between metabolic rate and the difference between ambient and body temperature.

In contrast to the work using steady state conditions, the experiments described here depend on intensive analysis of the unsteady thermal response. Perturbations to the steady state ambient conditions were introduced and the form and extent of transient thermal response were analyzed by several methods. Traditional unsteady state methods were used such as correlation and Fourier analysis. But in addition, the observed thermal response was compared to that computed from a mathematical model of the hibernator which had been developed in considerable detail (9). In this model the marmot was represented, when curled into the hibernating position, as three concentric spheres--a central core, a muscle layer and a layer of skin and fur. Thermodynamic unsteady state energy balances gave three coupled nonlinear differential equations which were solved numerically on a digital computer. The model was very successful in predicting the cooling term of the descent of a marmot into hibernation. Significantly, the model predicted a temperature differential of about 1.5° C between environment and core at 5° for the passive system. The observed average is nearer to 2.5 to 3.5° C suggesting non-minimal heat production rates with some degree of thermoregulation. The three experiments described here are from a series designed to describe, at least operationally, thermoregulation during hibernation and to demonstrate (a) existence of the control system and the type of controlling response that is evoked; (b) the degree and rate of controlled response; and (c) the long term effectiveness of the controller.

EXPERIMENT I. THERMAL RESPONSE TO CHANGES IN AMBIENT TEMPERATURE

Description. Previous to this experiment, the marmot had been in deep hibernation for over ten hours. Environmental temperature had been maintained at 5° C and the brain temperature was very steady at 8.5° C. At the start of the experiment, the room temperature was lowered at a fast rate as possible reaching 2° C in about 30 minutes (Fig. 5). This temperature was decreased to 1.6° and maintained at that value for about the next ten hours.

Because of thermal inertia, an immediate effect of the decrease in ambient temperature on the brain temperature could not be detected (Fig. 5a). However, approximately 24 minutes after the room temperature had dropped to 2° C, a sudden increase in heart rate occurred (Fig. 5b) from a previous average of about 5 b/m to a rate of 50 b/m over a period of 40 minutes. Simultaneously, an onset of EMG activity occurred (Fig. 5c). About 12 minutes after the heart rate increased the brain temperature also showed a rapid rise moving from an average of about 8.5° C to a peak of 11.6° C. The increase in heart rate and EMG activity preceded significant changes in brain temperature.

After peaking the heart rate decreased even more quickly than it had risen. The rate was back to 5 to 10 b/m even before the maximum brain temperature occurred. High EMG activity persisted even after the heart rate had returned to near minimal level. After about 3 minutes an abrupt EMG decrease to almost zero occurred. The brain temperature began decreasing immediately after the peak and followed an exponential Newtonian cooling term.

At about 200 minutes (Fig. 5b), increased irregularity of heart rate was observed and again at about 300 minutes. The second peak was lower

POSSIBLE MODEL FOR THERMOREGULATION

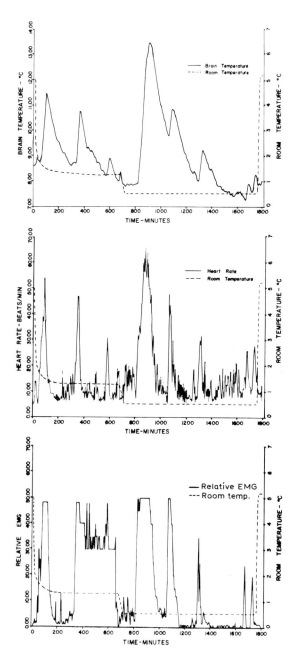

FIG. 5. Response of hibernating marmot to step changes in environmental temperature.

than that recorded earlier and suggests some "learning" or adaptation in the control system to produce a lesser temperature overshoot, i.e. the system "settled down." Brain temperature again lagged behind heart rate and it also had a smaller peak than previously.

Following the peaking, the heart rate returned very rapidly to 5 to 10 b/m while brain temperature followed an exponential cooling curve. The EMG activity decreased after 25 minutes as compared to about 40 minutes in the first episode. This time, however, it did not decrease back to nearly zero, but a moderate level of EMG voltage and hence presumably EMG activity, persisted even with low heart rate.

At about 540 minutes, an additional rapid rise in heart rate occurred. Short term increases in EMG activity were also evident. This time a peak heart rate of 32 b/m was obtained with a following peak brain temperature of 8.9° C. These reactions were clear improvements over those in previous episodes and suggest an adaptation of the control system.

In order to learn whether some degree of graded thermoregulatory response could be obtained, the ambient room temperature was decreased to 0.5° C. This further temperature reduction produced an almost immediate response in the form of increased irregularity of heart rate along with a higher average value of about 16 b/m. Thereafter, at 820 minutes another very rapid increase in heart rate occurred, this time peaking at 65 b/m, the highest observed. The brain temperature, lagging behind as before overshot to 13.5°C, two degrees higher than the previous maximum: the larger stress of the lower ambient temperature appears to have produced a more extreme response. Heart rate did not decrease immediately after peaking but maintained elevated rates for over 90 minutes as did the EMG activity, which was twice the duration of the first episode.

The pattern of sudden increases in heart rate and EMG activity followed by a settling period occurred twice more at 1030 and 1270 minutes. As before, substantial adaptation of the thermoregulatory regulating system is impressive with sequential reduction of overshoot. Finally, at about 1500 minutes, the brain temperature seemed to have stabilized at about 7.5° C with only moderate variability. The heart rate, though highly variable and at the higher average rate of about 15 b/m, did not show the gross overshoot that occurred previously.

Subsequently the experiment was terminated by adjustment of the room temperature back to 5° C. The brain temperature increased rapidly to 8.5° C with a small overshoot. The marmot resumed normal hibernation which was maintained for the following 16 hours, at which time monitoring was ended.

Analysis. Cross correlation between heart rate and brain temperature (Fig. 6) showed that the average brain temperature time lag was about 22 minutes. Inspection of the data indicated that the lag was somewhat greater in the second half of the experiment than in the first. This could be a cummulative effect due to the duration of the experiment or it could be due to the increased stress of the lower ambient temperature.

Fourier analysis of this data showed that the coherence between brain temperature and heart rate was high, indicating that a large fraction of the deviations in heart rate correlated with those in brain temperature.

The mathematical simulation for the passive system included a computed minimal metabolic rate which decreased with temperature according to an exponential relationship. A variable multiplicative factor onto this minimal metabolic rate was introduced into the simulation. This factor was adjusted during the numerical solution of the differential equations so that the computed output exactly matched the measured brain temperatures

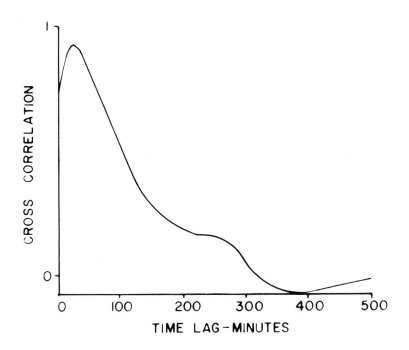

FIG. 6. Cross correlation of brain temperature and heart rate of hibernating marmots.

from this experiment. This computed metabolic factor shown in Figure 7 is nearly a perfect linear function of measured heart rate. This data indicates that an increased metabolic rate, either causing or caused by, the increased heart rate was the controlling mechanism for thermoregulation.

This experiment showed that there could be no doubt of the existence of some type of low temperature control. The temperature detection system demonstrated a strong sensitivity to the secondary change from 1.5° to 0.5° C environment. The control of brain temperature was rough with large overshoots occurring, followed by slow recovery time. However, an adaptive process was operative with reasonable control apparently being obtained after only three overshoot cycles. The control was characterized by incessant small variations in brain temperature resulting from intermittent bursts of increased heart rate.

It is clear that the initial responses were not the result of changes in brain temperature. There is a small EKG-EMG response near the start of the experiment that is almost certainly due to the rapid decrease in skin temperature and consequent to the reduced ambient temperature. Later responses seem to be triggered by a combination of decreasing skin temperature and low muscle and core temperature. The secondary adaptations might also be considered to be functions of these latter quantities. Attempts have been made to define these relationships mathematically but we have had little success in these efforts so far.

EXPERIMENT II. DEGREE AND RATE OF CONTROL RESPONSE

Description. The objective of this experiment was to expose a marmot in deep hibernation to a series of sudden changes in ambient temperature and to observe the response. The technique is related to the concept of frequency response testing which often is used to quantify the dynamic parameters in industrial process systems.

After a period of several hours of hibernation with ambient temperature at the normal 5° C, the room temperature was dropped to about 1° C as rapidly as possible (10-15 minutes). The low temperature was maintained for one hour and then the temperature was raised to about 6° C at the highest possible rate (also about 10 minutes). After one hour at 6° C, the cycle was repeated (Fig. 8). The procedure gave an approximate "square wave" input that is among the most useful for dynamic analysis (3,4).

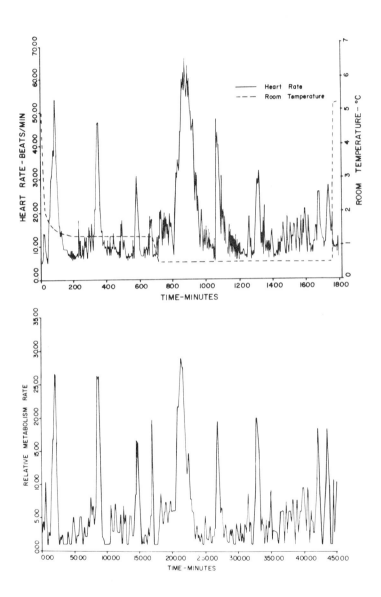

FIG. 7. Comparison of computed metabolic rate with measured heart rate.

After six cycles, the length of the cycle was doubled so that two hours were spent at both the high and low temperature level. This change added resolving power to the analysis of lower frequency components of the system. After three of these cycles, cycle time was again doubled so that four hours were spent at the high and low temperature levels.

Shortly after the first temperature decrease, heart rate and EMG activity increased rapidly as had been seen in other experiments. The brain temperature also increased although clearly lagging behind heart rate and EMG as in the previous experiments (Fig. 8). After the first hour at low temperature the ambient was raised to the higher level as scheduled. The EMG activity ceased almost immediately and heart rate slowed to basal levels shortly afterwards. The brain temperature continued upward however, with the overshoot being reinforced by the sudden change in ambient conditions. The natural frequency of this system was such that the brain temperature response was "out-of-phase" with the imposed input.

The brain temperature peaked near the middle of the high temperature cycle and then began to decrease in a Newtonian fashion. When the ambient was again lowered to 1° C, the trend of the brain temperature was again reinforced by this input. Thus the low temperature undershoot was also accentuated. Heart rate and EMG activity reacted immediately to the renewed lower ambient temperature but the degree of reaction was reduced from the first episode. This action is reminiscent of the adaptation in the earlier experiment except that this time there has been an intervening warmer period. The adaptation had been retained from the cold cycle one hour previous. Despite the overshoot, brain temperature eventually began to rise from the metabolic heat input. The rise was again reinforced as a consequence of the ambient temperature being driven upward on the next cycle. This time, however, the more moderate heart and EMG activity reduced the brain temperature

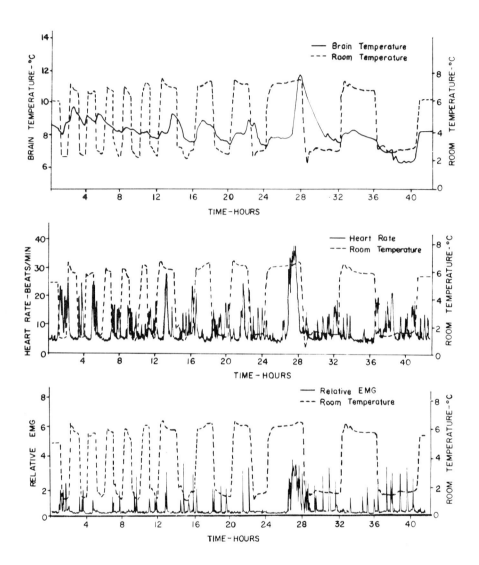

FIG. 8. Response of hibernating marmot to square wave inputs of environmental temperature.

overshoot. The pattern suggests that the control system was beginning to adapt itself to the ambient temperature switch and the cycling of brain temperature was damping out. At the end of the six cycles of one hour square wave input, there remained only a slow drift downward of brain temperature and the unstable, out-of-phase cycling had almost disappeared. This action of the control system is all the more interesting at these low temperatures when the length of the control time constant is considered.

After lengthening the cycle time, an extremely surprising phenomenon was observed. Midway through the first two hour period at 6° C, at the time when in the previous six cycles a sudden temperature decrease had occurred, there was a sudden onset of increased heart rate accompanied by increased EMG activity. No input perturbation causing this reaction was evident; is it possible that the control system not only "remembered" the previous input but also tried to improve the temperature control mechanism with an anticipatory response?

A rapid rise in brain temperature followed the rise in metabolic activity after which the heart rate and EMG rapidly subsided. This caused a peaking and subsequent fall in brain temperature at such a rate that it was again strongly reinforced by the cyclic reduction in ambient temperature. Out-of-phase control was again obtained as the regulating system began adapting itself to the altered schedule. Some improvement in control was observed - but only three cycles were allowed and complete adaptation, if forthcoming, was not observed.

At this point, cycle lengths were again increased so that four hours would be spent at both the high and low temperatures. During the first long cycle (at 6° C)), a most striking example occurred of apparent control system anticipation of cyclic disturbances. After a little over two hours at the higher temperature, at a time when in the previous cycle a sudden decrease in ambient temperature would have occurred, there was a sudden increase in heart rate and EMG activity. Having been in an ambient temperature of 6° C for the previous two hours, the marmot was actually being maintained at normal hibernating conditions. No external perturbation was evident. Thus, it appears that a knowledge of cyclic disturbances was stored in some way to be used in an anticipatory control. The previous cycle times were two hours at each temperature. The thermoregulatory model proposed here does not predict this phenomena. In fact, it is difficult to think of any physiological mechanism, particularly a physiological mechanism operative at 6° C, for this phenomenon.

Analysis. Again there is incontrovertible evidence for low temperature thermoregulation. This fact is more readily appreciated if the actual response is compared to the passive response to the same input as computed by the model (a series of descending rounded steps in temperature.) A passive stationary state is reached at approximately 4°C, a level considerably below the observed steady state.

There was a strong correlation between increasing ambient temperature and activation of heat generation mechanisms. The degree of response also showed adaptation so as to improve control of brain temperature. As in the previous experiment however, it was difficult to derive an exact mathematical control formulation that would predict the observed phenomenon. The precise point where thermogenesis was initiated was somewhat variable suggesting the necessity to formulate the control law in probabilistic terms.

EXPERIMENT III. LONG TERM CONSTANCY OF CONTROL

Description. This experiment was somewhat similar to the first. It was designed to evaluate the effectiveness of regulation for an extended period at low ambient temperature.

The reaction of the animal after the rapid decrease in ambient temperature from 5° C to 0.5° C was similar to that observed in the other experiments. Increases in heart rate and EMG activity were followed by a

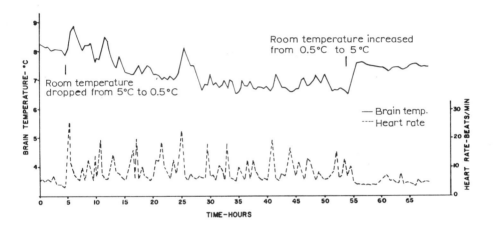

FIG. 9. Response of hibernating marmot to long term low temperature stress.

rapid rise in brain temperature (Fig. 9). Subsequent reduction of the rapid heart rate and of EMG activity was followed by Newtonian cooling of the marmot. This sequence occurred somewhat periodically with an average period of about 1.3 hours.

Controller magnitude adapted so that after three episodes with successively decreasing overshoot a moderately constant temperature was achieved. During the time period between 13 and 35 hours, the variation of brain temperature, except for one episode, was about $\pm 0.25°$ C around an average which itself declined about 1.0° C to a value of 6.7° C.

At about the 23rd hour an increase in cortical and subcortical activity occurred with subsequent increases in heart rate, EMG activity and, eventually, of brain temperature. No external perturbation or excitation of the marmot was noted at this time and this episode remains the result of an unknown factor. Regulation of brain temperature was regained in about four hours and with very little overshoot.

After the temperature slowed or ceased its slow decline, the average brain temperature was about 6.8° C with a persistent variability of $\pm 0.3°$ C. The heart rate also continually varied between five and 18 b/m. Bursts of EMG activity were infrequent and of short duration. In ten hours there were nine episodes of EMG activity with an average duration of about four minutes.

This experiment was terminated during the 49th hour by rapidly raising the cold room temperature from 0.5° to 5° C. There was a fast rise of brain temperature to about 7.5° C with negligible overshoot. This is a normal hibernating brain temperature in a 5° environment, although it is about 0.7° lower than the pre-experimental brain temperature for this animal. Normal hibernation continued for 30 hours after the experiment.

<u>Analysis</u>. While again demonstrating a clear thermoregulatory response to decreased ambient temperature, the control was only partially successful. The brain temperatures were as predicted, more variable under increased cold thermal stress. Also the average apparent low temperature setpoint was

displaced and appeared to drift to a level almost 1° lower than in the previous, less stressful environment. However no signs of premature arousal were noted.

THERMODE EXPERIMENTS

In order to be consistent with the model it would be expected that cooling the preoptic area of a hibernating animal would increase the heart rate and, at a somewhat lower temperature, initiate EMG activity. Furthermore, should one raise the preoptic temperature, either no change in the apparent physiological state of the animal should occur or, as a consequence of an increased output from the inhibitory function, a reduction in heart rate to minimal levels along with EMG silence would be expected. The results of our current experiments were in accord with these expectations (7).

SUMMARY

A thermoregulatory model is proposed where the normothermic regulatory responses are inhibited to permit attainment of low temperature hibernation. It is hypothesized that the inhibitory function fails near freezing permitting reactivation of normal low temperature thermolytic responses. This simple model seems to account for most of the observed phenomena. One important exception is the apparent anticipatory control that was observed during the square wave experiments.

ACKNOWLEDGEMENT

Supported by NASA grant number NGR 26-004-025.

REFERENCES

1. ARIS, R., and N. R. AMUNDSON. An analysis of chemical reactor stability and control, - I. Chem. Eng. Sci. 7: 121-147, 1958.

2. BENEDICT, F. G., and R. C. LEE. In: Hibernation and Marmot Physiology, Washington, D.C.; Carnegie Inst. of Washington Publ. 497, 1938.

3. CLEMENTS, W. C., and K. B. SCHNELLE. Pulse testing for dynamic analysis. I. & E. C. Proc. Design Devel. 2: 94, 1963.

4. DREIKE, G. E., and J. O. HOUGEN. Experimental determination of system dynamics by pulse methods. In: Joint Automatic Control Conferences, Minneapolis: University of Minnesota, 1963.

5. HAMMEL, H. T. Neurons and temperature regulation. In: Physiological Controls and Regulations, edited by W. Yamamoto and J. R. Brobeck. Philadelphia: W. B. Saunders, 1965.

6. HAMMEL, H. T. Temperature regulation and hibernation. In: Mammalian Hibernation III, edited by K. C. Fisher, A. R. Dawe, C. P. Lyman, E. Schönbaum and F. E. South. Edinburgh: Oliver and Boyd, 1967.

7. HARTNER, W. C., F. E. SOUTH, H. K. JACOBS, and R. H. LUECKE. Preoptic thermal stimulation and temperature regulation in the marmot (M. flaviventris) (Abstract). Cryobiology, 1971. In press.

8. HENSHAW, R. E. Thermoregulation during hibernation: application of Newton's Law of Cooling. J. Theoret. Biol. 20: 79-90, 1968.

9. LUECKE, R. H., E. W. GRAY, and F. E. SOUTH. Simulation of passive thermal behavior of a cooling biological system: entry into hibernation. Arch. Ges. Physiol. In press.

10. LYMAN, C. P. Circulation in mammalian hibernation. In: Handbook of Physiology. Circulation, edited by W. F. Hamilton and P. Dow. Washington, D.C.: Amer. Physiol. Soc., 1965, sect. 2, vol. III, p. 1967-1989.

11. SOUTH, F. E., and W. A. HOUSE. Energy metabolism in hibernation. In: Mammalian Hibernation III, edited by K. C. Fisher, A. R. Dawe, C. P. Lyman, E. Schönbaum, and F. E. South. Edinburg: Oliver and Boyd, 1967.

12. STRUMWASSER, F. Thermoregulatory, brain and behavioral mechanisms during entrance into hibernation in the squirrel, Citellus beecheyi. Amer. J. Physiol. 196: 15-22, 1959.

13. WILLIAMS, B. A., and J. E. HEATH. Responses to preoptic heating and cooling in a hibernator, Citellus tridecemlineatus. Amer. J. Physiol. 218: 1654-1660, 1970.

14. WILLIS, J. S. Cold adaptation of activities of tissues of hibernating mammals. In: Mammalian Hibernation III, edited by K. C. Fisher, A. R. Dawe, C. P. Lyman, E. Schönbaum and F. E. South. Edinburgh: Oliver & Boyd, 1967.

THE RESPONSIVENESS OF THE PREOPTIC-ANTERIOR HYPOTHALAMUS TO TEMPERATURE IN VERTEBRATES

JAMES EDWARD HEATH[1], BILL A WILLIAMS[2]
STEVEN H. MILLS[1], and MATTHEW J. KLUGER[1]

[1]Department of Physiology and Biophysics
University of Illinois, Urbana, Illinois 61801
and
[2]NASA, Ames Research Center
Moffett Field, California 94035

INTRODUCTION

The purpose of this paper is to draw together what is known of the central nervous system (CNS) control of thermoregulation among vertebrates with the view of identifying special features of thermoregulation among phylogenetically related, or similarly adapted groups of animals. To do this we must delineate the general features of thermal control common to vertebrates. Finally, we will review what is known of CNS control of temperature in animals that hibernate. Only a handful of species have been studied. Therefore, our comparisons will be hampered by crucial shortages of information. On the other hand, our speculations can be freer. We hope that the ideas that emerge from these comparisons will stimulate more comparative studies of CNS control of temperature.

Idealized Controller. Modeling the control system of temperature regulation requires solution of several simultaneous problems: (a) the heat exchange between the animal and its surroundings; (b) the responses of organisms to thermal inputs; (c) neuro-anatomical units involved in control and (d) the principles of controller action. These problems are assembled in ascending order of generality. That is, the solution of heat exchange requires studies upon individual organisms, the responses are generalizable at the species (population) level, anatomical organization is at higher levels of systematic order and the principle of controller action should be applicable to non-living as well as living systems. We will proceed from the most general level toward the most specific.

<u>Principles of Controller Action</u>. Hardy (24) outlined the principle means of control of body temperature. He compared the responses of organisms to the action of four industrial controllers. In each form of regulator the input, or error signal, is operated upon by the controller to produce an output, or thermoregulatory action, having simple mathematical relationships to the input. These are a non-linear or on-off response, a proportional response, a derivative or rate response and an integral or summating response. We can now generalize that on-off and proportional controls are characteristic of animal temperature control whether physiological or behavioral outputs are involved (19, 26). Convincing evidence of rate and integral control of body temperature is lacking.

On-off control is the simplest controller. Whenever the body temperature (T_B) exceeds a set level an effector output is turned on to a maximum degree. So long as T_B is displaced, the output remains maximal. When T_B returns to its original level the effector is shut off. This form of regulator models some thermoregulatory response remarkably well, for example, those behavioral regulatory responses which involve gross displacement of the animal in the environment (26). It also describes regulatory activities such as panting which depend upon mechanical resonant properties of the organism (15). On-off controllers appear proportional to an imposed thermal load over a wide range of loads if viewed over a long enough period. For example, a dog evaporates water proportional to environmental temperature by panting (28). The proportionality may depend largely upon intermittent bursts of patnting activity not upon intensity of panting. Finally, on-off control requires a reference point within the controller a T_{set}. Presumably, T_{set} is varied according to heat load on the animal. In industrial controllers a change in T_{set} may be some degree of flexion of a bimetallic bar, but in biological systems it is very difficult to model a reference which remains constant in spite of changes in its mileau.

Proportional control relates the change in effector output to a simple or complex internal temperature comparator. Hammel (19) argues that both the metabolic response to decreasing temperature and the evaporative response to increasing temperature can be modeled by the following equation:

$$R_1 - R_0 = \alpha (T_H - T_{set})$$

Where $R_1 - R_0$ is the effector response, T_H is hypothalamic temperature, T_{set} is a reference point whose value is altered by changes in heat load, and α is a constant. Hammel found that α is unaffected by changes in load. The value of α is about the same for heat loss and heat production in the dog. On the other hand, Jacobson and Squires (31) feel that α varies with thermal load in the cat.

There are difficulties in accounting for a T_{set} and of either a constant or uniformly changing α. Hammel has done a remarkable job of conceptualizing both quantities as inherent properties of a simple neuronal circuit. In his model T_{set} is the intercept of the firing rate of two rather ordinary neurons having differing responsiveness to temperature. External input affects the activity of one of the neurons (low Q_{10}) in a regular way, causing a shift of the intercept. The constant α expresses the regularity of the shift of the intercept when T_H is constant. Because α is linear, this system will work if the responsiveness of both neurons is linear. This calls for a special property, a linear rate-temperature response rather than a log response, for the neurons involved in thermoregulation. A similar but more compli-

TABLE 1. Responses of Vertebrates to POAH Heating and Cooling.

Animal	Metabolism	Body Temp.	Panting	Shivering	Other Effects	T_{Air}	Ref. No.
Fish						(T_{water})	
Sculpin							
cooling					$+T_B$ at escape from warm water	12-20	23
heating					$+T_B$ at escape from warm water		23
Reptile							
Lizard							
cooling					+"colonic exit" temp. from high T_{air}	40-48	20,35
heating					+"colonic exit" temp. from high T_{air}	40-48	20,35
Bird							
Sparrow							
cooling	+	+		+		22-23	34
heating	↓	↓	+			22-23	34
Mammals							
Opossum							
heating			+		licks, "sleepy"	Room T°	38
Bat							
cooling		↓					32
heating		↓					32
Heat stressed							
cooling		+					32
heating		+					32
Rabbit							
cooling	+	+				22-24	16
heating	+					35	16
Ground Squirrel							
cooling	+	+		+		8,22,33	47,48
heating	↓	↓	+			8,22,33	47,48
Hamster							
heating	↓	↓				10.5,34	33
White rat							
cooling		+		+	bar press for heat	21-27	43
					+bar press for heat	4,24,0	13,41
heating		↓				21-27	43
Cat							
cooling	+	+			vasoconstriction	5,23	1,2,4
heating	+	+		+		9-37	31
	↓	↓	+			9-37	31
	↓	↓			vasodilation	5,23	39
Dog							
cooling	+	+				15-35	28
	+	+	+	+		25	22
		+		+	+bar press for heat		22
heating	↓	↓				25	22
	↓	↓	+			15-35	28
Goat							
cooling		+			vasoconstriction	22,34	3
				+		<18	3
heating		↓				22	3
			+		polypnea	22	4
Ox							
cooling	+					10-35	29
heating					+evap. heat loss	10-20	49
			+		+periph. resist.	15	49
			+		polypnea		30
					changes of resp. freq.	40	17
Pig							
cooling					bar press for heat	0-25	5,6
					no bar pressing	30-35	5,6
	↑				+resp. rate	40	7
heating	↓				+resp. rate	>30	7
					+bar press for heat	<24	5,6
Monkey							
heating			+		sweating	27	9
Baboon							
cooling		+			+bar press for heat & vasoconstriction	24,40	18
heating		↓			+bar press for heat & vasodilation	12	18

cated model proposed by Jacobson and Squires (31) also implies linearity in the thermal responsiveness of neurons.

Another feature of Hammel's model is that artificial displacement of T_H generates the same effect or response as the shift of set-point by ambient temperature. We will explore this feature and its importance in understanding variability among species in apparent preoptic sensitivity.

Anatomical Structures Involved in Thermoregulation. The hypothalamus has long been implicated as a CNS control center for thermoregulation. The usual portrayal of hypothalamic thermoregulatory control is to assign heat loss regulation to a region in the anterior area of the hypothalamus. Heat production regulation in this scheme resides with the posterior hypothalamus. Further, the preoptic region is known to be thermosensitive and it is given the role of an "integrator" center. Spinal and midbrain regions are also involved in thermoregulatory control (11,12,37). A possible reticular involvement in this scheme might be to "activate" and "deactivate" the neurons in this chain, lowering or raising responsiveness to stimuli (19). This functional relationship of hypothalamic nuclei seems to be similar for most of the species studied.

In summary, then, normal control of temperature regulation in mammals is dependent upon the integrity of hypothalamic, midbrain, and spinal structures with a thermosensitive controller area located in the preoptic nucleus.

COMPARATIVE STUDIES

Very few studies of CNS control of temperature have been designed to produce data for comparison among species. More generally, workers have selected convenient animals at hand for study, in most cases large domestic mammals. Table 1 summarizes the species on which preoptic temperatures have been systematically manipulated and the qualitative response of the animal.

Most animals respond to artificial changes of preoptic temperature by compensatory changes in body temperature. Cooling the preoptic generally initiates responses which conserve or generate heat. Heating the preoptic initiates heat dissipation. Investigators have not found similar coherence in response to artificial changes in temperature of other parts of the brain.

Comparisons Among Classes. Very little is known of the control of temperature in groups other than mammals. Hammel et al. (23) placed arctic sculpins, Myoxocephalus sp. in a water shuttle box with hot and cold water compartments. The fish alternated between the two compartments to maintain body temperature at about 8°C. Heating the forebrain decreased the voluntary exposure of the fish to the warm compartment and lessened the deep body temperature before exit to the cold compartment. Cooling the forebrain increased the body temperature of escape from the warm compartment or suppressed escape entirely.

Reptiles regulate their internal temperature behaviorally (14). Heath (25) analyzed the behavior patterns resulting in thermoregulation in reptiles. He suggested that most of the patterns operate as on-off responses to some internal temperature, but in at least one response, lizards may show a proportional controller response.

Thermal stimulation of the preoptic and hypothalamic areas of reptiles increases heart rate and blood pressure (27,40). Hammel et al. (20) showed that warming the preoptic area of the skink, Tiliqua scincoides, caused the lizard to retreat from the warm end of a shuttle box, at lower body temperatures than controls, while cooling the preoptic caused a shift upwards. They also showed that both upper and lower set temperatures shift together with changes in preoptic temperature. Very high ambient temperatures caused the blue-tongued skink to retreat from the hot end of a shuttle box and at a lower body temperature than controls (35). This may indicate that peripheral input, possibly skin temperature, resets the hypothalamic responsiveness downward. Most recently Templeton (45) presented evidence that threshold temperature for panting in a lizard, Dipsosaurus dorsalis, shifts downward with increasing air temperature. Finally, extracellular recordings of neurons in the preoptic region of T. scincoides reveal units with high sensitivity to temperature change (10).

In summary, both the sculpin and skink have a thermally sensitive preoptic region, and temperature for behavioral responses are shifted downward by preoptic heating and upward by preoptic cooling. Peripheral temperature alters the responses of the lizard.

We group birds and mammals because little is available on birds and the short-term response of birds to cold resembles mammals closely. Further, too few mammals have been studied extensively enough to permit comparison between families. We have therefore concentrated our attention on the location of thermal responsiveness and functions of the controller in heat production.

Among birds and mammals the responses to changes in brain temperature include not only behavioral responses, but also adjustments of heat production and heat loss. For these animals the responsiveness to direct thermal stimulation of the CNS reflects the capacity of each animal to gain and lose heat. Comparisons between species should take into account differences in insulation, morphology, and behavior which markedly alter heat gain and loss. Even if enough information were available, development of this theme would be beyond the scope of this paper. We have limited our discussion to the simplest model of heat exchange.

Homeotherms increase metabolic rate in order to maintain thermal equilibrium and constant body temperature at air temperatures below the thermoneutral zone. The increased heat production is linear and extrapolates to body temperature where metabolism is zero (Figure 1). The slope of the metabolism on air temperature is the heat exchange coefficient, K, of the animal with its environment.

$$M = K (T_B - T_A)$$

where M = metabolism, kcal hr^{-1}

K = heat exchange coefficient, kcal hr^{-1} C^{-1}

T_B = body temperature (°C)

T_A = air temperature (°C)

In a way, the metabolic response to changes in air temperature depicts the sensitivity of the peripheral thermodetector system.

The hypothalamus is also thermally sensitive, at least in mammals. If the hypothalamic temperature (T_H) is raised, metabolism and body temperature decrease. Hellstrome and Hammel (28) plotted the relationship of metabolism to hypothalamic temperature in the dog and found a linear relationship similar to that illustrated in Figure 2.

The curves representing metabolic levels at various hypothalamic temperatures are shifted by air temperature. Hammel has used these shifts to describe proposed set-point shifts of metabolic rate. The thermosensitivity of the hypothalamus as illustrated by the curves is related in some way to air temperature. The slopes of these curves are equal as long as the air temperature does not fall within the thermoneutral zone. The slope is "α" and relates the sensitivity of the hypothalamus to changes in its own temperature (Fig. 2).

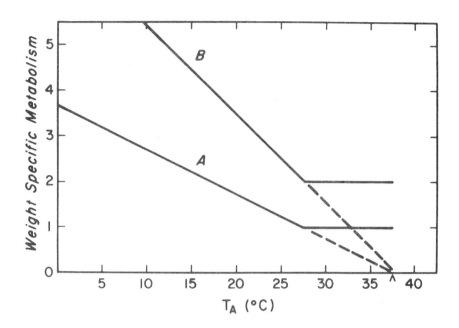

FIG. 1. Two idealized examples of the homeothermic response of metabolism to ambient temperature. The response of animal A shows that it is larger and had a lower basal metabolism than animal B. Both maintain 37.5 body temperatures. The slope of metabolism on ambient temperature is the coefficient of heat exchange, K.

$$M = \alpha (T_H - T_{set})$$

where M = change in metabolism (kcal hr^{-1} kgm^{-1})

α = hypothalamic sensitivity (kcal hr^{-1} kgm^{-1} °C^{-1})

T_{set} = set temperature (°C)

T_H = hypothalamic temperature (°C)

From these two values, then, we see that metabolism is altered as a function of air temperature, (K), and as a function of hypothalamic temperature, (α).

The relative weighting of K and α is described by experimentally measured changes of metabolism in response to changes in air temperature or

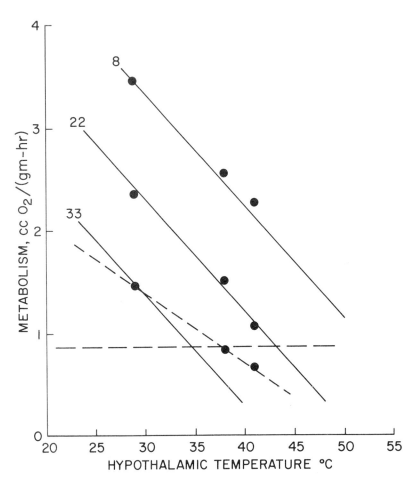

FIG. 2. Metabolic response of <u>Citellus tridecemlineatus</u> to POAH heating and cooling. At each air temperature (numbers on lines) when the POAH is cooled the metabolism increases from its value at 38C POAH temperature. When the POAH is heated, metabolism is decreased. Each point represents the mean response of nine animals. The slopes of the lines connecting points at each air temperature is hypothalamic thermosensitivity, α. At 33C air temperature the data may be interpreted as either a change of slope (dotted line), or that in thermoneutrality the animal adjusts its insulation when the POAH is heated rather than decreasing metabolism below minimal levels. (After Williams and Heath, personal communication.)

hypothalamic temperature. The ratio of α to K is the open-loop-gain OLG.

$$OLG = \frac{\alpha W}{K}$$

where W = body weight (kg)

The OLG thus relates hypothalamic thermosensitivity to peripheral thermosensitivity. OLG for metabolic responses have been calculated as 12.0 for the dog (19) and 8.0 for the cat (46).

Sensitivity to temperature change of the preoptic region varies among mammals and birds. Very small persistent changes cause relatively large changes in metabolic rate and T_B in dogs (28). However, very large changes in preoptic temperature in the house sparrow cause relatively small changes in metabolism and T_B (34). This is an unexpected result in that the response to air temperature is reversed between a dog and sparrow.

Table 2 shows the quantitative relationships among representative endotherms of the metabolic response to change in T_A when T_H is constant (k), and metabolic response to change in T_H when T_A is constant (α). This comparison requires that K and α be determined at T_A's, above the lower critical temperature.

The value of K diminishes with body size as expected (36), but the value of α, which is calculated as weight specific metabolism, is not strongly influenced by body size. Thus, the preoptic sensitivity is about the same in all of these animals. Having the same sensitivity does not mean that thermoregulation proceeds in the same way in all of these animals. Rather, it implies that smaller animals are more dependent on peripheral input in their thermoregulation than larger ones. In Table 3 we show the percent increase of metabolism caused by a 1° C decrease in preoptic-anterior hypothalamic (POAH) temperature at an air temperature at or below the lower critical temperature. The same responsiveness in a dog causes doubling or greater

TABLE 2. Preoptic thermosensitivity (α), heat loss coefficient (K) and open loop gain (OLG) of representative vertebrates.

Species	Wt (Kg)	α (Kcal Kg^{-1} hr C)	K† (Kcal hr^{-1} C)	OLG	Ref. No.
Dog	10	0.7-1.3	0.72	14-27	28
Rabbit	3.0	1.1	0.48	7.0	16
Cat	2.5	0.64-1.0	0.35	8.7	31
Ground Squirrel	.22	0.4*-.85	0.10	0.88	9
Sparrow	.025	0.77	0.0275	0.7	34

*mean of 8 animals
†standard metabolism, unrestrained animals - dog, rabbit, cat - Prosser and Brown, 1961; ground squirrel - Williams and Heath, 1970; sparrow - Hudson and Kimzey, 1964.

increase in metabolism, while the sparrow's increase is probably not experimentally detectable. Open loop gain is an expression for this effect. Figure 3 shows the relationship between OLG and body weight, and K is given for comparison.

Even with a similar sensitivity of the preoptic/anterior hypothalamus, the neural integration required to regulate temperature is greatly affected. In Figure 4 we compare two animals of different size. These idealized animals have the same thermoneutral point but differ greatly in size. Animal B has twice the basal metabolism of animal A and only one-half the insulation. Both have the same POAH sensitivity. When the preoptic temperature is displaced downwards 2.5°C the metabolism of each animal increases to the same extent. In animal A, this is the same metabolic increase as occasioned by a decrease in air temperature of 5°C, but in animal B it is the same as 2.5°C. Consider the shift in metabolism inducible by a 10°C decrease in air temperature. In animal A (Figure 4A), this change in metabolism is equi-

TABLE 3. Metabolic response to a 1°C decrease in POAH temperature in dog and sparrow.

	Dog	Sparrow
Weight (Kg)	10	.025
Standard Metabolism (Kcal Kg^{-1} hr)	2	20
(Kcal $Kg^{\alpha-1}$ hrC)	.7-1.3	.77
Δ Metabolism (%)	33-63	3.8
K (Kcal hr^{-1}C)	.72	.0275
ΔT_b (C)^b	>10	>1

*Calculated as $\frac{(\alpha W)}{K} \times 1C = T_b$

valent to a shift in set point of 5°C. In animal B (Figure 4B), this change in metabolism is equivalent to a shift in set-point of 10°C. Clearly these two animals are not integrating the thermal inputs in the same way.

It is tempting to suggest that variability of body temperature, a common occurrence in small vertebrates, is due to low open loop gain.

All vertebrates studied so far have temperature sensitive units in the POAH, and it is presumed that these units are involved in thermoregulation (see, however, Barker & Carpenter [8]). Some of these units respond to changes in ambient air temperature and to changes in the temperature of other parts of the body. Our observations here would suggest that in small animals, thermosensitive units in the POAH should respond more strongly to changes in ambient temperature than to changes in the temperature of the surrounding neural tissue.

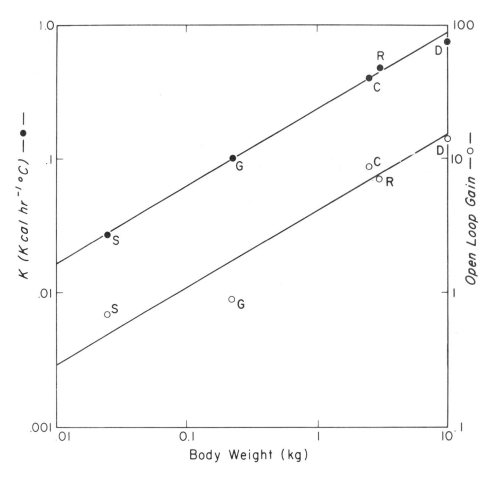

FIG. 3. Coefficient of heat exchange, (K), and open loop gain of the dog, (D), cat, (C), rabbit, (R), ground squirrel, (G) and house sparrow, (S). See Table 2 for sources of data. Both functions are dependent upon body weight. The heat exchange characteristics of each animal strongly perturbs the gain of the central nervous control system.

The response to heating of the POAH is to increase blood flow to the periphery and to initiate evaporative mechanisms of heat loss as well as changes in posture and behavior which facilitate heat loss. These responses are more diverse and therefore, more difficult to compare between species.

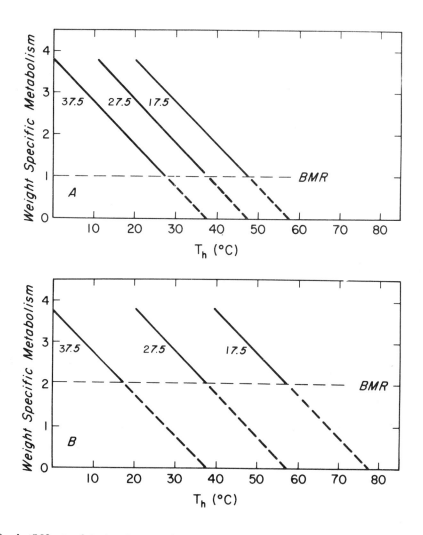

FIG. 4 Effect of body size on the set-points of thermoregulation. Both hypothetical animals, A and B, have the same central nervous thermosensitivity, slope of metabolism on T_h (peroptic-anterior hypothalamic temperature). A is larger than B. Each slope intersects the metabolic rate required to maintain a body temperature of 37.5C when brain temperature is not disturbed. The slopes extrapolate to 0 metabolism when $T_h = T_{set}$ in the controller equation. Thus set-points are shifted far more at the same air temperature in a small animal (B) than in a larger one (A).

The most thoroughly studied is the panting of the dog. Hellstrome and Hammel(28) found the heat dissipated by panting in response to preoptic heating to be proportional to the displacement of preoptic temperature. They calculated the open loop gain to be the same magnitude as for heat production. On the other hand, panting may be an on-off response in dogs and the proportionality emerges only by summation of intervals of presence and absence of panting. The neural message would be on-off, and the proportionality would emerge from the relationship of changes in deep body temperature to the capacity of the animal to lose heat. Thus, the value of α for evaporative heat loss is not necessarily a property of the central nervous network.

The changes in peripheral perfusion or sweating accompanying thermal stimulation of the brain alter the temperature of the surface of the animal. Because the surface temperature, or the temperature of receptors in the skin, is the source of nervous input estimating air temperature, T_{set} is changed by the effector response. Thus, the experimental animal will operate on a new but experimentally uncontrolled set of inputs. Further, these responses differ between species, and there seems little likelihood of useful quantitative comparisons at this time.

In summary, the amount of change of deep body temperature accompanying thermal stimulation of the POAH depends on the size of the bird or mammal studied. This suggests decreasing preoptic sensitivity with size. Since the basal metabolic rate increases with decreasing size, the relative metabolic response to preoptic cooling also diminishes with size. The rate of heat loss is greater for small animals. Nevertheless, the weight specific metabolic response to preoptic cooling is about the same in all animals studied. Therefore, although the preoptic response is the same, changes in air temperature must cause much greater shifts in the value of T_{set} in small animals than in large ones. Thus, neural integration of all relevant thermal inputs differs among animals in accordance with their heat loss characteristics.

The central nervous control of heat loss effectors has not been studied sufficiently among animals of differing size to permit a general statement.

Hibernating Animals. The thermal responsiveness of the central nervous system of hibernators has received little attention. The preoptic region is required for thermoregulation in Citellus tridecemlineatus, a ground squirrel. The thermoresponsiveness of this ground squirrel is organized like other mammals (48,49). Hammel, et al. (21) suggest that ground squirrels do not thermoregulate when in hibernation. Citellus tridecemlineatus shows a good deal of variability in the body temperature when awake. This may be due to the low OLG of animals of that size and may not be evidence of a special feature of the thermoregulator.

The bat, Eptesicus fuscus, has received study by Kluger (32). He found an apparent low preoptic sensitivity to temperature, but this is expected in an animal of only 15 gms. Thermosensitivity of the POAH is enchanced under artificially induced internal heat stress, but this bat does not resist lowering of body temperature by similar cold-stress. Electrolytic lesions in the POAH did not interfere significantly with the animal's ability to thermoregulate or to enter and arouse from torpor (Fig. 5).

Thus, the ground squirrel has a rather normal preoptic response to temperature and is dependent upon its integrity for thermoregulation, while the bat is not dependent upon POAH control. We see no general pattern that would indicate specialization of the thermal controller for hibernation.

Hibernation is a special form of thermoregulation which requires, in part, that the normal mode of regulation be "turned-off" or "turned-down" during the sojourn at lowered body temperature. This special ability to

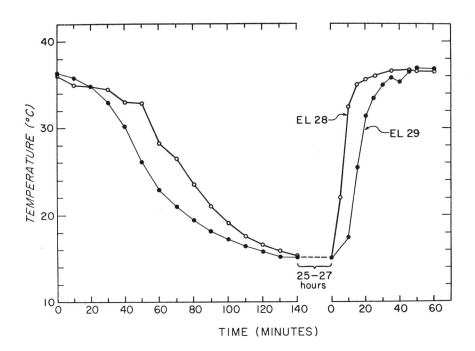

FIG. 5. Entry and arousal from hibernation by bats, Eptesicus fuscus, with massive lesions in the forebrain, (EL 28), and in the POAH, (EL 29). This bat is not dependent upon the integrity of the POAH for thermoregulation. (After Kluger, 32.)

turn-down regulation, then, must arise from a source which controls the level of activity (or responsiveness) of the hypothalamus, and may be the direct result of reticular formation interaction with the hypothalamus (21,49). Reticular input to the hypothalamus might alter the ability of the hypothalamic region to respond to changes in peripheral (air) temperature receptors. Reticular-hypothalamic inhibition might increase during hibernation to turn-down the responsiveness to air temperature and other stimuli and then decrease later to make the hypothalamus responsive once again. Such an interaction would explain the coincidence of arousal from hibernation on reticular oriented stimuli (noise, etc.) as well as upon other endogenous cues. Reticular control of hypothalamic responsiveness could also, in part, explain other conditions of lowered body function, such as sleep, torpor, estivation, and coma. One might speculate that decreased reticular influence on hypothalamic neuronal responsiveness is important in the development of a tighter control of body temperature. That is, as constant heat balance becomes more important, achievement of a finer body temperature control may occur as a result of less reticular influence. We have already mentioned that hibernators accomplish the hypothermic state with a seemingly typical mammalian CNS, so more reticular dependence (or less reticular inhibition) might be one way to explain this special form of thermoregulation. The hibernator might be more responsive to reticular activity levels and would therefore change its thermoregulatory patterns in accord with changes in reticular activity. Thus, one would expect hibernators to have more variable "normal" ranges of body temperature, varying with reticular function as seems to be the case.

ACKNOWLEDGMENTS

Parts of the work reported here were supported by NSF GB-6303 and GB-13797.

REFERENCES

1. ADAMS, T. Body temperature regulation in the normal and cold acclimatized cat. J. Appl. Physiol. 18: 772-777, 1963.

2. ADAMS, T. Hypothalamic temperature in the cat during feeding and sleep. Science 139: 609-610, 1963.

3. ANDERSON, H. T., B. ANDERSSON and C. C. GALE. Central control of cold defense mechanisms and the release of "endopyrogen" in the goat. Acta Physiol. Scand. 54: 159-174, 1962.

4. ANDERSSON, B. and B. LARSSON. Influence of local temperature changes in the preoptic area and rostral hypothalamus on the regulation of food and water intake. Acta Physiol. Scand. 52: 75-89, 1961.

5. BALDWIN, B. A., and D. L. INGRAM. The effect of heating and cooling the hypothalamus on behavioral thermoregulation in the pig. J. Physiol. 191: 375-392, 1967.

6. BALDWIN, B. A., and D. L. INGRAM. Effects of cooling the hypothalamus in the pig. J. Physiol. 186: 72P-73P, 1966.

7. BALDWIN, B. A., and D. L. INGRAM. The influence of hypothalamus temperature and ambient temperature on thermoregulatory mechanisms in the pig. J. Physiol. 198: 517-529, 1968.

8. BARKER, J. L., and D. O. CARPENTER. Thermosensitivity of neurons in the sensorimotor cortex of the cat. Science 169: 597-598, 1970.

9. BEATON, L. E., W. A. McKINLEY, C. M. BERRY, and S. W. RANSON. Localization of cerebral center activating heat-loss mechanisms in monkeys. J. Neurophysiol. 4: 478-485, 1941.

10. CABANAC, M., H. T. HAMMEL, and J. D. HARDY. Tiliqua scincoides: temperature-sensitive units in lizard brain. Science 158: 1050-1051, 1967.

11. CABANAC, M., and J. D. HARDY. Effect of temperature and pyrogens on unit activity in the rabbit's brain stem. Federation Proc. 26: 555, 1967.

12. CABANAC, M., J. A. STOLWIJK, and J. D. HARDY. Effect of temperature and pyrogens on single-unit activity in the rabbit's brain stem. J. Appl. Physiol. 24: 645-652, 1968.

13. CARLISLE, H. J. Behavioral significance of hypothalamic temperature-sensitive cells. Nature 209: 1324-1325, 1966.

14. COWLES, R. B., and C. M. BOGERT. A preliminary study of the thermal requirements of desert reptiles. Bull. Amer. Mus. Nat. Hist. 83: 261-296, 1944.

15. CRAWFORD, E. C. Mechanical aspects of panting in dogs. J. Appl. Physiol. 17: 249-251, 1962.

16. DOWNEY, J. A., R. F. MOTTRAM, and G. W. PICKERING. The location by regional cooling of the central temperature receptors in the conscious rabbit. J. Physiol. 170: 415-441, 1964.

17. FINDLAY, J. D., and J. R. HALES. Hypothalamic temperature and the regulation of respiration of the ox exposed to severe heat. J. Physiol. 200: 62P-63P, 1969.

18. GALE, C. C., M. MATTHEWS, and J. YOUNG. Behavioral thermoregulatory responses to hypothalamic cooling and warming in the baboons. Physiol. Behav. 5: 1-6, 1970.

19. HAMMEL, H. T. Regulation of internal body temperature. In: Annual Review of Physiology. edited by V. E. HALL, A. C. GIESE, and R. R. SANNERSCHEIN. Palo Alto: Annual Reviews, Inc. 1968, p. 641-710.

20. HAMMEL, H. T., F. T. CALDWELL, and R. M. ABRAMS. Regulation of body temperature in the blue-tongued lizard. Science 156: 1260-1262, 1967.

21. HAMMEL, H. T., T. J. DAWSON, R. M. ABRAMS, and H. T. ANDERSEN. Total calorimetric measurements on Citellus lateralis in hibernation. Physiol. Zool. 41: 341-357, 1968.

22. HAMMEL, H. T., D. C. JACKSON, J. A. J. STOLWIJK, J. D. HARDY, and S. B. STRØMME. Temperature regulation by hypothalamic proportional control with an adjustable set point. J. Appl. Physiol. 18: 1146-1154, 1963.

23. HAMMEL, H. T., S. B. STRØMME, and K. MYBRE. Forebrain temperature activates behavioral thermoregulatory response in Arctic sculpins. Science 165: 83-85, 1969.

24. HARDY, J. D. Physiology of temperature regulation. Physiol. Rev. 41: 521-606, 1961.

25. HEATH, J. E. Temperature regulation and duirnal activity in horned lizards. Univ. Calif. Publ. Zool. 64: 97-135, 1965.

26. HEATH, J. E. Behavioral regulation of body temperature in poikilotherms. Physiologist 13: 399-410, 1970.

27. HEATH, J. E., E. GASDORF, and R. G. NORTHCUTT. The effect of thermal stimulation of anterior hypothalamus on blood pressure in the turtle. Comp. Biochem. Physiol. 26: 509-518, 1968.

28. HELLSTRØM, B., and H. T. HAMMEL. Some characteristics of temperature regulation in the unanesthetized dog. Amer. J. Physiol. 213: 547-556, 1967.

29. INGRAM, D. L., J. A. McLEAN, and G. C. WHITTOW. Increase of evaporative loss of water from the skin of the ox in response to local heating of the hypothalamus. Nature 191: 81-82, 1961.

30. INGRAM, D. L. and G. C. WHITTOW. The effect of heating the hypothalamus on respiration in the ox (Bos taurus). J. Physiol. 163: 200-210, 1962.

31. JACOBSON, F. H., and R. D. SQUIRES. Thermoregulatory responses of the cat to preoptic and environmental temperatures. Amer. J. Physiol. 218: 1575-1582, 1970.

32. KLUGER, M. J. Thermoregulation in the bat, Eptesicus fuscus. (Ph.D. Thesis.) University of Illinois, Urbana, Ill., 1970.

33. MALAN, A. Controle hypothalamique de la thermorégulation et de l'hibernation chez le hamster d'europe, Cricetus cricetus. Arch. Sci. Physiol. 23: 47-87, 1969.

34. MILLS, S. H., and J. E. HEATH. Thermoresponsiveness of the preoptic region of the brain in house sparrows. Science 168: 1008-1009, 1970.

35. MYBRE, K., and H. T. HAMMEL. Behavioral regulation of internal temperature in the lizard Tiliqua scincoides. Amer. J. Physiol. 217: 1490-1495, 1969.

36. PROSSER, C. L., and F. A. BROWN. Temperature. In: Comparative Animal Physiology. W. B. Saunders Co., Philadelphia, 1961, p. 237-284.

37. RAWSON, R. O., K. P. QUICK, and R. F. COUGHLIN. Thermoregulatory response to intra-abdominal heating of sheep. Science 165: 919-920, 1969.

38. ROBERTS, W. W., E. H. BERQUIST, and T. C. L. ROBINSON. Thermoregulatory grooming and sleep-like relaxation induced by local warming of preoptic area and anterior hypothalamus in opossum. J. Comp. Physiol. Psychol. 67: 182-188, 1969.

39. ROBERTS, W. W., and T. C. L. ROBINSON. Relaxation and sleep induced by warming of preoptic region and anterior hypothalamus in cats. Exper. Neurol. 25: 282-294, 1969.

40. RODBARD, S., F. SAMSON, and D. FERGUSON. Thermosensitivity of the turtle brain as manifested by blood pressure changes. Amer. J. Physiol. 160: 402-408, 1950.

41. SATINOFF, E. Behavioral thermoregulation in response to local cooling of the rat brain. Amer. J. Physiol. 206: 1389-1394, 1964.

42. SATINOFF, E. Impaired recovery from hypothermia after anterior hypothalamus lesions in hibernators. Science 148: 399-400, 1965.

43. SPECTOR, N. H., J. R. BROBECK, and C. L. HAMILTON. Feeding and core temperature in albino rats: changes induced by preoptic heating and cooling. Science 161: 286-288, 1968.

44. STROM, G. Effect of hypothalamic cooling on cutaneous blood flow in the unanesthetized dog. Acta Physiol. Scand. 21: 271-277, 1950.

45. TEMPLETON, J. R. Panting and its control in the desert iguana, Dispsosaurus dorsalis. Amer. Zool. 10: 521, 1970.

46. VON EULER, C. The gain of the hypothalamic temperature regulating mechanisms. In: *Progress in Brain Research*, edited by W. Bargmann and J. P. Schade, Elsevier Publ. Co., 1964, p. 127-131.

47. WHITTOW, G. C. Cardiovascular response to localized heating of the anterior hypothalamus. *J. Physiol.* 198: 541, 1968.

48. WILLIAMS, B. A. Central nervous control of temperature regulation in a hibernator. (Ph.D. Thesis.) University of Illinois, Urbana, Ill., 1969.

49. WILLIAMS, B. A., and J. E. HEATH. Responses to preoptic heating and cooling in a hibernator *Citellus tridecemlineatus*. *Amer. J. Physiol.* 218: 1654-1660, 1970.

STATUS OF THE ROLE OF THE CENTRAL NERVOUS SYSTEM AND THERMOREGULATION DURING HIBERNATION AND HYPOTHERMIA

INTRODUCTION

After much discussion of the material presented in the preceding papers, it became apparent that it would be possible to arrive at a number of fairly definitive positions. Rather than merely summarize what has been presented already, some of these positions are outlined below as concise statements which are undocumented in the text and which are generally unqualified. The purpose of this approach was two-fold. In the first place we have attempted to define what we believe to be the most probable operational situation with respect to the central nervous system and to thermoregulation throughout the hibernation cycle. Secondly, these statements have been proposed both as a challenge to individual investigators and to provide focus for future studies. In some cases we have taken the liberty of suggesting areas for investigation.

CENTRAL NERVOUS SYSTEM IN NON-HIBERNATING HIBERNATORS

A. The evidence that hibernators have imperfectly developed thermoregulatory systems or that they may be considered as "primitive" organisms remains unconvincing. On the contrary, considerable evidence would indicate that when mammalian hibernators and non-hibernators are compared and allowance is made for size and shape factors, neither group appears to thermoregulate differently or less effectively than the other.

B. Hibernators as a whole are relatively more dependent upon behavioral rather than physical thermoregulation as a means of defense against heat and cold. This behavior would include not only postural adjustments but self-selection of appropriate environmental temperatures as well.

C. Gross morphological examinations of the central nervous systems have revealed no differences, thus far, between the hibernator and the non-hibernator which are relevant to an organism's ability to hibernate.

D. Although differences in central nervous system ultrastructure between hibernators and non-hibernators have been found, far too little work has been done in this field. It should prove a fruitful area of investigation in the future. This should be especially rewarding with respect to investigations centering on the ultrastructural features of the limbic system.

E. Spontaneous electrical activity of the central nervous system does not differ qualitatively between active hibernators and non-hibernators. Sleep patterns, durations and distributions may differ to a slight, but significant, extent.

ENTRY INTO HIBERNATION

A. The cues which initiate entry into hibernation are too varied and unrelated to one another to warrant the assignment of any single one as fundamental to the process. That is to say, certain external cues or behavioral patterns may be important for hibernation to occur in a given species - but not to hibernation as a general phenomenon; this includes food gathering, hoarding behavior, adiposity, season and even environmental temperature.

B. Animals enter hibernation from a state of sleep. Thereafter periods of apparent arousal and activity may intervene. The complexity and frequency of these periods are species dependent and probably not fundamental to the neural mechanisms of entry.

C. Entry into hibernation is accomplished by way of a state of inhibition originating from the limbic system. The process is characterized by a programmed successive suppression of activity at the main functional levels of the brain as well as in non-specific and specific systems. Higher cortical functions are not essential to the process.

D. Two stages of entry are proposed. Stage I: a resistance phase characterized by a cyclical interaction between excitatory and inhibitory mechanisms. Stage II: a compliance phase during which the inhibitory function is imperatively imposed. The initiation of the second phase is signalled by peculiar outbursts of electrical activity from portions of the limbic system including the hippocampus and preoptic area.

E. The following areas of investigation are important, profitable and may be considered as separate goals: (1) the elucidation of the mechanisms which initiate hibernation and, (2) the induction, or mimicking, of hibernation at will.

    1) In analyzing the functional attributes and morphological components of the nervous system of the hibernator, special attention should be given to the interactions between the structures of the limbic system. The simultaneous exploration of other subcortical structures is also needed; in which a variety of well-developed neurophysiological techniques including ablation, electrical stimulation, and recording should be utilized. A thorough investigation of neurohumoral factors involved in hibernation should be undertaken with emphasis placed on the hypothalamus, structures of the thalamus, and mesencephalon. Elevation of the local concentration of mono - and diamines, ACh and amino acids by refined methods of chemical stimulation is imperative. Further, demonstration of the release of neurohumoral factors as well as their efficacy in evoking coordinated responses following microinjection into different areas of the brain-stem

would be of great value. The effects of altering the ionic milieu, especially sodium and calcium in the brain of a hibernator also should be examined. Such studies as these should be correlated not only with the process of entry but with other parts of the hibernation cycle as well.

The functional relationship between nerve and glial cells is of significant interest. Might not these cells have some special relevance to hibernation?

2) It is expected that at some stage in future development we will be able to apply the results of such studies as these to successfully induce or mimic the hibernating state. Success in this not only would further our understanding of the hibernating state itself but would open the door to the possibility of utilizing this knowledge in the solution of applied problems.

DEEP HIBERNATION

A. Although the cortex is functionally deafferented during deep hibernation, the subcortical structures remain in active control throughout the entire period of deep hibernation. Typically, the electrical activity is characterized by intermittent localized bursts or spikes. Somatic activity, when it occurs in a species, is limited to pseudo-affective behavior.

B. Deep hibernation, in contrast to induced hypothermia, is an active state maintained by inhibitory control exerted through the limbic system.

C. Active thermoregulatory control is present throughout the hibernating period. It is at least a one-sided system (defense against cold) and possibly two-sided. A change in activity of "low-Q_{10}" neurones probably is involved, however.

AROUSAL

A. Although arousal from hibernation may be initiated readily by almost any external disturbance, the nature of the stimuli or cues responsible for triggering spontaneous arousal remains moot.

B. No specific signal or complex of electrical activity originating in the brain has been found to herald arousal.

C. The process, once initiated, is mediated and coordinated by the limbic system and expressed as a generalized sympathetic discharge.

D. Arousal proceeds, following in a reversed order, the pattern seen during entry. This is characterized by an overall increase in electrical activity and the appearance of oscillator-like potentials in the preoptic-anterior hypothalamic and midline thalamic nuclei - signaling the onset of the final stage of arousal.

Symposium Committee on CNS and Thermoregulation

Frank E. South, Chairman
James E. Heath
Richard H. Luecke
Lj. T. Mihailovic
Robert D. Myers
Joseph A. Panuska
Bill A. Williams
William C. Hartner
H. Kurt Jacobs

TIMING AND SYNCHRONY OF THE ENVIRONMENT

AN ANALYSIS OF THE MECHANISMS BY WHICH MAMMALIAN HIBERNATORS SYNCHRONIZE THEIR BEHAVIOURAL PHYSIOLOGY WITH THE ENVIRONMENT

E. T. PENGELLEY AND SALLY J. ASMUNDSON

Department of Life Sciences
University of California
Riverside, California 92502

INTRODUCTION

Modern biological research into the phenomenon of mammalian hibernation may safely be said to have started with a monograph on the marmot, written at the end of the nineteenth century by the French physiologist, Dubois (19). Since that time a vast literature on the subject has accumulated. In more recent years the proceedings of three international symposia on mammalian hibernation have been published (21,35,64) and certain rather well delineated specialties have developed within the field. One of the functions of the present symposium is to point out clearly the problems associated with these specialties and to give impetus to further research in an attempt to solve them. This paper is one of the series concerned with the various mechanisms by which animals synchronize their behaviour and physiology to changing environmental factors on a daily, seasonal and lifetime basis. The particular animals to be considered in this paper are species of hibernators which have been reasonably well studied and about which some conclusions can be drawn. From these it may be possible to speculate on others and put forth more general postulates and theories.

THE PROBLEM AND ITS DEVELOPMENT

Dubois (19), his compatriot Kayser (28), and later Lyman (31,32,33) all recognized clearly the temporal nature of hibernation, particularly that this was not a simple response of an animal to an environmental stimulus. However this was not their major concern, and it is only in the last decade or so that the temporal nature of hibernation has been given much attention (27,30,31,46,47,52-55,61,65,66).

The mechanisms by which animals respond to the changing environment associated with the 24 hour period of the earth's rotation on its axis have been well studied (4,9,10,12,36) and there seems no reason to doubt that hibernating mammals possess much the same mechanisms, at least when they are active (22,37,39,62,63). On the other hand the mechanisms by which animals synchronize the timing of various behavioural and physiological events with the changing seasons associated with the earth's 365 1/4 days orbit around the sun are relatively poorly studied and even less understood. The problem may be simply stated as follows; how do animals repeatedly perform the correct behavioral or physiological act at the specific time during the year when presumably it is most beneficial for the biological fitness of the species?

From the scientific point of view there would seem to be only three basic ways that an animal can synchronize its behavioral physiology to a changing environment (47). These are (a) a direct response to some changing geophysical stimuli; (b) an endogenous rhythm which programs the animal's behavior to the exogenous temporal period, i.e. 24 hours or 365 1/4 days; or (c) a combination of both. There seems little doubt that animals and plants, from single celled protists to complex multicellular animals, synchronize their activities to the changing environment associated with the 24 hour period of the earth's rotation on its axis. This is done by an endogenous circadian rhythm (24) modified by a complexity of Zeitgebers (1,2) which entrain the organism on a daily basis. Aschoff (3) has ably pointed out the survival value of such mechanisms, and it seems appropriate to ask whether the same basic mechanisms apply to the much longer annual period.

The seasonal timing of the reproductive activities of animals, as well as the migrations associated with them, are probably the major physiological events of the year, and have of course been noted by man before recorded history. However it is only in this century that any scientific understanding of the phenomena has developed. Rowan (57,58) clearly demonstrated for the first time that photoperiodism, reproductive periodism, and the annual migrations of birds were clearly related, though he also noted that photoperiodism could not be the only factor involved. Similarly Bissonette (8) showed that modification of some mammalian sexual cycles was possible by manipulation of the photoperiod. However, apart from increasing emphasis on the photoperiod as the apparently major controlling environmental aspect in seasonal reproductive periodism, little work had been attempted until very recently on a synthesis of the various other possible factors involved in the timing of mammals' physiological activities with the seasons of the year. Nevertheless it should be noted that a productive attempt at this was first made by the French workers, Benoit, Assenmacher and Brard (5-7).

In addition to the timing of reproductive activities, hibernating mammals display another major and easily observable event which they undergo on an annual basis, namely they hibernate for a certain period of time. This is a phenomenon comparable to the migration of birds, but with the special advantage that it is easily studied in the laboratory. It is therefore not surprising that careful studies of these mammals have enormously enhanced our knowledge of the probable means by which physiological and behavioral synchrony with the changing annual environment is achieved.

PRESENT STATE OF KNOWLEDGE IN SOME HIBERNATORS

The most recent studies which partially elucidate the problem are those of Pengelley and Asmundson (49,50), Heller and Poulson (25), Dawe et al. (17) and Mrosovsky and Fisher (45). In addition there have been two recent attempts to supply an overall theory (44,47).

As demonstrated in our laboratory (49,50) there seems good reason to suppose that in the extensively studied golden-mantled ground squirrel, Citellus lateralis, the underlying mechanism of the animal's ability to synchronize its behavioral physiology with the changing environment throughout the year is an endogenous rhythm, termed circannual (23,50,56), which approximates a year's duration. This has been confirmed by Heller and Poulson (25).

It is important to note that in order to demonstrate an endogenous circannual rhythm, the following criteria must be met (23,47): (a) The rhythm must persist year after year under constant environmental conditions; (b) the free-running period must approximate a year in duration, but should deviate from it, otherwise it is impossible to ensure that some unknown geophysical factor (67) is not disturbing the constant environment, and serving as a Zeitgeber; (c) the rhythm must be relatively temperature independent; and (d) the organism should exhibit a differential sensitivity throughout the year to one or more environmental Zeitgebers which will serve as an entraining agent for the organism to a period of one year.

Figures 1 and 2 show graphically most of the known facts which support the circannual rhythm theory. The representative animals in Figure 1 show in some detail the approximate annual rhythmic nature of two parameters over a period of nearly four years, and to our knowledge this is by far the longest time an experiment of this nature has been kept going. The two parameters are the body weight of the animals, plotted in grams, and what is technically termed the homothermic-heterothermic rhythm. This latter refers to the fact that during part of the year the animals are homothermic or active (open bars) and during the rest of the year they are heterothermic and capable of hibernation (solid black bars); a single period of homothermy together with a heterothermic period comprises the free-running period. The latter was measured in this experiment from onset to onset of each subsequent heterothermic period. It is clear that both these rhythms persist under constant environmental conditions (i.e. 12 hour artificial photoperiod and two ambient temperatures of 3°C and 12°C, with food and water ad libitum). The relationship between the two rhythms is such that the body weight reaches a peak at the onset of the heterothermic period and a minimum at the onset of, or shortly after, the homothermic period. These two rhythms would appear to be phase locked, though it is of theoretical importance that artificial abolition of the weight rhythm does not abolish the homothermic-heterothermic rhythm (48) nor vice versa (54). It may also be noted here that in addition to the ambient temperatures of 3° and 12°C, these two rhythms persist at ambient temperatures of 0° and 22°C, and the body weight rhythm, at least, persists at 35°C ambient (54). It is probable that the homothermic-heterothermic rhythm also persists at this high ambient temperature, but for obvious reasons it is difficult to observe, since the animals cannot reduce their body temperature below 35°C.

The animals in Figure 1 were merely two in an experiment with a total of twenty-three. These animals were caught in the wild in August and comprised both adults and juveniles. They were divided into two matched groups of eleven and twelve each, so far as sex and age (juveniles/adults) were concerned, and in addition one half of all males and females were castrated. Table 1 summarizes the results found for the free-running circannual homothermic-heterothermic periods for all the animals in the experiment. We were unable

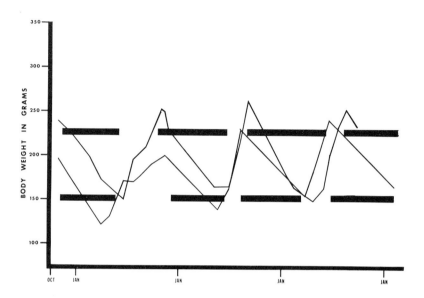

FIG. 1. Interrelationship of body weight (g), and whole hibernation periods (black bars), of two representative animals (C. lateralis) for nearly 4 years. Upper animal at 12°C ambient, lower at 3°C ambient, both with artificial photoperiod of 12 hrs (49).

to detect any significant differences in the circannual periods between sexes, adults and juveniles, or castrated and controls, and there is also no significant difference between those at ambient temperatures of 12° and 3°C. However the ambient temperature does affect the amplitude of the body weight rhythm (Fig. 1) the higher temperature giving rise to lower troughs and higher peaks (54). It may also be noted (though we do not feel conclusions are warranted) that of the sixty-one complete free-running periods (onset to onset of hibernation) which it was possible to measure, only 13 were

TABLE 1. Free running circannual periods (onset to onset of hibernation) in days of C. lateralis under constant environmental conditions. Animals 1-11, 12°C 12 hr photoperiod; animals 12-23, 3°C 12 hr photoperiod (49).

Animal	Circannual periods (days)			
	1st	2nd	3rd	4th
1	346	357		
2	356	324		
3	331			
4	303	374	346	338
5	299	279	277	229
6	290	337	390	
7	345	318	348	358
8	387	231	234	288
9	333	318	345	
10	380	345	393	
11				
12	385	366		
13	438			
14	388	312	325	
15	386	262	325	
16	353	379	413	
17	365	267	310	325
18	348	360		
19	322	445	354	
20	409	248	309	319
21	332	343	319	
22	325			
23	340	312		

longer than the annual period of 365 days. The longest period was 445 days and the shortest 229.

Figure 2 represents graphically the results from another similar experiment, but with important variations. All the animals in this experiment were laboratory born. Shortly after weaning they were divided into four matched groups, and set up under experimental conditions as follows. Group 1 comprised the controls with an artificial photoperiod of 12 hours. Group 2 were normal but had an artificial photoperiod of 20 hours. Group 3 were castrated and then placed with an artificial photoperiod of

12 hours. Finally group 4 were bilaterally enucleated. All the animals were at an ambient temperature of 3°C for nearly 4 years. In this experiment we measured the free-running period from termination to termination of each heterothermic period. The results of this experiment indicate a highly significant fact, namely that the animals in Group 4 (bilaterally enucleated) have more accurately timed successive free-running periods than

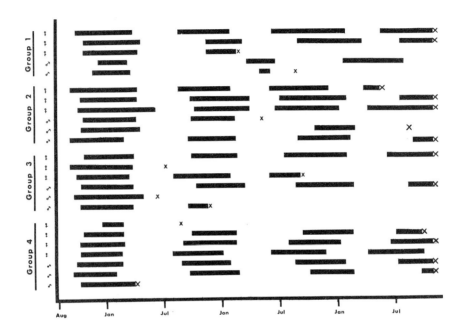

FIG. 2. Graphical representation of free-running circannual periods of 24 animals (C. lateralis) for 4 yrs. Black bars indicate heterothermic period, clear space homothermic period, X = death. All groups at 3°C ambient temperature. Group 1, controls with artificial photoperiod 12 hrs. Group 2, normal with artificial photoperiod 20 hr. Group 3, castrated with artificial photoperiod of 12 hr. Group 4, bilaterally enucleated. M = Male, F = Female (50).

any of the other groups or those in the previous experiment. Thus while the time of termination of the heterothermic period varies considerably between each animal in Group 4, the successive variations in the free-running periods of individual animals are remarkably small. In one animal of this group (the fourth animal) it is possible to measure three complete free-running periods. They are 313, 314 and 310 days. The largest variation of the free-running period of an individual animal in this group is 29 days, whereas variations of up to 75 days in the other groups are common. We concluded three things from this experiment. Firstly, that light is a factor involved in the determination of the free-running circannual rhythm of this species, and is therefore a potential <u>Zeitgeber</u>. Secondly, laboratory born juveniles establish circannual rhythms without detectable differences from wild caught animals. This includes the fact that the homothermic-heterothermic and body weight rhythms are synchronously established from the beginning of the animals' lives, and continue unto death. This seems to be clear evidence that no imprinting mechanism is involved, but rather that the circannual rhythm is genetically determined. Thirdly, it seems that the influence of light on the free-running period takes some time (over a year) to show appreciable affect, for it is easily discernable from Figure 2 that in Groups 1, 2, and 3, it is not until the third and fourth years that large discrepancies occur in the temporal sequence of events. In this connection it is interesting to recall that as far back as 1955, Hock (26) speculated that in the Arctic ground squirrel, <u>Citellus</u> <u>undulatus</u>, the photoperiod was probably a stimulus for the onset of hibernation in the wild, a fact since confirmed in the laboratory by Drescher (18). However Pengelley and Fisher (54) were unable to demonstrate this in the laboratory in <u>C</u>. <u>lateralis</u>. With some 10 years of hindsight and more recent experiments it would appear that this conclusion was hasty because the results of the experiments reported here do implicate light as a potential Zeitgeber. Table 2 summarizes the data from Figure 2, Group 4; with the annual cycle measured in A from onset to onset of heterothermy and in B from termination to termination of heterothermy. It can be seen that in this group the method of measurement makes little difference in the length of the free-running period, however when measuring from onset to onset of heterothermy three complete cycles can be measured for 5 animals.

TABLE 2. Free-running circannual periods in days of C. lateralis bilaterally enucleated and kept at 3°C. Data from Figure 2, Group 4.

A. Circannual periods in days (onset to onset of heterothermy)

Animal	1st	2nd	3rd
2	338	341	293
3	329	324	324
4	293	293	305
5	334	354	324
6	358	369	353

B. Circannual periods in days (termination to terminatation of heterothermy)

Animal	1st	2nd	3rd
2	348	377	
3	349	329	
4	313	314	310
5	337	359	
6	376	361	

(Animals 1 and 7 are not shown as neither completed an annual cycle before death).

Since it is known that all animals in this group were born within a period of only a few weeks it is clear their temporal physiological differences would be slight in relation to their initial entrance into hibernation. Experiments are currently underway in our laboratory to determine whether the spring breeding and raising of a litter will affect the temporal physiology of females in relation to the onset of hibernation and the length of the free-running period. It is not clear at this point which way of measuring the free-running period is more accurate.

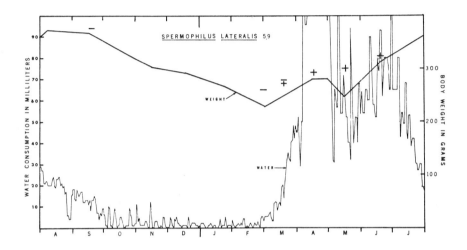

FIG. 3. Record of daily water consumption (fine line) and body weight (bold line) for C. lateralis male during its second year under constant laboratory conditions: + stands for reproductive competence: - stands for reproductive quiescence (25).

Recent work on the problem by Heller and Poulson (25) is summarized in Figures 3 and 4. This work, in the same species, Citellus lateralis, not only confirms the basic hypothysis of an endogenous circannual rhythm, but also adds new parameters and, importantly, another means of measuring the length of the free-running period. These facets are illustrated in Figure 3 where in addition to body weight, daily water consumption and reproductive competence are plotted. This is for a single representative animal during its second year under a constant environment of 16°C and a photoperiod of 12 hours. Their method of judging reproductive competence consisted of continuous observations on the testes of males and the vulvae

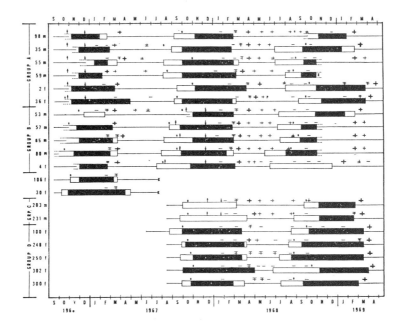

FIG. 4. Circannual rhythms of reproductive condition, hibernation and body weight in C. lateralis kept under constant laboratory conditions for 33 months, and the influence of prevention of torpor and dieting on these rhythms. Solid horizontal line is the life line and represents the time an animal was under observation under constant conditions. A solid black bar indicates the hibernation phase and a white bar indicates winter condition phase. A solid black circle above a record marks the occurrence of a body weight maximum. Full reproductive competence is denoted by +, reproductive quiescence by -, the transition into reproductive competence by ∓, and declining reproductive competence by ±. An arrow pointing up means dieting and prevention of torpor was begun and an arrow pointing down marks the return to ad lib. food and water consumption. The transition from either a solid bar or an open bar to a single life line is the date of terminal arousal. When the transition between active and inactive times was not clear, or when the record was incomplete, dashed lines are shown. In the record of animal No. 36, P (parturition) marks the birth date of a litter. An X indicates death (25).

of females. During most of the year males had abdominal testes and were considered reproductively incompetent, but this condition gradually changed at certain times to a competent condition where the testes became scrotal and the skin in that region became pigmented. Similarly for most of the year the females had closed vaginae and unswollen vulvae (reproductive incompetence) which gradually gave way to reproductive competence where the vaginae were open and the vulvae swollen. It is noteworthy that a "reproductively competent" pair did in fact mate in the laboratory and in due course the female produced a litter of six (end of May). In Figure 3 the declining weight phase corresponds to the previously mentioned heterothermic phase which is here characterized by extremely low (often zero) water consumption and reproductive incompetence. The increasing weight phase, on the other hand, corresponds to the homothermic phase with high water consumption and reproductive competence. Heller and Poulson took as their date of terminal arousal the first sharp increase of water consumption. At this time they also observed a very great increase in activity confirming the observations of Pengelley and Fisher (55). Using the first sharp increase of water consumption as the terminal arousal date, and the free-running period as terminal arousal to terminal arousal, the data for all C. lateralis in Heller and Poulson's work are plotted in Figure 4, and the evidence seems overwhelmingly in favor of a circannual rhythm. By the method previously described they were able to measure 17 free-running circannual periods, the extremes of which were 44 and 59 weeks (308 and 413 days), with a mean of 51 weeks (358 days). Heller and Poulson conclude "regardless of the experimental manipulations we imposed upon the animals, the free-running period length measured from terminal arousal to terminal arousal is generally only slightly less than a year and shows remarkably little variation." When Heller and Poulson's data are compared with Pengelley and Asmundson's (Figs. 1,2, Table 1 and 2) it is seen that there is a remarkable similarity in the free-running periods. Though Heller and Poulson's figures more closely approximate a year's duration, this is probably due to their different method of determining the free-running period. Which of the two is better is very difficult to judge.

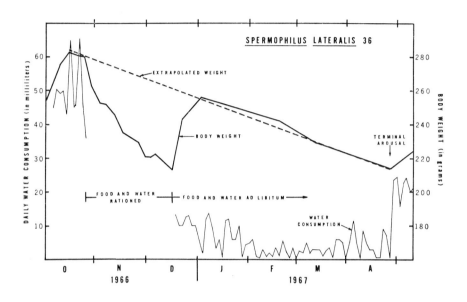

FIG. 5. Daily water consumption and body weight of a C. lateralis before, during and after dieting and prevention of torpor. This demonstrates the post-diet return of body weight to a programmed weight determined by a line extrapolated from weight peak to minimum weight at terminal arousal (25).

Further data from Heller and Poulson, shown in Figure 5, strongly support the endogenous rhythm concept. In this figure it may be seen that upon reaching a weight peak, the food and water supply of the animal was rationed in such a way that a decline in body weight inevitably followed. When food and water were restored ad libitum, it was found that the animal regained weight only to the point expected on the basis of a line extrapolated between weight peak and minimum weight at terminal arousal. Of seven animals in this experiment, six behaved in the predicted manner. This is clear evidence of a programming mechanism so far as body weight is concerned. This same phenomenon has recently been demonstrated in *Citellus tridecemlineatus* by Mrosovsky and Fisher (45). Heller and Poulson's data also demonstrate that excessive depletion of weight due to food and water rationing at the beginning of the heterothermic period has no effect on the time of terminal arousal, nor on the onset of the next heterothermic period, nor the length of the free-running circannual rhythm. However work in our laboratory (47) has shown that deprivation of food throughout the heterothermic period will prolong heterothermy. Heller and Poulson also confirm my conclusion (48) that artificially holding animals near their minimum weight during the homothermic period does not effectively alter the time of onset of the heterothermic period. Mrosovsky (44) has subjected this data to more sophisticated statistical analysis, and claims there is a slight difference.

To our knowledge the data for *Citellus lateralis* summarized here concerning the underlying mechanism of an endogenous circannual rhythm are far more complete than for any other species and seem to us reasonably conclusive. Furthermore the data fit all four criteria mentioned earlier for circannual rhythm with the exception that we have not yet been able to effectively entrain the rhythm by any combination of *Zeitgebers*. Nevertheless, the theory has recently been challenged by Mrosovsky (44). In addition the work of Dawe and Spurrier (16) and Dawe et al. (17) cast some doubts on it. Mrosovsky's criticisms of the endogenous circannual rhythm concept are (a) that the phase of a relatively constant weight found in the spring in natural populations is missing under laboratory conditions, (b) there are instances where the free-running periods are too short (251 days) to be considered circannual, (c) that the temperature independence is not as conclusive as appears at first sight, and (d) no one has yet discovered what the Zeitgeber is. Mrosovsky's alternative theory is best summarized in his own words. "An alternative system that could give rise to annual cycles is a sequence of linked stages, each one taking a given amount of time to complete and then leading into the next, with the last stage linked back to the first one again (43,63). These stages might be based largely on hormonal, nutritional, neural or other internal states and each last a relatively standard time, just as a pregnancy does. A cycle of this kind would be capable of persisting in constant conditions and in this sense would be endogenous, but it would be very different from the endogenous free-running oscillator mechanisms that seem to be involved in circadian rhythms." In our opinion the only one of Mrosovsky's criticisms that is valid is the last, namely that no one has yet discovered what the Zeitgeber is; to which it may be added that this is not surprising considering the appalling lengths of time involved in studying circannual rhythms. However, potential Zeitgebers have been implicated, both temperature (18,54) and light (18,50). His other three main criticisms do not seem to us to be warranted considering the nature of the phenomenon, and the very fundamental biological fact, so forcibly stressed by Charles Darwin (13), that animals and plants vary. Dawe and Spurrier (16) have reported that in Citellus tridecemlineatus they can induce hibernation during the normally homothermic period when recipient animals are transfused with blood from those in hibernation. Their data seems to us to be inconclusive in that they had not correctly judged the expected onset of hibernation in the recipient animals had they not been transfused. Our laboratory (unpublished data) has repeated their work using Citellus lateralis, and our results do not confirm theirs. McCarley (personal communication) using Citellus tridecemlineatus has also repeated their work and does not agree with their interpretation. Nevertheless more data on this aspect are clearly needed.

The foregoing analysis of the mechanisms by which mammalian hibernators synchronize their behavioural phsyiology with the environment is based on reasonably extensive studies in a few species of ground squirrels and it would be folly to try to apply the same theories to all hibernators, if for no other reason than that hibernation would appear to be of polyphylogenetic origin (11,56). Furthermore experimental evidence is available to show that the same mechanisms do not apply in such animals as the hamster, Mesocricetus auratus (20,28,32,33,40). However the very recent and excellent study by Smit-Vis and Smit (59) indicate that hibernation in the hamster is dependent partly on age, and possibly on an endogenous rhythm as well. Further experiments on this animal are badly needed. In addition, the dormouse, Glis glis (41,42,43) shows quite a different mechanism; there are no doubt other species also which show varying mechanisms. On the other hand the underlying factor of a circannual rhythm with entraining Zeitgebers is probably the basic mechanism in many species of Citellus (56), chipmunks (25), the woodchuck (14,15), the hedgehog (30,31), and some bats (37,38). The foregoing comprise representatives of all the major groups of mammalian hibernators.

DISCUSSION AND CONCLUDING REMARKS

The adaptive value of the mechanism of a circannual rhythm for such animals as hibernators has been pointed out by Pengelley (47) and more recently by Heller and Poulson (25). In general terms a circannual rhythm allows the animal to prepare well in advance for a future environmental condition, such as a combination of cold and food scarcity or, on the other hand, optimal breeding conditions. Thus for example if the animal waited until the onset of cold weather in the fall as a stimulus to increase its fat stores for the winter, it would be too late since the food supply would already be declining. But with the programming of an endogenous rhythm, the fat supply is automatically increased in the summer when the food supply is greatest. On the other hand a circannual rhythm allows the organism some flexibility in an environment that changes from year to year. This enables it to integrate a large number of environmental cues and Heller and Poulson (25) believe that environmental stimuli act to either shorten or lengthen some portion of the rhythm, i.e., to phase it without affecting its circannual period. In our opinion Heller and Poulson are partially correct in that phasing most certainly occurs, but it seems unlikely that the circannual period is exactly 365 days; in fact it is not, and thus one or more entraining _Zeitgebers_ are also necessary. Heller and Poulson also confirm our view (56) that the balance between phasing by environmental stimuli and annual endogenous control depends on the species ecology. Thus they argue that C. lateralis and C. beldingi, which are sympatric and thus experience the same physical weather conditions, nevertheless have very different food habits. The animal's adaptations are that C. lateralis is mainly under the control of a circannual rhythm, whereas C. beldingi is much more susceptible to phasing by environmental stimuli. They were able to confirm this in the laboratory. Similarly they argue that temperate zone hibernators are more susceptible to phasing than are Arctic zone animals, mainly because in the latter case the environment demands that the time of spring breeding be very accurately timed. Pengelley and Kelly (56) showed an apparent gradation of the balance between the circannual rhythm and the phasing by environmental factors in various other species of _Citellus_. There are some 28 species in the genus _Citellus_ ranging in habitat from the low deserts of the southwestern United States to the Arctic tundra, and thus the value of comparative studies is obvious.

The problems that future researchers face in this field are formidable. First, we know virtually nothing about the physiological or biochemical mechanism of a circannual rhythm. This is true also of circadian rhythms, and because of the time spans involved knowledge of this aspect must surely come first from those studying circadian phenomena. Such information is of prime importance, and it is hoped that such workers as Pittendrigh will shortly give us some clues. Second, a more accurate parameter must be found for measuring the free-running circannual period and perhaps this could be done by measuring an internal parameter (51,60), rather than an external one. Third, it must be admitted that we cannot yet apply a known Zeitgeber, or combination thereof, which will effectively entrain hibernators to an exact annual period. In our opinion the problem here is merely one of time; the biologist studying circadian rhythms has a time factor of over 300 in his favor. And fourth, we have no guarantee that the entraining mechanisms for circannual rhythms are the same as those for circadian rhythms. The phasing and entraining factors in circannual rhythms in hibernators vary from species to species and may be an incredibly complex interaction of environmental temperature, light (both length and intensity), food and water supply, and access to locomotor activity, to name but a few. However in our opinion nothing is so pressing as some understanding of the physiological and biochemical mechanisms giving rise to endogenous rhythms.

SUMMARY

A. Evidence for the mechanisms by which mammalian hibernators synchronize their behavioral physiology with the environment has been critically examined.

B. In most cases it appears that there is an endogenous circannual rhythm which allows the animal to program its behavior for future environmental conditions throughout the year.

C. The endogenous rhythm is more rigorous in animals living under harsh conditions (Arctic) than under moderate (temperate) conditions. This allows the organism some flexibility in its phasing of events to meet a changing environment.

D. The length of the circannual rhythm usually seems to be less than a year, and there is as yet no exact knowledge of the Zeitgebers which may entrain the rhythm.

E. It is pointed out that there are other hibernators, which do not have such mechanisms and about which very little is understood.

F. Various suggestions have been offered as to the nature of immediate problems in the field, and to their possible solutions.

ACKNOWLEDGMENTS

This paper has been prepared, and our research being continued, with the help of Grant GB 8605 from the National Science Foundation, and secondarily by a grant from the University of California.

REFERENCES

1. ASCHOFF, J. Exogenous and endogenous components in circadian rhythms. In: Cold Spring Harbor Symposium on Quantitative Biology. XXV. Biological Clocks. Long Island, New York: Biol. Lab., Long Island Biol. Assoc. Inc., 1960, p. 11-28.

2. ASCHOFF, J. Comparative physiology: diurnal rhythms. Ann. Rev. Physiol. 25: 581-600, 1963.

3. ASCHOFF, J. Survival value of diurnal rhythms. Symp. Zool. Soc. London 13: 79-98, 1964.

4. ASCHOFF, J., editor. Circadian Clocks. Amsterdam: North Holland Pub. Co., 1965, p. 479.

5. BENOIT, J., I. ASSENMACHER, and E. BRARD. Évolution testiculaire du Canard domestique maintenu à l'obscurité totale pendant une longue durée. Compt. Rend. Acad. Sci., 241: 251-253, 1955.

6. BENOIT, J., I ASSENMACHER, and E. BRARD. Étude de l'évolution testiculaire de Canard domestique soumis très jeune à un éclairement artificiel permanent pendant deux ans. Compt. Rend. Acad. Sci. 242: 3113-3115, 1956.

7. BENOIT, J., I. ASSENMACHER, and E. Brard. Apparition et maintien de cycles sexuels non saisonniers chez le Canard domestique placé pendent plus de trois ans à l'obscurité totale. J. Physiol., Paris, 48: 388-391, 1956.

8. BISSONETTE, T. H. Modification of mammalian sexual cycles. J. Exptl. Zool. 71: 341-373, 1935.

9. BUNNING, E. Die endonome Tagesrhythmik als Grundlage de photoperiodischen Reaktion. Ber. Deut. Botan. Ges. 54: 590-607, 1936.

10. BUNNING, E. The Physiological Clock. Endogenous Diurnal and Biological Chronometry. New York: Academic Press, 1964, p. 145.

11. CADE, T. J. The evolution of torpidity in rodents. In: Mammalian Hibernation II Ann. Acad. Sci. Fenn. Ser. A, IV, 71, edited by P. Suomalainen. Helsinki: Suomalainen Tiedeakatemia, 1964, p. 71-112.

12. Cold Spring Harbor Symposia on Quantitative Biology. XXV. Biological Clocks. New York: Biol. Lab. Long Island Biol. Assoc., Inc., 1960, p. 524.

13. DARWIN, C. On the Origin of Species By Means of Natural Selection, or The Preservation of the Favoured Races in the Struggle for Life. London: John Murray, 1859, p. 502.

14. DAVIS, D. E. Behavioural aspects of torpor in woodchucks. (Abstract) Amer. Zool. 5: 196, 1965.

15. DAVIS, D. E., Factors initiating torpor in woodchucks. (Abstract) Amer. Zool. 5: 208, 1965.

16. DAWE, A. R., and A. SPURRIER. Hibernation induced in ground squirrels by blood transfusion. Science 163: 298-299, 1969.

17. DAWE, A. R., A. SPURRIER, and A. ARMOUR. Summer hibernation induced by cryogenically preserved blood "trigger." Science 168: 497-498, 1970.

18. DRESCHER, J. W. Environmental influences on initiation and maintenance of hibernation in the Arctic ground squirrel, Citellus undulatus. Ecology 48: 962-966, 1967.

19. DUBOIS, R. Étude sur le mécanisme de la thermogenese et du sommeil chez les mammiferes. Physiologie comparée de la marmotte. Ann. Univ. Lyon 25: 1-268, 1896.

20. FAWCETT, D. W., and C. P. LYMAN. The effect of low environmental temperature on the composition of depot fat in relation to hibernation. J. Physiol. 126: 235-247, 1954.

21. FISHER, K. C., A. R. DAWE, C. P. LYMAN, E. SCHOENBAUM, and F. E. SOUTH, editors. Mammalian Hibernation III. Edinburgh: Oliver and Boyd, 1967, p. 535.

22. FOLK, G. E. Day-night rhythms and hibernation. In: Mammalian Hibernation, Proc. First Intl. Symp. Nat. Mammal. Hib., edited by C. P. Lyman and A. R. Dawe. Cambridge, Mass.: Mus. Comp. Zool. Harvard Coll., 1960, p. 209-232.

23. GWINNER, E. Circannuale Periodik als Grundlage des jahreszeitlichen Funktionswandels bei Zugvögeln. Untersuchungen am Fitis (Phylloscopus trochilus) und am Waldlaubsänger (P. sibilatrix). J. Ornithol. 109: 70-95, 1968.

24. HALBERG, E., F. HALBERG, C. P. BARNUM, and J. J. BITTNER. Physiologic 24-hour periodicity in human beings and mice, the lighting regimen and daily routine. In: Photoperiodism and Related Phenomena in Plants and Animals, edited by R. B. Withrow. Washington: A.A.A.S., 1959, p. 803-878.

25. HELLER, H. C., and T. L. POULSON. Circannian rhythms - II. Endogenous and exogenous factors controlling reproduction and hibernation in chipmunks (Eutamias) and ground squirrels (Spermophilus). Comp. Biochem. Physiol. 33: 357-383, 1970.

26. HOCK, R. J. Photoperiod as stimulus for onset of hibernation. Federation Proc., 14: 73-74, 1955.

27. HOFFMAN, R. A. Speculations on the regulation of hibernation. In: Mammalian Hibernation II, Ann. Acad. Sci. Fenn. Ser. A, IV, 71, edited by P. Suomalainen. Helsinki: Suomalainen Tiedaketamia, 1964, p. 199-216.

28. HOFFMAN, R. A., and R. J. REITER. Pineal Gland: influence on gonads of male hamster. Science 148: 1609-1610, 1965.

29. KAYSER, C. Essai d'analyse du mécanisme du sommeil hibernal. Ann. Physiol. 16: 314-372, 1940.

30. KRISTOFFERSSON, R., and A. SOIVIO. Hibernation of the hedgehog (Erinaceus europaeus L.). The periodicity of hibernation of undisturbed animals during the winter in a constant ambient temperature. Ann. Acad. Sci. Fenn., Ser. A, IV, 80: 1-22, 1964.

31. KRISTOFFERSSON, R., and P. SUOMALAINEN. Studies on the physiology of the hibernating hedgehog: changes of body weight of hibernating and non-hibernating animals. Ann. Acad. Sci. Fenn., Ser. A, IV, 76: 1-11, 1964.

32. LYMAN, C. P. The oxygen consumption and temperature regulation of hibernating hamsters. J. Exptl. Zool. 109: 55-78, 1948.

33. LYMAN, C. P. Activity, food consumption and hoarding in hibernators. J. Mammal. 35: 545-552, 1954.

34. LYMAN, C. P. Oxygen consumption, body temperature and heart rate of woodchucks entering hibernation. Amer. J. Physiol. 194: 83-91, 1958.

35. LYMAN, C. P., and A. R. DAWE, editors. Mammalian Hibernation Proceeding of the First International Symposium on Natural Mammalian Hibernation. Cambridge, Mass.: Mus. Comp. Zool. Harvard Coll., 1960.

36. MAYERSBACH, H. V., editor. The Cellular Aspects of Biorhythms. Berlin-Heidelberg: Springer-Verlag, 1967, p. 198.

37. MENAKER, M. The free running period of the bat clock: Seasonal variations at low body temperature. J. Cellular Comp. Physiol. 57: 81-86, 1961.

38. MENAKER, M. Hibernation-hypothermia: An annual cycle of response to low temperature in the bat Myotis lucifugus. J. Cellular Comp. Physiol. 59: 163-173, 1962.

39. MENAKER, M. Frequency of spontaneous arousal from hibernation in bats. Nature 203: 540-541, 1964.

40. MOGLER, R. K. Das endokrine system des Syrischen gold-hamsters, Mesocricetus auratus auratus Waterhouse, unter berücksichtigung des natürlichen und experimentellen Winterschlafs. Z. Morph. Okol. Tiere 47: 267-308, 1958.

41. MORRIS, L., and P. MORRISON. Cyclic responses in dormice, Glis glis, and ground squirrels, Spermophillus tridecemlineatus, exposed to normal and reversed yearly light schedules. Science in Alaska: Proc. Fifteenth Alaskan Sci. Conference A.A.A.S. 40-41, 1964.

42. MROSOVSKY, N. The performance of dormice and other hibernators on tests of hunger motivation. Animal Behaviour 12: 454-469, 1964.

43. MROSOVSKY, N. The adjustable brain of hibernators. Sci. Amer. 218(3): 110-118, 1968.

44. MROSOVSKY, N. Mechanism of hibernation cycles in ground squirrels: circannian rhythm or sequence of stages. Trans. Penn. Acad. Sci. In press.

45. MROSOVSKY, N., and K. C. Fisher. Sliding set points for body weight in ground squirrels during the hibernation season. Can. J. Zool. 48: 241-247, 1970.

46. PENGELLEY, E. T. Responses of a new hibernator, Citellus variegatus to controlled environments. Nature 203: No. 4947: 892, 1964.

47. PENGELLEY, E. T. The relation of external conditions to the onset and termination of hibernation and estivation. In: Mammalian Hibernation III, edited by K. C. Fisher, A. R. Dawe, C. P. Lyman, E. Schoenbaum, and F. E. South. Edinburgh: Oliver and Boyd, 1967, p. 1-29.

48. PENGELLEY, E. T. Interrelationships of circannian rhythms in the ground squirrel, Citellus lateralis. Comp. Biochem. Physiol. 24: 915-919, 1968.

49. PENGELLEY, E. T., and S. J. ASMUNDSON. Free-running periods of endogenous circannian rhythms in the golden-mantled ground squirrel, Citellus lateralis. Comp. Biochem. Physiol. 30: 177-183, 1969.

50. PENGELLEY, E. T., and S. J. ASMUNDSON. The effect of light on the free running circannual rhythm of the golden-mantled ground squirrel, Citellus lateralis. Comp. Biochem. Physiol. 32: 155-160, 1970.

51. PENGELLEY, E. T., S. J. ASMUNDSON, and C. UHLMAN. Homeostasis during hibernation in the golden-mantled ground squirrel, Citellus lateralis. Comp. Biochem. Physiol. In press.

52. PENGELLEY, E. T., and K. C. FISHER. Onset and cessation of hibernation under constant temperature and light in the golden-mantled ground squirrel, Citellus lateralis. Nature 180: 1371-1372, 1957.

53. PENGELLEY, E. T., and K. C. FISHER. Rhythmical arousal from hibernation in the golden-mantled ground squirrel, Citellus lateralis tescorum. Can. J. Zool. 39: 105-120, 1961.

54. PENGELLEY, E. T., and K. C. FISHER. The effect of temperature and photoperiod on the yearly hibernating behaviour of captive golden-mantled ground squirrels (Citellus lateralis tescorum). Can. J. Zool. 41: 1103-1120, 1963.

55. PENGELLEY, E. T., and K. C. FISHER. Locomotor activity patterns and their relation to hibernation in the golden-mantled ground squirrel, Citellus lateralis. J. Mammal. 47: 63-73, 1966.

56. PENGELLEY, E. T., and K. H. KELLY. A "circannian" rhythm in hibernating species of the genus Citellus with observations on their physiological evolution. Comp. Biochem. Physiol. 19: 603-617, 1966.

57. ROWAN, W. On photoperiodism, reproductive periodicity, and the annual migrations of birds and certain fishes. Proc. Boston Soc. Natur. Hist. 38: 147-189, 1926.

58. ROWAN, W. Light and seasonal reproduction in animals. Biol. Rev. Cambridge Phil. Soc. 13: 374-402, 1938.

59. SMIT-VIS, J. H. and G. J. SMIT. Hibernation in the golden hamster in relation to age and to season. Netherlands J. Zool. 20: 141-147, 1970.

60. SPAFFORD, D. C., and E. T. PENGELLEY. The influence of the neurohumor serotonin on hibernation in the golden-mantled ground squirrel, Citellus lateralis. Comp. Biochem. Physiol. In press.

61. STRUMWASSER, F. Factors in the pattern, timing and predictability of hibernation in the squirrel, Citellus beecheyi. Amer. J. Physiol. 196: 8-14, 1959.

62. STRUMWASSER, F., J. J. GILLIAM, and J. L. SMITH. Long term studies on individual hibernating animals. In: Mammalian Hibernation II, Ann. Acad. Sci. Fenn. Ser. A, IV, 71, edited by P. Suomalainen. Helsinki: Suomalainen Tiedakatemia, 1964, p. 399-414.

63. STRUMWASSER, F., F. R. SCHLECHTE, and J. STREETER. The internal rhythms of hibernators. In: Mammalian Hibernation III, edited by K. C. Fisher, A. R. Dawe, C. P. Lyman, E. Schoenbaum, and F. E. South, Edinburgh: Oliver and Boyd, 1967, p. 110-139.

64. SUOMALAINEN, P., editor. Mammalian Hibernation II, Ann. Acad. Sci. Fenn. Ser. A, IV, 71. Helsinki: Suomalainen Tiedakatemia, 1964.

65. TWENTE, J. W., and J. A. TWENTE. The duration of hibernation cycles as determined by body temperature. Abstract. Amer. Zool. 4: 295, 1964.

66. TWENTE, J. W., and J. A. TWENTE. Seasonal variations in the hibernating behavior of Citellus lateralis. In: Mammalian Hibernation III, edited by K. C. Fisher, et al., Edinburgh, Oliver and Boyd, 1967, pp. 47-63.

67. WEIHAUPT, J. G. Geophysical biology. Bioscience 14: 18-24, 1964.

CREDITS

The editors and authors are grateful to the following publishers, organizations and individuals for permission to use materials previously published:

Figure 1: Comparative Biochemistry and Physiology; E. T. Pengelley and S. J. Asmundson, Comparative Biochemistry and Physiology 30: 177-183, 1969.

Figure 3: Comparative Biochemistry and Physiology; H. C. Heller and T. L. Paulson, Comparative Biochemistry and Physiology 33: 357-383, 1970.

Figure 4: Comparative Biochemistry and Physiology; H. C. Heller and T. L. Paulson, Comparative Biochemistry and Physiology 33: 357-383, 1970.

Figure 5: Comparative Biochemistry and Physiology; H. C. Heller and T. L. Paulson, Comparative Biochemistry and Physiology 33: 357-383, 1970.

ASPECTS OF TIMING AND PERIODICITY OF HETEROTHERMY

GEORGE A. BARTHOLOMEW

Department of Zoology
University of California
Los Angeles, California 90024

INTRODUCTION

It is a truism that biology is indivisible. Nevertheless, from time to time it is helpful for us biologists to reaffirm that each of us is studying a very small segment of an immense continuum. This is particularly true if one is a physiologist, because physiology is an analytical discipline, and its methods demand the clear definition of a phenomenon, followed by the identification and isolation of its individual components for study under controlled conditions. Indeed, the great strength of physiology is that it can subdivide a complex system, identify the critical variables, hold constant most of the parameters affecting some selected component or components, and thus isolate a subsystem and make it amenable to analysis. Although the process of subdivision, isolation, and control makes for effective laboratory investigation, it sometimes hides the forest by the trees. Therefore, it may be the method of choice for defining the limits of biological phenomena, particularly those involving complex integrative levels. In fact, this approach can be positively misleading when applied to problems involving the performance of intact organisms, because it may result in the dissection of the details of mechanism before the limits of the problem itself have been determined. The resulting premature analysis, particularly if it yields novel results and involves technically advanced methods, can establish traditions of investigation and priorities of questions which produce a chaos of unstructured detail that literally defies ex post facto synthesis. The study of hibernation is a case in point.

A FRAME OF REFERENCE

If any single thing is clear from the history of the study of hibernation, it is that the term as traditionally used encompasses not just one phenomenon, but a whole class of phenomena. Although most of the persons who study hibernation are physiologists, hibernation is much more than a physiological process. It involves behavior, ecology, and most importantly, evolution. Nevertheless, physiologists who have studied hibernation have often been indifferent to the biological complexity of the subject. They have conceived of it primarily in terms of mammalian adjustment to seasonally cold temperatures, and their understandable fascination with the details of the mechanism of this spectacular variation on the theme of mammalian temperature regulation has led to an awesome accumulation of data. The volume of this information using mammalian homeothermy as its base of reference has, from my point of view, blunted our insights and deflected our thoughts from some of the central issues. By establishing an arbitrary frame of reference that encompasses only part of the problem, it has inhibited us from analyzing the phenomenon into its natural biological units and has perpetuated our tendency to frame our questions in terms of physiological tradition rather than biological appropriateness.

If we can free ourselves of the tyranny of standard points of view we may be able to see the phenomenon of hibernation from an angle that will broaden our biological perspective, sharpen our focus, and also indicate possible new ways to integrate and synthesize the available plethora of information.

Hibernation is most often considered in relation to body temperatures, but it may be helpful if we select a closely related but more basic frame of reference, namely animal energetics. It is obvious that the word hibernation describes a major pattern of manipulation of one of the most central elements in biology - free energy. It has been said with some validity that "the struggle for existence" is in fact a struggle for free energy for doing physiological work. Viewed thus, hibernation becomes not just a special case of temperature regulation, but a major strategy in the struggle for existence. Moreover, it becomes enormously complicated - too complicated to be amenable to strictly physiological analysis. It ceases to be a finite discipline defined in terms of established preoccupations of physiologists. Indeed, it becomes a major subsection of animal biology.

By changing our frame of reference from thermoregulation to energetics we can begin to understand the long, tortuous, and frustrating history of the study of hibernation, particularly the depressing dearth of generalizations with a satisfactory breadth of applicability. Perhaps by looking at hibernation within the framework of energetics we can pose questions the answers to which will have more generality than has often been the case when other frames of reference are used. Such an effort is far from novel, but I should like to interpret the theme more broadly than is customary. From the standpoint of energetics the hibernation of mammals can conveniently be thought of as a special case of the more widespread and more general phenomenon of adaptive hypothermia, a pattern of dormancy which occurs under a variety of conditions in a number of taxa of birds and mammals. The adaptive hypothermia of birds and mammals in turn is functionally a special case of the heterothermy (which for present purposes can be defined as intermittent endothermy) that occurs in not only birds and mammals but also in insects belonging to several different orders.

ENERGETICS AND HETEROTHERMY

The availabity of free energy for doing physiological work varies spatially and temporally. The relevant scales on which to examine adaptations of individuals in different taxa to this variability range from millimeters to kilometers and from minutes to seasons (for some populations, of course, the scale is global and the time is measured in centuries or millenia).

The patchy availability of free energy, which of itself presents many challenges to its biological exploitation, is further complicated, particularly in middle and high latitudes, by seasonal, daily, and ahemeral fluctuations in environmental temperatures. As everyone knows temperature is a primary determinant of rates of biological reaction, including energy utilization. Because of the familiar van't Hoff effect, the advantages of the sustained high levels of elevated body temperature characteristic of mammals and birds, have from the evolutionary point of view paid their own way, despite the exorbitant rates of energy utilization required. An obligatorily high metabolic rate, of course, exacerbates the problems inherent in the temporal variability of the availability of free energy in the environment.

The patchy distribution of available energy in time and space is further complicated by seasonal variability of water supply, since water, like energy, is a primary requirement for living processes.

Viewed in this context, it is to be expected that natural selection will repeatedly have favored the establishment of temporal patterns of energy utilization, including dormancy, that will allow animals to survive shortages of energy and water, and that the timing of the periods of dormancy will in some ultimate sense correlate with recurrent environmental periods during which temperature and water availability are limiting.

Furthermore, it would be surprising if some of the physiological mechanisms that were advantageous in meeting recurrent periods of environment stress were not closely related to those involved in minimizing the exorbitant energy demands of endothermy when it is associated with small size, and with adjustments to chronic rather than short-term food scarcity. Some of the more obvious ecological and evolutionary aspects of the temporal patterning of energy utilization in endothermic animals are examined below.

VARIATIONS ON THE THEME OF DORMANCY

Hibernation. Although endothermy almost certainly evolved independently in birds and mammals, it shows many convergent mechanisms and patterns in the two groups. However, hibernation does not appear to be among them, or at least is only marginally among them. Birds belonging to at least three orders (Apodiformes, Caprimulgiformes, Coliiformes) undergo diurnal cycles of torpidity, but with the possible exception of some members of the family Caprimulgidae (goatsuckers and nightjars) no incontrovertible cases of avian hibernation are available (cf. 5 for a recent review). The almost complete absence of demonstrable instances of hibernation in birds is particularly impressive in view of the common occurrence of hibernation in mammalian taxa, some members of which utilize daily torpor. Apparently because of the conspicuous mobility of birds, seasonal migration, rather than seasonal dormancy, has been the evolutionary strategy favored by natural selection.

Estivation. Although estivation is a less common pattern of adaptive hypothermia than hibernation, it exists in several mammalian orders (10). However, its occurrence under natural conditions in birds has not so far been demonstrated.

Birds and mammals of low and middle latitudes appear able to adjust to any naturally occurring conditions of heat by physiological and behavioral mechanism as long as water and food are available. However, high air temperatures and intense solar radiation when combined with seasonal drought present problems to mammals which only a few large species have become adapted to meet head-on, and which some small species meet only by prolonged periods of dormancy.

There is no reason to doubt that the most important ultimate (evolutionary) cause of mammalian estivation is seasonal drought. However, as discussed by MacMillen (14), no proximate (stimulatory) cause other than food shortage has been demonstrated in the laboratory for any species.

Estivation affords a much smaller energy saving than hibernation, because adaptive hypothermia at high ambient temperatures results in a smaller reduction in metabolism than it does at low ambient temperatures. In fact, unless a mammal's lower critical temperature is well above the ambient temperature of the site in which it estivates, the energy savings afforded by allowing T_B to approach T_A are relatively modest. With few exceptions, the lower critical temperature is inversely related to body weight, which assists us in understanding why estivation occurs most frequently in small species. When the thermal neutral zone extends as low as 18° to 20° C, as it does in most larger mammals, the energetic savings that result from estivation are so small that this pattern of adaptive hypothermia has rarely beem favored by natural selection.

Daily torpor. The phenomenon of daily torpor in birds and mammals has received much attention during the past several decades. From the diversity of taxa in which it occurs and from the variations in pattern which have been found, it is reasonable to treat it as a polyphyletic phenomenon, involving a complex array of metabolic, thermoregulatory, nervous, and cardiovascular responses which have been assembled in a variety of adaptive patterns that allow different species to cope with a wide range of ecological situations.

Moreover, as will be discussed below, daily torpor intergrades with more prolonged periods of dormancy.

In the present context a detailed review of the occurrence of daily torpor is not needed. Suffice it to say that among mammals, the phenomenon is particularly well developed in temperate zone insectivorous bats and is well documented in several families of rodents, and that among birds it is best known in hummingbirds, but also occurs in swifts, caprimulgids, and colies.

In small, temperate zone bats of the suborder Microchiroptera a clear cut pattern of daily torpor appears to be the general rule (cf. 15 and 19 for discussion and references). Except during incubation, a similar situation sometimes exists in some temperate zone hummingbirds, although torpidity appears to be infrequent or absent when their energy balance is positive (13).

Both bats and hummingbirds, however, are primarily tropical in distribution and in the tropics the situation with regard to torpidity in these two groups is variable. For example, McNab (15) has found in tropical members of the Microchiroptera that daily torpidity is correlated with food habits, being common in insectivorous forms and rare or absent in frugivorous forms. Size also correlates with food habits in the Microchiroptera, insectivorous forms being smaller than the frugivorous forms. McNab's assignment of a primary role to food habits in the heterothermy of bats is reasonable, and the correlation between food supply and size shown in bats emphasizes the central role of the interplay between energy demands and energy availability in adaptive hypothermia.

The weight-specific energy demands of bats and hummingbirds are great not only because of their small size, but because they feed on the wing, and flight is energetically expensive. It is of interest that shrews, the other smallest amniote endotherms, are at least semifossorial and apparently do not employ daily torpor.

Swifts and goatsuckers are much larger than most hummingbirds. Hence their weight-specific energy demands are less severe, but they also are aerial feeders and their food supply is subject to drastic reduction during inclement weather. Their capacity for short periods of torpidity should clearly be of selective advantage in that it would tide them over irregularly occurring periods of bad weather (2,11,16). The ecological significance of the facultative hypothermia of colies which are frugivorous and weight 40 to 50 grams (3) remains to be assessed.

From the preceding discussion we may conclude that even in the absence of stressfully low temperatures facultative daily torpor represents a strategy of energy conservation which has evolved in situations where high weight-specific energy demands exist either because of small size or aerial feeding. When both conditions obtain, as they do with many microchiropterans and some hummingbirds, daily torpor has been the strategy of choice. Where only one of these conditions obtains (sometimes it has been selected for, sometimes not) the smallest megachiropterans are frugivorous but show daily torpor (1); the smaller insectivores are largely insectivorous but apparently do not experience daily torpor. Medium-sized aerial feeders such as some swifts (Aeronautes, _Apus_) and goatsuckers (_Caprimulgus_, _Phalaenoptilus_) under conditions of mild inanition enter a state of torpor from which they can arouse by the production of endogenous heat, but other insectivorous birds of equal or smaller size, do not have such a capacity.

When we consider the many rodents which are known to undergo daily torpor, it is clear that a different ecological situation is involved. For purposes of the present discussion I shall confine my remarks to members of the family Heteromyidae, the group in which this phenomenon has been most intensively studied. Although some heteromyids, such as Perognathus longimembris, which undergo torpor weigh only about 10 grams, most members of the group are several fold heavier than the smaller bats and hummingbirds, and of course all of them are primarily granivorous. Consequently, the family Heteromyidae must have evolved the capacity for torpidity under selective forces different from those shaping the analogous response in the aerial feeders previously discussed.

Tucker (17) in his excellent study of the thermodynamics of torpor in Perognathus californicus found that even if this pocket mouse enters torpor at 15° C and immediately arouses, it used only 55% of the energy that would be required to maintain normal body temperature for the same amount of time. Hence, even the shortest possible torpor cycle results in a significant energy savings, and of course, the longer the period of torpidity and the lower the ambient temperature, the greater the energy savings.

The pocket mice, kangaroo mice, and kangaroo rats of the family Heteromyidae are pre-eminently adapted to conditions of aridity and are among the most successful mammals in the arid regions of western North America. In these desert areas leaf and seed production is low, seasonally restricted, and sometimes essentially lacking for several consecutive years. Consequently, the small herbivorous mammals that occupy them must be able to carry out their essential functions on a fixed and limited energy supply. Under such conditions strict homeothermy could be self-defeating. They circumvent this problem by estivation, hibernation, and daily torpor (cf. 9 for a review), but there is a dearth of quantitative information on the patterns of timing which allow them to budget their metabolic expenditures against this limited energy supply. It is known (4) that in heteromyids torpor is absent or rare when food is available in excess, but that a modest restriction in the availability of food during any season induces daily torpor in

Perognathus, and Microdipodops and at least one species of kangaroo rat, Dipodomys merriami. Tucker (18) has found that the daily torpor of P. californicus shows a clear-cut circadian rhythm and that the duration of the daily torpor is inversely related to the daily food ration.

In this instance, the timing of torpidity appears to be a special case of circadian rhythmicity, with the animal's clock showing a pattern of temperature independence reminiscent of comparable rhythms in plants and ectothermic animals. However, some heteromyids show patterns of recurrent dormancy intermediate between diurnal torpor and the prolonged periods of seasonal dormancy shown by hibernating and estivating rodents such as ground squirrels. The relation of these recurrent non-seasonal patterns of torpor to food supply and to environmental temperature has been analyzed for the kangaroo mouse, Microdipodops pallidus by Brown and Bartholomew (4). In M. pallidus the length and periodicity of torpor are related to both environmental temperature and to food availability in such a way that the animals can maintain weight and also accumulate food stores under a variety of conditions. By varying the ratio between the time spent in torpor and the time spent in a homeothermic condition, they are able to match their energy demands with their energy resources, even though their energy resources are stored in part in their own tissues (particularly in the proximal part of the tail), and in part in underground food caches. The effective utilization of these two separate energy supplies requires that the animal measure the total energy resources it can tap and balance this against the energy cost of foraging for more food (which will depend on food availability) and the energy cost of regulating its body temperature near 35° C as compared with becoming torpid (which will depend on ambient temperature). Moreover, in addition to measuring the food resources available, these mice are apparently able to respond differentially to the rate at which new food becomes available. In an ambient temperature of 6° C they do not become torpid when a slight excess of food is supplied daily, but they do show torpor when a large quantity is supplied at weekly intervals. Under natural conditions this pattern of response could allow them to adjust the duration of torpor to foraging success.

The selective advantages of such a complex integration of inputs to the control of the timing of dormancy is readily apparent with reference to both energy availability and energy demands. By the precision of its control of the timing of torpor this species should be able to adjust its energy expenditure so as to compensate for food shortages associated with either chronic food shortage or exhaustion of the seed crop or with acute shortages that might develop because of brief periods of inclement weather. It can also minimize the big energy costs which a small endotherm must pay in cold weather, because as ambient temperature goes down the cost of homeothermy increases while the cost of torpidity decreases.

Thus, Microdipodops pallidus, faced with an erratic environment with limited energy availability has evolved a timing system which apparently is neither circadian nor circannual, nor does it depend on some regular astronomical clue such as day length. Instead, it integrates a complex of factors, ambient temperature, fat stores, food caches, and the rate that new food becomes available, in such a manner that it achieves a maximum of environmental independence.

BODY WEIGHT AND PATTERNS OF HETEROTHERMY

Insect heterothermy. Many insects cannot fly unless the temperature of their wing muscles is 30° C or higher. Those forms, such as sphinx moths that are crepuscular or nocturnal cannot depend on external sources for heat, but most produce it endogenously like birds and mammals. Some sphingid and saturniid moths have thoracic temperatures of 40° C or more while they are in flight and cannot take off until they warm up to approximately 35° C (cf. 6,7,8 for a guide to the literature). They warm up by vibrating their wing muscles, a process that is analogous to shivering in birds and mammals. In fact, the preflight warm-up of large moths is strikingly similar to the performance of birds and mammals during arousal from the hypothermia associated with daily torpor, estivation, and hibernation. However, in these endothermic insects, high body temperatures occur only in association with flight and are ordinarily restricted to the thorax. They carry out all their terrestrial activities at body temperatures near ambient. Thus, except during flight and preflight warm-up they are poikilotherms. Thus, their pattern of heterothermy can be referred to as <u>facultative endothermy</u> to distinguish it from the <u>adaptive hypothermia</u> of heterothermic vertebrates. The energetic utility of restricting endothermy to periods of intense activity is obvious when one considers the very small size of these insects. The energy expenditure necessary for an animal weighing a gram or less to maintain body temperature 20° C or more above ambient is too great to be practical for more than brief periods - less than an hour per day in some sphingids, and only for a few hours in a life time for those saturniids which do not feed as adults. In this regard the periodicity and timing of the endothermic state employed by some bumble bees is particularly instructive.

Bumble bees do not hover while gathering nectar, but they must be warm to fly from flower to flower. When <u>Bombus terricola</u> (mean weight of workers, 200 mg) feeds on a flower panicle of golden rod (<u>Solidago</u>) which has 1000 or more closely spaced florets, it crawls from one floret to another. Its thoracic temperature falls below flight temperature and varies directly with ambient. Before it can fly to another plant it must warm itself up at least to 30° C. When the same species feeds on milkweed (<u>Asclepias</u>) the inflorescence of which contains only a few florets, it continuously maintains its thoracic temperature between 34 and 36° C over a wide range of ambient temperatures; it remains on each panicle briefly and flies from one plant to the next without having to go through a preflight warm-up. Thus, in this species the temporal pattern of thoracic temperature, and hence energy expenditure, depends on the spacing, density, structure, and presumably nectar content, of the flowers on which it feeds (7). Obviously the time constant for the alternating periods of high and low thoracic temperatures in this bumble bee contrasts markedly with that of even the smallest

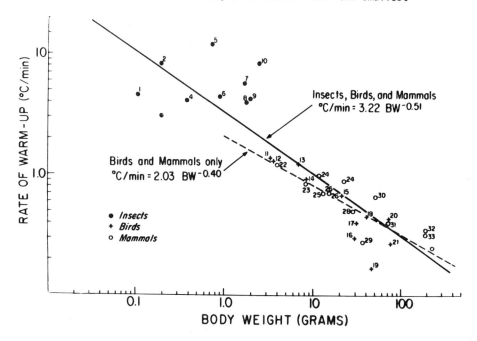

FIG. 1. Rates of increase in body temperature during warm-up in insects, birds, and mammals. In all cases ambient temperatures and initial body temperatures were between 20 and 25° C. For names of species and references, cf. ref. 8 from which the figure is taken.

avian or mammalian heterotherms, since the bumble bee may cycle through several warmings and coolings per hour. Our understanding of patterns of timing of the endothermic state is assisted by examining the rates of warm-up in a variety of heterotherms.

Body weight and rate of warm-up. It has long been known that small heterothermic mammals warm up more rapidly than large ones. The same relationship exists in heterothermic birds (2,13). The generalization that rate of warm-up is inversely related to body weight can be extended to include representatives of several orders of insects (Fig. 1). The correlation between body weight and warm-up rate is notably similar in birds and mammals. The apparent difference between vertebrates and insects is probably explained in part by the fact that the insects warm up only the thorax, but the rates of warm-up are plotted in terms of total body weight. For present purposes the salient feature is that in ambient temperatures of 20° to 25° C a 100 mg insect warms up about 10° C per minute while a 100g mammal warms up about 0.3° C per minute. Thus, there is time for an insect to pass through many heating-cooling cycles per day while few mammals pass through more than one.

From the equation for the lower curve in Figure 1, one and five kilo mammals would warm at rates of 0.13° and 0.07° C per minute respectively - too slow to fit easily into a daily cycle. It is of interest that adaptive hypothermia with a daily periodicity is known to occur only in birds and mammals weighing much less than 100 g.

CONCLUSIONS

If one uses animal energetics rather than mammalian temperature regulation as a frame of reference, hibernation can be treated as one of many mechanisms which allow animals to match their energy demands to their energy resources through manipulation of rates of energy expenditure. Rates of energy expenditure are strongly affected by temperature, and natural selection has acted on this relationship in a variety of ways in many different taxa. The results are visible in a fascinating array of adaptations which share a common feature, they exploit the relationship between rate of energy utilization and body temperature. The net result of all these adaptations is that the animals which possess them acquire increased independence of the environment, particularly of temporal variation in environmental conditions.

These adaptations are found in several different groups on different phyletic lines so their underlying similarities almost certainly result from convergence. Nevertheless, a reasonably complete continuum of temporal heterothermy is visible in this polyphyletic series extending from the intermittent endothermy of large insects, through the short-term adaptive hypothermia of small birds and mammals, to the long-term dormancy of estivation and hibernation in mammals with a variety of intermediate stages. The organisms which show one or another of these patterns of temporal heterothermy range in size from 100 milligrams to several kilograms and belong to several different major taxa, insects (beetles, bees, moths), mammals (monotremes, marsupials, and placentals of several orders), and birds (hummingbirds, swifts, goatsuckers, colies, and possibly swallows).

In addition to affording energy conservation to very small forms, and to flying forms dependent on erratic food supplies, temporal heterothermy allows animals that have high metabolic rates during normal activity to tide themselves over periods when energy resources are scarce or absent because of seasonal drought or seasonal cold. It also makes it possible for them to accommodate their energy demands to such subtle variables as size of energy stores and foraging success, as well as to chronic food shortages of more than seasonal duration.

In view of the variety of ecological situations and the range of time scales to which temporal heterothermy can be adaptive, it is obvious that no single pattern of timing, or single set of timing mechanisms, will be found. Underlying endogenous rhythms of annual, lunar, daily, and shorter periods are to be expected, and imposed on or integrated with these will be timing cues of many sorts depending on the ecology and physiology of the species in question - sunrise, sunset, photoperiod, seasonal rain, seasonal drought, environmental temperature, nutritional state, negative water balance, size of food cache, and foraging effort per unit food, to name but a few of the more obvious.

ACKNOWLEDGEMENT

The preparation of this report and some of the projects discussed in it were supported in part by NSF Grant GB-18744.

REFERENCES

1. BARTHOLOMEW, G. A., W. R. DAWSON, and R. C. LASIEWSKI. Thermoregulation and heterothermy in some of the smaller flying foxes (Megachiroptera) of New Guinea. Z. Vergl. Physiol. 70: 196-209, 1970.

2. BARTHOLOMEW, G. A., T. R. HOWELL, and T. J. CADE. Torpidity in the White-throated Swift, Anna Hummingbird, and Poor-will. Condor 59: 145-155, 1957.

3. BARTHOLOMEW, G. A., and C. H. TROST. Temperature regulation in the Speckled Mousebird, Colius striatus. Condor 72: 141-146, 1970.

4. BROWN, J. H., and G. A. BARTHOLOMEW. Periodicity and energetics of torpor in the kangaroo mouse, Microdipodops pallidus. Ecology 50: 705-709, 1969.

5. DAWSON, W. R. and J. W. HUDSON. Birds. In: Comparative Physiology of Thermoregulation I, edited by G.C. Whittow. New York: Academic Press, 1970, p. 223-310.

6. HANEGAN, J. L., and J. E. HEATH. Mechanisms for the control of body temperature in the moth, Hyalophora cecropia. J. Exptl. Biol. 53: 349-362, 1970.

7. HEINRICH, B. Temperature regulation in bumblebees: biology and ecology. Z. Vergl. Physiol. In press.

8. HEINRICH, B., and G. A. BARTHOLOMEW. An analysis of preflight warm-up in the sphinx moth, Manduca sexta. J. Exptl. Biol. In press.

9. HUDSON, J. W. Variations in the patterns of torpidity of small homeotherms. In: Mammalian Hibernation III, edited by K.C. Fisher, A.R. Dawe, C.P. Lyman, E. Schönbaum, and F.E. South. Edinburgh: Oliver and Boyd, 1967, p. 30-46.

10. HUDSON, J. W., and G. A. BARTHOLOMEW. Terrestrial animals in dry heat: estivators. In: Handbook of Physiology. Adaptation to the Environment, edited by D.B. Dill, A.E. Adolph, and C.G. Wilber. Washington: American Physiological Society, 1964, sect. 4, p. 541-550.

11. KOSKIMIES, J. 1961. Fakultative Kältelethargie beim Mauersegler (Apus apus) im Spätherbst. Vogelwarte 21: 161-166.

12. LASIEWSKI, R. C. Oxygen consumption of torpid, resting, active, and flying hummingbirds. Physiol. Zool. 36: 122-140, 1963.

13. LASIEWSKI, R. C., and R. J. LASIEWSKI. Physiological responses of the Blue-throated and Rivoli's Hummingbirds. Auk 84: 34-48, 1967.

14. MACMILLEN, R. E. Aestivation in the cactus mouse, Peromyscus eremicus. Comp. Biochem. Physiol. 16: 227-248, 1965.

15. MCNAB, B. K. The economics of temperature regulation in neotropical bats. Comp. Biochem. Physiol. 31: 227-268, 1969.

16. PEIPONEN, V. A. On hypothermia and torpidity in the nightjar (Caprimulgus europaeus L.). Ann. Acad. Sci. Fenn. Ser. A. IV 87: 4-15, 1965.

17. TUCKER, V. A. The relation between the torpor cycle and heat exchange in the California pocket mouse Perognathus californicus. J. Cellular Comp. Physiol. 65: 405-414, 1965.

18. TUCKER, V. A. Diurnal torpor and its relation to food consumption and weight changes in the California pocket mouse Perognathus californicus. Ecology 47: 245-252, 1966.

19. TWENTE, J. W. and J. A. TWENTE. An hypothesis concerning the evolution of heterothermy in bats. In: Mammalian Hibernation II, edited by P. Suomalainen. Helsingfors: Suomalainen Tiedeakatemia, 1962, p. 435-442.

PROBLEMS IN EXPERIMENTATION ON TIMING MECHANISMS FOR ANNUAL PHYSIOLOGICAL CYCLES IN REPTILES

PAUL LICHT

Department of Zoology
University of California
Berkeley, California 94720

INTRODUCTION

Reptiles, especially in temperate regions, typically exhibit pronounced seasonality in many aspects of their physiology. This is hardly a surprising observation for a poikilothermous animal in which most physiological functions are temperature dependent. No doubt many of the seasonal physiological changes in reptiles may be explained by a relatively direct relationship to seasonal temperature changes (i.e., "Q_{10}-effect"). However, both phenological and experimental data for a diversity of temperate and tropical species indicate that such a simple temperature relationship cannot account for all aspects of annual cyclicity in Reptilia; and other types of mechanisms for regulating cycles must exist. Experimentation on these mechanisms in reptiles is relatively limited in depth and scope. However, sufficient work has been done to provide some valuable insights into the general mechanisms for the timing of annual cycles in vertebrates. The literature dealing with reptiles is examined here with the objective of evaluating the major problems involved in experimenting with annual cycles and interpreting the ecological significance of such laboratory experimentation.

Seasonal physiological changes clearly can be adaptive in a seasonal habitat if they are synchronized with the appropriate climatic conditions. Of particular importance in this connection are those timing mechanisms that enable the species to "anticipate" and thereby be prepared for the arrival of a particular season. A wide array of both reproductive and non-reproductive functions appear to depend on such regulatory mechanisms.

Both endogenous (e.g., circannual rhythms) and exogenous control of annual cycles have been proposed for reptilian cycles. However, few studies have been specifically designed to distinguish between these two alternatives. Studies with reptiles have typically been of relatively short duration and have had the aim of altering (accelerating or slowing) some aspect of a season cycle by manipulating one or more environmental conditions. I hope to show that such short term laboratory studies frequently lead to erroneous conclusions about the mechanisms involved in the control of the annual cycle in nature, especially when the conditions used are unrealistic with reference to natural conditions.

MECHANISMS FOR TIMING ANNUAL CYCLES

Endogenous rhythms. In the present context of seasonal cycles, "endogenous" timing is used to indicate that a change in physiological state may occur autonomously of external conditions. Evidence for such an internal rhythm is derived from the demonstration that the physiological change may occur under presumably constant external conditions, even though the exact timing of these changes may be modified by external conditions. In fact, external cues (Zeitgebers) may be very important for the phasing of an internal rhythm.

The first good experimental evidence for a circannual rhythm in a reptile was the data on appetite and growth in the American alligator. Hernandez and Coulson (21) observed that their captive alligators underwent seasonal "...hypoglycaemia and anorexia coincident with the approach of winter regardless of temperature in the animal's habitat" (unfortunately, they did not report what range of temperatures existed). When kept at a constant temperature (28°C) with either 10 or 14 hours of light daily, starting with the beginning of a period of anorexia in the fall, the alligators showed a complete cycle of growth and blood glucose. Even under these constant conditions, growth and blood glucose showed elevations at about the normal time (spring) and then declined again at about the normal time the next autumn (Fig. 1). Interestingly, a captive tropical crocodilian (Caiman) did not show these cycles. Photoperiod can apparently be ruled out as having an effect on this rhythm in the alligator but it seems likely that, in nature, the annual changes in temperature would modify growth, i.e., temperature might be a Zeitgeber, since changes in temperature greatly influence many aspects of the alligator's physiology (9).

In the horned lizard, Phrynosoma m'calli, anorexia typically occurs in conjunction with a depression in the standard rate of oxidative metabolism associated with winter dormancy or brumation. (Mayhew [41] proposed the term brumation for certain types of winter dormancy in reptiles to avoid the difficulties involved in applying the definition of mammalian hibernation to a poikilotherm. Brumation is intended to indicate a change in physiological state that differs from the temporary "numbness" or cold torpor that typically occurs in response to unfavorably low temperatures; brumation may occur even under constant warm temperatures). However, the loss of appetite seemed relatively independent of both temperature and photoperiod even though metabolic rate appeared to be regulated by temperature (41). These data suggest that at least part of the appetite cycle - the onset of anorexia - might be controlled endogenously. The termination of brumation ("arousal") normally depends on high temperatures following a cool period (i.e., heat acts as a cue). Lizards kept at continuously high temperatures (35°C) died, probably from starvation, before appetite returned

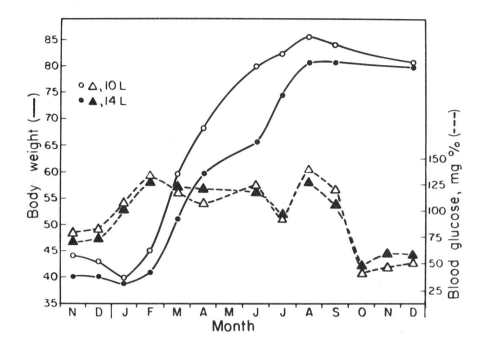

FIG. 1. Annual cycles in growth and blood glucose in juvenile alligators kept under constant temperatures (28°C) and 10 or 14 hr photoperiods. Based on data presented by Hernandez and Coulson (21: Table 1).

to normal, indicating a period of thermal refractoriness to high temperatures. However, on the basis of several studies dealing with reproduction discussed below, one might expect that if inanition were prevented, these lizards might have eventually aroused even under constant high temperatures. One lizard, aroused spontaneously from low temperatures in complete darkness. Unfortunately, in no case was an individual kept under any constant condition long enough to determine whether it would exhibit a complete activity cycle (i.e., entry, emergence and subsequent re-entry into brumation).

Stebbins (45) observed several complete "annual" activity cycles in lizards, Sceloporus, kept under a constant 9 hour photoperiod without seasonal temperature changes. The length of the cycle under these conditions tended to be shorter than in nature but data were insufficient for extensive analyses.

Studies on reproductive cycles in reptiles are replete with suggestions of endogenous rhythmicities. For example, Fitch (16) discussed several snakes that underwent annual egg-laying cycles for several years in captivity, but insufficient environmental data were given to evaluate these observations; no attempts to maintain constancy were mentioned. More careful experimentation has provided evidence for endogenous control of several aspects of reproductive activity in a diversity of reptilian species.

The lizard, Lacerta sicula, undergoes a complete testis cycle if kept under constant warm temperatures and photoperiod starting shortly after the testes have regressed in the fall (15). Testes remained completely inactive during the winter and then developed at approximately the normal time in spring followed by regression in early summer. More detailed examination of this species provides additional evidence for the various endogenous components of the cycle and their Zeitgebers (cf. Fig. 7 for summary). For example, in nature, testes develop slowly through the fall and winter and warm temperatures are required to complete development in spring (35). These same high temperatures that stimulate in the spring will suppress spermatogenesis during late summer, fall and winter. Thus, there is a period of thermal refractoriness to high temperatures starting in July which may be responsible for regression (35). The length of this refractory period is normally influenced by temperature: low temperatures accelerate the termination of refractoriness (11,12). However, refractoriness will eventually cease "spontaneously" even under constant high temperatures (15,35).

Endogenous rhythms may be particularly important for the regressive phase of testes cycles. The onset of testis regression in L. sicula can be delayed by short day lengths; however, regression cannot be prevented by photoperiod manipulation (15). Attempts to prevent testis regression by manipulating temperature and photoperiod over wide ranges were also unsuccessful in two other lizards, Dipsosaurus dorsalis and Xantusia vigilis (Licht, unpublished data). Thus, the cessation of reproductive activity may be "obligatory" in many lizards. In contrast, in several species of Anolis lizards, testis regression is largely "facultative" and depends on photoperiod (reduced day lengths), although testes may regress "endogenously" after a long period even under long days (31, and unpublished data). Several species of scincid lizards have also been maintained in constant reproductive condition in the laboratory indicating that regression is facultative (1,2).

There is also evidence for endogenous rhythmicity in some female lizards. Tinkle and Irwin (46) and Marion (37) showed that ovarian development was dependent on warm temperatures in two species of lizards but there was a seasonal cycle in responsiveness that resembled the situation in male Lacerta. Both species (Uta stansburiana and Sceloporus undulatus) were refractory to these warm temperatures in fall and winter but this refractoriness eventually terminated in mid-winter even if the lizards were kept continuously at warm temperatures for many months starting in the fall; no photoperiodic effects were evident.

I recently observed that a number of individual male Dipsosaurus dorsalis underwent complete testis cycles under constant conditions of temperature and photoperiod (unpublished data, Fig. 2). The period of active spermatogenesis occurred at approximately the normal time of year (April) in only a few animals and was delayed by a few months in most.

In at least one lizard (#224), the duration of the breeding season was extended by several months due to relatively late regression. About half of the colony underwent one complete cycle but a very small percentage (only one animal -#304) showed a second cycle in the next year. These rhythms were observed only in animals kept at relatively warm temperatures which is consistent with the observation that high temperatures are normally required to trigger testis growth in the spring (Licht, unpublished). Low temperatures during the winter probably serve as a Zeitgeber to synchronize the population since all lizards are responsive when they emerge from hibernation in the spring and the time of emergence is relatively uniform.

Although Fischer (15) has used the term circannual periodicity in connection with his studies on Lacerta, it is not clear that such a conclusion is warranted yet. It seems reasonable to demand as the minimal proof for the existence of a circannual rhythm that at least one complete period and the initiation of a second be observed under constant conditions (23). With the exception of Hernandez and Coulson's (21) study on growth in the alligator, some of Stebbins (44) observations on Sceloporus activity, and the observations on the testis activity in Dipsosaurus, none of the above studies were sufficiently extended to provide this minimal proof. Furthermore, none of these studies have used "free-running" conditions (i.e., continuous light or complete darkness). The majority of data for reptiles warrant only the conclusion that some phase of the annual reproductive and appetite cycles (e.g., initiation or termination) may be regulated endogenously and in almost all cases, the existence for external Zeitgebers is apparent.

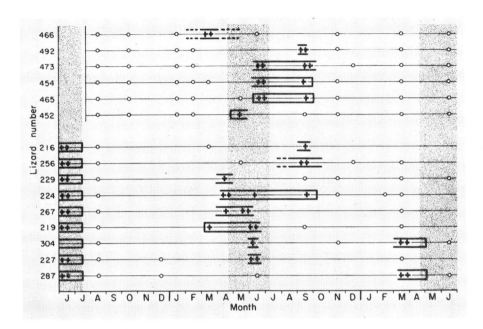

FIG. 2. Spermatogenic activity in captive desert iguanas (<u>Dipsosaurus dorsalis</u>) kept under constant conditions of temperature (<u>ca</u>. 38°C) and photoperiod (12 hrs) starting near the time of testes regression in the summer of 1968. Individuals were examined by laporatomy at the times marked by symbols: O, no indication of spermatogenesis; +, testes partially developed; ++, testes fully enlarged. Bars delineate approximate "breeding" season for each. Only lizards #227 and #304 showed any indication of activity throughout 1970. Stippled areas demarcate the normal "breeding" season. Based on Licht, unpublished data.

EXOGENOUS STIMULI

In addition to the apparent need for Zeitgebers for endogenous rhythms, there is evidence that many aspects of the annual cycle are essentially entirely under exogenous control. The two environmental factors that have received the most attention as possible cues for timing annual cycles in reptiles are temperature and photoperiod. Phenological data for reproduction in some lizards, especially for females in tropical regions, have strongly implicated seasonal rainfall cycles as a basis for reproductive seasonality (reviewed in Licht and Gorman, 33). However, there are no experimental

TABLE 1. Experimental evidence for photoperiodism in reptiles[a]

Physiological Cycle	Observed Effects of Photoperiod	Species[b]	Reference
Reproduction	1. Long days initiate or accelerate testis recrudescence	Lacerta sicula Anolis carolinensis Xantusia vigilis Uma spp. Phrynosoma cornutum Chrysemys picta	11-15,20 8,18,25, 26,29,30,31 4,5 39 43 7
	2. Long days accelerate testis regression	Lacerta sicula	15
	3. Long days prevent testis regression	Anolis carolinensis	7,18,30
	4. Long days promote ovarian growth	Uma notata Callisaurus draconoides Phyrnosoma m'calli Scincus scincus Chalcides occelatus	38 39 39 1 2
Appetite - growth	1. Long days cause hyperphagia and body growth	Anolis carolinensis Dipsosaurus dorsalis	17,31 40
Thermal relations	1. Long days increase preferred body temperature	Sceloporus undulatus	3
	2. Long days increase heat resistance	Anolis carolinensis Chrysemys picta Sceloporus occidentalis	28 22 24
Brumation	1. Short day lengths initiate winter dormancy	Phrynosoma m'calli	41
Thyroid activity	1. Long days increase height of thyroid epithelium	Anolis carolinensis	19

[a] This compilation includes all experimental studies that have been interpreted as demonstrating photoperiodism by their authors or by subsequent reviews.

[b] All are lizards except for one turtle. Pseudemys scripta.

studies on this mechanism and only temperature and photoperiod effects will be examined in detail here. Since there seems to be a strong bias towards demonstrating a photoperiodic dependence in reptilian studies, attention will be focused on this phenomenon.

Table 1 summarizes all of the studies in which it was concluded that photoperiodism exists in connection with some aspect of an annual physiological cycle in a reptile. Early studies dealt with reproductive cycles and paralleled those in birds and mammals. In 1937, Clausen and Poris (8) reported that artificially increased day lengths stimulated testicular growth in the lizard Anolis carolinensis and Burger (7) reached similar conclusions for a turtle, Pseudemys scripta. These reports suggested that increasing day lengths might be responsible for timing the onset of breeding in the spring, but this conclusion is tenuous: temperature was poorly regulated in both studies; Burger used extremely small sample sizes and his turtles were in poor health; and Clausen and Poris' control animals were smaller than experimentals and showed abnormal testis activity (34). Many of the subsequent studies also dealt with reproduction, but there is some evidence for photoperiodism in connection with several non-reproductive cycles.

Although it would seem that the use of photoperiod in timing seasonal activities may be an important and widespread phenomenon among reptiles, I propose that some of these studies do not warrant any statement about the importance or even the existence of photoperiod in reptiles. Criticism revolves around two major questions: (a) Were the observed effects due to photoperiodic manipulation as proposed or to some other controlled variable (e.g., temperature)? (b) Even if photoperiodism was demonstrated experimentally, does it necessarily follow that photoperiod is responsible for the synchrony of the annual cycle in nature; i.e., is photoperiodism ecologically relevant?

Regulation of environmental conditions and experimental design.

a. Were all variables adequately regulated? Perhaps the most common error connected with the study of photoperiodism in reptiles (and many other poikilotherms) is the lack of adequate temperature regulation. It is well established that many reptiles, especially lizards, exhibit thermoregulatory behavior. When provided with a temperature gradient, each species will "strive" to maintain body temperatures at a characteristic preferred level. Typically, this thermoregulation depends on an exogenous heat source such as solar radiation in nature or, under laboratory conditions, the lamps are used to provide illumination. Standard measurements of air temperatures - which are generally the ones reported - may fail to show that body temperatures are warmed by radiant heating. Unless special precautions are taken, it is likely that manipulation of photoperiod will simultaneously vary body temperature. If temperature has not been ruled out as an important variable in connection with the cycle under consideration, it seems reasonable to assume, a priori, that temperature will have some effect; temperature effects have been clearly demonstrated in many cases. When authors fail to report body temperatures of their experimental animal, potentially important experimental errors are difficult to evaluate.

The generally poor regulation of body temperature in the early experiments on photoperiodism in turtles (7) and lizards (8) has already been mentioned. In 1951, Galgano examined several different temperatures and photoperiods in connection with testis growth in Lacerta sicula (20). However, his experimental conditions were poorly defined and temperature was not well regulated. Further, the small sample sizes and variability in results do not allow meaningful evaluation about which variable was responsible for observed testis changes. There is also some doubt surrounding other work on testis activity in Lacerta sicula. For example,

Fischer (14) reported that increasing photoperiods could stimulate the early stages of testis growth at very low temperatures (ca. 5°C) that normally suppressed spermatogenesis in the dark (cf. 35). However, Figure 1 in Fischer's paper demonstrates a rise of several degrees in air temperature while the lights were on, and the possibility that the lizards were heated was not ruled out. In another study Fischer (12) commented that it was difficult to rule out the possibility that responses were due partly to the heating effects of the lamps.

Similar lack of temperature control may underlie some of the non-reproductive responses to photoperiod. For example, Mayhew (40) compared the growth rates of young Dipsosaurus dorsalis exposed to constant illumination from an incandescent lamp with those illuminated for only 11 hours per day. While the greater growth rate of the former was attributed to light, measurements revealed that these animals also were warmer on the average since they did not cool at night (see Mayhew, 41; Table 1). The fact that growth rates also seemed to vary with the quality of light used is not evidence for photoperiodism.

    b. Were adequate experimental controls employed? As important as the lack of temperature regulation is the apparent lack of aquate experimental control when both temperature and photoperiod are knowingly varied. For example, Mayhew (38,39) subjected female Uma notata, Callisaurus draconoides and Phrynosoma m'calli and males of several Uma species to continuous light at elevated (summer-like) temperatures during the winter. He concluded that the resultant stimulation of the gonads was due to the added illumination. Mayhew (42) also drew similar conclusions from the studies of Badir (1), Badir and Hussein (2) and Mellish (43) in which several scincid lizards (Scincus scincus and Chalcides occellatus) and in iguanid (Phrynosoma cornutum) were stimulated to "out-of-season" reproductive activity by continuous light and high temperatures. None of the above studies employed a "short-day" control at elevated temperatures or a "long-day" at lower temperatures. Since elevated temperatures alone have been shown to stimulate ovarian and testicular development in six saurian species (31,35,37,45 and Licht, unpublished), Mayhew's conclusions regarding photoperiodism hardly seem justified in the absence of appropriate controls. In fact, all of these data are equally consistent with the conclusion that gonadal stimulation depends solely on warm temperatures.

In a series of studies on testis growth in Lacerta sicula, Fischer (11,12,13,15) employed a complex experimental regime in which lizards were transferred from low temperatures (5°C) and darkness to warm temperatures (28-30°C) with gradually increasing photoperiods. The stimulation of testis growth after this transfer was attributed to the effects of increased illumination; i.e., photoperiodism. However, separate studies on the same species (35) showed that testis growth could be stimulated by elevating temperatures independent of day length. Fischer's (12) data also show that testis growth is not wholly dependent on any particular day length. Thus, there is little evidence for the conclusion that testis growth is regulated by photoperiod.

Ecological relevance of photoperiodism.

    a. Reliability of photoperiod as a proximate factor. The marked thermal dependence of behavior in reptiles may introduce considerable complexity into their photoperiodic relations and it has been argued (6) that such complications may greatly diminish the reliability of photoperiod as a seasonal indicator for these poikilotherms. For example, reptiles may not emee from nocturnal retreats until after sunrise, they may retreat again before sunset, and midday activity may be interrupted by excessively high temperatures (Fig. 3).

Unfortunately, little is known about conditions in the "retreats." Some species remain relatively exposed; e.g., they utilize deep shade to cool and spend the night in the open. In one such lizard, Anolis carolinensis, the illuminance threshold for photo-sexual responses is about 1-footcandle (29). This level of sensitivity tends to negate twilight as an effective part of the photoperiod. However, after sunrise, sufficient light may reach a reptile to promote photoperiodic responses even in a "retreat." On the other hand, body temperature may be too low to facilitate a photoperiodic response (cf. part c below).

Recent avian studies (reviewed by Farner, 10; Lofts et al., 36) indicate that a circadian component may be involved in the measurement of photoperiod. "Interrupted-night" or "scotophase-scan" experiments with several avian species suggest that the temporal relation between the first and last exposure to light in each day may be more important in determining photoperiodic responses than the light during the intervening hours. Such a phenomenon would have special ecological significance for reptiles since it could mean that the retreat into darkness for several hours during midday might have little consequence for photoperiodic responses. However, this possibility is not supported by recent interrupted-night experiments dealing with photo-sexual responses in Anolis carolinensis (Licht, unpublished). A 14 hour photoperiod maintains spermatogenesis while a short (13 hours or less) causes testes to regress in late summer. A basic 6 hour photoperiod combined with an hour of light at various times during the night did not have consistent effects (Fig. 4). Testes remained active in part of each group and regressed in the rest, independently of when the light pulse was given. In a second experiment designed to examine photostimulation of testis recrudescence in September, the night was interrupted by a 2 hour light pulse at various times after a 6 hour light period. Testis growth was

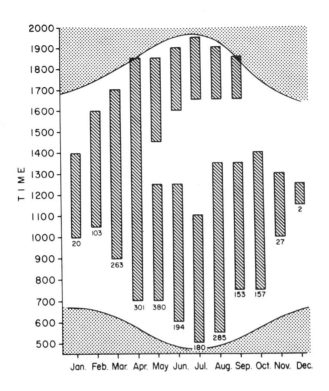

FIG. 3. Time of day when three species of Uma were active each month. Numbers beneath bars are sample sizes. Data include records for 727 U. notata, 783 U. inornata, and 555 U. scoparia; not all species were equally active at all times of day. The lower and upper curves delimit the times of sunrise and sunset, respectively, for the first day of each month; shaded areas indicate night time. Modified from Figure 3 in Mayhew (39).

successfully stimulated in this way. Thus, if the lizard is in subliminal light during the middle of the day, testis responses would probably not correspond to that expected for the full photoperiod on that day.

    b. Selection of "meaningful" photoperiods in experimental design. Although the complications described above may cause great difficulty in defining effective day lengths for reptiles, some estimates of such values are not impossible. For example, the maximum effective day length clearly cannot exceed the normal period between morning and evening twilight and for reptiles, it is probably determined by sunrise and sunset (29). Since reptiles tend to exhibit photoperiodic thresholds or "critical day lengths" which mark the division between stimulatory and non-stimulatory photoperiods (31), differences on only a fraction of an hour can have profound effects. Failure to take these factors into account when selecting photoperiods for experimental study may hinder evaluation of the ecological relevance of the results.

In the lizard, Anolis carolinensis, the critical day length for photoperiodic stimulation of testis development lies between 13 and 14 hours (30, 31), yet testes recrudescence is normally initiated and completed when photoperiods are below 13 hours. Thus, this aspect of the testis cycle cannot be timed by photoperiod as proposed by earlier studies based largely on relatively long (> 14 hours) day lengths (8,17). Likewise, Batholomew (6) reported that testis growth in Xantusia vigilis was accelerated by a 16 hour photoperiod in midwinter and Miller (44) reported the same effects with continuous light, suggesting that "long" days might be responsible for the stimulation of the gonads in the spring. However, photoperiods are below 14 hours when the testes normally begin to grow in nature (44, and Licht, unpublished data) and I found no difference between the effects of 8 and 14 hours photoperiods in this lizard (i.e., there was no evidence of photostimulation)(unpublished data). Even is increased illumination does accelerate testis growth, long days are clearly not required for normal recrudescence, since artificially shortened days in no way inhibit development in the spring (4).

The ecological significance of apparent stimulation of reproductive activity by continuous light (2,38,39,44) - assuming that light and not temperature was responsible for stimulation - is impossible to interpret.

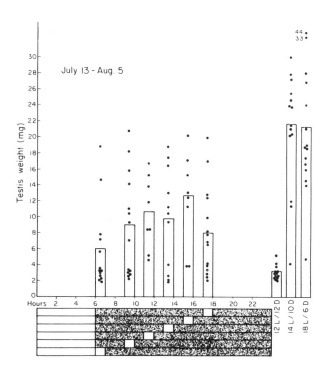

FIG. 4. Testis weights in the lizard Anolis carolinensis exposed to various "interrupted-night" photoperiods in late summer. Lizards had fully developed testes (averaging 21 mg) when treatment started on July 13; this is shortly after the seasonal onset of photosensitivity (cf. Figs. 6 & 7). The basic photoperiod consisted of 6 hrs light alternating with 18 hrs darkness with an hour of light given at various times during the night as shown below the graph. The individual (solid points) and average (vertical bars) testis weights for each of the five experimental lighting regimes (designated by Roman numerals) are shown directly above the time when the 1-hr light pulse was given. The results obtained with a standard 7, 12, 14, and 16 hr photoperiod are shown to the right. In general, testes weighing under 8 mg were regressed (spermatogenically inactive) (35).

These examples demonstrate the value of precise phenological data for designing and interrupting experimental studies on photoperiodism. These same arguments of course apply to considerations of all other possible timing mechanisms for annual cycles.

c. Temperature dependence of photoperiodism. Photoperiodism may be one of the most temperature sensitive of the physiological responses in reptiles. In Anolis carolinensis (25,27,31,34) photo-sexual responses of males occur over a temperature range of only a few degrees C. Photoperiodism is evident at a constant body temperature between 28° and 32°C but not at 25°C (Fig.5). With more natural thermocycles, photo-sexual responses occur only if body temperatures are raised to 32°C for about 8 hours daily (depending slightly on nocturnal temperatures), and this heating must occur during the day. Photoperiodism in females of this species requires even more heat than in males (Licht, unpublished data). In Anolis lineatopus from Jamaica, testis activity is dependent on day length (short days can induce regression) only if temperatures are moderately cool (e.g., 20°C for half the day); at constant warm temperatures (32°C) testes remain enlarged and active indefinitely at all photoperiods (Licht, unpublished data). Thus, high temperatures completely annul the effects of short photoperiods.

Bartholomew's (5) report that photo-sexual responses of male Xantusia vigilis were only slightly modified by temperature between 9° and 30°C conflicts with the results obtained in Anolis. However, his study was based on a single (16 hour) photoperiod without appropriate short-day controls. Studies in my lab indicate that the effects observed in Xantusia were due largely to temperature conditions and that photoperiodism was not involved.

Temperature dependent responses are also evident for photoperiodism in non-reproductive functions. For example, the photoperiodic related modification of heat resistance in the lizard A. carolinensis (28) and in the turtle, Chrysemys picta (22) occurred only at cool temperatures. In these cases, heat resistance was lower at short than at long day lengths. The photoperiodic effects on appetite and growth in Anolis carolinensis (first observed at 28°C by Fox and Dessauer (17) also depend on temperature (31)

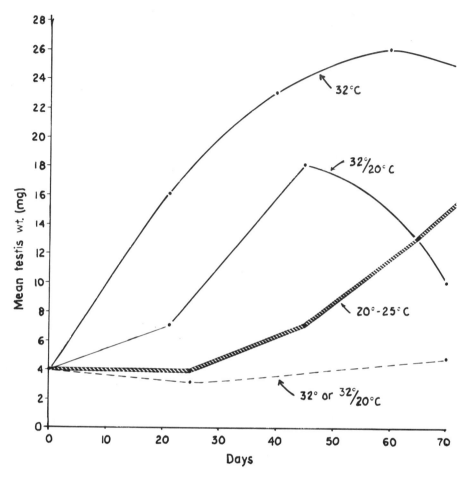

FIG. 5. Effects of body temperature on the photo-sexual response of male lizards (Anolis carolinensis) between Sept. 15 and Nov. 23. The two fine solid curves show testis growth with a "long" photoperiod (14 hrs) and the dashed line shows suppression of testis growth with "short" photoperiods (6 or 12 hrs) when body temperatures are continuously at 32°C or fluctuating between 32°C and 20°C daily (12 hrs at each). The heavy hatched curve shows testis growth when body temperature is maintained at 20 or 25°C or if there is less than 8 hrs heating to 32°C with photoperiods of 0, 6, 12, 14 or 16 hrs: the same growth rate is observed in nature. Modified from a figure in Licht (30).

although in a slightly different manner from the photo-sexual responses in this species.

Thus, each photoperiodic system may have its own characteristic temperature relations. The generally high temperature coefficient of photoperiodic responses raises the possibility that the failure to demonstrate a photoperiodic effect in the laboratory may have been due to the use of inadequate temperatures. On the other hand, the demonstration of photoperiodism in the laboratory might not be relevant to the field if temperatures are not normally in the appropriate range to facilitate photoperiodism. The latter possibility has been suggested in connection with photoperiodically-induced testis recrudescence in Anolis carolinensis. Although testis activity can be experimentally modified by photoperiod in the fall, it is unlikely that temperatures are high enough for this effect to occur in nature (29,30).

d. Seasonality in photoperiodism. Another prerequisite for the conclusion that photoperiodism controls a particular aspect of the annual cycle is the demonstration that the animal is photo-sensitive at the appropriate time of year. There is good evidence that photoperiodic responses may vary seasonally in some reptiles. For example, in Anolis carolinensis, there is a clear photoperiodic response in connection with testis activity, but only for about four months (from July through October). In other months, the male reproductive system is essentially independent of photoperiod (Figs.6,7). Since breeding normally occurs in spring (May), photoperiodism cannot be responsible for this phase of the cycle as suggested by the early studies on this lizard (8,18). The timing of the photosensitive period is consistent with the hypothesis that the regressive phase of the testis cycle (in August) is timed by photoperiod (31). A similar seasonality in photoperiodism is also evident in connection with the ovarian cycle (Licht, unpublished data) and with the appetite-growth cycle in this lizard (31), although the exact timing of photosensitivity in these functions differs slightly from that for the photo-sexual responses of the male. The female remains photosensitive later than the male, until about December, but photosensitivity is still terminated well before natural photoperiods reach potentially stimulatory levels.

Unfortunately, the seasonal aspect of photoperiodism has not been extensively explored in connection with most reptilian studies. Thus, the

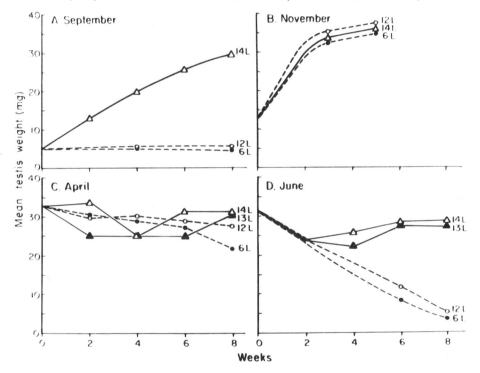

FIG. 6. Seasonal variations in the responsiveness of the testes in Anolis carolinensis to photoperiod. Body temperatures were regulated at 32°C in all experiments and daily photoperiods ranged from 6 to 14 hours (6L-14L) of light alternating with 18 to 10 hrs darkness, respectively. Dates show the months when experimental treatment was started, observed between early Sept. and mid-Oct.; in "B" early Nov. to March; in "C" from April to late May; in "D" from June through mid-Aug. Significant differences among photoperiods occur only in "A" and "D." Data are derived from: "A" (26), "B" (27), "C" and "D" (31).

experimental demonstration of a photoperiodic response in one season does not constitute sufficient proof that photoperiod controls the cycle in a different season. Alternatively, the failure to demonstrate photoperiodism in one season cannot be used to dismiss the possibility that it may account for physiological changes in other seasons.

CONCLUDING REMARKS

Interspecific variation in the control of a particular type of cycle and differences among the controls of diverse cycles in a single species are evident in reptiles. It should be noted that the vast majority of experimental studies on reptiles deal with lizards; exceptions include only two turtles and an alligator. However, several trends are now emerging with regard to the major mechanisms used for synchronizing annual cycles in reptiles. Perhaps most striking is a form of thermoperiodism that appears in connection with the initiation of various physiological activities in the spring. Despite numerous attempts to demonstrate photoperiodism in reptiles, there is considerable doubt about whether photoperiod represents an important _proximate_ factor for springtime activities. Certainly the emergence of an animal from a dark hibernaculum cannot depend directly on photoperiod and experimental studies indicate that temperature or an endogenous rhythm is the main cue. It also seems likely that after spring emergence, elevated temperatures represent the major cue for initiating or completing gonadal development and inducing hyperphagia. This form of temperature control is relatively complex since many of these activities exhibit varying degrees of refractoriness to high temperature stimulation in other seasons. This refractoriness apparently prevents premature activity in the late summer, fall or midwinter. If photoperiod (i.e., increasing day length) does stimulate certain activities in the spring, there is no evidence that it is essential for the initiation of the activity.

Another generalization about timing mechanisms is that the termination of activities in the fall (e.g., gonadal regression and anorexia) is controlled by mechanisms other than the thermal one involved in the initiation of the cycle. Both endogenous rhythmicity and photoperiod seem important for the termination of breeding. Of course, some activities may simply be terminated by the physiological depression due to low temperatures in winter.

The major timing mechanisms for some major cycles are still unclear. For example, in the only study regarding the regulation of the brumation cycle in a lizard (Phrynosoma m'calli) Mayhew (41) postulated that the onset of winter dormancy is controlled by reduced photoperiod. However, his own data do not support this conclusion since animals exposed to both long (15 hour) and short (6 hour) day lengths failed to enter dormancy.

One final aspect of annual cycles that has not been considered, and that may have especial relevance for the present consideration of hibernation, is the possibility of interrelated systems. For example, almost all lizards studied show a marked fat storage that typically occurs before winter. While it is often associated with winter dormancy, recent studies have shown that tropical lizards also show pronounced fattening at about the same time of year even though they are continually active (33). Little direct experimentation has dealt with the control of the fat cycle but a review of field studies (33) and preliminary experiments with some lizards in our laboratory suggest that it is closely linked to the reproductive cycle. There is no evidence that hyperphagia (in fact, appetites may be relatively low) or increased food availability account for the sudden fattening. However, there is a strong inverse relation between reproductive activity and fattening. It appears that the normally high energy demands of reproduction prevent fattening and that the latter is a consequence of a shift in energy partitioning when reproduction ceases. Thus, the survival of the species over winter may depend on a cessation of reproduction allowing energy storage. The timing of the reproductive cycle thereby simultaneously times the fat cycle or is at least closely related to the fat cycle.

Figure 7 illustrates two of the complex patterns of seasonally changing sensitivities to light and temperature that are responsible for the control of the annual testicular cycle in lizards. These exemplify a "facultative" cycle; i.e., one that is almost wholly dependent on exogenous cues, and a cycle that is "obligatory" because of an endogenous component. Although these represent the two best studied cases, similar physiological response patterns are emerging for several other reptilian cycles. These should emphasize the need for caution in interpreting the results of laboratory studies based on a few experimental regimes at only one or a few times of year. The arguments against some of the existing photoperiodic studies apply equally to studies of any other single factor that is suspected of being an important timing cue for some seasonal change in activity.

In addition to the experimental studies discussed herein, a number of field (phenological) studies on reptiles have included efforts to elucidate the mechanisms for environmental control. Such studies typically seek correlations between physiological and climatic changes. Some of these field studies have provided useful insights into potential timing mechanisms, and the importance of these phenological data for designing and evaluating experimental data should serve to illustrate the enormous difficulties involved in attempting to understand the nature of control mechanisms on the basis of climatic correlations alone.

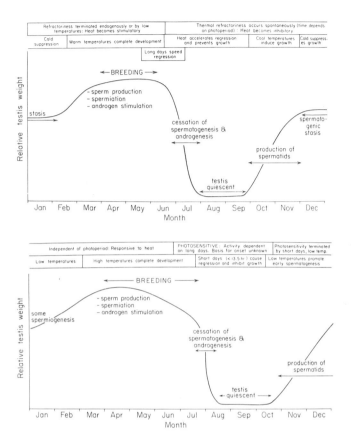

FIG. 7. Examples of two distinct patterns for timing annual reproductivity in male lizards. Curves show the yearly cycle in testis weights and descriptions of the major activities associated with them are superimposed. The top line of each graph indicates the physiological state of "responsiveness" in different seasons and lower lines summarize the major environmental influences that normally account for seasonal changes in testis activity. Data for <u>Lacerta sicula</u> (upper) illustrate an "obligatory" cycle and in which case seasonal changes in temperature sensitivity are prominent in the response to environmental conditions. Data for <u>Anolis carolinensis</u> (lower) illustrate a "facultative" cycle and one that is dominated by seasonal shifts in photosensitivity. Data for <u>Lacerta</u> are based on reports by Fischer (11,12,13,15) and Licht <u>et al.</u> (35). Data for <u>Anolis</u> are based on a review for this species in Licht (31).

SUMMARY

A. Not all aspects of the seasonality evident in the physiology of reptiles can be accounted for by simple temperature (i.e., "Q_{10}") effects. The present paper examines the problems involved in elucidating the mechanisms involved in the seasonal synchrony of these physiological cycles.

B. There is some evidence for an endogenous (internal) annual rhythm in connection with growth and blood glucose cycles in alligators and for spermatogenesis in some lizards. Other studies indicate that at least some phase of an annual cycle may be timed endogenously; e.g., the termination of reproductive refractoriness and testis regression.

C. Although photoperiod has been stressed as a proximate factor for various annual cycles in reptiles, there is considerable doubt surrounding the conclusion that photoperiodism in important in reptiles. Doubts regarding photoperiodism arise from poor experimental design and because it is unclear that photoperiodism is ecologically "relevant."

D. Regulation of temperature has been poor in some studies and others have failed to employ appropriate experimental controls when several variables were involved. For example, although both light and temperature were varied in some cases, it was concluded that light alone was responsible for observed effects. However, separate studies have demonstrated that temperature alone is often an important timing cue for annual cycles in reptiles.

E. The ecological relevance of photoperiodism is considered from the standpoint that thermoregulatory behavior of reptiles tends to obfuscate their photoperiod relations in nature; photoperiodic responses in reptiles are highly temperature sensitive and photoperiodism tends to be seasonal.

Also many studies failed to utilize "meaningful" photoperiods (i.e., day lengths that might actually exist when the particular physiological change normally occurs).

F. Considerable interspecific variation in timing mechanisms is evident among reptiles even though only a relatively few species (mostly lizards) have been examined. Nonetheless, some trends are evident: temperature appears to be a major cue for timing physiological changes in the spring after emergence from winter dormancy, and non-thermal mechanisms, photoperiod or endogenous rhythms appear primarily responsible for altering activities in later parts of the cycle.

ACKNOWLEDGMENT

The preparation of this report and many of the projects discussed herein were supported in part by Grant GB-7306 from the National Science Foundation.

REFERENCES

1. BADIR, N. Seasonal variation of the male urogenital organs of Scincus scincus L. and Chalcides ocellatus Forsk. Z. Wiss. Zool. Abt. A 160: 290-351, 1958.

2. BADIR, N, and M. F. HUSSEIN. Effect of temperature, food and illumination on the reproduction of Chalcides ocellatus (Forsk.) and Scincus scincus (Lin.). Bull. Fac. Sci. Cairo Univ. 39: 179-185, 1964.

3. BALLINGER, R. E., J. HAWKER, and O. J. SEXTON. The effect of photoperiod acclimation on the thermoregulation of the lizard, Sceloporus undulatus. J. Exptl. Zool. 171: 43-47, 1969.

4. BARTHOLOMEW, G. A. The effects of artificially controlled temperature and day length on gonadal development in a lizard, Xantusia vigilis. Anat. Rec. 106: 49-59, 1950.

5. BARTHOLOMEW, G. A. The modification by temperature of the photoperiodic control of gonadal development in the lizard Xantusia vigilis. Copeia 1953: 45-50, 1953.

6. BARTHOLOMEW, G. A. Photoperiodism in reptiles. In: Photoperiodism and Related Phenomena in Plants and Animals, edited by R. B. Withrow. Washington, D. C.: American Association for the Advancement of Science, 1959, p. 669-676.

7. BURGER, J. W. Experimental sexual photoperiodicity in the male turtle Pseudemys elegans (Wied). Amer. Nat. 71: 481-487, 1937.

8. CLAUSEN, H. J., and E. G. PORIS. The effect of light upon sexual activity in the lizard, Anolis carolinensis, with special reference to the pineal body. Anat. Rec. 69: 39-53, 1937.

9. COULSON, R. A., and T. HERNANDEZ. Biochemistry of the Alligator. A Study of Metabolism in Slow-Motion. Baton Rouge: Louisiana State University Press, 1964, p. 138.

10. FARNER, D. S. Predictive functions in the control of annual cycles. Environ. Res. 3: 119-131, 1970.

11. FISCHER, K. Untersuchungen zur Jahresperiodik der Fortpflanzung bei männlichen Ruineneidechsen (Lacerta sicula campestris Betta). I. Die Refraktärperiode. Z. Vergl. Physiol. 60: 244-268, 1968.

12. FISCHER, K. Untersuchungen zur Jahresperiodik der Fortpflanzung bei männlichen Ruineneidechsen (Lacerta sicula campestris Betta). II. Einflüsse verschiedener Photoperioden und Temperaturen auf die Progressions- und die Regressionsphase. Z. Vergl. Physiol. 61: 394-419, 1968.

13. FISCHER, K. Untersuchungen zur Jahresperiodik der Fortpflanzung bei männlichen Ruineneidechsen (Lacerta sicula campestris Betta). Zool. Anz. Suppl. Bds. 31: 325-340, 1968.

14. FISCHER, K. Einflüsse von Licht und Temperatur auf die Jahresperiodik der Fortpflanzung bei männlichen Ruineneidechsen. Verh. Dtsch. Zool. Ges. Innsbruck: 278-283, 1968.

15. FISCHER, K. Untersuchungen zur Jahresperiodik der Fortpflanzung bei männlichen Ruineneidechsen (Lacerta sicula campestris Betta) III. Spontanes Einsetzen und Ausklingen der Gonadenaktivität; ein Beitrag zur Frage der circannualen Periodik. Z. Vergl. Physiol. 66: 273-293, 1970.

16. FITCH, H. S. Reproductive cycles in lizards and snakes. Univ. Kans. Mus. Nat. Hist. Misc. Publ. 52: 1-247, 1970.

17. FOX, W., and H. C. DESSAUER. Photoperiodic stimulation of appetite and growth in the male lizard, Anolis carolinensis. J. Exptl. Zool., 134: 557-575, 1957.

18. FOX, W., and H. C. DESSAUER. Response of the male reproductive system of lizards (Anolis carolinensis) to unnatural day-lengths in different seasons. Biol. Bull. 115: 421-439, 1958.

19. FOX, W., and H. C. DESSAUER. The response of the thyroid of male lizards, Anolis carolinensis to artificial manipulations of the photoperiod. (Abstract) Anat. Rec. 133: 277-278, 1959.

20. GALGANO, M. Prime richerche intorno all'influenza della luce e della temperatura sul ciclo sessuale di Lacerta sicula campestris (Betta). Boll. Zool. 18: 109-115, 1951.

21. HERNANDEZ, T., and R. A. COULSON. Hibernation in the alligator. Proc. Soc. Exptl. Biol. Med. 79: 145-149, 1952.

22. HUTCHISON, V. H., and R. J. KOSH. The effect of photoperiod on the critical thermal maxima of painted turtles (Chrysemys picta picta). Herpetologica 20: 233-238, 1965.

23. JEGLA, T. C., and T. L. POULSON. Circannian rhythms - I. Reproduction in the cave crayfish, Orconcetes pellucidus inermis. Comp. Biochem. Physiol. 33: 347-355, 1970.

24. LASHBROOK, M. K., and R. L. LIVEZEY. Effects of photoperiod on heat tolerance in Sceloporus occidentalis occidentalis. Physiol. Zool. 43: 38-46, 1970.

25. LICHT, P. Reproduction in lizards: influence of temperature on photoperiodism in testicular recrudescence. Science 154: 1668-1670, 1966.

26. LICHT, P. Environmental control of annual testicular cycles in the lizard Anolis carolinensis. I. Interaction of light and temperature in the initiation of testicular recrudescence. J. Exptl. Zool. 165: 505-516, 1967.

27. LICHT, P. Environmental control of annual testicular cycles in the lizard Anolis carolinensis. II. Seasonal variations in the effects of photoperiod and temperature on testicular recrudescence. J. Exptl. Zool. 166: 243-253, 1967.

28. LICHT, P. Response of the thermal preferendum and heat resistance to thermal acclimation under different photoperiods in the lizard Anolis carolinensis. Amer. Midl. Natur. 79: 149-158, 1968.

29. LICHT, P. Illuminance threshold and spectral sensitivity of photosexual responses in the male lizard (Anolis carolinensis). Comp. Biochem. Physiol. 30: 223-246, 1969.

30. LICHT, P. Environmental control of annual testicular cycles in the lizard Anolis carolinensis. III. Thermal thresholds for photoperiodism. J. Exptl. Zool. 172: 311-322, 1969.

31. LICHT, P. Regulation of the annual testes cycle by photoperiod and temperature in the lizard Anolis carolinensis. Ecology. In press.

32. LICHT, P. Response of the male reproductive system to interrupted-night photoperiods in the lizard Anolis carolinensis. Z. Vergl. Physiol. In press.

33. LICHT, P., and G. C. GORMAN. Reproductive and fat cycles in Caribbean Anolis lizards. Univ. Calif. Publ. Zool. 95: 1-52, 1970.

34. LICHT, P., and L. K. PEARSON. Failure of parietalectomy to affect the testes in the lizard Anolis carolinensis. Copeia 1970: 172-173, 1970.

35. LICHT, P., H. E. HOYER, and P. G. W. J. VAN OORDT. Influence of photoperiod and temperature on testicular recrudescence and body growth in the lizards, Lacerta sicula and Lacerta muralis. J. Zool. 157: 469-501, 1969.

36. LOFTS, B., B. K. FOLLETT, and R. K. MURTON. Temporal changes in the pituitary-gonadal axis. Memoirs Soc. Endocrinol. 18: 545-575, 1970.

37. MARION, K. R. Temperature as the reproductive cue for the female fence lizard Sceloporus undulatus. Copeia 1970: 562-564, 1970.

38. MAYHEW, W. W. Photoperiodic response of female fringe-toed lizards. Science 134: 2104-2105, 1961.

39. MAYHEW, W. W. Photoperiodic responses in three species of the lizard genus Uma. Herpetologica 20: 95-113, 1964.

40. MAYHEW, W. W. Growth response to photoperiodic stimulation in the lizard Dipsosaurus dorsalis. Comp. Biochem. Physiol. 14: 209-216, 1965.

41. MAYHEW, W. W. Hibernation in the horned lizard, Phrynosoma m'calli. Comp. Biochem. Physiol. 16: 103-119, 1965.

42. MAYHEW, W. W. Biology of desert amphibians and reptiles. In: Desert Biology I, edited by G. W. Brown. New York: Academy Press, 1968, p. 195-356.

43. MELLISH, C. H. The effects of anterior pituitary extract and certain environmental conditions on the genital system of the horned lizard (Phrynosoma cornutum, Harlan). Anat. Rec. 67: 23-33, 1936.

44. MILLER, M. R. The seasonal histological changes occurring in the ovary, corpus luteum, and testis of the viviparous lizard, Xantusia vigilis. Univ. Calif. Publ. Zool. 47: 197-224, 1948.

45. STEBBINS, R. C. Activity changes in the striped plateau lizard with evidence on influence of the parietal eye. Copeia 1963: 681-691, 1963.

46. TINKLE, D. W., and L. N. IRWIN. Lizard reproduction; refractory period and response to warmth in Uta stansburiana females. Science 148: 1613-1614, 1965.

CREDITS

The editors and authors are grateful to the following publishers, organizations and individuals for permission to use materials previously published:

Figure 4: Zeitschrift für vergleichende Physiologie, Springer-Verlag, Berlin and Dr. h.c.H Autrum, München; P. Licht, H. Hoyer and P. G. W. J. van Oordt, Journal of Zoology. London. 157: 311-322, 1969.

PROBLEMS IN THE CHRONOBIOLOGY
OF HIBERNATING MAMMALS

This paper attempts to analyse the weak points in our knowledge of the mechanisms by which hibernating mammals synchronize their physiology and behavior with variations in the physical environment, and to suggest to present and future investigators ideas which may help to solve the complex problems involved. The ideas presented here are in part the outcome of the three papers given by the authors in the Hibernation-Hypothermia IV Symposium, but also include the points of view of many participants who presented their ideas in various discussion sessions.

It is clear that the timing of hibernation is inseparable from the total ecology of the hibernating species. Therefore it is essential that long term studies of the natural history of hibernators be carried out. Both from discussions at this symposium and from the general literature, it seems obvious that we know pitifully little about the natural history and ecology of the hibernating mammals which have been so extensively studied in the laboratory. This fact was stressed time and again and it can hardly be denied that the comparison of data obtained in the laboratory with that of reliable field information would be of inestimable value in understanding how these mammals adjust the temporal pattern of their physiology and behavior to the environment. How precisely do we know when animals in nature begin and end hibernation from year to year? When does preparation for hibernation and exactly how is this related to breeding? It was also pointed out that such fundamental field information as life span, feeding habits, nesting and breeding behavior, energetics, etc. is very incomplete and all of these are bound to affect the timing of hibernation.

From the extensive laboratory studies which have been made on a few species, the timing of hibernation and dormancy appears to involve some combination of an endogenous circannual rhythm and seasonal factors in the environment. However, knowledge of the nature of the interaction between the endogenous and exogenous factors is almost totally lacking. There has been an understandable tendency to assume that the zeitgeber mechanisms which entrain circadian rhythms also entrain circannual rhythms but documentation is lacking. It is possible that completely different timing mechanisms are involved in daily dormancy as compared with seasonal dormancy. Moreover one part of a rhythm may be endogenous while another may be exogenous.

Simple answers to biologically complex questions need not be expected. Thus not only may some parts of the rhythm be refractory to various environmental stimuli while other parts are not, but there may be discrete short periods of time, "windows", when the central nervous system responds to special stimuli. Furthermore, an environmental factor such as photoperiod may produce an effect at one environmental temperature but not at another, at one season and not another, or on one sex and not the other, or any environmental factor may interact with another in such a way that both may be required for the animal to respond. Yet again, environmental temperature may be a critical factor at one season while photoperiod is critical at another. It is important also to remember in comparing the results obtained in the laboratory to the situation occurring in the natural environment that even biologically reasonable experimental stimuli may not be ecologically meaningful. For example, even though "short days" may affect the onset of dormancy, the effective day length to which a hibernating mammal is exposed may be determined not by astronomical factors but by the daily period that an animal remains on the surface of the ground. Therefore, unless the behavior of a species under natural conditions is accurately known, ecologically irrelevant factors may be examined in the laboratory. Thus, the theoretically possible total exposure to light may be insignificant compared to the time between first exposure and last exposure; and this may vary with the intensity of the stimulus, the environmental temperature, or foraging success.

Although meaningful laboratory experiments which can be compared with data from the field are urgently needed, we nevertheless need to study the mechanisms of long term endogenous rhythms independently. This topic has been considered for only a few species and the ratio of ignorance to knowledge is impressive. Comparative studies are needed because different species may well have evolved different mechanisms. A long term rhythm may be dependent upon several different types of oscillators, or there may be rhythms within rhythms, or one rhythm may drive another. Moreover, the trigger which starts a rhythm may occur long before the event. That is, a stimulus may have a latent response which becomes overt only after the passage of several months, or after other and highly specific conditions are met.

It would probably be a fruitful exercise to try to get key physiological events out of cycle with the others which it normally accompanies. For example, by endocrine injections it might be possible to induce experimental animals to breed out of season and then to determine whether this affects the time of onset of hibernation. One could also try to prevent young animals from entering hibernation in the first fall and then see if this affects the timing of reproductive competence. In this type of manipulative investigation the importance of comparative studies of different species under the same controlled conditions can scarcely be overestimated. These experimental conditions should not only be carefully controlled, they must be accurately described, because minor differences in experimental protocol may have major effects on biological rhythmicity.

Other major problems which await a solution include not only an understanding of the internal mechanisms involved in endogenous rhythms, but also the identification of the various internal physiological changes which occur as a result of endogenous factors, those which occur as a result of environmental stimuli, or a combination of both.

Variability is one of the most conspicuous attributes of biological phenomena. We strongly urge that more attention be paid not only to the inevitable variations that animals display but also to those that deviate conspicuously. Analysis of the variability of patterns of dormancy may reveal essential features which will allow us to understand its more regular cyclical aspects. For example, it has long been known that a few animals of a hibernating species never hibernate - why? Thorough studies of both the behavior and physiology of such aberrant forms would surely be of great value in furthering our knowledge of how hibernating mammals keep in temporal adjustment with their environment.

Despite the enormous amount of energy that has been devoted to the seasonal dormancy of animals and the many words that have been written about it, we still do not really know what initiates, or what terminates, the hibernation of any species of mammal.

Symposium Committee on Timing and Synchrony of the Environment

E. T. Pengelley, Chairman
George A. Bartholomew
Paul Licht

AUTHOR INDEX

ABBOTT, B. C., 446, 451
ABRAMS, R. M., 608, 610, 621, 622, 625
ADAIR, G. S., 362, 371
ADAMS, T., 608, 624
ADELSTEIN, S. J., 7, 8, 9, 14
ADOLPH, E. F., 169, 188
AGER, M. E., 202, 218
AIGNER, P., 168, 169, 176, 189
AKERT, K., 505, 529
ALANIS, J., 307, 352
ALBERS, R. W., 136, 144, 158, 164, 165
ALBIN, M. S., 508, 511, 530
ALEXANDER, A. E., 253, 274
ALFREY, T., 434, 451
ALICAN, F., 101, 119
ALLARD, A. A., 152, 165
ALLEN, J. C., 155, 158, 159, 160, 164, 166
ALPERT, N. R., 434, 436, 446, 452, 453, 484
AMANTEA, L., 508, 528
AMUNDSEN, N. R., 577, 604
ANDERSON, H. T., 608, 621, 622, 624, 625
ANDERSON, P., 490, 527
ANDERSSON, B., 608, 624
ANDERSSON-DEDERGREN, E., 294, 353
ANDJUS, R. K., 110, 116, 172, 182, 188, 515, 527
ANDREOLI, T. E., 18, 19, 20, 49
ANGELAKOS, E. T., 436, 451
APPEL, P., 373, 374, 398
APTER, J. T., 435, 451
ARIS, R., 577, 604
ARMOR, J. A., 421, 452
ARMOUR, A., 640, 651, 658
ARNTZEN, C. J., 304, 350
ASCHOFF, J., 638, 657
ASMUNDSEN, S. J., 637, 638, 640, 642, 643, 644, 652, 655, 660, 661
ASSAF, S. A., 66, 77
ASSENMACHER, I., 639, 657
ATKINSON, D. E., 12, 13, 59, 64, 77
AVI-DOR, Y., 133, 136, 145, 219, 235

BABA, M., 109, 119
BACH, L. M. N., 535
BADER, H., 137, 144
BADIR, N., 686, 689, 692, 696, 708
BAETHMANN, A., 168, 176, 180, 188, 189
BAILEY, K., 357, 362, 368, 371
BAKAY, L., 108, 188, 189
BAKER, D. E., 84, 98
BAKER, H., 358, 370, 373, 374, 398
BAKER, M. A., 569, 573
BAKER, O., 460, 477
BAKER, P. F., 127, 136, 144, 203, 215
BALDWIN, B. A., 608, 624
BALDWIN, J., 61, 65, 77
BALE, W. F., 110, 118
BALLEW, D. W., 460, 477
BALLIN, H. M., 509, 511, 529
BALLINGER, R. E., 689, 708
BANGHAM, A. D., 228, 235
BAR, R. S., 269, 275
BARANY, K., 39, 395
BARANY, M., 390, 391, 395
BARGOOT, F. G., 260, 279
BARKER, J. L., 617, 624
BARNES, H. W., 511, 530
BARNUM, C. P., 638, 658
BARR, L., 298, 350
BARR, R. E., 108, 118
BARRON, D. H., 511, 527
BARTHOLOMEW, G. A., 663, 667, 670, 671, 672, 674, 675, 676, 679, 689, 693, 696, 698, 708, 716
BASKIN, R. J., 305, 350
BATTISTA, A. F., 101, 116, 488, 489, 501, 514, 527, 528
BAUER, A. A., 110, 119
BECKETT, A. H., 246, 274
BEDUSCHI, A., 507, 527
BEECHER, H. K., 182, 190
BELESLIN, B., 528, 530, 531, 533
BELESLIN, D., 557, 560, 573, 574
BELESLIN, D. B., 560, 574
BENDAT, J. S., 435, 451
BENEDITTI, E. L., 296, 336, 350
BENEDICT, F. G., 580, 604
BENITEZ, D., 307, 352
BENOIT, J., 639, 657

BENOIT, O., 501, 508, 510, 527, 528
BENSON, E. S., 295, 353, 354
BENTAL, E., 540, 549
BENTON, L. E., 608, 616, 624
BERGMEYER, H. U., 188
BERGREN, A. B., 202, 215
BERING, E. A., 508, 530
BERNAL, J. D., 82, 97, 110, 116
BERNARD, L. D., 244, 246, 274
BERNHARD, W. F., 508, 530
BERQUIST, E. H., 608, 626
BERRY, C. M., 608, 618, 624
BESCH, H. R., 155, 164
BETZ, E., 190
BEUTLER, E., 133, 135, 136, 137, 147
BEYER, R. E., 17, 21, 22, 49, 51, 54, 113, 115
BIAJI, G. L., 269, 275
BIDET, R., 128, 133, 144
BIGELOW, W. G., 507, 527
BIHARI, B., 540, 549
BIRNBAUM, E. R., 364, 368
BISHOP, V. S., 460, 467, 477, 478
BISSONETTE, T. H., 639, 657
BITTAR, E. E., 198, 208, 215
BITTNER, J. J., 638, 658
BJÖRK, V. O., 167, 188
BLAKE, A., 198, 199, 218
BLAND, J. H., 81, 96, 97, 110, 115, 116
BLATTEIS, C. H., 101, 116
BLAUSTEIN, M. P., 127, 136, 144, 201, 215
BLY, C. G., 110, 118
BLINKS, D. C., 431, 436, 453
BLOND, D. M., 203, 204, 215
BLUMENFELD, O. O., 260, 270, 274
BOARD, R. D., 92, 97
BOETTINGER, S. D., 434, 451
BOGERT, C. M., 610, 624
BOGOROCH, R., 219, 235
BOOIJ, H. L., 239, 274
BOUCHER, L. J., 357
BOUCKAERT, J. P., 451, 453
BOWEN, T. E., 431, 432, 433, 453
BOWERS, N. P., 125, 147
BOWIE, W. C., 457, 460, 468, 477
BOWLER, K., 11, 12, 133, 136, 141, 144, 168, 188, 219, 220, 235
BOWMAN, R. H., 101, 116

BORUT, A., 203, 206, 215
BOYD, E., 109, 116
BRADFORD, H. F., 239, 274, 279
BRARD, E., 639, 657
BRADY, A. J., 428, 429, 430, 431, 446, 451
BRAUER, R. W., 101, 116
BRAUN, L., 435, 454
BRAUNS, R., 240, 274
BRAUNWALD, E., 109, 118, 458, 459, 460, 479
BREAZILE, J. E., 490, 493, 515, 516, 522, 525, 532, 534
BREEMAN, V. L. V., 294, 350
BREMER, F., 527
BRENDEL, W., 167, 168, 169, 172, 174, 180, 182, 188, 189, 190
BRENDEL, W. B., 501, 531
BRENNEMAN, C., 19, 50, 42
BRETSCHNEIDER, H. C., 178, 188
BREUER, H., 193, 215
BRICKER, N. S., 200, 215
BRIDGMAN, P. W., 85
BRIGGS, F. N., 360, 361, 369
BRILLER, S. A., 508, 531
BROBECK, J. R., 608, 626
BROBERG, S., 10, 14
BROOKS, C. M., 507, 511, 512, 527, 529, 533, 534
BROOKS, J. C., 21, 52
BROWN, D. H., 25, 50
BROWN, F. A., 615, 626
BROWN, F. F., 431, 432, 433, 454
BROWN, J. H., 671, 672, 679
BROWN, W., 358, 360, 370, 385, 386, 397
BRUSH, A. H., 11, 13
BRUTSAERT, D. L., 431, 432, 451, 454
BUCHER, T., 213, 217
BUCHTAL, F., 434, 451
BULLARD, R. W., 172, 188, 472, 475, 477
BULLIVANT, S., 298, 350
BUNGENBERG DE JONG, H. G., 239, 260, 268, 270, 274
BUNNELL, I. L., 460, 477
BUNNING, E., 638, 657
BURGER, J. W., 689, 690, 611, 708
BURLINGTON, R. F., 3, 4, 8, 9, 10, 11, 13, 14, 115, 436, 451
BURRIS, R. H., 19, 49

CABANAC, M., 609, 610, 624
CADENAS, E., 109, 118

CADE, T. J., 653, 657, 670, 676, 679
CAHN, J., 508, 527
CALDWELL, F. T., 608, 610, 625
CALDWELL, R. S., 73, 77
CALDWELL, W. M., 460, 480
CALLAGHAN, J. C., 507, 527, 531
CAPELLEN, L., 423, 451
CAPULONG, A. L., 381, 382, 383, 395
CARLISLE, H. J., 608, 624
CARLISLE, P., 460, 477
CARLSEN, T., 306, 352
CARLSON, F. D., 388, 395
CARMICHAEL, E. A., 557, 573
CARPENTER, D. O., 617, 624
CARROL, H. W., 101, 116
CARROLL, W. R., 374, 385, 397
CARTER, J. D., 7, 13
CATTELL, M., 421, 452
CASWELL, A. H., 266, 275
CAZZULO, C. L., 505, 507, 527
CECIL, R., 270, 274
CERBÓN, J., 241, 242, 261, 274
CESSAC, G. L., 92, 98
CHAFFEE, R. R. J., 4, 13
CHAI, C. Y., 547, 549
CHALCROFT, J. P., 298, 354
CHAMBERLAIN, N. F., 94, 97, 115, 117
CHAMDLER, B. M., 109, 118
CHAN, S. I., 270, 276
CHANCE, B., 202, 204, 205, 215
CHANG, H. T., 499, 508, 527
CHANGEAUX, J. P., 258, 266, 268, 274, 276
CHAPMAN, C. B., 460, 477
CHAPMAN, D., 241, 242, 244, 261
CHAPMAN, I., 133, 145
CHATFIELD, P. O., 101, 116, 488, 489, 493, 499, 501, 509, 514, 515, 527, 528, 530, 542, 549
CHAVEZ-DIAZ, G., 255, 257, 275
CHEZ, R. Z., 127, 146, 233, 236
CHIGNELL, C. F., 158, 164
CHRISTENSEN, H. N., 219, 223, 235, 236
CHRISTENSEN, J. J., 366, 367, 368
CHUNG, C. - S., 373, 374, 398
CHURAEV, N. V., 92, 97
CID, M. E., 248, 255
CIER, J. F., 507, 528
CINGOLANI, H. E., 460, 479, 468
CLARK, R., 359, 362, 369

CLAUSEN, H. J., 689, 690, 691, 700, 708
CLAUSEN, T., 124, 145, 215, 218
CLEMENTE, C. D., 538, 550
CLEMENTS, F. G., 595, 604
COATS, J. H., 19, 49
COHEN, C., 362, 368, 377, 398
COHEN, P., 205, 215
COLLIER, R. H., 198, 204, 208, 217
COLLIN, D. T., 241, 274
COLLINS, W. F., 510, 532, 533
COLOMBO, V. E., 220, 236
COOKE, P. H., 393, 399
CONN, P. A., 109, 117
CONNELLY, C. M., 203, 215
CONSTANTINIDES, S. M., 19, 27, 28, 29, 43, 49, 53
CONWAY, B. E., 12, 97
COPE, F. W., 93, 97, 115, 116
COPLAND, C., 102, 117
CORDES, E. H., 262, 264, 267, 275
CORI, C. F., 19, 25, 49, 50
CORI, G. T., 19, 49
CORNWELL, D. J., 269, 275
COSENZA, B. J., 242, 278
COTANCH, W. W., 109, 117
COTHRAN, L. N., 457, 460, 468, 477, 478
COUGHLIN, R. F., 609, 626
COULSON, R. A., 682, 683, 684, 687, 708, 709
COURJON, J., 508, 510, 528
COVELL, J. W., 458, 459, 460, 479
COVINO, B. G., 99, 100, 102, 103, 108, 116
COWEY, C. B., 66, 77
COWLES, R. B., 610, 624
COWMAN, R. A., 19, 49
CRABBE, J., 219, 235
CRANE, F. L., 260, 275, 304, 350
CRANE, R. K., 127, 144
CRAWFORD, E. C., 109, 116
CRIMSON, J. M., 606, 615, 624
CRONIN, R., 101, 117
CROSS, M. F., 421, 452
CRYSTAL, D. K., 460, 479
ČUPIĆ, D., 528, 529, 530, 531
CURRAN, P. F., 127, 146, 233, 236

DANIELS, V. G., 219, 235
DANILYUK, V. P., 499, 500, 514, 532
DARNALL, D. W., 364, 368
DARWIN, C., 652, 657

DAS, M. L., 260, 275
DATTA, A., 18, 52, 259, 278
DATTA, P., 19, 49
DAUDOVA, G. M., 9, 11, 13
DAVEY, B. G., 84, 98
DAVIES, R. E., 388, 390, 395
DAVILA, J. C., 459, 460, 477, 479
DAVIS, D. E., 653, 657
DAWE, A. R., 637, 640, 651, 652, 658, 659
DAWSON, A. G., 219, 235
DAWSON, T. J., 621, 622, 625
DAWSON, W. R., 667, 670, 679
DEAL, W. C., 19, 27, 28, 29, 42, 49, 53
DEAMER, D. W., 305, 350
DEAN, J. M., 74, 77
DEBAKEY, M. E., 101, 118
DEBLENDE, J., 423, 451
DEDUVE, C., 127, 144
DEHL, R. E., 85, 97
DELGADO, J. M. R., 557, 573
DEMENT, W. C., 493, 528
DEMUS, H., 257, 275
DENYES, A., 7, 13
DENT, C. E., 224, 235
DEROSIER, D. J., 360, 370, 398
DERYAGEN, B. V., 92, 97
DESSAUER, H. C., 689, 700, 709
DETTINGER, E., 190
DEWEY, M. M., 294, 298, 350, 351
DIANOUX, A. - C., 242, 243, 245, 279
DILLEY, R. A., 304, 350
DIORIO, A. F., 374, 388, 398
DIVAC, I., 531
DODGE, H. T., 460, 477, 479
DONBROW, M., 246, 278
DOTY, P., 434, 451
DOUDOROFF, M., 19, 53
DOW, J., 383
DOWNEY, J. A., 608, 616, 625
DOWNING, S. E., 459, 460, 477
DRABIKOWSKI, W., 359, 368
DREIKE, G. E., 595, 604
DREIZEN, P., 360, 363, 373, 374, 376, 377, 378, 379, 381, 382, 383, 385, 386, 387, 389, 395, 396, 400, 484
DRESCHER, J. W., 645, 652, 658
DREW, C. E., 507, 528
DRYER, R. L., 3, 14

DRYER, W. J., 257, 277
DUA, R. D., 19, 49
DUBOIS, R., 637, 658
DUDEL, J., 405, 419
DUMONT, J. E., 198, 208, 218
DUNCAN, C. J., 11, 13, 133, 136, 141, 144, 188, 219, 220, 235
DYDYNSKA, M., 202, 211, 215

EBASHI, F., 360, 361, 362, 368
EBASHI, S., 299, 350, 357, 359, 360, 361, 362, 368, 369, 373, 395
ECKEL, R. E., 202, 215
EDELMAN, I. S., 219, 235
EDMAN, K. A. P., 425, 428, 430, 431, 432, 452
EDNER, O. J., 260, 279
EGERTON, N., 167, 188
EGERTON, W. S., 167, 188
EISENBERG, D., 88, 97
EISENBERG, E., 358, 368, 390, 396
EISENMAN, G., 226, 235
EKBERG, D. R., 56, 57, 74, 77
ELIHSSEN, E., 125, 133, 144
ELLIOTT, K. A. C., 13, 188
ELLORY, J. C., 233, 237
ELSHOVE, A., 203, 211, 215
EMMELOT, P., 296, 336, 350
ENDO, M., 299, 350, 357, 359, 360, 362, 368, 369, 373, 395
ENGELKEN, E. J., 460, 467, 477
ENGLEMAN, D. M., 242, 269, 275
ERASMUS, B., 109, 116
ERDELT, H., 24, 53
ERLANDER, S. R., 83, 86, 88, 89, 90, 91, 97, 98
ESTRADA, O., 262, 265
EVARTS, E. V., 540, 549
EXTON, J. H., 110, 116
EYRING, H., 65, 78, 243, 276

FAHN, S., 136, 144
FANG, L. S. T., 123, 133, 136, 137, 141, 142, 144, 185, 288
FARQUHAR, M. G., 294, 350
FARNER, D. S., 694, 708
FARNER, K., 685, 686, 687, 689, 693, 705, 708, 709
FAWCETT, D. W., 295, 304, 350, 352, 653, 658
FAY, T., 511, 528
FEATES, S. F., 85

AUTHOR INDEX

FEDYAKIN, N. N., 90
FEINBERG, H., 109, 116
FELDBERG, R. S., 19, 42, 43, 44, 50
FELDBERG, W., 508, 528, 553, 557, 559, 561, 565, 573, 575
FELTS, P. W., 110, 119
FELDMAN, S. M., 541, 549
FENN, W. O., 423, 424, 426, 452
FERGUSON, D., 610, 626
FERGUSON, S., 4, 14
FERRANS, V. J., 351
FERRARI, E., 507, 528
FICHAM, J. R. S., 19, 50
FIELD, J., 101, 117
FIFKOVA, E., 543, 550
FINDLAY, J. D., 608, 625
FISCHER, E. H., 24, 26, 50
FISHER, B., 102, 117
FISHER, E. R., 102, 117
FISHER, K. C., 81, 97, 126, 144, 637, 638, 640, 641, 642, 645, 649, 651, 652, 658, 660
FITCH, H. S., 685, 709
FLEISCHAUER, K., 508, 528
FOLK, G. E., 638, 658
FOLLETT, B. K., 694, 710
FORBES, A., 501, 528
FORSSMANN, W. G., 304, 350, 351
FOSTER, R. F., 123, 126, 147, 185
FOWKES, F. M., 244, 252, 278
FOWLER, R. H., 82, 91
FOX, C. F., 242, 280
FOX, W., 689, 698, 709
FOZZARD, H. A., 403, 414, 419
FRANK, H. S., 86, 87, 97
FRANKLIN, D. L., 459, 460, 477, 479
FRANKENHAUSER, B., 168, 184, 188
FRANZEN, J. S., 251, 275, 276
FRANZINI-ARMSTRONG, C., 300, 351
FREHN, J. L., 134, 144
FRENCH, J. D., 509, 528
FREUNDLICH, M., 19, 42, 50
FRIEDRICHS, D., 110, 119
FRONEK, K., 460, 479
FROWEIN, R. A., 169, 174, 176, 190
FUCHS, F., 360, 361, 369
FUHRMAN, F. A., 101, 117
FUHRMAN, G. H., 101, 117
FUISZ, R. P., 127, 146, 233, 236
FUNKHAUSER, G. E., 472, 477
FURCHGOTT, B. F., 406, 407, 419

GAENSHIRT, H., 506, 528, 533
GAINES, G. L., 244, 275
GALAMBOS, R., 514, 529
GALE, C. C., 608, 624, 625
GALGANO, M., 689, 691, 709
GALSTER, W. A., 8, 9, 13
GALSWORTHY, P. R., 161, 164
GALLIMORE, J. K., 109, 119
GAMBLE, W. J., 109, 117
GAMMACH, D. B., 239, 274
GARCIA, J. P., 101, 116, 488, 489, 501, 514, 527, 528
GARRAHAN, P. I., 136, 145
GASCA-RIVAS, J. M., 256
GASDORF, E., 610, 625
GASSER, H. S., 423, 447, 452
GAY, W. A., 293, 351, 429, 433, 452
GENEL, M., 128, 129, 130, 146
GENT, W. L. G., 261, 275
GERGELEY, J., 359, 362, 363, 368, 369, 390, 396, 399
GERSHMAN, L. C., 374, 375, 376, 377, 378, 379, 381, 382, 383, 384, 385, 386, 387, 395, 396, 399, 400
GEST, H., 19, 42, 49, 50
GIAJA, J., 182, 188
GIESE, A. C., 56, 57, 79
GILBOE, D. D., 109, 117
GILLES-BAILLIEN, M., 220, 235
GILLIAM, J., 492, 505, 532, 638, 661
GILMAN, S., 167, 189
GIRADIER, L., 124, 145, 210, 215, 304, 350, 351
GIRETTI, M. L., 528
GITLER, C., 239, 240, 242, 243, 245, 252, 255, 257, 262, 263, 264, 266, 267, 271, 275, 278, 279, 280
GLASER, M., 270, 276
GLICK, G., 155, 164
GLOVER, M. B., 107, 117
GLYNN, I. M., 136, 145
GODDARD, E. D., 276
GOLD, H., 421, 452
GOLDFARB, A. R., 247, 276
GOLDMAN, S. S., 126, 133, 136, 141, 142, 145, 147
GOLDSTEIN, J., 109, 188
GONZE, J., 198, 208, 218
GORDON, A. M., 292, 351, 389, 396, 426, 452
GORDON, E. E., 202, 216

GORMAN, G. C., 688, 703, 710
GORTNER, R. A., 93
GOTTERER, G. S., 260, 276
GRANT, C., 460, 477
GRANT, E. H., 261, 275
GRANT, W. H., 92, 98
GRAVES, D. J., 19, 25, 26, 50, 53, 66, 77
GRAY, E. W., 589, 604
GRAY, I., 101, 117
GREASER, M. L., 362, 363, 369
GREEN, D. E., 260, 270, 278
GREEN, I., 381, 398
GREENE, D. G., 460, 477
GREGORY, K. F., 65, 79
GRIBBE, P., 460, 477
GRIESE, D. W., 425, 428, 430, 431, 432, 452
GRIGG, G. C., 71, 77
GRISOLIA, S., 19, 52, 57, 77
GROOT, G. S. P., 22, 51
GROSSMAN, A., 407, 419
GRUENER, N., 133, 136, 145, 219, 235
GRUNDFEST, H., 511, 528
GUARESCHI, A., 505, 507, 527
DE GUBAREFF, T., 406, 410, 411, 412, 414, 419
GUIDOTTI, G., 260, 270, 278
GUTFREUND, H., 362, 368
GWINNER, E., 640, 658
GWYNNE, J. T., 257, 276

HABEEB, A. F. S. A., 247, 276
HACKENBROCK, C. R., 308, 351
HAGBARTH, K. E., 497, 528
HAGYARD, C. J., 374, 377, 378, 379, 397, 398
HALBERG, E., 638, 658
HALBERG, F., 638, 658
HALE, J. R., 625, 608
HALPERN, W., 421, 434, 436, 452
HAMILTON, C. L., 608, 626
HAMMEL, H. T., 493, 528, 547, 549, 577, 579, 581, 584, 604, 606, 607, 608, 609, 610, 612, 615, 616, 620, 621, 622, 624, 625, 626
HAMRELL, B. B., 421, 446, 453
HANAHAN, D. J., 250, 279
HANEGAN, J. L., 674, 679
HANNON, J. P., 99, 100, 102, 103, 108, 115, 116, 120

HANSON, J., 351, 358, 359, 360, 369, 370, 373, 385, 386, 389, 396, 432, 453
HARARY, I., 200, 216
HARDY, D. J., 388, 395
HARDY, J. D., 606, 608, 609, 610, 624, 625
HARRINGTON, W. F., 374, 397, 399
HARRIS, E. J., 202, 215, 211, 405, 419
HARRIS, R. A., 158, 164
HARTLEY, G. S., 240, 241, 276
DE HARTOG, M., 202, 216
HARTNER, W. C., 582, 603, 604, 633
HARTREE, W., 422, 453
HARTSHORNE, D. J., 357, 358, 359, 360, 361, 362, 363, 369, 370, 375, 377, 395, 396, 484
HARVA, O., 276
HARVEY, R. B., 101, 102, 117
HASSELBACH, W., 299, 351
HASSELBERGER, F. X., 169, 172, 174, 176, 189
HAVER, E. A., 19, 37, 38, 39, 40, 50, 54
HAWKER, J., 689, 708
HAWTHORNE, E. M., 457, 460, 464, 468, 469, 477, 478, 484
HAYES, F. R., 57, 74, 78
HAZELWOD, C. F., 94, 97, 115, 117
HAYWARD, J. N., 569, 573
HEATH, J. E., 524, 533, 582, 604, 605, 606, 608, 610, 615, 616, 622, 624, 625, 626, 627, 633, 674, 679
HEBB, D. O., 536, 549
HEFNER, L. L., 431, 432, 433, 453, 454
HEGVARY, C., 140, 145
HEIKKILA, R. E., 269, 275
HEINRICH, B., 674, 675, 679
HELDT, H. W., 22, 50
HELLER, H. C., 640, 647, 648, 650, 653, 654, 658, 661
HELLSTRØM, B., 606, 608, 612, 615, 616, 620, 625
HEMPLING, H. G., 192, 198, 200, 203, 210
HENDRICKSON, E. M., 19, 51
HENSHAW, R. E., 589, 604
HERD, P. A., 124, 145
HERMAN, L., 389
HERNANDEZ, T., 682, 683, 684, 687, 708, 709

HERNANDEZ-MONTEZ, H., 255, 276
HERNANDEZ-PÉON, R., 542, 549
HERZ, R., 299, 355, 359, 361, 371
HESS, W. R., 541, 542, 549
HIBBS, R. C., 351
HICKMAN, C. P., 71, 77
HIEBEL, C., 101, 118, 488, 489, 491, 514, 515, 529, 531
HILBER, C., 168, 169, 174, 176, 180, 190
HILDEBRAND, J. H., 43,50
HILL, A. V., 297, 351, 388, 389, 397, 421, 423, 424, 425, 426, 428, 431, 434, 447, 452, 453
HIMMELFARB, S., 399
HINDS, J. E., 457, 460, 464, 468, 469, 477, 478
HIRAKOW, R., 304, 306, 351
HIRSCHOWITZ, B. I., 204, 217
HIRVONEN, L., 460, 477
HOCH, F. L., 4, 13
HOCHACHKA, P. W., 17, 50, 57, 58, 60, 61, 62, 63, 65, 70, 73, 74, 75, 77, 78, 79
HOCK, R. J., 645, 658
HODGKIN, A. L., 127, 144, 168, 184, 189, 215, 402, 409, 411, 419
HOFFMAN, R. A., 638, 653, 658
HOFSTEE, B. H. J., 19, 50
HOHORST, H. J., 190
HOKIN, L. E., 152, 161, 164, 165
HOLLOWAY, R. J., 101, 116
HOLMES, T. G., 308, 529
HOLT, J. P., 459, 460, 478
HOLTZAPPLE, P., 128, 129, 130, 146
HOLTZER, A., 359, 362, 369
HONALD, G. R., 7, 8, 15
HOOD, W. P., 460, 478
HORNE, R. A., 85
HORNIG, D. F., 87, 98
HOROWICZ, P., 409, 419
HOROWITZ, J. M., 124, 140, 145
HOROWITZ, L. D., 460, 467, 477, 478
HORSTEN, G. P. M., 505, 533
HORVATH, A., 99, 101
HORVATH, S. M., 101, 116
HORWITZ, B. A., 3, 4, 13, 14, 124, 140, 145
HÖSLI, L., 540, 550
HOUGEN, G. E., 595, 604
HOUSE, W. A., 3, 6, 8, 9, 10, 14, 584, 604
HOUSTON, A. H., 71, 78

HOWELL, T. R., 670, 676, 679
HOWSE, H. D., 351
HOYER, H. E., 685, 692, 693, 697, 705, 710
HUDSON, J. W., 667, 671, 679
HUERTAS, J., 540, 549
HUGHES, M., 19, 42, 50
HULTQUIST, G., 167, 188
HUMMELER, K., 128, 146
K.-HUNTER, F., 457, 478
HUSSEIN, M. F., 686, 689, 692, 696
HUSSON, F., 240, 244, 276, 277
HUSTON, R. L., 7, 8, 15
HUTCHINSON, V. H., 689, 698, 709
HUTTENLOCHER, P. R., 540, 549
HUTTER, O. F., 405, 419
HUXLEY, A. F., 301, 351, 358, 369, 389, 396, 402, 411, 419, 426, 438, 452, 453
HUXLEY, H. E., 292, 351, 358, 359, 360, 369, 370, 373, 374, 385, 386, 388, 389, 396, 397, 398, 432, 453

IGGO, A., 564, 573
ILLANES, A., 436, 453
ILANI, A., 269, 277
ILLINGWORTH, B., 25, 50
INGRAM, D. L., 608, 624, 625
IRIAS, J. J., 19, 33, 34, 35, 41, 42, 50, 54
IRWIN, L. N., 686, 711
ISRAEL, Y., 145
ITO, A., 253, 276
IVES, D. J. G., 85
IWAKURA, H., 357, 368
IZATT, R. M., 366, 367, 368

JACKSON, D. C., 608, 625
JACOBS, H. K., 589, 604, 633
JACOBSON, F. H., 607, 608, 609, 616, 626
JAMIESON, J. D., 294, 309, 351
JANSZ, H. S., 25, 50
JARABEK, J., 19, 41, 42, 50
JARETT, L., 200, 216
JASPER, H., 13, 188
JEGLA, T. C., 687, 709
JEFFRY, G. A., 82, 97
JENCKS, W. P., 251, 276
JEWELL, B. R., 425, 438, 439, 453
JEWETT, P., 291, 295, 299, 303, 304, 308, 352, 354
JOBSIS, F., 205, 215, 216

JOHANSEN, K., 490, 527
JOHANSSON, B. W., 392, 397
JOHNSON, D. S., 85
JOHNSON, E. A., 291, 293, 294, 295, 296, 298, 299, 300, 301, 303, 304, 306, 307, 308, 309, 318, 351, 352, 354, 429, 433, 452
JOHNSON, F. H., 65, 78
JOHNSTON, P. V., 72, 78, 79
JONES, T. G., 276
JOUVET, M., 508, 510, 528, 541, 543, 549
JULIAN, F. J., 292, 359, 396, 426, 452
JURTSHUK, P., 260, 270, 278

KAADA, B. R., 505, 529
KAHLENBERG, A., 161, 164
KAISER, E., 434, 451
KAISER, G. A., 479
KAKINUMA, K., 198, 216
KALLEN, F. C., 101, 117
KALLOW, W., 101, 117
KALSER, S. C., 101, 117
KAMB, B., 82, 88, 99
KANTER, G. S., 101, 102, 118
KANTHOR, H. A., 101, 117
KANUGO, M. S., 56, 57, 74, 78
KAPLAN, A. E., 253, 276
KARNOVSKY, M. J., 298, 336, 353
KASAI, M., 266, 268, 276
KATZ, A. M., 357, 370
KATZ, B., 402, 419
KATZUNG, B., 436, 454
KAUZMANN, W., 17, 43, 51, 88, 243, 276
KAVANAU, J. L., 82, 98
KAWA, A., 560, 574
KAWAMURA, K., 352
KAY, J. H., 167
KAYSER, C., 101, 118, 128, 133, 144, 482, 489, 491, 505, 508, 514, 515, 531, 637, 658
KEECH, D. B., 31, 53
KEILLEY, W. W., 358, 368, 374, 397
KELLY, D. E., 300, 352
KELLY, K. H., 640, 653, 654, 660
KELVINGTON, E. J., 101, 117
KEMP, A., 22, 51, 209, 216
KEMP, P., 227, 235, 237
KENT, K. M., 172, 189
KERNAN, R. P., 206, 216

KERR, D. I. B., 528
KEY, C. M., 397
KEYNES, R. D., 136, 144, 168, 184, 189
KIELY, B., 383, 385, 397
KIMZEY, S. L., 133, 145, 146
KIPNIS, D. M., 200, 216
KIRKMAN, H. N., 19, 51
KISELEVA, L. M., 490, 531
KITTEL, C., 258, 274
KLAHR, S., 200, 215
KLAIN, G. J., 7, 8, 9, 10, 11, 13, 15
KLEIN, R. A., 228, 229, 236
KLEINEKE, J., 110, 119
KLEINZELLER, A., 127, 129, 130, 146
KLEITMAN, N., 538, 549
KLINE, M. H., 152, 165
KLINGENBERG, M., 22, 24, 51, 53, 208, 209, 216
KLOTZ, I. M., 251, 276
KLUGER, M. J., 605, 608, 621, 622, 626
KNAPEIS, G. S., 300, 352
KNAUSS, T., 538, 550
KOCH, A., 168, 189
KODAMA, A., 357, 359, 360, 361, 362, 368
KOELLA, W. P., 509, 511, 529
KOHANA, L., 514, 529
KOIZUMI, K., 507, 511, 512, 529, 533, 534
KOMINZ, D. R., 359, 370, 374, 385, 397
KOOPMAN, L. J., 505, 533
KOSH, R. J., 689, 698, 709
KOSHLAND, D. E., 32, 51, 156, 164, 272, 277
KOSKIMIES, J., 670, 679
KOVAL, G. J., 136, 144, 158, 165
KRAZLIC, L., 528, 529, 530, 531
KREBS, E. G., 24, 26, 50
KREBS, J. S., 101, 116
KREISBERG, R. A., 110, 119
KRENKEL, W., 506, 528, 533
KRESPI, V., 406, 419
KREUTIGER, G. O., 295, 298, 352
KRISTOFFERSON, R., 10, 14, 637, 638, 653, 659
KROG, J., 490, 527
KRUGER, F. A., 109, 119
KUCZENSKI, R. T., 19, 42, 51
KUFFLER, S. W., 189
KUME, S., 136, 138, 146

AUTHOR INDEX

KUMAR, A. E., 109, 117
KURZ, J. L., 277
KUSHMERICK, M. J., 390, 395

LAICO, M. T., 257, 277
LAKI, K., 359, 370, 374, 388, 398
LAM, K. W., 18, 19, 20, 49
LAMY, F., 198, 208, 218
LANDON, E. J., 204, 216
LAPORTA, A. D., 4, 14
LAROCCA, M., 528
LARSON, R. E., 390, 395
LARSSON, B., 608, 624
LASHBROOK, M. K., 689, 709
LASIEWSKI, R. C., 669, 670, 676, 679
LASIEWSKI, R. J., 669, 676, 679
LAUGHTER, A. H., 158, 164, 165
LAW, P. F., 84, 98
LAWRENCE, A. S. C., 240, 277
LEAF, A., 182, 189, 219, 236
LEAK, L. V., 304, 352
LECLERC, M. M., 128, 133, 144
LEE, J. C., 168, 188, 189
LEE, K. W., 84, 98
LEE, R. C., 580, 604
LEHMANN, D., 538, 550
LEHMANN, O., 277
LEIGHTON, R. F., 109, 119
LEIVESTAD, H., 125, 133, 144
LEMAITRE, M., 505, 529
LENARD, J., 240, 270, 277
LEONG, G. F., 101, 116
LEVIN, A., 423, 453
LEVIN, V. A., 109, 117
LEVINSON, C., 203, 216
LEVY, H. A., 88, 98
LEW, V. L., 136, 145
LEWIS, J. K., 70, 78
LI, N. M., 133, 136, 147, 185, 190, 237
LI, C. - L., 216
LIANIDES, S. P., 17, 51
LIBERSON, W. T., 505, 529
LIBET, B., 536, 549
LICHT, P., 17, 51, 681, 685, 686, 688, 689, 690, 692, 693, 694, 696, 697, 698, 699, 700, 701, 703, 705, 709, 710, 711, 716
LIND, J., 460, 477
LINDBERG, O., 3, 14

LINDENMAYER, G. E., 158, 159, 160, 161, 162, 164, 166
LING, G. N., 93
LINNE, A., 19, 51
LIPMANN, F., 299, 350
LIPP, J. A., 508, 529
LIPPINCOTT, E. R., 92, 98
LISK, R. D., 540, 549
LITTLE, G. R., 407, 419
LIU, C. - C., 4, 14
LIVESZEY, R. L., 689, 709
LOCKER, R. H., 374, 376, 377, 378, 397, 398
LOGINOVA, E. V., 490, 531
LODISH, H., 202, 215
LOFTS, B., 694, 710
LORD, J. D., 460, 477
LOUGHEED, W. M., 505, 529
LOURIE, H., 508, 529
LOWENSTEIN, L. M., 128, 146
LOWEY, S., 358, 359, 362, 369, 370, 373, 374, 377, 383, 398, 399
LOWRY, O. H., 169, 172, 174, 189
LOWS, C. F., 270, 274
LOWY, J., 359, 360, 369
LUBIN, M., 127, 146
LUCK, S. M., 387, 398
LUECKE, R. H., 577, 582, 589, 603, 604, 633
LUSTY, C. J., 37, 38, 39, 40, 52, 53, 54
LÜTHI, U., 136, 145
LUZZATI, V., 240, 241, 244, 276
LYMAN, C. P., 4, 7, 8, 9, 13, 14, 101, 108, 116, 118, 126, 146, 431, 436, 453, 457, 478, 488, 489, 493, 499, 501, 509, 514, 515, 530, 549, 579, 580, 604, 637, 653, 658, 659

MA, N., 133, 136, 147, 185, 190
MADDEN, J. A., 71, 78
MACPHERSON, L., 428, 453
MADDY, A. H., 270, 277
MAGOWN, H. W., 501, 509, 528, 531
MAHER, J. T., 436, 451
MAIZELS, M., 193, 216
MAKINOSE, M., 299, 351
MALAN, A., 101, 118, 491, 529, 571, 573, 608, 626
MALCOLM, J. L., 511, 527, 529
MALCOM, B. R., 270, 277

MANASEK, F. J., 7, 14
MANDELKERN, L., 374, 388, 398
MANERY, J. F., 81, 97
MANIL, J., 136, 144
MARION, K. R., 686, 693, 710
MARQUEZ, E., 435, 451
MARSH, B. S., 424, 426
MARSHALL, J. M., 108, 118, 436, 453, 454
MARTINEZ DE MUÑOZ, D., 262, 263, 266, 278, 280
MARTINEZ-PALOMO, A., 307, 352
MARTINEZ-ROJAS, D., 239, 255, 257, 275
MARTINEZ-ZEDILLO, G., 239, 255, 257, 275
MARTONOSI, A., 266, 268, 279, 359, 370, 383, 385, 397
MARUYAMA, K., 359, 370
MASAKI, T., 359, 360, 369, 370
MASSOPUST, L. C., 108, 118, 508, 511, 514, 530, 547, 549
MATSUI, H., 156, 157, 164, 165
MATTHEWS, B. H. C., 511, 527
MATTHEWS, M., 608, 625
MAYERSBACH, H. V., 638, 659
MAYHEW, W. W., 789, 692, 695, 696, 703, 710
McCARLEY, 652
McCASKILL, E. S., 109, 116
McCLENNAN, H., 557, 573
McDONALD, I. G., 460, 478
McDONALD, R. H., 460, 468, 479
McELHANY, R. N., 72, 79, 242, 244, 245, 260, 269, 278, 279
McGINTY, D. J., 543, 550
McILWAIN, H., 198, 203, 216, 218, 239, 279
McKINLEY, W. A., 608, 616, 624
McLEAN, J. A., 608, 625
McMILLAN, J. K., 540, 549
McMILLEN, R. E., 668, 679
McMULLAN, R. K., 82, 97
McMURREY, J. D., 508, 530
McNAB, B. K., 669, 680
McNABB, R. A., 71, 77
McNUTT, N. S., 295, 298, 304, 350, 352
McOWENS, D. A., 507, 527, 531
MEDER, J., 101, 118, 508, 514, 530, 547, 549
MEDER, R., 511, 530

MEDZIHRADSKY, F., 152, 165
MEHL, E., 257, 275
MELLISH, C. H., 689, 692, 710
MELMAN, A., 259, 277
MENAKER, M., 108, 118, 119, 638, 653, 659
MENDLER, N., 167, 168, 171, 172, 189, 288
MEPHAM, T. B., 219, 236
MURILLEES, N. C. R., 309, 352
MERZBACHER, L., 487, 530
MESSMER, K., 167, 168, 168, 176, 188, 189
MICHAEL, C. R., 108, 118
MIHAILOVIĆ, L. T., 487, 490, 497, 498, 499, 501, 505, 518, 519, 522, 523, 525, 528, 529, 531, 533, 633
MIHALYI, E., 373, 398
MILLER, A. H., 19, 51
MILLER, F. S., 182, 189
MILLER, J. A., 182, 189
MILLER, L. L., 110, 118
MILLER, M. R., 687, 696, 711
MILLS, S. H., 605, 608, 615, 616, 626
MILSUM, J. H., 434, 454
MINIKAMI, S., 198, 216
MISHKIN, E., 435, 454
MITCHELL, E. R., 374, 385, 397
MITCHELL, J. H., 457, 460, 464, 465, 469, 477, 478
MOKHIN, K. M., 490, 531
MOLYNEAUX, P., 245, 246, 269, 277
MOGLER, R. K., 653, 659
MOMMAERTS, W. F. H. M., 381, 398, 446, 451
MONNERIE, L., 266, 268, 276
MONNIER, M., 540, 550
MONROE, R. G., 109, 117
MONTAL, M., 273
MORELAND, J. E., 4, 11, 15
MOORE, L., 168, 184, 188
MOORE, M. J., 229, 236
MOORE, P. B., 360, 370, 385, 398
MOORE, S., 221, 223, 224, 236
MOOS, C., 390, 396
MORALES, M., 358, 371
MORAN, A., 269, 277
MORGAN, H. E., 109, 118
MORINO, Y., 19, 51
MORRS, G., 101, 118

MORRIS, H. P., 192, 216
MORRIS, L., 653, 659
MORRISON, P., 8, 9, 13, 653, 659
MORUZZI, G., 501, 531, 542, 550
MOSELY, M., 297, 354
MOTTRAM, R. F., 616, 625
MOUNTCASTLE, V. B., 536, 550
MOYER, J. H., 101, 118
MROSOVSKY, N., 126, 144, 545, 550, 640, 651, 652, 653, 659, 660
MUDGE, G. H., 131, 146
MUELDENER, B., 169, 174, 190
MUELLER, C., 167, 169, 176, 188
MUELLER, H., 357, 358, 359, 361, 362, 369, 377, 398
MUKERJEE, P., 252, 261, 277
MULLINS, C. B., 460, 465, 478
MUNCK, B. G., 219, 236
MUNN, E. A., 220, 236
MUNTUS, R., 364, 371
MURTON, R. H., 694, 710
MUSACCHIA, X. J., 108, 118
MUTCHMOR, J. A., 17, 51
MYBRE, K., 608, 610, 625, 626
MYERS, R. D., 551-570, 573-575, 633

NAGAI, K., 147, 161, 165
NAKAJIMA, H., 357, 368
NAKAGAWA, Y., 147
NAKAMURA, R., 357, 368
NARTEN, A. H., 98
NAUTA, W. J. H., 542, 550
NAYLOR, W. G., 309, 352
NEET, K. E., 32, 51, 272, 277
NEHEI, T., 397
NELSON, D. A., 295, 353
NELSON, J. S., 71, 77
NELSON, L., 4, 13
NEMETHY, G., 43, 51, 52, 87, 98
NESTOR, E. W., 19, 49
NEVILLE, M. C., 288
NEWMAN, B. L., 168, 189
NEWEY, H., 219, 235
NICHOLS, J. G., 189
NICHOLS, B. L., 94, 97, 115, 117
NIEDERGERKE, R., 292, 351, 358, 369
NILSSON, E., 425, 428, 430, 431, 432, 452
NINOMIYA, I., 460, 478, 480
NOBLE, M. I. M., 431, 432, 433, 454
NOELL, W. K., 508, 531
NOLAN, A. C., 434, 454

NONOMURA, Y., 359, 360, 369, 370
NORDMANN, K., 176, 189
NORTHCUTT, R. G., 510, 625
NUMA, S., 19, 41, 51

O'BRIEN, R. C., 101, 118, 126, 146
OCHOA-SOLANO, A., 252, 264, 267, 275
O'CONNOR, G., 110, 116
OGSTON, A. G., 362, 368
OHASI, T., 147
OHTSUKI, I., 359, 360, 369, 370
OKAMOTO, H., 309, 353
OKUYAMA, T., 247, 277
OLCOTT, H. S., 373, 374, 378
O'LEARY, J. L., 508, 529
OLMSTED, M. R., 19, 33, 34, 35, 41, 42, 50, 54
ONISHI, T., 257, 277
OOI, Y., 357, 368
ORCUTT, B., 136, 138, 146
OVERATH, P., 242, 278
OXENDER, D. L., 219, 223, 236

PAGE, E., 301, 308, 353
PAKTER, A., 246, 278
PALADE, G. E., 294, 296, 299, 309, 350, 351, 353
PANUSKA, J. A., 633
PAPAHADJOPOULAS, D., 249, 278
PAPAS, T. S., 19, 51
PAPERMASTER, D. S., 257, 277
PAPPIUS, H. M., 180, 189
PARIS, E. G., 689, 690, 691, 701, 708
PARK, C. R., 109, 110, 116, 118
PARKER, L., 358, 370
PARMEGGIANI, A., 109, 118
PARMLEE, W. W., 431, 432, 433, 454, 455
PASSONNEAU, L. V., 169, 172, 174, 176, 189
PAULSON, T. L., 640, 647, 648, 650, 653, 654, 658, 661
PEARSON, L. K., 690, 698, 710
PEARSON, R. G., 364, 369, 370
PEIPONEN, V. A., 670, 680
PELOUCH, V., 108, 119
PENEFSKY, H. S., 18, 19, 20, 52, 54, 259, 278
PENGELLEY, E. T., 637-645, 649-655, 660, 661, 716
PEPE, F. A., 387, 398
PERRY, S. L. C., 460, 464, 478
PERRY, S. V., 361, 371, 377, 383, 398, 399

PERRY, T. O., 84, 98
PETROVIĆ-MINIĆ, B., 531
PFAFF, E., 208, 209, 216
PFAFF, E. M., 22, 51
PFLEIDERER, G., 169, 174, 176, 190
PHILLIPS, C. M., 460, 477, 479
PHILLIPS, C. S. G., 366, 370
PICKERING, G. W., 608, 616, 625
PIEPER, H. P., 459, 460, 478
PIERRE, R., 508, 527
PIERSOL, A. G., 435, 451
PLENGE, R., 109, 117
PLUM, F., 536, 550
PODOLSKY, R. J., 434, 454
POLISSAR, M. J., 65, 78
POLLACK, G. H., 430, 431, 454
POLLACK, J. D., 242, 278
POOL, P. E., 109, 118
POPESKOVIC, D., 496, 501, 522, 530, 533
POPOVIC, V., 514, 531
POPOVIC, V. P., 4, 13, 172, 189, 472, 475, 478
PORTER, G. A., 219, 235
PORTER, K. R., 296, 299, 351, 352
PORTZEHL, H., 387, 388, 399
POSIVIATA, M. A., 7, 14, 125, 147
POSNER, A. S., 374, 388, 398
PŎSNER, J. B., 536, 550
POST, R. L., 136, 138, 140, 145, 146, 151, 165
POULSON, T. L., 687, 709
POUPA, O., 108, 109
PRAGAY, D., 390, 396
PROTIĆ, S., 531
PRECHT, H., 71, 78, 123, 136, 146
PREISER, H., 409, 419
PROCHAZKA, J., 108, 119
PROSSER, C. L., 56, 57, 74, 78, 615, 626
PRUSINER, S., 168, 169, 174, 176, 180, 190
PULLINAN, M. E., 18, 52, 259, 278
PURKINJE, J. E., 305, 353
PURPURA, D. P., 488, 499, 501, 509, 514, 515, 527
PUTKONEN, P., 490, 531
PYUN, H. Y., 358, 359, 360, 362, 363, 368, 370

QUICK, K. P., 609, 626

RACKER, E., 18, 19, 24, 51, 52, 259, 278
RACKLEY, C. E., 460, 478
RADER, R. L., 242, 244, 245, 260, 269, 279
RAHN, H., 70, 78
RAIJMAN, L., 19, 52
RANCK, J. B., 168, 189
RANDALL, W. C., 4, 21, 452
RANDOLPH, M. M., 101, 117
RANDT, C. T., 510, 532, 533
RANSON, S. W., 608, 616, 624
RANG, H. P., 203, 217
RAO, K. P., 71, 78
RATHS, P., 490, 515, 517, 524, 525, 531, 534
RATNER, S., 19, 37, 38, 39, 40, 50, 52, 53, 54
RAVENTOS, J., 101, 119
RAWSON, R. O., 609, 626
RAY, A., 252, 261, 277
RAYNS, D. G., 301, 353
RAZEN, S., 242, 278
READ, K. R. H., 17, 52
RECHARD, T., 390, 395
REES, M. K., 359, 370
REGEN, D. M., 109, 118
REINERT, J. C., 72, 78, 79, 242, 244, 245, 260, 269, 279
REISS, I., 359, 371
REITER, R. J., 653, 658
REITSMA, H. J., 22, 51
REMINGTON, M., 193, 216
RENNIE, D. W., 109, 116
REULEN, H. J., 167, 168, 169, 174, 176, 180, 188, 189, 190
REUTER, H., 201, 217
REVEL, J. P., 297, 298, 336, 353
REX, R. L., 435, 454
RHODES, C. T., 245, 246, 269, 277
RICE, R. V., 359, 371, 373, 374, 398
RICHARDS, D. H., 379, 380, 383, 396
RICHARDS. E. G., 373, 374, 398
RICHTER, C. P., 538, 550
RINGLEMAN, E., 19, 41, 51
RINK, R. R., 101, 117
RITCHIE, J. M., 203, 217, 425, 428, 451
RIZZO, S. R., 202, 215
ROBERTS, G. T., 435, 454
ROBERTS, W. W., 608, 626
ROBERTSON, J. D., 298, 353

AUTHOR INDEX

ROBERTSON, R. C., 110, 119
ROBINSON, T. C. L., 608, 626
RODBARD, S., 430, 433, 454, 610, 626
ROENTGEN, W. K., 92, 98
ROGERS, G. B., 7, 14
ROHMER, F., 101, 118, 488, 489, 515, 529, 531
ROLDAN, E., 543, 550
ROSENBLITH, W. A., 514, 529
ROUSSEAU, D. L., 92, 97, 98
ROLETT, E. L., 460, 478
ROOTS, B. I., 72, 78, 79
ROSE, P., 242, 280
ROSEMBERG, S. A., 260, 270, 278
ROSENBAUM, R. M., 259, 277
ROSENTHAL, D., 433, 454
ROSS, I., 169, 174, 176, 190
ROSS, J., 458, 459, 460, 479
VAN ROSSUM, G. D. V., 191, 193, 198, 200, 201, 203, 206, 207, 211, 212, 215, 217, 218, 288
ROWAN, W., 639, 660
RUBALCAVA, B., 262, 263, 265, 266, 269, 271, 275, 278, 280
RUBENSTEIN, L., 557, 573
RUCCI, F. S., 528
RUDY, B., 256-273
RUECKERT, R. R., 101, 117
RUEGG, J. C., 357, 371
REISS, I., 299, 355
RUOSLAHTI, E. J., 257, 277
RUSHMER, R. F., 459, 460, 468, 477, 479

SAAKOV, B. A., 490, 531
SACHS, G. R., 198, 204, 208, 217
SAMPSON, J. H., 11, 13
SAMSON, F., 610, 626
SANADI, D. R., 18, 19, 20, 49
SANDLER, H., 457, 460, 477, 479
SANDOW, A., 409, 419
SANMARCO, M. E., 459, 460, 477, 479
SANTOMENA, D. M., 101, 117
SARAJAS, H. S., 490, 531
SARKAR, S., 393, 399
SATAKE, K., 247, 277
SATINOFF, E., 545, 550, 608, 626
SATO, R., 253, 276
SAURWEIN, T. D., 510, 532, 533
SAWYER, C. H., 540, 549
SCARPELLI, D. G., 109, 119
SCHAFFER, W. I., 110, 116

SCHAIRER, H. U., 242, 278
SCHATZMANN, H. J., 150, 165, 201, 217
SCHAUB, M. C., 361, 371
SCHERAGA, H. A., 43, 51, 52, 87, 98
SCHICK, M. G., 244, 252, 278
SCHIEBLER, T. H., 295, 353
SCHIFFER, J., 87, 98
SCHIMKE, R. T., 52, 57, 58, 79
SCHIPP, R., 309, 353
SCHLECHTE, F. R., 492, 493, 505, 532, 533, 638, 652, 661
SCHMAL, F. W., 190
SCHMIDT-NIELSEN, K., 203, 205, 215
SCHNELLE, K. B., 595, 604
SCHOENBAUM, E., 637, 658
SCHOENER, B., 205, 215
SCHOLZ, F., 390, 396
SCHOLZ, R., 196, 213, 217, 218
SCHOOL, H., 169, 174, 190
SCHRADER, L. E., 7, 8, 15
SCHRODT, G. R., 300, 355
SCHULMAN, J. H., 246, 278
SCHULTZ, D. W., 169, 172, 174, 176, 189
SCHULTZ, S. G., 127, 146, 233, 236
SCHULTZE, W., 304, 353
SCHULZE, T. I., 37, 38, 39, 40, 52, 53, 54
SCHUSTER, C. W., 19, 53
SCHWARTZ, A., 149, 155-162, 164-166
SCHWARTZ, H. G., 508, 529
SCHWARZ, E., 213, 217
SCHWERT, G. W., 19, 42, 52
SCOTT, J. W., 507, 527, 531
SCRUTTON, M. C., 19, 31, 32, 33, 34, 41, 53, 54
SEALOCK, R. W., 19, 25, 26, 50, 53
SEEDS, A. E., 19, 41, 42, 50
SEGAL, S., 128, 129, 130, 146
SEIDEL, J. C., 390, 396, 399
SEITZ, N., 201, 217
SEKUZU, I., 260, 270, 278
SELVERSTONE, B., 182, 190
SEN, A. K., 136, 137, 138, 144, 146, 151, 165
SENIOR, A. E., 21, 52
SENTURIA, J. B., 108, 119
SEXTON, O. J., 689, 708
SEYDEUX, J., 124, 145, 210, 215
SHARP, G. W. G., 219, 236
SHARPE, L. G., 553, 555, 559, 560, 574

SHAW, D. O., 246, 278
SHAW, T. I., 136, 144
SHEETZ, M., 270, 276
SHINITZKY, M., 242, 243, 245, 279
SHIPP, J. C., 109, 116
SHIRACHI, D. Y., 152, 165
SHOEMAKER, R. L., 198, 204, 208, 217
SHRAGO, E., 4, 14
SHTARK, M. B., 490, 495, 496, 499, 500, 501, 504, 505, 531, 532
SHUG, A. L., 4, 14
SHUKUYA, R., 42, 52, 119
SIEGEL, G. J., 158, 165
SIMPKINS, H., 270, 276
SIMPSON, F. O., 301, 353
SINGER, S. J., 240, 253, 254, 269, 270, 273, 276, 277, 279
SJOSTRAND, F. S., 294, 353
SKOU, J. C., 150, 165, 219, 236
SLATER, E. C., 196, 197, 199, 200, 216, 217
SLAUGHTERBACK, D. B., 308, 353
SLAYTER, H. S., 358, 370, 373, 374, 398, 399
SLEATOR, W. W., 401, 403, 406, 407, 410, 411, 412, 414, 419, 484
SLEIN, M. W., 19, 49
SMALL, D., 240, 241, 279
SMALLEY, R. L., 3, 14
SMIT, G. J., 653, 661
SMIT-VIS, J. H., 653, 661
SMITH, A. U., 182, 188
SMITH, D. E., 436, 454
SMITH, E. N., 374, 385, 397
SMITH, I., 128, 129, 130, 146
SMITH, J., 492, 505, 532
SMITH, J. L., 638, 661
SMITH, L. W., 511, 528
SMITH, M. W., 219, 220, 221, 227, 229, 233, 235, 236, 237, 288
SMITH, R. E., 3, 14, 124, 140, 145
SMYTH, D. H., 219, 235
SMYTHE, M. Q., 305, 306, 355
SNELL, E. E., 19, 51
SOIVIO, A., 638, 653, 659
SÖLING, H. D., 110, 119
SOMERO, G. N., 17, 50, 55, 56, 57, 58, 60, 61, 62, 63, 65, 67, 68, 69, 74, 75, 78, 80, 115
SOMMER, J. R., 291, 294, 295, 296, 299, 300, 303, 304, 306, 307, 308, 309, 318, 336, 352, 353, 354, 484

SONNENBLICK, E. H., 109, 118, 430, 431, 432, 433, 434, 446, 451, 454, 455, 458, 459, 460, 477, 479
SOUTH, F. E., 3, 6, 8, 9, 10, 14, 101, 108, 119, 140, 146, 436, 455, 490, 493, 514, 515, 516, 522, 525, 532, 534, 577, 582, 589, 603, 604, 633, 637, 658
SPACH, M. S., 304, 354
SPAFFORD, D. C., 655, 661
SPANN, J. F., 109, 118
SPECK, M. L., 19, 49
SPECTOR, N. H., 608, 626
SPIRO, D., 293, 297, 355, 432, 455, 479
SQUIRES, R. D., 607-609, 616, 626
SPURRIER, A., 640, 651, 652, 658
SRETER, F. A., 390, 396, 399
STALEY, N. A., 295, 309, 354
STANDISH, M. M., 228, 235
STEBBINS, R. C., 684, 693, 711
STECK, T. L., 270, 299
STEERE, R. L., 291, 297, 298, 304, 336, 354
STEGALL, H. F., 460, 467, 477, 478
STEIM, J. M., 72, 78, 79, 242, 244, 245, 260, 269, 279
STEIN, W. H., 221, 223, 224, 236
STEINHARDT, R. A., 127, 136, 144
STENGER, R. J., 297, 355
STEPANOVA, N. G., 9, 11, 13
STEPHENS, R. E., 251, 275
STERMAN, M. B., 538, 543, 550
STEUDE, U., 168, 169, 174, 176, 180, 190
STEVENSON, G. C., 510, 532, 533
STEWART, S., 108, 119
STOFFEL, W., 242, 278
STOLWIJK, J. A., 608, 609, 624, 625
STONE, H. L., 460, 467, 477, 478
STRACHER, A., 374-379, 381-383, 395, 396, 399
STRAPRANS, I., 362, 370
STRAUSS, J. H., 270, 279
STREETER, J., 492, 493, 532, 533, 638, 652, 661
STROM, G., 626
STROMBERG, R. R., 92, 98
STROMER, M. H., 359, 371
STRØMME, S. B., 608, 610, 625
STRUMWASSER, F., 490-493, 497, 505, 532, 533, 545, 550, 581, 604, (cont.)

AUTHOR INDEX

STRUMWASSER, F., (cont.) 638, 652, 661
SUDA, I. K., 507, 511, 533
SUELTER, C. H., 19, 42, 51
SUOMALAINEN, P., 490, 493, 531, 533, 637, 638, 653, 659, 661
SWAISGOOD, H. E., 19, 49
SWANSON, P. D., 133, 136, 146, 219, 237, 239, 274, 279
SWARBRICK, J., 245, 246, 269, 277
SZENT-GYÖRGYI, A. G., 93, 362, 368, 373, 398

TAIT, J., 101, 119
TANFORD, C., 257, 276
TALALAY, P., 19, 41, 42, 50
TAMIR, H., 19, 37-40, 50, 54
TANCHE, M., 507, 508, 510, 528
TANZINI, G., 109, 116
TAREN, J. A., 508, 530
TASHIMA, L. S., 8, 14
TAYLOR, R. E., 301, 351
TAYLOR, R. R., 459, 460, 468, 479
TAYLOR, S. R., 409, 419
TEMPLETON, J. R., 610, 626
TEN CATE, J., 505, 533
TER WELLE, H. F., 197, 199, 217
THAL, N., 460, 479
THAUER, R., 167, 172, 182, 190
THEINER, M., 361, 362, 369
THEWS, G., 182, 190
THIERY, J., 258, 274
THOMAS, J. A., 4, 13
THOMPSON, A. M., 110, 119
THOMPSON, W. J., 460, 478
THORN, W. H., 169, 174, 176, 190
THURMAN, R. G., 196, 218
TINKLE, D. W., 686, 711
TITUS, E., 158, 164
TOBIN, R. B., 198, 203, 218
TOBIN, T., 136, 138, 146
TONOMURA, Y., 358, 371
TOURTELOTTE, M. E., 72, 79, 242, 244, 245, 260, 269, 278, 279
TRAUTWEIN, W., 405, 419
TRAYER, I. P., 383, 399
TREVOR, A. J., 152, 165
TRIGGLE, D. J., 279
TROSHIN, A. S., 93
TROST, C. H., 670, 679
TROTTA, P. P., 374, 375, 378, 381, 395, 399

TRUEY, R. C., 305, 306, 355
TRUSCOE, R., 193, 216
TSAO, T. - C., 362, 371, 374, 399
TUCKER, S. W., 261, 275
TUCKER, V. A., 671, 672, 680
TUNG, Y., 258, 274
TURNER, M. D., 101, 119
TWENTE, J. A., 8, 14, 126, 146, 493, 530, 661, 669, 680
TWENTE, J. W., 8, 14, 126, 146, 493, 530, 661, 669, 680
TYLER, D. D., 198, 208, 218

UHLMAN, C., 638, 655, 660
UMBARGER, H. E., 19, 42, 50
ULLRICK, W. C., 421, 455
USHIYAMA, J., 511, 512, 529, 534
UTTER, M. F., 19, 31-35, 41, 42, 50, 52, 53, 54

VAN BREEMAN, E. D., 71, 77
VAN CITTERS, R. L., 459, 460, 477
VANDERKOOI, J., 266, 268, 279
VAN OORDT, P. G. W. J., 685, 692, 693, 697, 705, 710
VEALE, W. L., 561, 562, 564, 565, 574, 575
VERLOT, M., 182, 190
VERNBERG, F. J., 72, 73, 77
VERZEANO, M., 509, 528
VILLARICO, E. A., 388, 398
VOLPE, A., 390, 395
VON EULER, C., 615, 627

WAGNER, B. P., 192, 216
WALKER, M. L., 457, 479
WALKER, S. M., 300, 355
WALLACH, D. F. H., 241, 242, 244, 261, 270, 274, 279
WALLER, H. J., 541, 549
WANG, J. H., 19, 25, 26, 50, 53
WANG, S. C., 547, 549
WARNER, R. C., 18-20, 37-40, 50, 52, 259, 278
WATANABE, S., 358, 362, 371
WATKINS, J. C., 228, 235
WATSON, M. L., 110, 118
WAYS, P., 250, 279
WEBER, A., 299, 355, 357, 359, 361, 371
WEBER, G., 242, 243, 245, 279
WEBER, H. H., 387, 388, 399
WEDENSKY, N. E., 501, 533

WEEDS, A. G., 358, 370, 373, 374, 398, 399
WEHREN, A. B. - V., 309, 353
WEIDEMANN, M. J., 24, 53
WEIDMANN, S., 402, 403, 419, 420
WEINSTEIN, R. S., 270, 279, 298, 352
WEINSTEIN, W., 508, 529
WEISHAAR, W., 168, 189
WEISS, L., 249, 278
WEISS, T., 543, 550
WEISSLER, A. M., 109, 119
WEN, W. - Y., 86, 87, 97
WHEELER, K. P., 140, 146, 196, 198, 199, 218
WHITE, D. C. S., 432, 434, 455
WHITE, J. C., 182, 190
WHITEHORN, W. V., 421, 455
WHITTAKER, E. J. W., 364, 371
WHITTAM, R., 140, 146, 193, 196, 198, 199, 202-204, 215, 218
WHITTEN, B. K., 7, 8, 9, 10, 11, 14, 15, 125, 147
WHITTOW, G. C., 608, 625, 627
WIEBERS, J. E., 10, 13
WIGELIUS, C., 400, 477
WILEY, J. S., 202, 218
WILKIE, D. R., 388, 395, 425, 428, 438, 439, 453
WILLIAMS, B. A., 524, 533, 582, 604, 605, 608, 621, 622, 624, 627, 633
WILLIAMS, G. R., 204, 205, 215
WILLIAMS, J. P., 366, 370
WILLIAMS, R. J. P., 364, 371
WILLIAMSON, J. R., 110, 119
WILLMS, B., 110, 119
WILLIS, J. S., 3, 15, 108, 118, 123-133, 135-137, 140-142, 144-147, 185, 199, 219, 220, 234, 237, 239, 280, 288, 436, 454, 584, 604
WILSON, G., 242, 280
WILSON, M. F., 460, 464, 468, 478, 479, 480
WOHLSCHLAG, D. E., 56, 57, 79
WOLFF, H. H., 353
WOLIN, L. R., 101, 118, 508, 514, 530, 547, 549
WOLLENBERGER, A., 304, 353
WOOD, L., 133-137, 147
WOODS, E. F., 359, 362, 371
WOODWARD, R. J., 246, 274
WROBLEWSKI, F., 65, 79
WYMAN, J., 423, 453

YAKSH, T. L., 551, 553, 554, 556, 558, 564-567, 568, 574, 575
YAMAMOTO, T., 295, 355
YEATMAN, L. A., 433, 454
YOSHIDA, H., 147
YOSHIKAWA, H., 198, 216
YOUNG, D. M., 399
YOUNG, M., 359, 371
YOUNG, J., 608, 625

ZAHLER, P. H., 270, 279
ZAVELAR, S. A., 434, 454
ZIMNEY, M. L., 4, 11, 15
ZYLKA, W., 506, 528, 533

SUBJECT INDEX

Acclimation
   brain of hibernators, 283
   capacity adaptation, 123
   fish, 55-80
   in vitro and rewarming, 131
   membrane function, 125
   non-homeothermic, 123
   resistance adaptation, 123
Acetylcholinesterase, 61
Action potential, 149
Adaptation
   adaptive hypothermia, 665
   capacity adaptation, 123
   cellular, 3
   cold and membrane function, 132
      225, 239
   compensation, 55-80
   environmental, 55-80
   myosin ATPase, 392
   resistance adaptation, 123
   temperature, 55-80, 219, 237
   temperature control system, 592
   tissue, 3
   warm and cold fish, 58-80
Adenine nucleotides
   adenosinediphosphate, 12, 21, 102,
      202, 383
   adenosinemonophosphate, 12, 103
   adenosinetriphosphate, 12, 17-54,
      102, 135-143, 150-163, 169,
      174, 192-218, 219, 234, 259,
      283, 299, 304, 357, 373-400
ADP (see Adenine nucleotides)
Alkalosis, 184
Alligator
   circannual rhythm, 682
Allosteric
   effector, 25-48
   inhibitor, 24, 160
   site, 25
Amino acids, 127
   and ion-exchange resin, 224
   and TNBS, 247-251
   efflux rate, 231
   electrostatic, hydrophobic
      interactions, 233
   transport, 221-237
AMP (see Adenine nucleotides)

Amphibian
   heart, 295, 300
Anionic sites, 95
Arctic ground squirrel, 8
Arctic sculpin, 608, 610, 611
ATP (see Adenine nucleotides)
Avian salt gland, 205

Bat
   body temperature, 579
   brain, 490
   hibernating, 259
   POAH heating and cooling, 608, 621
   southeastern Myotis, 4
   torpor in, 669, 670
Behavior
   physiological, 637-661
   synchronization with environment,
      637-661
Bird
   caprimulgids, 669
   chicken heart, 301, 303
   chicken liver, 31
   colies, 669, 670
   diurnal torpidity, 667
   finch heart, 303
   goatsuckers, 667, 670
   heart, 293, 295, 300, 301, 303
   hibernation, 667
   hummingbird, 545, 669, 670
   nighthawk, 544
   nightjars, 667
   photoperiod, 694
   salt gland, 205
   sparrow, 608, 616, 617
   swifts, 669, 670
Blood
   cells and potassium, 126-135
   erythrocyte membranes, 242-244,
      255, 256, 257, 258
   glucose, 8, 9
   hemoperfusion, 176-179
   red cell ghost, 151, 249
   red cell transport, 201
   unidirectional flux in erythrocytes,
      123
Bonding
   high energy phosphate, 152

Bonding, cont'd.
  hydrophobic, 42-47
Brain (see Nervous system)
Brown fat, 124
Brumation, 683
Bullheads, 71

Calcium
  and ATPase activity, 357
  and muscle activity, 357-371, 408-409
  sodium balance, brain, 561
  heart muscle, 299, 401-410
  hypothermia, 551-576
  perfusal brain, 560-568
Caprimulgids, 669
Carassius auratus, 56
Cat, 539, 553, 557, 570, 608
Cations, 125
Cell
  blood, 126
  calcium uptake, 127
  cholesterol and erythrocytes, 243-245
  culture studies, 110, 111
  damage in hypothermia, 281
  enzyme system, 149-166
  eucaryotic, 240
  intestinal mucosa, 127
  liver, 101
  membranes and ionic gradients, 124
  renal tubule, 127
  repolarization, 149
  respiration and glycolysis, 191-201
  sensitivity to low temperature, 99, 100
  skeletal muscle, 127
  skin, 101
  subcellular studies, 111
  survival, 127
  survival and cold, 128
  transport, 201
Chimpanzee, 410-420
Chicken, 31, 301, 303
Chipmunk, 653
Chronobiology, 713-716
Circadian rhythm (also see Rhythms)
  electrical activity, CNS, and 492
  sleep-wake cycles, 538
Circannual rhythm (see Rhythms)

Circulation
  artificial, 171-176
  arrest, 169, 175
Citellus beecheyi, 492
Citellus beldingi, 654
Citellus citellus L., 489, 491, 515
Citellus lateralis, 4, 640, 644, 645, 646, 647, 648-655
Citellus tridecemlineatus, 4, 421-455, 457, 472-480, 524, 582, 651, 652
Citellus undulatus plesius, 99, 100, 645
Citric acid cycle, 203
Clathrate hydrates, 82
Cod, 66
Cold
  acclimation, 17
  adaptation in hibernators, 239
  ATPase activity, and, 20
  exposure, 142
  inactivation, 17-54
  labile enzyme, 37
Colies, 669, 670
Collagen, 93
Corypheroides sp., 391
Crab, 67, 150, 198, 208
Cricetus cricetus L., 490, 571
Crocodile, 683
Crystal structure, 88
Cyanide, 194
Cytochromes, 205

Dog
  brain, 167-190
  decapitate preparations, 507
  heart, 303-304, 457-480
  POAH heating, cooling, 608, 615, 617, 620
Dormancy
  energy utilization, and, 666
  tree, 84
  variations of, 667-673
Dormouse, 283, 579, 653

Ecology
  relevance of photoperiodism to, 693-702
Electric eel, 73, 202
Electrolytes
  transcellular, 175
Electron carriers
  cytochromes, 205
  nicotinamide nucleotides, 205
  redox levels, 204

SUBJECT INDEX 735

Electron microscopy
   cardiac muscle, 291-355, 373-381
   freeze-etch image, 298, 304
   transitional fibers, 306
Elephant
   heart, 306
Endergonic processes, 101, 102
Entrainment, 638, 640
Environment
   behavioral physiology, and, 637-661
   "catalytic", 70
   plants, 83
Enzymes
   acetyl-CoA carboxylase, 40-41
   activity, hibernation and arousal, 11, 111-114
   allosteric regulator, 30-54
   and ATP, 137
   and metabolic homeostasis, 10
   and ouabain interaction, 155-163
   arginosuccinase, 35-40
   bonding, ionic and hydrophobic, 20
   catalytic efficiencies, 58-73
   chicken liver, 31
   cold-labile, 18, 19
   cold-sensitive, 18-54, 58, 95
   concentration, changes in, 57-58
   conformation, 25, 26, 32, 35, 37, 42, 44, 67
   effectors, 12
   effects of low temperature on, 17-54
   energy of activation, 10, 26, 64
   glycogen phosphorylase, 27-30
   inhibitors, 196-201
   kinetic behavior, 12
   lipid component, 140
   new variants, 58, 67-80
   poikilothermic, 55-80
   pyruvate carboxylase, 30-35
   pyruvate kinase, 61, 67
   quarternary structure, 20, 21, 26, 67
   synthesis, 11, 12
Erinaceus europaeus, 10, 11, 490
Estivation, 667-668
European dormouse, 185
European hedgehog, 10, 11, 490

Fatty acids
   and adaptation, 227-230
   in growth media, 242

Fatty acids, cont'd.
   heat stress, brain, 560
   synthesis of, 40, 41
Finch, 303
Fish
   Arctic sculpin, 608, 610, 611
   bullheads, 71
   cod, 66
   cold-adapted, intestine, 221
   crab, 67-70
   goldfish, 56, 219-237
   heart, 295
   lobster, 66-70
   lungfish, 73
   tropical, 62, 64
   trout, 56, 61, 66, 68, 70
Frog, 202, 206, 211, 219, 293, 301, 402, 425

Glis glis, 185, 579, 653
Glycolysis, 27-30
   control, 201
   substrates, 172
Glycoproteins, 84, 93
Glycosaminoglycans, 84, 93
Goat, 608
Goatsuckers, 667, 670
Goldfish, 56, 219-237
Ground squirrel, 126, 128
   Arctic, 99, 100, 645
   beldingi, 654
   erythrocytes, 284
   European, 489, 491, 515
   golden-mantled, 4, 640, 644, 645, 646, 647, 648
   heart, 102
   POAH heating, cooling, 608, 614
   red blood cells, 135
   thirteen-lined, 4, 421-455, 457, 472-480, 524, 582, 621, 651, 652
"Guest" molecules, 82
Guinea pig
   atria, 403-405
   red blood cells, 135, 284

Hamsters, 4, 126, 128, 138, 140, 141, 185, 219, 283, 488, 489, 490, 515, 571, 579, 608, 653
Heart
   action-potential, 402-410
   adenine nucleotides, 153, 299, 373-400
   Aschoff-Tawara node, 305

Heart, cont'd.
  bundle of His, 305
  cardiac dimensions, 457-480
  cardiac dynamics, 457-480
  cardiac glycosides, 150-153
  cardiac output, 457-480
  cell and DNP, 200
  cells and bundles, 294-297
  conductiong fibers, 305-308
  configuration, muscle, 421
  contraction strength and action
    potential, 405-410
  couplings, 296, 297, 300, 301,
    401-410
  elasticity, muscle, 421, 436
  excitation and contraction
    hibernators, 481-484
  fascia adherens, 297
  force generator, muscle, 421-455
  force velocity, heart, 421-455
  force velocity, muscle, 421-455
  freeze-etch preparations, 298,
    304
  Golgi complexes, 308
  human, 306, 405
  human and chimpanzee atrial muscle
    410-418
  intercalated disc, 297-298
  junctional complexes, 297-299
  lanthanam space, 298
  left ventricular diameter,
    457-480
  length-tension, muscle, 421-455
  macula adherens, 297
  membrane processes and activation,
    401-420
  mitochondria, 308
  M-lines, 293
  muscle, configuration, 421
  muscle, ultrastructure, 291-355,
    373-381
  myocardium in cooling, 178
  nexus, 297-298
  open heart surgery, 167, 411
  papillary muscle, ground squirrel,
    421-454
  Purkinje fibers, 295, 305-308,
    402, 403
  rat and brain temperature, 593
  sarcoplasmic reticulum, 299-305
  stroke volume, 457-480
  structure function correlation,
    292
  trabecular carnae, 421

Heart, cont'd.
  transverse tubules, 295
  ventricular output, 457-480
  ventricular volume, 457-480
  working fibers, 294
  Z-lines, 295
  Z-tubule, 301
Heat production, 124
Hedgehog
  brain, 141, 220, 283, 490
  circannual rhythm in, 653
  European, 10, 11, 490
Hens, 238
Hepatic polyribosomes, 8
Hepatoma, 192, 193, 201
Heterothermy
  as form of hibernation, 663-680
  body weight and patterns of,
    673-676
  insect, 673-676
Hibernation
  and hypothermia, comparative
    analysis, CNS, 514-526
  and non-hibernators, chemical
    control, CNS, 569-572
  and poikilotherms, 219-221
  and sleep, 493-495
  arousal, 545
  arousal mechanism, 571-572, 633,
    649
  as form of heterothermy, 663-680
  as variation of dormancy, 667
  behavioral physiology of, 637-661
  behavioral thermoregulation, 629,
    630
  brain temperatures and electrical
    activity, 490-496
  chronobiology of mammals in,
    713-716
  CNS in non-hibernating hibernators,
    629, 630
  CNS, role of in thermoregulation,
    629-634
  electrical activity in brain
    during, 487-534
  energetics, animal, 665-667
  entrance and arousal, 110
  entry, 630-632
  evoked electrical activity,
    496-505
  food supply in, 653-654
  glial cells and nerves in, 632
  hibernating position, 589
  mammals, intermediary metabolism, 3

SUBJECT INDEX

Hibernation, cont'd.
    membrane, function in, 123-148
    model for thermoregulation in hibernation, 584-588
    myosin-ATPase, and, 390-394
    photoperiod in ground squirrel, 645, 646
    set-point and entrance, 570-571
    sleep, as contrasted to hypothermia, 632
    spontaneous electrical activity and, 487-496
    structural water and, 81, 93, 95, 96
    temporal nature of, 638
    thermal responsiveness of CNS in, 621-623
    thermoregulation during normothermia, 581-583
    thermoregulation in 577-604
    timing and periodicity, 663-680

Hibernators
    and cold swelling, 184-187
    and membrane function, 281-288
    and metabolic "restructuring", 76
    behavioral physiology in, 637-661
    brain lesions and eating, 545
    brain, in acclimation, 283
    cardiac function, 373, 421-455, 481-484
    CNS in non-hibernating hibernators, 629, 630
    enzyme metabolism, 18-54
    isolated hearts, 436
    kidney, 128
    respiration, 546
Hummingbird, 545, 669, 670
Hypercapnia, 182
Hyperthermia
    acetylcholine, and, 554
    5-HT, and, 553-560
    serotonin, and, 553
Hypothermia
    adaptive, 665, 674
    and brain tissue, 167-190
    and cell function, 132
    and hibernation, comparative analysis, CNS, 514-526
    and neurological damage, 167
    calcium hypothermia, 551-576
    cardiac function in, 481-484
    CNS, role in thermoregulation, 629-634

Hypothermia, cont'd.
    electrical activity in brain during, 487-543
    evoked electrical activity, CNS, 509-514
    in man, 167
    membrane failure in, 281-288
    perfusion, 172
    spontaneous electrical activity, CNS, 505-509

Ion
    alterations in nervous system, 632
    balance, 551-576
    balance in hibernators, 125
    cytoplasmic, 286
    detergents, 239-280
    distribution and swelling, 126
    ionic gradients, 124
    ionic hydration shells, 86-87
        inner A region, 86-87
        outer B region, 86-87
    in onset and arousal, hibernation, 572
    interactions, 245-251
    intracellular, 94
    isomorphous replacement, 364
    permeability, 240-280
    pumps, 95
    transport, 127, 135, 136, 191-218, 240, 282, 283
Inotropic effect, 155
Insects
    facultative endothermy in, 674
    heterothermy in, 673-676
Intestine
    cold-adapted, fish, 221-237
    everted intestinal sac, 108
    mucosa, 127, 197, 208, 220, 282
Isolated head
    perfusion, 176
Isozymes, 11, 17, 65

Kidney
    adenine nucleotides, 153
    and potassium 126
    cortex, hibernators, 185
    cortex, human, 128
    cortex, rats, 128
    homogenates, hamster, 219
    hypothermic, 101, 102
    ion transport, 197
    perfused, 110

Kidney cont'd
  potassium contents, 128-143
Krebs cycle, 30-50, 203

Lactate, 172
Lipid
  and cold sensitivity, 140
  components, 240-253
  hydrogen bonds, 251-253
  interaction with protein, 259, 260
  melting behavior, 285
  seasonal changes, metabolism, 7
Liver
  and potassium, 126
  citrate synthase, 61
  hypothermic, 101
  lung fish and eel, 73
  metabolism, 191-218
  perfused, 110
  slice, rat fetus, 192
Lizard, 608, 610, 611, 683
Lobster, 66-70
Lungfish, 73
Lysosome
  fragility, 127

Macromolecules
  collagen, 95
  globular protein, 95
  glycosaminoglycans, 95
  polyelectrolyte, 93, 95
  polyionic sites, 94
Magnetic inhomogeneities, 94
Mammals
  and metabolic control systems, 12
  chronobiology in, 713-716
  circadian rhythm in, 538
  enzymes of, 66
  heart, 293-295, 300, 304, 374, 291, 401-420
  metabolic pathways, cold-sensitive, 46
Marmot, 489, 580-582, 588-603, 637, 653
Marmota flaviventris, 580-582, 588-603
Marmota monax, 489, 637, 653
Membrane
  amphipathic interfaces, 262-269
  cell and ionic gradients, 124
  cell or plasma, 151, 201
  cellular, 71-73
  composition, 240
  depolarization, 124

Membrane cont'd
  detergents, 239-280
  function in hibernation, 123
  hydrogen bonds and lipids, 251-253
  lipid and protein interaction, 259, 260
  micelles, 243
  microvilli and amino acids, 233
  model, 269, 270
  passive permeability, 132
  processes and activation, cardiac muscle, 401-420
  protein components, 253-259
  restructuring, 72
  role of in hibernation and hypothermia, 281-288
  structure, 239-280
  surface, properties of, 262-269
  water, interactions with, 260, 261
  zeta potential, 225
Mesocricetus auratus, 4, 126, 128, 138, 140, 141, 185, 219, 283, 488, 489, 490, 515, 571, 579, 608, 653
Metabolism
  acclimation and, 74, 75
  and cellular function, 101
  "basal" metabolic rate, 135
  brain, 169
  brown fat, 3
  carbohydrate, 6
  cold swelling and, 184
  control, enzymes and temperature, 10
  coupling of ion transport, 191-218
  enzyme systems, 191
  function in low temperature, 99-120, 169
  heart, perfused, 108, 109
  inhibition, 180
  in intact animal, 105-108
  intermediary, 3-15, 182
  intermediates, 12
  isolated organ, 108-110
  lipid, 6, 7
  mitochondrial, 3
  pathways "reorganization", 73-79
  protein, 6
  rate compensation, 57-73
  regulation, 17-54
  response to air temperature change, 612-613

SUBJECT INDEX

Metabolism, cont'd.
   role of in hibernation;
     hypothermia, 282, 283
   temperature compensation to, 55
   temperature effects, 3
Migration, 639
Mitochondria
   heart, 4
   isolated, 204
   liver, 4
   membrane, 21
   oxidative phosphorylation, 192
   respiration rate, 4
Monoamines
   and eating, drinking, drowsiness, 553-560
   in hibernation and hypothermia, 551-576
Mouse
   kangaroo, 671, 672
   pocket and torpor, 671
Muscle
   absorption peaks, using NMR, 94
   and heavy water, 93
   action, 292, 358, 373
   action potential, 401-410
   actomyosin system, 357
   birefringence, 387
   calcium and muscle activity, 357-371
   cardiac, ultrastructure of, 291-355, 373
   configuration, heart, 421
   contraction, 124, 358, 401, 422-455, 458
   cross-bridges, 358, 373, 385, 429, 434, 448
   diaphragm and potassium, 126
   diaphragm, excitability and contractility, 108
   elastic model, 421-436
   force generator, heart, 421-455
   G-actin, 359, 373
   heart and potassium, 126
   heart, in hibernators, 481-484
   length-tension, heart, 421-455
   mammalian cardiac, 401-420
   muscle model, 422-436
   myosin, 292, 373-400
   "native tropomyosin", 357
   papillary, ground squirrel, 421-455
   pseudo-random white noise (PRWN), 434-455
   relaxation, 358

Muscle, cont'd.
   sarcomeres, 292, 293
   shortening heat, 388
   skeletal, 126, 127, 357, 374, 408, 422-428
   sliding filament model, 358
   thin filaments and Z-line, 359, 373
   tropomyosin, 357-367, 373
   troponin, 357-367, 373
   X-ray diffraction, 385, 388
Mycoplasma laidlawii, 242, 244, 245, 260, 269
Myosin, 292, 373-400
   -ATPase and adaptation, 392
   ATPase hydrolysis and hibernation, 390-394
   dissociation, 374-381
   linear model for contraction, 386
   meromyosin, 373-381
   molecular weight, 374
   myofibrillar organization, 385-390
   -nucleotide interaction, 381-385
   proteolytic fragments, 377-378
   subunit composition, 374-381
   torsional model for contraction, 387-390
Myotis a. austroriparius, 4
Myotis, Southeastern, 4

Nerve
   and glial cells in hibernation, 632
   crab, 150
   diaphragm preparation, 108
   neurological damage, surgery, 167
Nervous system
   acetylcholine as neurotransmitter, 555
   "behavioral vigilance", 535
   Brain
     acetylcholinesterase, 61
     amygdala, 493, 504
     and heavy water, 93
     and long-term temperature regulation, 600-604
     and urea, 180
     brainstem, 501, 521, 526, 552, 565, 571, 572
     cannulae, 562-568
     cerebral ventricles, 553, 556, 562, 563, 564, 566, 567
     cerebral edema, 167, 174
     cholinergic injections, 533-560

Nerve, cont'd.
  Brain, cont'd.
    chloralase injections, 557
    cochlear nucleus, 510
    "coding", chemo-specific, 553
    cold-swelling, rats and dogs, 167-190
    conciousness, 537, 623
    conciousness and activity levels, 535-550
    cortex, 487-521, 525, 526, 581, 602
    electrical activity, cortical and subcortical, 487-534
    electrical stimulation, human, 536
    electrical stimulation, spontaneous and evoked, 487-534, 630, 631
    electroencephalogram, 505, 537
    energy metabolism, rats and dogs, 167-190
    heat and cold stress and transmitters, 557-560
    heat stress and fatty acids, 560
    hippocampo-amygdaloid complex, 492, 496, 505, 521-523, 543
    hypothalamic control, 553
    hypothalamic control model, 569, 577, 604
    hypothalamus, 492, 494, 495, 505, 508, 517, 518, 520-524, 551-576, 581, 582, 605-623, 631
    hypoxia, 174
    ion transport, 197
    intracranial pressure, 183
    isolated, 109, 110
    isolated head perfusion, 176
    "learning" by temperature control system, 592
    lesions, in hibernators, 545
    limbic system, 489, 490, 496, 503, 504
    mesencephalic reticular formation, 492, 496, 508, 514, 518
    phospholipid content, membranes, 72
    potassium in hibernation, 126
    preoptic nuclei, 494, 515, 516, 522, 524, 551-576, 582, 603, 605-627
    preoptic-anterior hypothalamus, 553-572, 605-627

Nerve, cont'd.
  Brain, cont'd.
    reticular activating system, 488, 501, 503, 505, 521, 541, 547, 584, 609, 622, 623, 631, 632
    septum pellucidum, 492
    "silent areas", 536
    temperature, hibernators, 581
    thalamus, 496, 499, 501, 508, 520, 541, 551-576, 605-627, 631
    thermal response to ambient temperature, 590-595
    volume, 172
    cerebral spinal fluid, 565
    decapitate preparation, dog, 507-508
    electrocorticograms, 488
    epileptiform seizures, 504-508, 514
    hyperreflexia, 511-512
    isolated head, cat, 506
    non-hibernating hibernators, 629, 630
    role of in hibernation and hypothermia, 629-634
    sciatic nerve, 499
    sensory-motor cortex, 490, 492, 496
    thermal responsiveness in hibernators, 621
    thermoregulation, 551-627
    ultrastructure, 630
Nighthawk, 544
Nightjars, 667
Nocardia asteroides, 242
Nuclear magnetic resonance, 93-95

Oppossum
  heart, 295
  POAH heating, cooling, 608
Ouabain
  binding, 283, 284
  enzyme interaction, 155-163, 201
  insensitive activity, 152, 199
  K-influx, 135
  Na-efflux, 135
  Na:Na exchange, 136
Ox, 608

Paramagnetic ions, 94
Pasteur effect, 192

SUBJECT INDEX

pH, 70
Phosphate
 high energy, 10, 182
Phospholipids, 140, 240, 266
Phosphorylase
 cold inactivated, 25, 26
 cold sensitive, 25
 conformational form, 26
 glycogen, 23-27, 66
Phosphorylation
 oxidative, 4, 17, 18, 22, 24, 30, 192-204
Photoperiodism
 reptile, 689-702
Pig, 608
Plants
 "winter hardy", 83
Pocket mouse
 torpor in, 671
Poikilotherms
 aquatic, 55-80, 111, 113
 circannual cycles, 681-716
 temperature adaptation, transport, 219-237
 timing mechanisms in, 681-716
Polyelectrolytes, 86
Potassium, 125-132, 141, 168-190, 193-218, 402-418
Primate
 baboon, 608
 hypothalamus, 553
 monkey, 414, 507, 552-576, 608
Protein
 binding sites, 267, 268
 conformation, 17-54
 interaction with lipid, 259, 260
 membrane components, 253-259
 myofibrillar proteins, 360
 phosphorylation, 136, 137
 "protein component", 357
 synthesis, 125, 127
 tropomyosin, 357-367, 373
 troponin A & B, 361-364, 357-367, 373

Rhythm
 and exogenous periods, 638
 circadian, 638, 672
 circannual, 640-661, 681-711
 chronobiology in mammals, 713-716
 endogenous, 638, 650, 651, 682-687
 exogenous, 688-702
 homothermic-heterothermic rhythm, 641, 644

Rhythm, cont'd.
 in reptile, 681-717
 timing mechanisms in reptiles, 681-711

Salt gland, avian, 205-214
Sarcoplasmic reticulum, 299-305
Seminal vesicle, 192
Serotonin, 553
Shivering, 492
Sleep
 and entrance into hibernation, 630
 and wakefulness, 537-548
 "hibernal", 487, 492-495
 hibernation, hypothermia, and, 544-547
 initiator of, 540
 insomnia, 542
 patterns, durations, 630
 REM, 539-544
 reticular control, 623
 slow wave, 539
Snake, 685
Sodium
 cardiac conductance, 401-418
 current, 149, 402-405
 ion balance, 125
 perfusal, brain, 560-568
 potassium balance, 127, 168-190
 transport, 192-218, 234
Space travel, 107
Sparrow, 608, 616, 617
Spectroscope
 studies of enzyme, 163
Squid, 402
 axon, 201
Stoichiometry, 150-152
Structure
 of water, 81-98
Strychnine, 501
Succinate oxidation
 ADP, and, 5
Swift, 669, 670
Synchronization
 physiological, with environment, 637-661

Temperature
 adaptation of transport, poikilotherms, 219-237
 alteration of set-point, 560-568, 583
 ambient, brain response to, 590-595

Temperature, cont'd.
  brain, 490, 523
  compensation, molecular
    mechanisms of, 55-80
  control mechanisms, 12
  core, in hibernator, 17
  -dependance of myosin ATPase, 390
  environmental fluctuations,
    666-667
  isozymes, effect on, 217
  low and cold sensitive enzymes,
    17-54
  low and metabolism, 10, 102-105
  low and neural activity, 101
  low and thermoregulation, 588-604
  oxidative phosphorylation, and,
    17-54
  preoptic, 605-627
  renal tubular functions, and, 101
  skeletal muscle contraction, and,
    101
  ventilation, and, 101
Thermoregulation
  alteration of temperature
    set-point, 560-568
  animal energetics, as, 665-667
  brain response to ambient,
    590-595
  chemical control, 569-572
  comparative studies, CNS, 609-623
  controller, 523
  degree and rate, control response,
    595-600
  deep hibernation, 577-604, 632
  feedback system, 57
  hibernants, during normothermia,
    581-583
  hypothalamic responsiveness,
    605-627
  "limit cycle" in, 588
  long term control, 600-604
  low-temperature, 577, 588-604
  model for thermoregulation in
    hibernation, 584-588, 605-609
  neurochemical, background, 552-560
  preoptic, role in, 551-604,
    605-627
  responsiveness in hibernators,
    621-623
  thermal drifting, 579
Thyroid, 208
Timing mechanisms, 681-711
Tissue water, 94
Toad, 219

Torpidity
  daily 668, 669
  liver mitochondria in, 5
  reticular control, 623
Tortoise, 220
Transport
  activation of energy, 168
  ion, 191-218, 219-237
  membrane damage in, 281, 282
  microvillar membrane, 233
  rate of in adaptation, 132
  sodium, 192-201
Tree
  dormancy, 84
Tropical fish, 62, 64
Trout, 56, 61, 66, 68, 70
Tumor
  ascites and Pasteur effect, 192
  ion transport, 197-202
Turtle, 200, 201, 689, 691

Rabbit, 65, 99, 102, 249, 255, 304,
  357, 374, 391, 428, 608
Rat, 138, 155, 167-190, 182, 185,
  198, 608, 671-716
Regulation
  and glycolysis, 29
  enzyme and hibernation, 17-54
  pH, 70
Renal tubules, 127
Reptile
  alligator, 682
  circannual cycles, 681-711
  crocodile, 683
  heart, 295
  lizard, 608, 610, 611, 683
  snake, 685
  timing mechanisms in, 681-711
  tortoise, 220
  turtle, 200, 201, 689, 691
Rewarming
  and potassium, 126

Ungulate
  heart, 301, 306
Urea cycle, 35-40

Water
  absorption peaks, 94
  anomalous, 82, 90-96
  "bound water", 85
  "chair" structures, 86-90
  clathrate hydrates in, 82
  clusters, 89, 90

SUBJECT INDEX

Water, cont'd.
   cluster-aggregate model, 88
   crystal structure, 88
   cytoplasmic ion and adsorption, 286, 287
   D_2O (heavy water) in tissue, 93-95
   "guest" molecules, 82
   hibernation, and, 81, 93, 95, 96
   ice I, II, III, 87-93
   interaction of membrane with, 260-273
   ionic hydration shells, 86
   NMR, 93-95
   poly, 82, 90-96
   polyelectrolytes, 86
   polyions, 84
   structural models, 87-90
   structured, 81-98, 102, 114, 115
   super, 82, 90-96
   tissue, 94
   true water polymer, 85
Whale
   heart, 306

Zeitgeber, 638, 640, 645, 646, 651, 652, 655, 682-716
Z-tubule, 301

THE LIBRARY

285205

THE LIBRARY